AF530714

# Recent Trends in Plant Pathology

*Editors*


**Bishwanath Chakraborty**

Professor, Immuno-Phytopathology Laboratory
Department of Botany, University of North Bengal
Siliguri – 734 013, West Bengal

**Usha Chakraborty**

Professor, Plant Biochemistry Laboratory
Department of Botany, University of North Bengal
Siliguri – 734 013, West Bengal

*Published by:*

Scientific Publishers (India)
5 A, New Pali Road, P.O. Box 91
Jodhpur 342 001 (India)
E-mail: info@scientificpub.com
Website: www.scientificpub.com

Scientific Publishers (India)
4806/24, Ansari Road,
Darya Ganj,
New Delhi=-110 002 (India)

Print 2026

ISBN: 978-81-7233-921-0



Reprint from Review of Plant Pathology,
Vol. 6 (2014)

ISSN: 0972-9712; pp. 1-608

ypeset by : Rajesh Ojha
ublished by : Tanay Sharma, Scientific Publishers (India), Jodhpur

and Printed at Hinglaj Offset, Jodhpur (India)

# PREFACE

Sustainable agriculture needed for feeding millions of hungry mouths is the real challenge of the present day. With rapid changes in the environment leading to more and more adverse conditions for plant growth, crop productivity is severely affected. Plants are subjected directly to different abiotic stresses such as increasing temperature, water deficit, salinity, toxic chemical accumulation etc which adversely affect crop productivity. Besides, such environmental conditions also favour the growth and infectivity of pests and pathogens which lead to increased biotic stress conditions. It is well established that encounter of all organisms with temperatures higher than their optimal growth temperature triggers the evolutionarily conserved, ubiquitous phenomenon, the Heat Shock (Stress) Response. Metabolomics is emerging as a powerful post-genomic approach which is now being developed to be used in plant breeding strategies through the use of metabolite marker–assisted selection, acting as an early indicator of crop or environmental stress or to improve agricultural practices, for example allowing the improvement of formulations of treatment with plant growth promoting microbes. Properly applied therefore, metabolomics stands ready to play an important role in securing plant productivity against the effects of adverse environmental changes. There are a number of osmoprotectants in plants which play an immense role in stress protection among different organisms, stabilizing proteins and biological membranes from a large number of stresses, including oxidative stress, hydrostatic pressure, desiccation, osmotic stress, and exposure to toxic chemicals. These include compounds like proline, glycine, betaine, trehalose, mannitol etc. Biotechnological trials have been made for the production of stress tolerant crops, by incorporating genes for synthesis of osmolytes in stress susceptible plants. It is well known that a single disease has the potential to wipe out the total crop in a particular area and hence it is very essential to have detailed information on the epidemiology and prevalence of diseases in different areas. Host-resistance to plant pathogens has been more systematically investigated, because, unlike non-host resistance, it is genetically attainable. Perhaps the best characterized mechanism of plant host resistance is mediated by resistance (*R*) genes. Modern technology has made it easy to develop data banks and predict the possible outbreak of several diseases.

Plant Pathology has an important role to play in devising strategies suitable for sustainable agriculture. Some of the important steps to be taken by plant pathologists include the development of eco-friendly mechanisms of disease control through the use of biological resources, enhancing the resistant mechanisms of the plant through molecular means as well as studies on the interactions of biotic and abiotic stressors. The diversity of Agriculturally Important Microorganisms is far more than what we have found out till date and the function of these important microbes in agro-ecosystems is also equally diverse. The ability of plant growth promoting rhizobacteria, actinomycetes, plant growth promoting fungi, mycorrhiza, to influence plant growth depends upon the diverse mechanisms like phosphate solubilization, biological nitrogen fixation, phytohormone production, siderophore production, biological control of plant pathogens and direct or indirect induction of disease resistance. Application of these beneficial microorganisms in enriching soil and enhancing crop production will not only change the scenario of using harmful chemical fertilizers but will also open up new dimensions for utilizing microbial resources for sustainable agricultural practices.

The present review covers a wide spectrum of articles which are pertinent in the present day context and gives an indication to readers of the role of plant pathology in the current agricultural scenario. We take this opportunity to thank all scientists for their contributions and at the same time express regret at the unforeseen delay in bringing out this volume. We would also like to acknowledge Indian Society of Mycology and Plant Pathology for entrusting us with this task and Scientific Publishers for publishing this volume.

**Bishwanath Chakraborty**
**Usha Chakraborty**
**Editors**

# CONTENTS

*Rev. Plant Pathol. Vol. 6, 2014*
*Indian Society of Mycology and Plant Pathology*
*Scientific Publishers (India), Jodhpur pp. 1-57*

# A BIRD'S EYE-VIEW OF MOLECULAR CHAPERONES AND SOME FACETS OF THE HEAT SHOCK RESPONSE OF *NEUROSPORA CRASSA*

***M. Kapoor* and S.S. Roy***

Department of Biological Sciences, University of Calgary, Calgary, Alberta, Canada T2N 1N4
*E-mail: mkapoor@ucalgary.ca

**KEYWORDS:** Heat shock response; Hsps; Oxidative stress; *Neurospora* Hsp70, Hsp80; Molecular Chaperones; Homology-based Protein Structure Modeling.

## ABSTRACT

Exposure to sub-lethal temperature, toxic chemicals, metabolic poisons, glucose deficit, anoxia, reactive oxygen species or any other agents leading to protein unfolding, stimulates the rapid and transient over-expression of evolutionarily conserved genes encoding molecular chaperones, including heat shock proteins (Hsps). Accumulation of unfolded or misfolded aggregation-prone proteins is the principal signal that triggers the heat shock-stress response. Molecular chaperones play vital roles under normal growth conditions in controlling steps in metabolic pathways, genome stability, membrane integrity, proliferation, differentiation, signal transduction and apoptosis. The build-up of molecular chaperones under adverse conditions is critical for cellular survival to shield the target organism from damaged macromolecules. A large number of molecular chaperone families have been detected in the cytosol, mitochondria and the nucleus in species from all three biological kingdoms. In functional terms, the basic categories of molecular chaperones include the folding capable entities ('foldases') and the 'holdases' that sequester unfolded or partially folded proteins and transfer them to the former group. The ubiquitous class of small Hsps comprises the ATP-independent holdases. The foldase class includes the chaperonins, Hsp70, Hsp90 and Hsp110 families. Members of Hsp70 and 90 families are the most important, highly conserved, ATP binding/hydrolysis-dependent molecular chaperones. Hsp70s, with assistance from co-chaperones, are the major players in folding of nascent polypeptides emerging from the polysomes. The Hsp90s chaperone a wide spectrum of proteins: transcription factors, hormone receptors, viral replication proteins, plant pathogen sensors, mammalian innate immune response-related proteins and factors controlling apoptosis. A combination of biochemical methodology, X-ray crystallographic analysis and cryo-EM imaging techniques has led to the elucidation of structure-function relationships of the major molecular chaperones. Homology-based models of *Neurospora* molecular chaperones, Hsp70 and Hsp80 (90 family member), show an overall structural similarity with other eukaryotic counterparts.

## INTRODUCTION

It is well established that encounter of all organisms with temperatures higher than their optimal growth temperature triggers the evolutionarily conserved, ubiquitous phenomenon, the Heat Shock (Stress) Response. An increment of a few degrees is sufficient to disturb protein homeostasis by causing unfolding and aggregation of enzymes and structural proteins. In addition to hyperthermia, water logging, drought, salinity, herbicides, high-intensity light, and bacterial, fungal and viral infections induce a powerful stress response in plants. Accumulation of unfolded or misfolded proteins, which are intrinsically aggregation-prone, spells dire consequences for the host organism. The abrupt appearance of unfolded proteins in the cytosol is the principal signal that triggers the heat shock-stress response, characterized by the over-expression of numerous stress-inducible genes and the rapid synthesis of defensive proteins, the best known being the heat shock proteins (Hsps). The build-up of stress-responsive proteins is crucial for survival as it fortifies the target organism against adverse environmental conditions. This defense mechanism is eclectic in nature; it can react speedily to a wide variety of physiological and chemical challenges—heavy metals, glucose deprivation, oxidants, ethanol, metabolic poisons—and any other factors leading to protein unfolding.

The major stress-inducible, defense-capable proteins are encountered universally in the cytosol of species belonging to all three biological kingdoms: the Bacteria, Archaea and Eukarya. The heat shock proteins (Hsps)—also known as molecular chaperones—are designed for shielding the cellular machinery from damaged macromolecules. Although most Hsps are present at low levels in all cells during normal growth and metabolism, their synthesis is rapidly and dramatically up-regulated under conditions that specifically result in massive protein aggregation or misfolding. As chaperones are required in stoichiometric amounts vis-à-vis the population of unfolded polypeptides, they need to be synthesized in colossal quantities.

In the eukaryotes, pathogen attack, depletion of nutrients and a plethora of genetic/physiological factors can also cause a substantial rise in the population of unfolded proteins in the endoplasmic reticulum (ER). The lumen of ER is the seat of synthesis of secretory and membrane-specific proteins. Any type of perturbation involving changes in the redox status, calcium levels or deficiency of post-translational modifications of secretory proteins, can result in a deficit of local folding capacity. As a direct consequence, unfolded or misfolded ER proteins will accumulate. A potent surveillance system conserved in plants, fungi and mammalian species—designated the Unfolded Protein Response (UPR)—is implemented to control the build up of hazardous protein products. In mammalian cells, the immediate action of the UPR system is to promote cessation of normal protein synthesis and to accelerate the transcription of a select set of genes encoding chaperone proteins. The latter provide protection to the ER from the cytotoxicity of aberrant proteins by induction of the ER-specific degradation system and, in the event of a systems failure, apoptosis. In addition these chaperones aid in re-establishing normal ER function by resumption of proper folding (34).

A number of recent reviews with insightful analyses of the mechanistic parameters of the heat shock response, protein folding, aggregation, intricacies of macromolecular assemblies, and structure and function of molecular chaperones are available

(47, 68, 23, 44). In the following only a brief overview of the general aspects of the nature of environmental stress, defensive response and molecular chaperones will be considered. The major emphasis in this article is on the heat shock response and the molecular chaperones, Hsp70 and Hsp80(90) (Hsp70Nc and Hsp80Nc), of the model filamentous fungus, *Neurospora crassa*.

## MOLECULAR CHAPERONES: DISTRIBUTION, FUNCTION AND PROPERTIES

In recent years a large number of molecular chaperone families (approaching >100), with functions in the cytosol, mitochondria, chloroplasts and the nucleus, have been demonstrated in various eukaryotic, prokaryotic and archaeal species. Molecular chaperones encompass a generic class of proteins several of which are indispensable for normal growth as well as necessary for survival under unfavourable situations. Many of them are not heat shock proteins. While a majority of the molecular chaperones are present in the cell at a basal level throughout the life cycle, their rate of synthesis is accelerated manifold, immediately following exposure to stress-inducing conditions. As partially unfolded and aggregated proteins accumulate during exposure to higher-than-normal temperatures or a significant build-up of Reactive Oxygen Species (ROS), elevated synthesis of defensive proteins (including Hsps and other stress proteins) is critical for survival. In the short term, Hsps are instrumental in repair/reversal of cellular damage and in the longer term their availability ensures survival upon subsequent challenges with potentially lethal threats.

In addition to their well-documented role in defense, molecular chaperones play vital roles in normal metabolism by controlling critical steps in virtually all biological activities within the cell: DNA replication and repair; maintenance of genome stability, chromatin modeling, membrane integrity; the conduct and regulation of metabolic pathways; cell proliferation; differentiation; signal transduction and apoptosis (18, 13). Intervention by chaperones is witnessed at every step of the protein biogenesis pathway, beginning with the emergence of the nascent polypeptide chain from the ribosome to the completion of its synthesis and acquisition of a stable three-dimensional native structure. Stress-inducible proteins/molecular chaperones help to maintain the conformational integrity of structural proteins and enzymes by directing requisite folding of nascent and partially unfolded polypeptides or, alternatively, by promoting timely degradation of grossly damaged, irreparable proteins.

For eukaryotic cells chaperones are crucial for survival as the effects of hyperthermia can be particularly far reaching. Some of the major consequences include defects in the cytoskeleton such as aggregation of filament-forming proteins, disruption of intermediate actin and tubulin networks, incorrect localization of organelles, fragmentation of ER and the Golgi, deficit of mitochondria, impaired RNA splicing, aggregation of ribosomal proteins, anomalous processing of ribosomal RNA, changes in membrane fluidity and alteration of protein:lipid ratio (39, 47, 66).

As stated in the foregoing Hsps and their constitutively expressed cognates are known to perform vital functions in cellular physiology even under stress-free conditions. Thus a key role of Hsps/molecular chaperones is witnessed in the assembly of macromolecular complexes, trans-membrane transport of proteins, events critical to development and differentiation, cell cycle signaling pathways, regulation of gene expression and, last but not least, apoptosis. One of the well-studied attributes of molecular chaperones is the prevention of aggregation of misfolded polypeptides that may be capable of forming insoluble polymers which can subsequently lead to devastating disorders in humans. For instance, the so-called "conformational diseases" including autoimmune disorders and neurodegenerative conditions—such as Alzheimer's and Parkinson's disease—are linked to malfunction or deficiency of molecular chaperones. Furthermore, Hsps play an important role in oncogenesis, primarily by control of apoptosis. Sustained elevated expression of some Hsps is believed to be linked to metastasis and development of resistance to anti-cancer drugs (62).

## PROTEIN FOLDING AND AGGREGATION

During normal protein synthesis, hydrophobic segments of newly generated polypeptide chains display a strong tendency to associate with each other, prior to the completion of synthesis, thereby precluding subsequent proper folding and assembly. The intracellular environment in the cytosol is extremely crowded due to the high concentration of macromolecules—estimated at ~300-400 mg/mL—furnishing eminently suitable conditions for aggregation of proteins. In this milieu, molecular chaperones are absolutely essential. Their primary function is to retard intrinsic intramolecular interactions by sequestering the newly synthesized hydrophobic patches and, secondly, to guide the proper folding of the polypeptide and subsequent assembly of folded monomeric units into oligomers or macromolecular complexes. Exposed hydrophobic surfaces also have a strong propensity for inter-molecular aggregation.

In general, chaperones intervene to ameliorate the devastating effects of this type of damage either directly by prevention of aggregation and refolding of misfolded, non-native proteins or, by reducing their levels by promoting the degradation of severely malformed proteins via energy-dependent protease systems (78). As mentioned in the foregoing, some molecular chaperones mediate transport of macromolecules across membranes while others control remodeling of damaged multi-component assemblages. In addition eukaryotic chaperones also promote genetic diversity by providing a buffering system vis-à-vis deleterious mutations of potential evolutionarily adaptive value (51).

It is important to emphasize that in all encounters, the molecular chaperones engage only in short-lived, transient interactions with the polypeptide/substrate and promptly dissociate once the folding/refolding or disaggregation reaction is accomplished. Several of the chaperones are endowed with the capacity to distinguish between the native and non-native state of proteins, a system of "quality control" is thus poised to cleanse the cell of potentially hazardous, aberrant proteins (8). Molecular chaperones are often found to act in cooperative assemblages with other regulators or co-chaperones, forming complex interacting networks (reviewed in 61). In many instances the role of co-chaperones involves facilitating substrate recognition and productive binding to the substrate binding domain (SBD) of the chaperone protein. As the function of many of the chaperones (e.g. Hsp70 and 90 families) depends on ATP binding and hydrolysis, some co-chaperones are also involved in activation or inhibition of ATPase activity of the nucleotide binding domain (NBD) as well as downstream processing of substrates. Co-chaperones confer on the chaperone the ability to interact with multiple substrates selectively, thereby fine-tuning the system. During the past decade great strides have been made in understanding of structure-function relationships and molecular mechanism of action of chaperones in various organisms and a great deal of information is available on their nature and properties. Several excellent reviews on the subject of environmental stress response and molecular chaperones are available (18, 8, 67).

## ENDOGENOUS AND EXOGENOUS STIMULI CAUSING PROTEIN AGGREGATION

Both the intracellular physiological environment and exogenous factors contribute to protein misfolding and aggregation. Internal conditions that support aggregation comprise products of normal metabolism, a prominent example being reactive oxygen species. Another uncontrollable factor is the occurrence of spontaneous mutations, particularly those culminating in a natural tendency of affected proteins to misfold and aggregate, as exemplified by conformational diseases like type II diabetes, Huntington's disease, familial forms of Parkinson's disease and Alzheimer's. Moreover, defects in protein biogenesis following mis-incorporation of amino acids during translation and/or defective assembly of ribosomal protein complexes often result in aggregation susceptible products.

External stimuli responsible for generation of anomalous proteins comprise, in addition to heat, oxidants, prolonged exposure to UV and natural or synthetic toxic chemical agents. Under conditions where generation of misfolded proteins exceeds the refolding or degradation capacity of a cell, protein aggregates tend to accumulate. Hyperthermia is one such condition that directly leads to unfolding of cellular proteins on a global scale. The immediate action of the Heat-Stress response, therefore, is to manipulate the relative level of chaperones and proteases to cope with the added burden of unfolded/misfolded proteins. While heat-induced unfolding may be readily reversible by the action of molecular chaperones, proteins damaged by oxidative stress, amyloidogenic proteins and those already tagged with ubiquitin for degradation, can not be restored to their native state by molecular chaperones.

In view of the workings of intermediary metabolic pathways, oxidative stress presents a major, unavoidable, potentially damaging agent effecting cellular macromolecules that the organism is faced with on a regular basis. Over the course of evolution, highly sophisticated systems for protection from and repair of oxidative damage have been elaborated. The major ROS species include singlet oxygen ($^1O_2$), the hydroxyperoxyl radical ($HO_2^{\cdot}$), superoxide free radical ($O_2^{\cdot}$), hydrogen peroxide ($H_2O_2$), hydroxyl radical ($OH^{\cdot}$) and nitric oxide (NO). The significance of oxidative stress can be appreciated by the fact that ROS are produced continuously at a basal rate by the electron transport chain, during the operation of TCA cycle in the mitochondria and are involved in regulation of cell proliferation, differentiation and signaling (38). In plants ROS also play a central role in pathways involved in defense against pathogens (55).

Incomplete reduction of oxygen in oxidative phosphorylation reactions produces superoxide free radicals. The latter are converted by the ubiquitous enzyme superoxide dismutase (SOD) into relatively stable $H_2O_2$—which also acts as a regulator of some of the redox sensitive proteins—but that in turn, can be converted to the highly reactive, far more detrimental species, the hydroxyl radical. High levels of mitochondrial ROS may also initiate tumorigenesis. To combat this type of attack, a powerful cohort of antioxidant enzymes—catalases, glutathione peroxidases and peroxiredoxins—furnishes the second line of defense by catalyzing the reduction of $H_2O_2$ to water, thereby eliminating it as a source of downstream hazard. Therefore, under normal circumstances, ROS concentration is maintained strictly within a tolerable range by the concerted action of these enzymes. However, augmented intracellular ROS levels brought about by hypoxia or mutations in genes encoding oxidative phosphorylation proteins, are severely damaging to cellular macromolecules—proteins, membranes, lipids, DNA, RNA—and can be potentially lethal. Reduction in antioxidant defensive enzymes, responsible for enhanced oxidation of DNA—and the resulting increased mutation rate—is associated with many types of cancers (reviewed in 73). Hence aerobic organisms are particularly vulnerable to oxidative stress.

Oxidative stress, in contrast with heat stress, can also perpetrate irreversible modifications of proteins, such as free radical-induced fragmentation of the polypeptide backbone and replacement of side chains of specific amino acids by carbonyl groups. Carbonyl derivatives can accrue from direct oxidative modification of Proline, Arginine, Lysine and Threonine residues or from reactions of Lysine, Cysteine and Histidine residues with reactive carbonyl compounds or glycoxidation products, lipids and so-called advanced glycation end products (AGEP). Furthermore, carbonyl groups can react with α-amino group of Lysines giving rise to cross-linked derivatives that are completely resistant to degradation by the proteasome apparatus (63, 40). Such irreversible modifications lead to misfolding and, eventually, aggregation of the target protein. Protein aggregation also increases during cellular ageing on account of progressive deterioration of the quality control system (37).

## STRUCTURAL FEATURES OF AGGREGATES

Following is a brief comment on the structural characteristics of protein aggregates as a broader treatment of this subject is beyond the scope of this article. For a detailed, in-depth coverage of this topic an excellent recent review may be consulted (68). Extensive physico-chemical studies in conjunction with cryo-electron microscopy (cryo-EM) and other modern imaging techniques show that protein aggregates are either apparently amorphous, typically found in bacterial inclusion bodies, or amyloid-like in nature. Bacterial inclusion bodies, too, have been shown to contain amyloid-like structures. Aggregates of endogenous proteins often form in bacteria, notably under heat or oxidative stress conditions. It is interesting to note that the number of bacterial protein species susceptible to thermal unfolding and aggregation is reported to be as high as 150 to 200, perhaps even higher (75). Furthermore, as attested by a common frustrating laboratory experience in proteomics work, insoluble inclusion bodies are frequently observed in bacterial strains employed as host organisms for over-expression of heterologous proteins; individual inclusion bodies are likely to contain up to 16,500 protein molecules. It has been estimated that in *E. coli* as much as ~1.5 to 3% of total cytosolic protein contingent has the structural features suitable for forming aggregates (75). A typical structural feature of amorphous and amyloid-like aggregates comprises the formation of intramolecular β-sheets and amyloid-like fibrils. The degree of β-sheet organization is variable in different aggregates, the highest being recorded in amyloid fibrils where ß-sheets are perpendicular to the fibril axis (12).

## HEAT SHOCK PROTEINS (HSPS): A UBIQUITOUS CLASS OF MOLECULAR CHAPERONES

Exposure to sub-lethal temperature, toxic chemicals, metabolic poisons, glucose deficit, anoxia and ROS stimulates the rapid and transient over-expression of a class of evolutionarily conserved genes, encoding molecular chaperones, collectively referred to as the heat shock proteins, Hsps (15). The resulting massive increase in the cellular concentration of Hsps is the principal force in development of thermotolerance: resistance to subsequent encounters with lethal temperature. Investigations of the heat shock response by genome-wide profiling, bioinformatics analyses, differential display, transcriptional arrays and proteomics methodology have revealed the induction of >200 genes in different model organisms ranging from the bacteria, Archaea, fungi and plants to mammalian cells.

Stress-inducible proteins, in general, are classifiable into several functionally distinguishable categories (47). The best characterized and historically the most prominent set is, of course, represented by the Hsps. The other, equally crucial groupings include: proteolysis systems, required for cleansing the cells of residual misfolded and irreversibly aggregated proteins after the refolding/de-aggregation action of Hsps; catalytic proteins (enzymes) implicated in repair of DNA/RNA damage resulting from covalent modifications of nucleic acids; enzymes in metabolic pathways directly connected to cellular energy generation; transcriptional factors—activators, repressors, mediators—and other ancillary regulators of gene expression; proteins involved in the maintenance of cytoskeleton structure, the molecular transport machinery, and membrane integrity and stability.

Heat shock proteins are traditionally designated on the basis of their approximate molecular mass to the following protein families: small Hsp family (sHsps) including Hsp10, Hsp12, Hsp26 and other members, Hsp40, Hsp60, Hsp70, Hsp90 and the Hsp70-related Hsp110 family (35). Although the characteristic features of the heat shock response are preserved universally in species belonging to all three kingdoms, notable differences in the distribution of individual Hsps have been well documented. For instance, it is known that while bacterial and lower eukaryotic genomes harbour loci encoding all of the Hsp families, Hsp70 is apparently absent in many extremophilic species. Evidently Hsp90 and Hsp100 family proteins have not been detected in the Archaea, and cytosolic orthologs of Hsp100 also appear to be lacking in the nematodes, arthropods and the vertebrates. It has been noted that compared to prokaryotes and the Archaea, higher eukaryote genomes encode far more numerous loci for chaperones and regulators (47).

As mentioned in the preceding section, Hsps are required for vital functions during normal growth, differentiation and cell-cycle progression for stabilizing unfolded/unstable conformers of proteins and to facilitate proper folding/oligomeric assembly, translocation to the requisite sub-cellular organelles or targeting for degradation. The 70-kDa, 90-kDa and the chaperonin (~60-kDa) families, capable of folding nascent polypeptides or refolding unfolded proteins, are categorized as "foldases". They encompass an intensively investigated group, comprising some members that exhibit a generalized mode of action, operating on a wide spectrum of substrates as well as more specialized entities that recognize either a single substrate or a narrow, structurally delimited range thereof. Small Hsps, on the other hand, are aptly described as "holdases". They are not capable of folding polypeptides *per se*, but are crucial in prevention of protein aggregation by binding to unfolded intermediates that can be delivered subsequently to "foldase" complexes. Hence an appropriate balance must be maintained between the intracellular levels of holdases *versus* foldases under stress conditions. It is logical that holdases should be expressed only under stress, while foldases are encoded as both stress-induced and constitutively expressed counterparts. It is possible that some chaperones may perform a dual function as a holdase and a foldase depending upon the prevailing growth environment.

## REGULATION OF THE HEAT SHOCK RESPONSE

Heat shock gene expression is a tightly regulated process. In eukaryotes the most prominent and intensively investigated regulator of Hsp gene expression is the heat shock factor, Hsf1, a major transcriptional activator that binds specifically to cis-acting motifs, the heat shock elements (HSEs), located in the promoter region of all heat shock genes, to assemble the transcriptional machinery. In the absence of stress, Hsf1 is maintained in a biologically inactive, monomeric state in close association with components of the Hsp90/Hsp70 machinery. Under hyperthermia, however, unfolded or partially folded proteins abound and interact with Hsp70/Hsp90 chaperones with high affinity, exceeding by far that for Hsf1. Hence in the presence of massive quantities of unfolded polypeptides, Hsf1 monomers readily dissociate from Hsp70/90 and the freed subunits now assemble to form biologically active homotrimeric entities. The latter are transported to the nucleus, poised to catalyze transcription of heat shock genes. Upon return to normal conditions after disposal of unfolded proteins, the Hsp70/90 chaperones divested of their substrates, once again bind Hsf1 and entrap it in its quiescent monomeric state. The activity of Hsf1 is also regulatable by post-translational modifications such as phosphorylation, nitrosylation and acetylation as well as by oligomerization (1, 74).

In *E. coli* the transcriptional cofactor sigma-32 is responsible for stimulating the over-expression of heat shock genes (21). It is an alternative cofactor of bacterial RNA polymerase that replaces the standard sigma-70 specifically for transcription of heat shock genes under stress conditions. However, under stress-free conditions, σ32 is sequestered in a complex with the heat shock proteins, DnaK and DnaJ (bacterial equivalents of the eukaryotic Hsp70 and Hsp40, respectively). It is remodeled for degradation by these chaperones in order to ensure its maintenance at a low intracellular level so that the heat shock genes remain quiescent during physiological conditions. Accumulation of increasing amounts of unfolded proteins upon heat shock—that bind to the Hsps with high affinity—results in the ejection of σ32 from the RNA polymerase complex. Return to the usual growth conditions, once again, permits the occupancy of RNA polymerase by the canonical cofactor σ70, and resumption of quotidian transcription.

## PROPERTIES OF HEAT SHOCK PROTEINS

### Small heat shock proteins (sHsps), the ubiquitous holdases

Included among the small heat shock proteins are the ubiquitous families of ATP-independent molecular chaperones containing a diverse array of members, displaying a low degree of conservation as well as heterogeneity in size and sequence (33). sHsps are multi-subunit proteins capable of forming dynamic oligomeric units. These proteins interact with a wide spectrum of structurally unrelated, partially folded substrates and help to maintain them in a healthy state, fully capable of undergoing proper folding when the requisite equipment becomes available. This attribute of sHsps ("holdases") is significant in pre-empting the aggregation and irreversible denaturation of target proteins. The function of holdases is primarily to prevent the untimely entry of unfolded proteins into degradation pathways and to ensure their conversion to the active, native state by forwarding them to folding-capable partners, the "foldases".

The sHsps participate in exquisite chaperone networks in all three kingdoms. All small Hsps are structurally related: they display a common domain of 100 amino acids—the conserved α-crystallin domain, initially recognized in the eye lens α-crystallin protein. In individual sHsps, these domains are flanked on the N- and C-terminal ends by unrelated sequences that are variable in length and are believed to be involved in client recognition and the formation of larger, dynamic oligomeric assemblies. The latter may contain as many as 24 monomers (64). This group of chaperones is unique in that at normal temperatures, sHsps exist mainly as inactive, low-number oligomeric units—dimers to dodecamers, depending upon the individual sHsp species—that lack the capacity to bind client proteins. At higher temperatures, however, in conjunction with the presence of unfolding client proteins, conformational changes in the subunits expose hydrophobic surfaces permitting the formation of higher-order oligomers, as well as stable binding to the client protein. Unlike the "foldase" class of chaperones (e.g. Hsp70 and Hsp100) the activity of sHsps does not rely on ATP hydrolysis-dependent repetitive cycles of substrate binding and release, but is directly responsive to conformational modulation induced by temperature (64). Partially unfolded client proteins, produced in large quantities by heat shock and other stresses, are sequestered by sHsp chaperones thereby protecting the cell from potentially toxic aggregation-prone species. When the co-aggregates of sHsps and client polypeptides are subsequently resolved, the larger oligomers are converted back to their original inactive, low-number state after the foldase machinery steps in. An example of a unique small

Hsp is yeast Hsp12, a naturally unstructured protein that attains α-helical structure upon interaction with membrane lipids and stabilizes membranes by decreasing their fluidity (72)

## Chaperonins: circles and rings

The Hsp60 class of chaperones, also described as chaperonins, comprises a set of ring-shaped multi-subunit proteins that encapsulate non-native proteins in an ATP-dependent manner. In the bacteria, the best characterized chaperonin system, the GroE machinery (Group I chaperonins), consists of 14 GroEL monomers of ~57-kDa organized as two heptameric circles to which GroES (monomer molecular mass of ~10-kDa), also a heptameric ring, binds. The central cavity of GroEL is large enough to physically accommodate a non-native protein/polypeptide of up to ~60-kDa. Upon contact with the substrate, the receptor cavity is closed by binding of GroES, acting like the proverbial 'lid' in the presence of ATP (25). It has been suggested that as much as ~50% of *E. coli* proteins can potentially bind to GroEL. As immense amounts of GroE proteins will be required to trap significant quantities of proteins unfolded during heat stress, an enormous up-regulation of GroE genes is witnessed. Experimental evidence indicates that in *E. coli* GroEL/ES chaperone system is absolutely essential for survival under all conditions (65).

There are no GroE homologs in the eukaryotic cytosol. Instead a functionally similar chaperonin (assigned to Group II), resembling GroE in activity, CCT or TriC (TCP1 ring complex)—composed of eight heterogeneous subunits—is present. In the archaea, chaperonins referred to as "thermosomes", are octameric or nonameric rings formed by subunits of the same type or two or three different gene products. CCT protein, however, is not a conventional heat shock protein and is actually down regulated during hyperthermal stress. The best known substrates of CCT are actins and tubulins, but it also recognizes a substantial proportion of newly synthesized proteins. It has been shown to interact with several other chaperones including Hsp70 but apparently does not require any co-chaperonins for its activity, unlike the interaction of GroEL with GroES. The building blocks of all chaperonins, the subunits, display a three-domain structure consisting of a central equatorial domain containing the ATP-binding site, an apical domain with the substrate-binding site and the so-called intermediate domain, bridging the two. ATP binding to GroEL subunits and hydrolysis are governed by complex and highly sophisticated allosteric interactions, manifested by positive cooperativity within members of each ring and negative cooperativity between the two rings (24). TriC and GroEL are similar in this respect. The unfolded substrate is recognized by an "open" ATP-bound state and upon ATP hydrolysis, is lodged within a "closed" conformation where the folding reactions are to take place (reviewed in 76).

## Hsp100 family proteins: dynamic hexameric ATPases

Hsp100 family of molecular chaperones is a conserved group of AAA+ ATPases (AAA+, acronym for ATPases Associated with diverse cellular Activities). In bacteria this group includes members referred to as ClpA, B, C, E, X, Y; similar proteins are present in mitochondria as well as in plants, fungi and mammals (4). Two major classes are recognized among the Hsp100 family based on the number of AAA domains: class 1 includes entities with dynamic, hexameric structures containing two nucleotide-binding sites per monomer, exemplified by the bacterial ClpA, B, C and E, yeast Hsp104 and homologs in plants; class 2 proteins harbor a single nucleotide binding domain as seen in bacterial ClpX and ClpY. Similar AAA+ ATPase domains, exhibiting homology to Hsp100, have also been described in mammals and in ATPases associated with the proteasome assembly (53). Hsp100 chaperones are thought to function by drawing unfolded polypeptides through the central aperture of the hexameric ring and subsequently refolding them. In addition, Hsp100 chaperones of class I are instrumental in supporting protein disaggregation, as illustrated by the system composed of ClpB, Hsp70- and Hsp40-like proteins that can reclaim substrates from aggregates and refold them. Interestingly, some higher eukaryotes including nematodes, arthropods and mammals apparently lack cytosolic isoforms of Hsp100 class 1 proteins. The exact mechanism of the refolding process is not completely understood.

## Hsp70 family and co-players: major allosteric foldases

Hsp70 and Hsp90 families of heat shock proteins comprise the most important molecular chaperone families in both prokaryotes and eukaryotes. Heat-induced and constitutively expressed Hsp70 cognates are the chief players in the folding process of nascent polypeptides and are implicated in protection of proteins under stress conditions, as well as in regulating trans-membrane trafficking of proteins. Hsp70s may also mediate chaperoning of some regulators of apoptotic events, thereby diminishing the efficacy of apoptosis-inducing anti-cancer drugs. Hsp70 family is one of the most structurally and functionally conserved of molecular chaperones: members of this family are found in the bacteria and in cytosol, endoplasmic reticulum and mitochondria of fungi and higher eukaryotes, as well as in chloroplasts in plants. A remarkable overall amino acid sequence similarity/identity is witnessed between Hsp70 orthologs across biological kingdoms, conservation of the three-dimensional structure being even more striking. The common structural features of Hsp70 family comprise two major domains: the ~40 kDa N-terminal nucleotide (ATP)-binding /ATPase domain (NBD) and a C-terminal ~25-kDa peptide binding domain (SBD) connected by a short, hydrophobic flexible linker (16). The NBD is believed to regulate the conformation of SBD allosterically, which then binds to short substrate stretches enriched in hydrophobic residues. Further details of structure are presented in the section on *Neurospora* Hsp70.

During growth under permissive conditions, Hsp70 and constitutively expressed isoforms are associated with normal folding of nascent polypeptides. Exposed hydrophobic surfaces of unfolded or partially folded nascent polypeptides are captured by Hsp70s and thereby prevented from self- or cross-association with similar segments that could lead to aberrant folding pathways.

According to the current understanding of the mechanism of action of Hsp70, substrates are bound by Hsp70s in an ATP-dependent manner and the binding site is believed to be capable of admitting a stretch of 5-7 hydrophobic residues *in vitro* (50). Hsp70 alternates between ATP- and ADP-bound forms by virtue of ATP hydrolysis and nucleotide exchange reactions. ATP hydrolysis results in high-affinity substrate binding as the rates of binding and dissociation are rapid in the ATP-bound state and slow in the ADP-bound state. However, as most Hsp70s have a very low intrinsic ATPase activity, hence interaction with co-chaperones is often required to raise it to a workable level.

Structural studies conducted on DnaK (bacterial Hsp70) show that substrates are enclosed within a cavity formed by β-sheets that is covered by a lid-like structure composed of α-helices. For the substrate to gain access to the cavity it is necessary for the lid to attain an "open" state; the lid is believed to be capable of acquiring different conformational states. Recently it has been shown that *in vivo* DnaK can also bind to extended stretches of polypeptides as well as to regions containing some tertiary structure (52).

### *Co-chaperones of Hsp70 family*

Hsp70 family of chaperones acts in collaboration with a class of co-chaperones—collectively referred to as J-proteins (bacterial equivalents of Hsp40 in eukaryotes)—that are implicated in the initial binding to the unfolded substrate polypeptide that is conveyed to Hsp70 (bacterial equivalent is DnaK) in the next step for folding. According to the current models proposed on the mechanism of action of Hsp70 molecular chaperones, J proteins first bind the unfolded/partially folded client protein by their peptide binding domains and interact with [Hsp70-ATP] via one of their signature domains, the so-called J domain. The latter interaction results in stimulation of the ATPase activity of Hsp70.

The substrate is proposed to bind to the "open" peptide binding site of Hsp70. As a consequence of ATPase action a conformational change is propagated in Hsp70, leading to a movement of the helical "lid" over the catalytic site with the resident substrate in place, concurrently with dissociation of the J protein. Another component of the so-called Hsp70 "core" machinery is the nucleotide exchange factor (NEF) that now interacts with the ADP-bound Hsp70. In the next step the client protein dissociates from the complex, having undergone a cycle of folding. ATP can re-bind to the ligand-free, apo state of the chaperone at this stage and the cycle of substrate binding and release can be repeated multiple times until complete folding is achieved (29).

### *Structure and properties of Hsp40 (J proteins)*

The eukaryotic Hsp40 or bacterial J protein families encompass a diverse group of individual proteins exhibiting marked sequence diversity and molecular mass that may vary considerably from the typical 40 kDa. Based on domain structure encountered in bacterial DnaJ, three distinct classes of J proteins have been recognized: in class I, each monomer has an N-terminal J domain, a flexible Gly-Phe rich region, four repeats of a $Zn^+$-finger motif and a C-terminal sequence with two similar domains (CTD I and CTD II) mainly composed of ß-strands. A hydrophobic pocket in CTD I harbors the critical site for binding client proteins, and a $Zn^+$-finger emanating from it possibly assists in substrate binding. Terminal sequences in CTD II constitute a dimerization domain presumably involved in enhancing the binding affinity for substrates. J proteins of class II diverge slightly in possessing an N-terminal J domain and a Gly-Phe region but are devoid of the $Zn^+$-finger domain; class III J proteins, on the other hand, constitute a collage of entities with structural features that are inconsistent with the above criteria. Interestingly, some J proteins are molecular chaperones in their own right and may be involved in maintaining intrinsically aggregation-prone peptides in a form competent for degradation (29).

## Hsp90 family: chaperones of diverse cellular processes

The Hsp90 class chaperones also recognize hydrophobic patches on non-native, unfolded or partially folded polypeptides as well as near-native proteins and regulate a variety of critical cellular processes—cell proliferation, signal transduction, biogenesis of miRNAs and damage repair—through their extensive repertoire of client proteins. The spectrum of their substrates or client proteins is enormous, spanning transcription factors, cell cycle/checkpoint kinases, nuclear hormone receptors, nitric oxide synthase, telomerase, plant pathogen resistance gene products, mammalian innate immunity response-related proteins and numerous others (http://www.picard.ch/downloads/hsp90interactors.pdf). In the fungi and higher eukaryotes the presence of an active cytosolic Hsp90 isoform is believed to be indispensable for cellular survival. Like the Hsp70 family, interaction of Hsp90 family chaperones with substrates is dependent upon cycles of ATP binding and hydrolysis (18; 78). Although ATPase activity is imperative for function of Hsp90s, ATP hydrolysis is a very slow process: turnover numbers for yeast and human Hsp90 ATPase activity are 1 $min^{-1}$ and 0.04 $min^{-1}$, respectively. Eukaryotic Hsp90s form complexes with several co-chaperones of which Aha1 (Activation of Hsp90 ATPase) is an activator of its ATPase. In addition several other co-chaperones are known to be inhibitors of ATPase activity and/or downstream steps in binding and processing of bound substrates (48, 44).

It has been convincingly demonstrated that mammalian Hsp90, in concert with other chaperones, is involved in cytoprotection and maintenance of cellular integrity. Inhibition of the ATP-dependent chaperoning activity of Hsp90 by agents such as geldanamycin accelerates detergent-induced lysis of human, yeast and bacterial cells. Furthermore, it is believed to contribute to an epigenetic mechanism; the deficiency of Hsp90 elicits a heritably altered chromatin state in *Drosophila* cells. As Hsp90 contributes to the conformational maturation of certain oncogenic signaling proteins, its inhibitors are known to selectively destroy cancer cells. The compound, 17-allylaminogeldanamycin (17-AAG), an analogue of geldanamycin, is bound by Hsp90 from tumour cells with a 100-fold higher affinity, compared to that from normal cells (28). In tumour cells Hsp90 with high ATPase activity is found associated with multi-protein complexes involved in chaperoning oncoproteins, thereby facilitating malignant progression. The use of Hsp90 and its inhibitors, in the quest for targets of anti-cancer agents and immunotherapeutic vaccines against infectious diseases, is a highly promising approach with some recent successes.

General features of Hsp90 structure have been uncovered by application of biochemical, biophysical and imaging technologies including limited proteolysis, x-ray crystal structures of isolated domains, NMR spectroscopy and more recently cryo-EM, rotary shadowing EM and small-angle x-ray scattering (SAXS). All Hsp90s are dimers in their active state, the monomer being composed of three major structural domains, with secondary and tertiary structure that is virtually constant in diverse species in all three kingdoms: a conserved ~25 kDa N-terminal domain (NTD) containing the ATP-binding pocket; a highly conserved middle domain (MD) of ~40 kDa and a partially conserved C-terminal domain (CTD) of ~12 kDa. Several crystal structures of isolated N-terminal, middle and C-terminal domains of Hsp90s are available in the PDB database (44, 45). The NTD and MD in the eukaryotic Hsp90s are connected by a charged linker of length varying from 30 to 70 residues in different species, containing alternating stretches of charged residues, both acidic and basic. The linker segment is absent in HtpG, the bacterial version of Hsp90, and the mitochondrial protein. Another important difference between the bacterial and eukaryotic cytosolic Hsp90s is the presence of a ~35 residue C-terminal sequence terminating in the motif MEEVD, in the lower and higher eukaryote homologs, which is absent in the bacterial and mitochondrial Hsp90s. This motif is also present in the C-terminal end of Hsp70s and is the binding site for co-chaperones containing tetratricopeptide repeats (TPR), degenerate 34-amino acid motifs, mediating protein-protein interactions. The C-terminal provides the surface for dimerization in the unliganded state while residues in the extreme end of NTD are involved in dimerization upon binding of ATP. Following which, a crucial segment in the NTD, the "lid" region (residues 94-125 in yeast), moves over the active site cavity, freeing a short sequence in the N-terminal domain. The latter interacts with the corresponding region in the other subunit of the dimer and the accompanying conformational changes result in a characteristic appearance of the dimer, a twisted structure (22, 3).

Recent structural studies of full-length *E. coli* HtpG (Hsp90) and the native form of pig brain Hsp90 using small-angle X-ray scattering and single-particle cryo-EM revealed that in the nucleotide-free state, apo-Hsp90 dimers, in solution, exist in an equilibrium between two 'open' conformations (56,7). Furthermore species-specific dynamic equilibrium between three conformational states, namely apo and nucleotide-bound (ATP and ADP) of yeast Hsc82, HtpG and human Hsp90α was demonstrated by cryo-EM and three-dimensional reconstructions (60).

### *Co-chaperones of Hsp90*

Several co-chaperones—known to associate with the Hsp90 homodimer and acting on its ATPase activity—have been characterized. This list includes negative regulators of ATPase activity as well as cofactors that influence substrate binding and downstream processing without interference with ATP hydrolysis (44, 70). For instance, as shown for yeast Hsp82, binding of co-chaperones Cdc37, Sba1, Sti1 diminishes ATPase activity. The latter is modified by selective interaction of co-chaperones with a specific conformational state of the Hsp90 dimer. Crystal structure of a complex between yeast Hsp82 and Sba1 showed that this co-chaperone binds to the N-terminal domain and stabilizes its 'closed' (ATP-bound) dimerized state (3). On the other hand Sti1 is proposed to bind by its TPR domain to the CTD of Hsp90 as well as to a second site in the NTD/MD region and inhibit the ATPase activity. The action of Cdc37 involves interaction with the 'lid' segment in the NTD, thereby interfering with the dimerization at the N-terminal end. As mentioned in a preceding section, the only activator of Hsp90 ATPase activity is the co-chaperone Aha1 which binds to the MD of the Hsp90 dimer and alters its conformation to empower the catalytic site. This action results in approximately 10-fold stimulation of ATPase activity (70, 44).

### *The role of Hsp90 and co-chaperones in plant and animal pathology*

Hsp90 is a key component of the system involved in recognition of potential pathogens in higher eukaryotes and induction of the disease-defense response. In plants specific resistance to pathogens is conferred by the so-called resistance (R) genes. A number of R genes have been isolated from a variety of plant species and most encode cytosolic sensors referred to as NLR proteins (NLR, an acronym for nucleotide-binding site [NB] and leucine-rich repeat [LRR]). Close to 150 NLR protein-encoding genes have been documented in the *Arabidopsis thaliana* genome and up to 600 in rice genome. In humans, 21 NLR proteins are believed to be involved in sensing pathogen products as well as in the regulation of innate immune responses. In higher plants NLR sensors recognize specific pathogen effector proteins that enhance the virulence of the invading organism. Genetic screens identified key immune regulating molecules in plants, the most prominent being the R gene products, RAR1 (Required for MLA12 resistance 1) and SGT1 (suppressor of the G2 allele of Skp1) that bind to NTD of Hsp90 and act as co-chaperones. SGT1 is highly conserved in eukaryotes and it contains three distinct structural domains one of which is the TPR domain but the latter is apparently not involved in interaction with Hsp90 (57). Evidence of direct association between Hsp90 and plant NLR proteins *in vivo* was furnished by co-immunoprecipitation analyses (26).

## HEAT SHOCK AND OXIDATIVE STRESS RESPONSE OF *NEUROSPORA*: GENERAL ASPECTS

Normal growth temperature employed in the laboratory for culturing *Neurospora crassa* is 28°C although conidial germination and mycelial growth can tolerate a broader range (15°C to 35°C). Transfer of normally-growing cultures to sub-lethal temperatures, heavy metals or oxidizing agents for a short duration elicits the heightened expression of a set of stress-inducible genes and a concomitant repression of a range of non-stress genes. In cultures administered [$^{35}$S]-methionine during stress-inducing treatments, several newly synthesized proteins, or those with increased rate of synthesis, are detectable within 10-15 min and their synthesis continues for approximately another 45 – 60 min. A dramatic increase in corresponding mRNA levels is evident for up to 60 min, declining thereafter. Like all other eukaryotes, the most prominent cytosolic chaperones in *Neurospora* are the high molecular mass Hsps, Hsp70Nc and Hsp80Nc, functional equivalents of the mammalian 70- and 90-kDa stress family chaperones. The process of recovery commences around 90-120 min after initial exposure as witnessed by degradation of Hsp-specific mRNA, cessation of stress protein synthesis and restoration of normal transcription and translation. However, heat shock proteins persist for several hours after the onset of the recovery process.

A multiplex purification protocol enabled the recovery of near-homogeneous preparations of four high-molecular mass *Neurospora* proteins from the same sample of heat-shocked mycelium: Hsp70Nc, Hsp90Nc, a heat- and oxidative stress-induced peroxidase, cobalamin-independent methionine synthase and EF2 (30). The identity of these proteins was determined by western blotting, employing specific polyclonal antibodies and verified by MALDI-TOF mass spectrometry (Fig. 1B).

Two-Dimensional electrophoretic analysis of crude and partially fractionated extracts of mycelial cells, subjected to hyperthermal and oxidative stress conditions, resolved ~600 polypeptide spots in each case; only a partial set is shown (Fig. 1A). A computer-assisted comparison of typical two-dimensional SDS-PAGE profiles of heat-shocked and non-shocked cell extracts showed a decrease in relative intensity in 45% and an increase in 34% of spots in heat-shocked *versus* unstressed samples, attesting to the complexity of the heat shock response. The high-molecular-mass Hsps (Hsp70Nc and Hsp80Nc), methionine synthase, stress-induced peroxidase and stress-inducible glyoxalase I were located by western blotting using specific antibodies. *Neurospora* genome also encodes Hsp60 and equivalents of Hsp110 as well as several small Hsps ranging in molecular mass from 25 to 35 kDa (11).

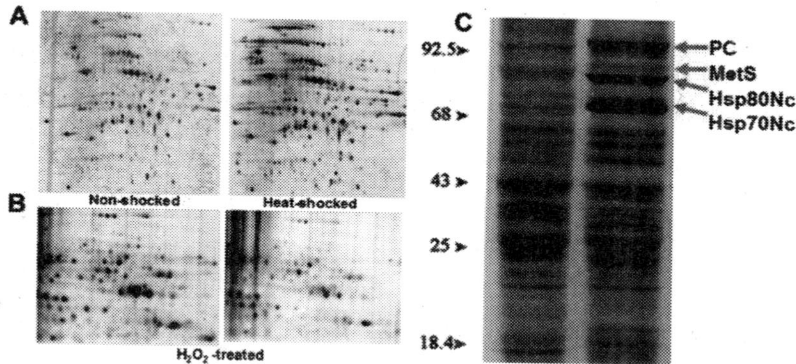


**Fig. 1**. Heat shock proteins of *N. crassa*. **A**. Differential expression of proteins in heat-shocked and normally-grown cells. Two-dimensional electrophoretic resolution of cell extracts prepared from mycelial cells grown at 28°C for 15 hours; 14-hr-old cultures heat-shocked at 48°C for 1 hr and 14-h-old cultures treated with 1 and 3 mM $H_2O_2$ for 1 h. **B**. SDS-PAGE profile of cell extracts prepared from heat-shocked and non-shocked cells, labeled with $^{35}S$-methionine for 1 hour. The enlargement of a section in the right-hand panel shows the high $M_r$ proteins identified by MALDI-TOF MS; PC, peroxidase-catalase; MetS, methionine synthase; Hsp80Nc and Hsp70Nc.

## STRESS-INDUCED THERMOTOLERANCE

To evaluate the potential for acquisition of thermotolerance by *Neurospora*, colonies grown on sorbose-agar plates at 28°C were subjected to heat shock and/or oxidative stress followed by 1h of exposure to lethal temperature. Stress treatments such as a single dose of heat shock (1 hr at 48°C) or 1 mM $H_2O_2$, 100μM cadmium chloride or sodium arsenite for one hr resulted in development of thermotolerance, as witnessed by survival upon subsequent exposure to the lethal temperature of 51.5°C. The surviving fraction was estimated by replica-plating of colonies on to fresh medium and growth for 24 h. As shown in Fig. 2, colonies that survived the lethal exposure grew readily on the replica plates. Interestingly, >90% of the previously heat-shocked colonies survived the lethal temperature. No survivors were recovered after direct lethal-temperature treatment of non-shocked colonies (A-D). Heat-shocked colonies tested positive for peroxidase activity (E, F); pre-treatment with 1 mM $H_2O_2$ elicited survival of 45% of colonies on subsequent exposure to 51.5°C for 1 h. Significant, percentage survival was also recorded upon lethal temperature treatment of colonies pre-treated with sodium arsenite (52%) and cadmium chloride (69%) (32). To test the effect of hyperthermal exposure on conidial viability and the general growth pattern, repeated doses of heat shock were administered to mycelial cells growing in liquid medium. After three successive episodes of heat shock, alternating with recovery for brief intervals, marked changes were observed in mycelial morphology (G, H) but the overall long term effect appeared to be marginal.

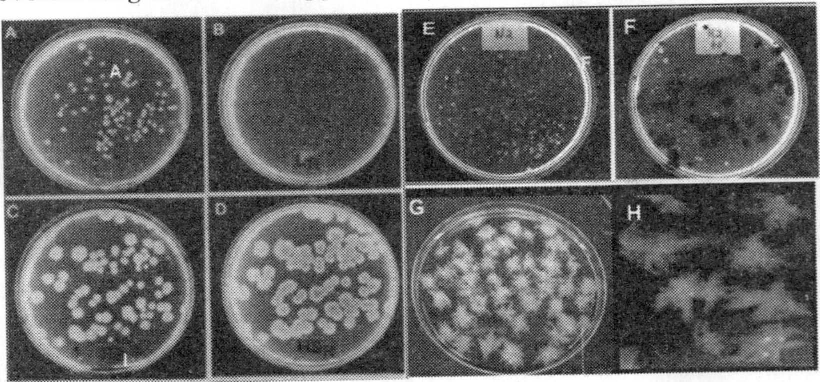


**Fig. 2**. Induction of thermotolerance by heat shock and hydrogen peroxide. **A**, 60-h-old colonies on a sorbose-agar plate after "lethal" shock; **B**, replica plate, prepared from plate in panel **A** and incubated at room temperature for 24 h. No surviving colonies are seen. **C**, 60-h-old colonies subjected to heat shock for 1 h and lethal temperature for 1 h: **D**, replica plate of panel **C** after 24 h; **E** and **F**, show peroxidase-positive colonies in non-shocked and heat-shocked plates, respectively; **G**, photograph of mycelial growth pattern in liquid culture after 3 consecutive heat shock episodes; **H,** enlargement of G.

Extracts of wild-type mycelium grown in liquid culture medium and subjected to heat shock or 1mM $H_2O_2$ for 1 h, revealed strong induction of a large high molecular mass (~95-115 kDa) protein exhibiting peroxidase activity (subsequently identified as peroxidase-catalase) which was also induced in cadmium- and arsenite-treated cells, but virtually undetectable in stress-free cells (36). A correlation between thermotolerance and relative level of stress-induced peroxidase led us to consider its role as a contributing factor in development of thermotolerance. Support for this notion emerged from a serendipitous finding of a second, constitutively expressed peroxidase activity, detected in a mutant strain of *Neurospora*, harbouring a defective heat-inducible peroxidase. This mutant strain, paradoxically, appeared to be highly thermotolerant. Purification and characterization of both of these proteins demonstrated that the constitutive and heat-inducible peroxidase activities were, in fact, attributable to two proteins, distinct in molecular size and catalytic properties (54). Thus, thermotolerance in *Neurospora* appears to be dependent, at least partially, upon a critical level of total intracellular peroxidase activity, irrespective of whether it was heat-induced or constitutive in origin.

## HSP70NC AND HSP80NC: GENES AND PROTEINS

The genes encoding the heat-inducible molecular chaperones, Hsp70Nc and Hsp80Nc, were cloned, sequenced and mapped by Restriction Fragment Length Polymorphism (RFLP). The genetic locus for Hsp70Nc (*hsp*70-1) maps to the left arm of Linkage Group (LG) II, while that for Hsp80Nc (*hsp*80) maps to the right arm of Linkage Group V, between the loci *cya*-2 and *inl* (31, 49). The coding sequence of *hsp*70-1, with four introns, translates to a 646-residue polypeptide, with a molecular mass of 70, 562 Da. The coding sequence of *hsp*80 is interrupted by 2 short introns and translates to a polypeptide of 695 residues ($M_r$ 78,894 Da) (20).

**Hsp70Nc:** The amino acid sequences of Hsp70Nc displays a high degree of homology/identity to the corresponding eukaryotic family members with strongly conserved segments in the N-terminal half. The C-terminal half exhibits a significantly lower degree of sequence conservation but has the terminal VEEVD motif, similar to the canonical MEEVD, characteristic of eukaryotic cytosolic Hsp70s implicated in recognition of co-chaperones. Using a multiple sequence alignment program, comparison of the amino acid sequence with a set of other species showed close to 80% identity with *Saccharomyces cerevisiae* Hsp70s, notably in the N-terminal half (31). The amino acid sequence encompasses the highly conserved, Hsp70 family "signature" motifs IDLGTTYS (residues 7-14) and IFDLGGGTFDVSLL (195-208).

The C-terminal end of Hsp70Nc is particularly interesting as it contains five tetrapeptide repeat units rich in Glycine and Proline residues, namely GAPP, GGMP, GAAP, GGFP and GGAP. Interestingly, similar repeats have been reported in Hsp70s of the phytopathogenic fungus, *B. lectucae* and human disease-causing parasitic organisms *Trypanosoma cruzi*, *Leishmania major* and *Plasmodium falciparum*. The number of such repeat units, however, varies in different organisms: for instance, the Hsp70 of *T. cruzi* has 10 consecutive GGMP units following a single GMP, whereas the *B. lectucae* sequence has 7 reiterations of GGMP. This motif has been suggested to have a possible role in binding specific proteins/peptides associated with aspects of pathogenesis (58). The significance of this motif in Hsp70Nc is not clear. As *N. crassa* is not a pathogenic organism, this motif may represent an evolutionary relic. It would be interesting to see if it acts as a binding site for some proteins or co-chaperone-like molecules.

## HOMOLOGY-BASED STRUCTURAL MODELING OF HSP70NC

The experimentally derived three dimensional structure of intact Hsp70Nc protein is currently not available to us as difficulties were experienced in crystallization of the full-length native protein. As discussed in a preceding section, the structure of Hsp70 class of proteins consists of two major domains: an N-terminal ~40 kDa nucleotide (ATP)-binding /ATPase domain (NBD) and a C-terminal ~25-kDa peptide binding domain (SBD) connected by a short, hydrophobic flexible linker (16, 71). Limited trypsinolysis of Hsp70Nc released peptide fragments of a similar size. Judging by the high degree of sequence conservation among eukaryotic Hsp70s, it is a reasonable hypothesis that the overall structure of the native Hsp70Nc would be similar to that of other eukaryotic Hsp70s, particularly that of *S. cerevisiae* Hsp70s. Hence we decided to employ the methodology of homology-based modeling to arrive at a structure for Hsp70Nc.

The amino acid sequence of Hsp70Nc is shown in Fig. 3 with secondary structure assignments for constituent residues. In view of the available Hsp70 sequences stored in the protein structure database, the largest domain in the Hsp70Nc, the N-terminal, ATP binding domain (NTD) comprises the first 380 residues and contains the 'lid' region. Residues between 380 and 400 form a linker—likely without any significant secondary structure—between the NTD and the Middle Domain (MD). The MD is proposed to start at residue 400 and has a secondary structure that is mostly composed of □-strands, in contrast with the highly □-helical CTD starting from residue 533.

A search of the PDB database revealed that complete or nearly complete (>90%) experimentally derived structures of eukaryotic Hsp70 were not available. Therefore, constructing a homology based model of Hsp70Nc required the construction and subsequent splicing of two partial, but overlapping, models based on appropriate Hsp70 templates. The majority of structures suitable for that role fell in three categories: a) constructs containing the first ~380 residues or the N-terminal domain; b) constructs containing ~550 residues encompassing the region containing the N-terminal domain + the middle domain and c) constructs containing most of the C-terminal domain.

Initially two partial models of Hsp70Nc were constructed. The first template used is based on the structure of bovine Hsc70 construct covering residues 1-553: PDB identifier **1yuw** (27). The second model was based on **3lof**, containing the C-terminal domain of human Hsp70 (41) spanning residues 529-612. Both structural templates had >60% sequence homology to corresponding regions of the target sequence in Hsp70Nc.

Files containing structural information about these two constructs were downloaded from NCBI as *.pdb files and used as starting material for homology-based modeling. MOE 2008.10 (Molecular Operating Environment. Chemical Computing Group Inc., Montreal. http://www.chemcomp.com) software was used for generating and refining the homology-based models of Hsp70Nc. The homology modeling algorithm in MOE consists of the following steps: 1) generation of initial partial structures, where Cartesian coordinates from identical residues, in the 'template' molecule, are conserved; 2) Boltzmann-weighted randomized sampling of non-conserved and partially conserved residues, consisting of a data-collection step and model building step. Data collection involves acquiring backbone fragments from a high-resolution structural database and assembly of alternative side chain conformations for non-identical residues, derived from an extensive rotamer library. During model building, 10 independent intermediate models were created based upon loop and side chain placements and then scored by a contact energy function. The final model was an energy minimized version of the best intermediate model.

As noted previously, the partial Hsp70Nc models had an overlapping region of 25 residues from 529-553. Visual examination of both models revealed that the residue stretch in question lacked an ordered secondary structure in **1yuw**-based model, while it was part of a highly ordered α-helical bundle in the **3lof**-based model. The **1yuw**-based model was therefore cut at residue 528 and spliced with the **3lof**-based model. The relative positioning of the C-terminal domain in the Hsp70Nc model was based on the position of a homologous domain in Hsp110s—a protein family with considerable evolutionary and structural similarity to Hsp70s. The species specific C-terminal tail sequence (613-646) was reconstructed using the secondary structure prediction libraries within the MOE homology modeling script.

The splicing and manipulations described in the previous steps had the potential to introduce undesirable strains, improper angles and side-chain placements. A final homology modeling step, with the spliced model as template, was therefore used to revisit side-chain placement, rotamer selection, followed by an additional round of localized energy optimizations. The resultant model was then scrutinized for flaws, both visually and through a series of computational filters. The phi-psi plots (Ramachandran plots), rotamer profiles, lack of residue clashes and contact energy profiles of the model displayed values within acceptable limits, indistinguishable from those of experimentally derived structures of Hsp70s (i.e. **1yuw** and **3lof**).

## Hsp70Nc model

The reconstituted homology-based model of Hsp70Nc, is a tri-lobed 'Y' or 'T' shaped molecule. Each of the three lobes is connected to the adjacent one by short (<20 residues long) unstructured linkers. The three structural domains that make up the bulk of Hsp70Nc—the N-terminal (NTD), Middle (MD) and C-terminal domains (CTD), respectively—are presented schematically in Fig. 4.

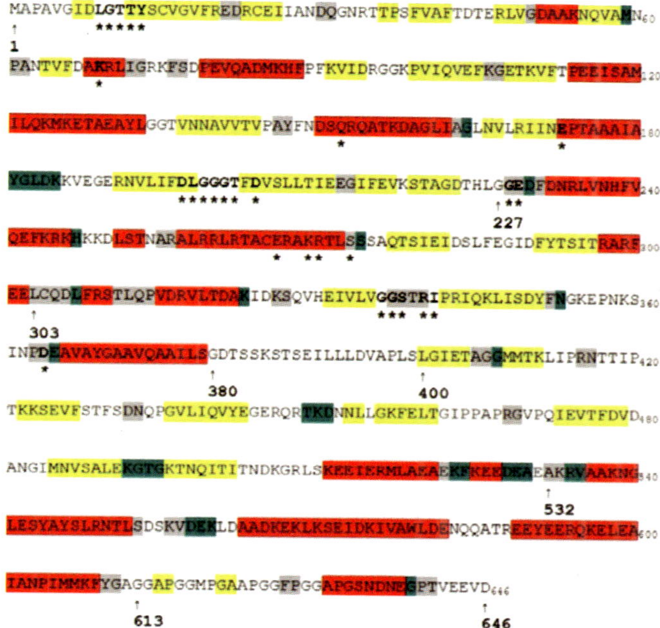


**Fig. 3**. The annotated sequence shows the location of different domains and important features in the amino acid sequence of Hsp70Nc. The N-Terminal Domain (NTD) spans residues **1-380** and contains the 'lid' region covering residues **227-303**. The Middle Domain (MD) extending from residue **400-531** is characterized by a dominance of β-strand type secondary structure. The C-Terminal Domain (CTD) covers residues **532-646**. It can be sub-divided into two parts: the highly structured and predominantly α-helical region common to all Hsp70s which extending from residue **532-612** and the species-specific and relatively unstructured C-terminal tail covering the rest of the sequence. Red, α-helix; yellow, β-strand; grey, π-helix; green, turn residues**;** *, residues lining the receptor pocket.

The NTD shows a strong structural resemblance to the equivalent domain in Hsp110, even though the sequence homology between the two is only ~30%. The NTD in Hsp70Nc comprises ~380 residues and contains the nucleotide (ATP/ADP) binding site, similar to that in other high-$M_r$ Hsps; the shape and volume of the receptor pocket (cavity) resembles that seen in Hsp110. While residues from more than one β-sheet have an edge-on interaction with the receptor cavity wall, only one of them (R191-D223) exhibits a face-on interaction with it. A few residues from five-helices form the rest of the cavity. Four residues from the 'lid' segment (E266, K269, R270, S273) provide crucial stabilizing interactions with the ligand. Three categories of residues are witnessed in the cavity: 1) interacting with the ligand (T11, Y13, K69, T202, E266, K269, S273, R340, D364); 2) interacting with each other to form a structural scaffold (T12, Y13, E173, D197, D204, D229, E266, R270, R340, D364) and 3) form the hydrophobic surface (L9, G10, P145, L198-G201, G228, G336, G337, I341). It is noteworthy that there are some common residues amongst the first two categories.

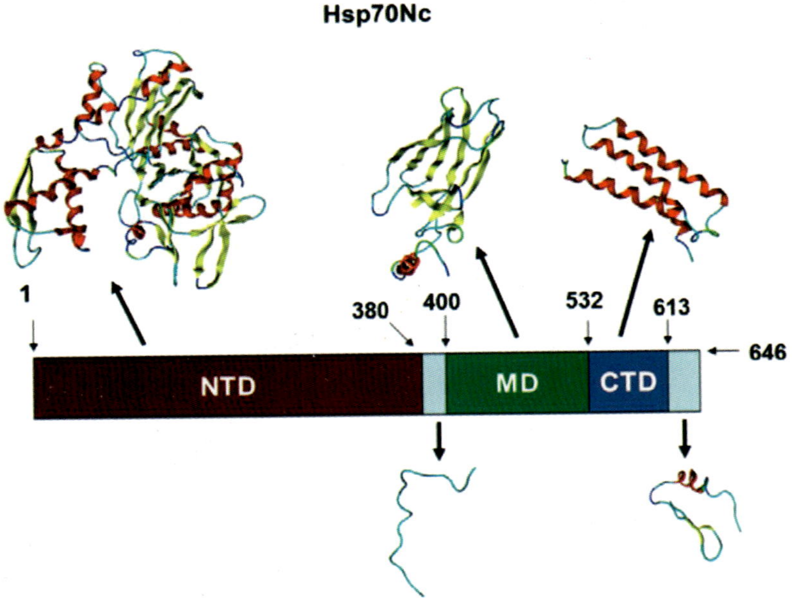

**Fig. 4.** Homology-based structural model of Hsp70Nc monomer. The reconstructed homology-based model of nHsp70 monomer is shown in **A** and **B**. The NTD (red) spans residues 1-380 and contains the active site; MD (orange) with a preponderance of β-strand secondary structure extending from residue 400-532; CTD has two regions-- a highly structured segment from residue 533-612 (violet), common to all Hsp70s and a species specific C-terminal tail (blue) from residue 613-646. The 'lid' region in the NTD (green), depicted in **B,** spans residues 227-303 and forms part of the lining of the active site, shown by rendering the water-accessible (Connolly) surface of the cavity.

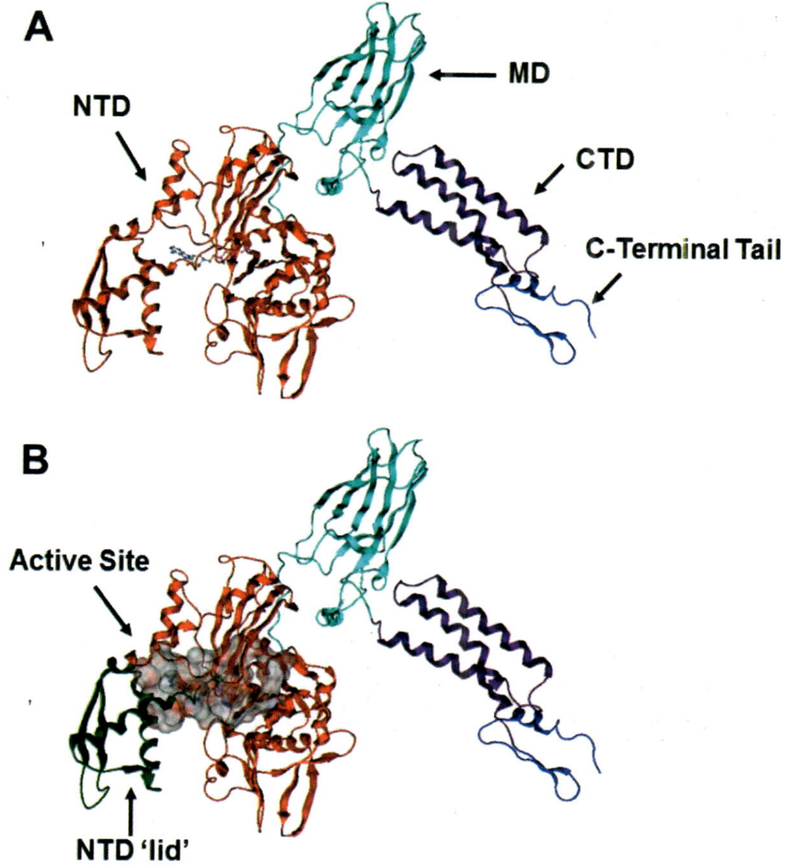


**Fig. 5.** Active site in Hsp70Nc model. A close-up of the active site in the NTD of Hsp70Nc showing the relative placement of the ligands, ADP and $PO_4$, inside the active site cavity, rendered as the water-accessible (Connolly) surface. Some residues in the 'lid' region (green) form part of the active site's lining.

The MD region, extending from residues 400-532, contains an anti-parallel β-sheet sandwich (L400-I502) — along with an α-helical region (K509 - A530)—which appears to be fairly rigid. The β-sheet sandwich is composed of three smaller sheets: two smaller on one side and a large one with 4 strands on the opposite side. It is likely that exposed residues in the sandwich form recognition motifs for co-chaperones and other accessory proteins. The CTD

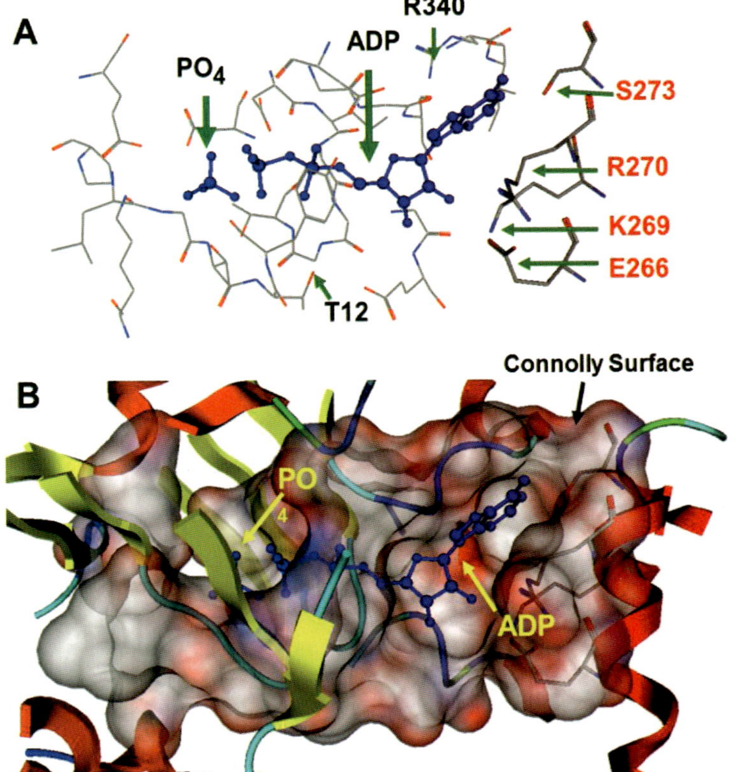


**Fig. 6.** Features of the active site in Hsp70Nc model. Two styles of rendering are used to illustrate specific features of the active site in Hsp70Nc. **A,** the line, stick and ball-and-stick style shows the relative positions of various residues near the active site. Residues lining the active site cavity and capable of forming hydrogen bonds with the ligands (ADP and $PO_4$) are identified by name and number. These include E266, K269, R270 and S273 from the 'lid' region, and R340 and T12. Hydrophilic residues near the active site but apparently not involved in ligand binding include T11, Y13, K69, E173, D197, T202, D204, E229, S338, and D364. They might be involved in interactions that stabilize the shape of the active site. Hydrophobic residues lining the active site include L9, G10, P145, L198, G199, G200, G201, G228, G336, G337 and I341. **B,** secondary structure around the active site shows that the more rigid parts of the cavity have contain predominantly β-strands with some random coil, while the somewhat flexible part derived from the lid is predominantly α-helical.

spans residues 532 - 646 and is notable for its predominant α-helical content, in contrast with the majority β-sheet content of the MD. It contains a bundle of three α-helices, from A536 - F609, modeled to be perpendicular to the main axis of Hsp70, based on a similar arrangement of equivalent domains in Hsp110. The three helical regions (A536 - L552; A563 - E582 and E589 - F609) are likely stabilized by their considerable internal interactions and, thereby, present a stable arrangement of surface residues for interactions with co-chaperones. The final ~30 residues in the CTD were modeled *ab initio* and while they do exhibit some secondary structure, especially the helical content of the stretch A631 - G639, such structural assignments are to be considered tentative.

The model of Hsp70Nc monomer (Fig. 5) graphically illustrates some of the information shown in Fig. 4. The general shape, size and secondary structural content of the three domains of the protein are depicted in A and B. The NTD is the largest domain in the protein, in terms of sequence length and size. It bears reiterating that the C-terminal tail of the Hsp70Nc model was modeled *de novo*, and should therefore be seen as one of the more likely conformations of this flexible and species-specific region. The placement of the 'lid' region and receptor cavity in the NTD relative to the CTD (Fig. 5B), suggests that allosteric interaction between those two regions is probably unlikely without assistance from co-chaperones.

Different aspects of the environment in, and around, the active site are illustrated by two representations of the residues surrounding the active site (Fig. 6). The stick-and-ball representation of the ligands in Fig. 6A shows the receptor cavity to comprise two segments: the larger one derived from non-hinge regions of the NTD with a significant β-strand type secondary structure and a smaller one from the 'lid' region containing a few crucial hydrogen-bond forming residues (E266, K269, R270, S273) backed by an α-helix (Fig. 6B). The cavity in the model is shown with ADP and $PO_4$, instead of ATP; the overall shape and fit of the cavity is not likely to undergo significant shifts between binding to the physiological ligand (ATP) and its products. The water-accessible (Connolly) surface of the receptor cavity provides a depiction of the features of the cavity surface as 'perceived' by any molecule in that spatial volume (Fig. 6B).

## Hsp80Nc: sequence and structure

The coding sequence of Hsp80Nc comprises a polypeptide of 695 residues ($M_r$ 78, 894 Da), with a high level of homology with the Hsp90s of fungal, plant and animal species: 84% identity (and 90% similarity) with modE gene product of *Podospora anserina*, 76% and 72%, with Hsp82 of *Ajellomyces capsulatus* and Hsp90 of *Candida albicans*, and a much lower degree of conservation with respect to Hsp82 of *Saccharomycces cerevisiae* (71%) and Hsp90 of *Schizosaccharomyces pombe* (67%) (20). Hsp80Nc also displays the typical C-terminal sequence MEEVD, characteristic of cytosolic localization, the latter being independently confirmed by immuno-electron microscopy (49).

The crystal structure of the complete, native form of Hsp80Nc is currently not available to us but judging by sequences and structures in databases, the characteristic Hsp90 family signature sequence YSNKEIFLRE is located at residues 25 – 34; and residues 7 - 226 comprise the putative ATPase domain, judging by the striking sequence identity/similarity with ATPase domains of other eukaryotic Hsp90s. A highly charged region (DEEEEEK-KKRRR), also reported in other Hsp90s, is seen at position 230 - 241 (Fig. 7). In view of published structural studies with yeast and other eukaryotic Hsp90s, the C-terminal region contains the dimerization domain and the putative binding site(s) for proteins with multiple TPR domains. The C-terminal MEEVD motif is thought to be implicated in substrate binding/selectivity. However, it is not clear as to how many TPR proteins are present in *Neurospora* that could interact with Hsp80Nc: one TPR domain containing protein, a cyclophilin of 40-kDa that can potentially interact with Hsp80 has been described (14). But as shown by genome sequence analysis, none of the other, mitochondrial, immunophilins of *Neurospora* contain TPR domains (5) resembling the mammalian entities that mediate formation of Hsp70/Hsp90 multi-protein complexes.

## Homology Based Modeling of Hsp80Nc

A search of the PDB database revealed that complete or almost complete (>90%) experimentally derived models of eukaryotic Hsp90 were, again, not available. The truncated dimeric protein construct (PDB identifier **2cg9**) (3), containing the major domains in *S. cerevisiae* Hsp82 (Hsp82Sc) is the closest structural facsimile for the major domains in Hsp80Nc. Therefore developing a homology based model of Hsp80Nc initially required construction and splicing of three partial, but overlapping, models based on Hsp90 templates from eukaryotic organisms, chiefly *S. cerevisiae* (Hsp82Sc). Moreover two regions in the protein, namely the 'charged' linker and C-terminal tail, were constructed by us *de novo* as no suitable structural templates for these segments were available.

Homology-based models of domains used to reconstruct Hsp80Nc included: a) templates with the N-terminal 205 residues of yeast Hsp82Sc (PDB identifier **1amw**; 46) and the N-terminal domain in **2cg9** (residues 1-677) of Hsp82Sc (3). In the latter construct the charged linker residues (221-255) were deleted and substituted with the sequence LQHMASVD; b) templates containing residues 257-513 (first domain in **2cge**, comprising the middle and C-terminal residues and second domain in **2cg9**) and c) templates with residues 514-664 (the last domain in **2cge** and **2cg9**). All structures selected as templates displayed a relatively high homology to the relevant regions of the target sequence in Hsp80Nc. Files containing structural information about the templates were downloaded from NCBI as *.pdb files and used as the starting point for homology modeling.

As for homology-based structural modeling of Hsp70Nc, MOE 2008.10 software was employed for generating and refining the Hsp80Nc models. The rationale for template selection, steps in data collection and analysis were exactly as outlined in the preceding section for Hsp70Nc. The final model was an energy minimized version of the best intermediate model.

## Hsp80Nc model

The annotated sequence in Fig. 7 shows the secondary structure assignments for residues in the homology-based model of Hsp80Nc. The schematic representation and rendering used graphically illustrate the shape, size and secondary structural features of major and minor structural domains (Fig. 8). The NTD is the second largest domain in the protein and contains the primary active site for ATP. It covers the first ~200 residues and is a fairly rigid structure but for two mobile regions— the N-terminal

(M1 - E12) and the 'lid' region (N94 - G121). These two regions are involved, respectively, in dimer formation and encasing ATP in the ligand-binding site. Experimental evidence from site-directed mutagenesis suggests that these two regions are mechanistically coupled (69). A comparison of the ADP/ATP bound structures of Hsp82Sc (**2cge** vs **2cg9**) and homology-based Hsp80Nc models suggests a possible mechanism for such an effect. The NTD of Hsp80Nc contains 6 major α-helices (I13 - N22; K28 - Y48; K89 - T98; T104-A113; G126 - L132; E187 - K196) and 3 minor ones (K57-L60; I117 - G121; L182- L185) and one large 7-stranded anti-parallel β-sheet (α strands: E5-E8; I67-D72; T77- D82; R136-K142; Q146-S152; T158-P163; I174-L177). Upon dimer formation the E5-E8 strand of one monomer interacts with the T158-P163 strand from the other monomer. The NTD also contains the main ligand-binding cavity (site) in Hsp80Nc. It has a distinct and fairly mobile 'lid' (N94-G121) unlike the semi-rigid ones seen in Hsp70 and 110. In its inactive ADP-bound form the predominantly α-helical lid exposes the entrance to the ligand-binding cavity. ATP binding appears to make the 'lid' lose some of its α-helical content, thereby trapping the ligand in the active site cavity. As seen in Hsp70Nc, the latter contains three categories of residues: 1) that interact with the ligand (N38, N95, S102, T104, *G124, V125, G126, F127*); 2) interacting with each other to furnish a structural scaf-fold (E34, D41, R101, K105, Q122, Y128, *G124, V125, G126, F127)* and 3) those contributing to the hydrophobic surface of the cavity (L35, A39, A42, I85, G86, M87, L96, I99, G103, G121). There is some overlap between the first two categories as residues *G124, V125, G126, F127* engage in backbone amide interactions with the ligand while contributing to the structural scaffold as well.

The 'charged' linker (I201-L255) connecting the NTD and the MD is traditionally modeled as a collapsed random coil, however analysis of Hsp80Nc Circular Dichroism (CD) spectra suggests that it has significant α-strand content. Therefore, we modeled its stretches of alternatively charged residues as a collection of loose but interacting α-strands. The MD (N256-T510) is the largest and relatively rigid domain in Hsp80Nc. Its distinguishing feature is two large anti-parallel β-sheets linked by a two-part helix between Q370 and E416. A short, unstructured linker (K511-T518) connects the MD with the CTD.

**Hsp80Nc**

**Fig. 7.** Amino acid sequence and secondary structure of Hsp80Nc. The NTD covers the first 205 residues of the sequence. It has a significant amount of α-helical content and smaller but crucially important β-strands which support part of the active site. The 'lid' extends from residue 94-124 and is one of two regions in the NTD which change structure during dimer formation. The region with residues 1-15 crosses over to bind with the NTD from the other chain during dimer formation. The 'charged' linker covers residues 206-256 and though it lacks a defined secondary structure, phi-psi angle analysis of its constituent residues suggests a β-strand like structure. The MD covers residue 257-513 and is noteworthy for the presence of two long α-helices between residues 360-420. The CTD covers residues 514-695, and consists of a highly structured core region extending from 514-663 and a mostly unstructured C-terminal tail from 664-695. Red, α-helix; yellow, β-strand; grey, π-helix; green, turn residues; *, residues lining the receptor pocket.

The CTD spans residues 519-695 and is characterized by 5 major α helices (E520-N541, M575-A581, P606-V615, V625-V639, S649-L660) and one minor one (P675-A677). It also contains one 2 stranded β-sheet (C562-R665(( T598-I601) which likely forms part of the secondary ligand-binding site in the CTD. The final ~30 residues, comprising the C-terminal tail in the CTD were modeled *ab initio* (like the 'charged' linker) and while they do exhibit some secondary structure, such assignments are considered tentative.

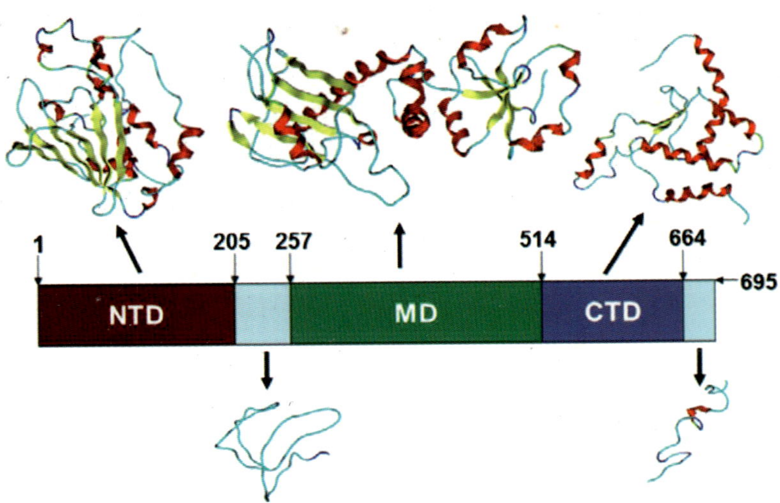


**Fig. 8**. A schematic depiction of structural domains of Hsp80Nc. The relative size, position, arrangement and conventional delineation points for various domains in the residue sequence are indicated. The N-terminal domain (NTD) which covers the first 205 residues, is the second largest domain in the protein and contains the well-characterized active site. It has overall structural similarity to the active site from nHsp70. The linker between the NTD and Middle Domain (MD) is about 52 residues long and contains a series of short stretches of charged residues. The MD extends from residue 257 to 513 and is the largest domain in nHsp80. The CTD starts at residue 514 and is predominantly α-helical with some β-strands. The relative positioning of strands and helices in it support experimental results suggesting that the CTD contains a second nucleotide binding site. The C-terminal tail, consists of the last 30-odd residues of CTD and lacks a defined secondary structure.

As stated in the foregoing, there are two regions in Hsp80Nc that could not be modeled by homology-based methodology since suitable structural templates were not available. The first of these regions, the 'charged' linker, was particularly challenging to model as it is relatively long (~50 residues) and in all likelihood contains some intrinsic secondary structure, but is also known to be flexible during the course of normal protein function. It also has a very characteristic pattern of short and alternating stretches of charged residues. We hypothesized that the combination of these features could allow intra-region interaction such that it could adopt two structural states: a 'compact semi-structured' form and an 'open unstructured' form. The compact semi-structured form was modeled as a loose α-strand sandwich based on *de novo* structural assignments and interpretation of CD spectra of Hsp80Nc. Its position was based on the corresponding region in its truncated version in **2cg9**. The C-terminal tail was modeled by *de novo* methods as a random coil with a short stretch of α-helical content.

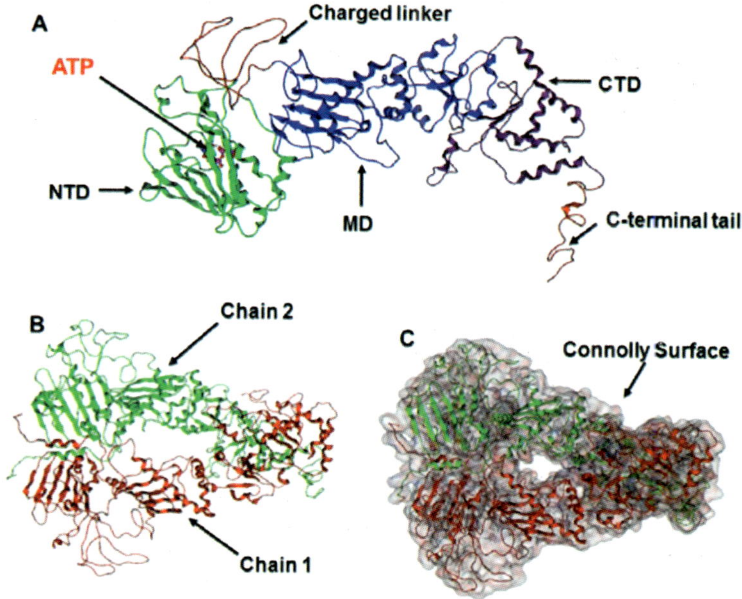


**Fig. 9.** Depiction of a reconstructed Hsp80Nc 'half-dimer'. **A,** a color-coded rendering of a reconstructed nHsp80 'half-dimer' based on a scaffold derived from one monomer in the **2cg9** complex. The NTD (green), MD (blue) and CTD (violet) are based on homology-based models while the 'charged' linker and C-terminal tail (both red) were reconstructed using *de novo* modeling techniques. **B** and **C,** a reconstructed and optimized model of the dimer. The rendering style in **B** highlights the relative positions and interactions between subunit 1 (red) and subunit 2 (green), especially near the N- and C-terminal ends. The semi-transparent Connolly surface used in **C** depicts the external morphology of the dimer, similar to published models of Hsp90s and images of Hsp90 dimers obtained via single particle cryo-EM and small-angle X-ray scattering (56, 7).

The splicing and manipulations described in the previous steps had the potential to introduce undesirable strains, improper angles and side-chain placements in the model. A final homology modeling step, with the spliced model as template was, therefore, used to re-assess side-chain placement, rotamer selection and to perform an additional round of localized energy optimizations. The resultant final model (Fig. 9A), was then scrutinized for flaws, as described for Hsp70Nc. The phi-psi plots, rotamer profiles, lack of residue clashes and contact energy profiles of the model yielded values indistinguishable from those of experimentally derived structures of Hsp90s (i.e. **2cg9** and **2cge**).

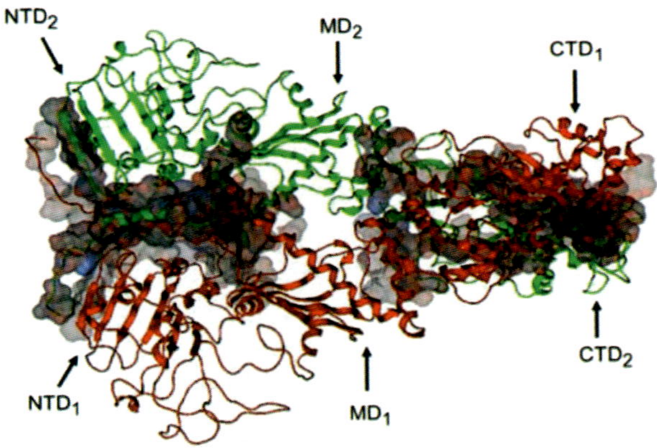


**Fig. 10.** The Connolly interaction surface between the two constituent monomer chains in the Hsp80Nc dimer model. Monomer 1, red; monomer 2, green. Hydrophilic regions are rendered as red or blue, and hydrophobic areas as grey. The NTD and CTD regions show significant interaction with the corresponding domain in the opposite monomer. The major pattern of inter-monomer interactions are: $NTD_1 \leftrightarrow NTD_2$, $NTD_1 \leftrightarrow MD_2$, $MD_1 \leftrightarrow CTD_2$ and $CTD_1 \leftrightarrow CTD_2$. Intra-monomer interactions between NTD $\leftrightarrow$ MD and MD $\leftrightarrow$ CTD are also evident.

**Hsp80Nc dimer construction**: Next, the Hsp80Nc dimer was reconstructed by positioning two chains of Hsp80Nc such that their relative position and contact pattern resembled the dimer construct in **2cg9**. A subsequent round of backbone-constrained energy minimization was used to remove clashes and strains from residue side-chains at the interface to generate the final structure. A reconstructed and optimized model of the Hsp80Nc monomer, with ATP in the primary active site in the NTD is presented in Fig. 9A. The major domains are: NTD (green), MD (blue) and CTD (purple). Segments that were built by *de novo* methods—the 'charged' linker and C-terminal tail—are rendered as red coils. A model of the Hsp80Nc dimer is illustrated in Figs 9B and C. Interaction between the two monomers—chain 1 (red) and chain 2 (green)—is evident. The overall shape of this complex is visualized by rendering their Connolly surfaces (Fig. 9C). It is noteworthy that our model shows extensive interaction between corresponding domains from the two monomers in the dimer (Fig. 10). It appears that the major domains, engage in significant inter- and intra-chain interactions. The observed pattern of inter-monomer interaction is as follows: $NTD_1 \leftarrow\rightarrow NTD_2$, $NTD_1 \leftarrow\rightarrow MD_2$, $MD_1 \leftarrow\rightarrow CTD_2$ and $CTD_1 \leftarrow\rightarrow CTD_2$.

## Co-chaperones of Hsp70Nc and Hsp80Nc?

Accurate information on possible co-chaperones of Hsp70Nc and Hsp80Nc is currently not available. In view of the higher intrinsic ATPase activity of these chaperones and their chaperoning activity described in the succeeding sections, they appear to function *in vitro* without assistance from co-chaperones. However, the *Neurospora* genome contains 18 predicted ORFs, capable of encoding potential DnaJ- Hsp40-like proteins (5) which could, theoretically, act as independent chaperones or Hsp70Nc-cochaperones under specific circumstances. Mammalian 70-kDa stress proteins, Hsp70 and Hsc70, are known to bind the TPR (tetratricopeptide repeat units) domains of cochaperones, Hsp40, BAG-1, HIP (Hsp70-Interacting Protein) and HOP (Hsp70-Hsp90-Organizing Protein), which serve as regulators of their chaperone function. One can only speculate that the C-terminal region may contain binding site(s) for *Neurospora* proteins with multiple TPRs. It is not entirely clear as to how many TPR proteins are present in *N. crassa* cells: one TPR domain containing protein, a cyclophilin of 40-kDa that can reportedly bind to Hsp80Nc has been described (14). But as shown by genome sequence analysis, none of the other, mitochondrial, immunophilins of *Neurospora* appear to contain TPR domains analogous to the mammalian entities that mediate formation of Hsp70/Hsp90 multi-protein complexes (5). A search of genome database shows that *Neurospora* genome does indeed code for putative, uncharacterized, homologs of yeast Hsp90 cochaperones Cdc37, Aha1, Sba1 and Sti1 (19).

## SOME DISTINCTIVE FEATURES AND PROPERTIES OF HSP70NC AND HSP80NC

Although similar in many respects to their eukaryotic and prokaryotic counterparts, Hsp70Nc and 80 exhibit several distinctive features. As outlined in the following, nHsp70 and 80 differ from their eukaryotic counterparts in two significant properties: in comparison with mammalian Hsp90 and Hsp70 representatives, Hsp70Nc and 80 exhibit a higher level of intrinsic ATPase activity (42, 43) and the ability to interact directly with each other, without intervention of additional protein cofactor(s). Both proteins act as independent molecular chaperones.

### Complex formation between Hsp70Nc and Hsp80Nc

Molecular chaperones often function in cooperation, classical examples being the bacterial GroEL/ES and the DnaK/DnaJ/GrpE systems, the eukaryotic Hsp70-Hsp40 complexes and multi-protein assemblages containing Hsp90, Hsp70, cochaperones and ancillary protein cofactors, along with a large entourage of "client" substrates. In mammalian systems, in complexes involving Hsp70 and Hsp90—important in promoting the maturation of hormone receptors—a bridging protein connects the two via its TPR domains. Binding of cochaperones is thought to confer selectivity to the complex, thereby permitting a wide range of structurally and functionally distinct client proteins to be recognized and processed by Hsp90 in an ATP-binding and hydrolysis-dependent manner (70). In contrast, Hsp70Nc and Hsp80Nc interact with each other apparently independently of ancillary proteins. By means of biochemical, immunological and physico-chemical methodology we demonstrated the formation of a hetero-oligomeric complex between purified, native forms of Hsp70Nc and Hsp80Nc.

#### *1. Chemical Crosslinking and Dynamic Light Scattering (DLC) measurements*

Native subunit structure of Hsp70Nc and Hsp80Nc was inferred from crosslinking with bifunctional reagents and dynamic light scattering experiments. The data from the latter indicated that both Hsp70Nc and Hsp80Nc form soluble homo-oligomers, even in dilute solutions (6). The inter-protomer crosslinking of Hsp80Nc with glutaraldehyde (GA) and dimethyl suberimidate

(DMS)—upon resolution by SDS-PAGE—had shown cross-linked subunits indicative of a quaternary structure consistent with a dimer-of-dimers (17); binding of CTP yielded higher-order assemblies >tetramers. Unliganded Hsp70Nc, cross-linked with GA, yielded a profile consistent with a tetrameric or higher-order structure. Nucleotides, ATP, ADP, NAD and CTP, on the other hand, inhibited cross-linking between Hsp70Nc protomers by GA and DMS (42). Interaction of Hsp70Nc and Hsp80Nc to form a complex was readily demonstrable by chemical cross-linking (17). Treatment of 1:1 mixtures of Hsp70Nc and Hsp80Nc with GA, followed by SDS-PAGE, western blotting and screening of the same membrane with anti-Hsp80Nc- and Hsp70Nc-IgG in succession, revealed protein bands. corresponding to hetero-oligomers composed of 1:1 and 2:1 of nHsp70:nHsp80, respectively, along with the products of inter-protein cross-linking (Fig. 11A, B).

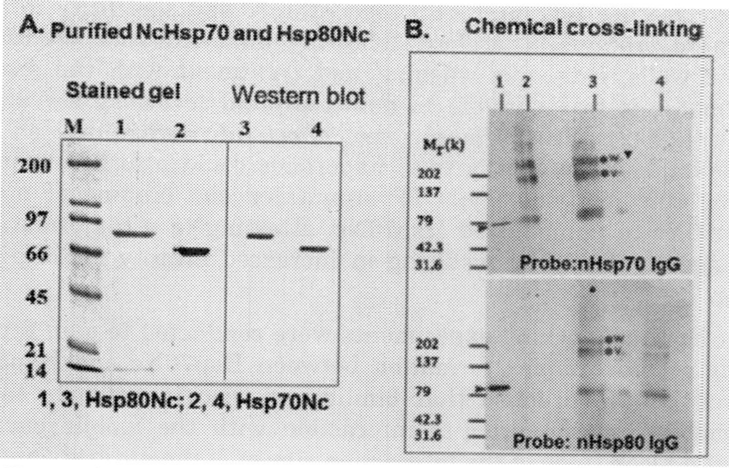


**Fig. 11**. Identification and chemical crosslinking of Hsp70Nc and Hsp80Nc purified to near homogeneity. Panel **A**. Samples of crude extract (10µg) of the two proteins were resolved by SDS-PAGE and gels stained with Coomassie blue (1, 2) and identical samples were electro-blotted onto nitrocellulose membranes and treated with specific antibodies. Panel **B**. Western blot of chemically cross-linked Hsp70Nc, Hsp80Nc and a mixture of the two, probed successively with specific IgG for each of the proteins. Lane 1, crude extract of heat-shocked mycelium; lanes 2, 3 and 4, cross-linked Hsp70Nc (2), cross-linked Hsp70Nc + Hsp80Nc (3) and cross-linked Hsp80Nc (4). Hetero-oligomeric complexes, W and V, reactive with antibodies to both proteins are shown in lane 3. Arrows in lane 1 indicate Hsp70Nc (top gel) and Hsp80Nc (bottom gel) polypeptides, respectively.

## 2. Affinity Chromatography and Immunological Analysis

Co-purification of Hsp70Nc and Hsp80Nc in the initial steps of our purification procedure first suggested the possibility of direct association between the two. However, Hsp80Nc does not bind to ATP-agarose wherein ATP is linked to the matrix via C-8 with a spacer of nine atoms while Hsp70Nc binds strongly and specifically. A complete retention of Hsp80Nc by the same agarose matrix was witnessed if Hsp70Nc was pre-bound to it. We used immunoprecipitation and structural ELISA (Enzyme-Linked Immunosorbent Assay) in conjunction with specific polyclonal antibodies against Hsp80Nc and Hsp70Nc to investigate this further. In the ELISA protocol, one of the proteins is immobilized and the binding of the target, mobile protein to it is monitored with antibodies against the mobile protein. Anti-Hsp70Nc and anti-Hsp80Nc IgG bind with high affinity and specificity to their cognate proteins. Therefore, following immobilization on ELISA plate wells, extensive washing and treatment with the specific antibody, it was possible to detect protein levels as low as 0.5 μg/well (42). Furthermore, the effect of nucleotides on the interaction between the two chaperones could also be examined using this methodology; CTP stimulated the binding of mobile Hsp80Nc to immobilized Hsp70Nc, attesting to a conformational change in Hsp80Nc, resulting in increased affinity for Hsp70Nc (Fig. 12A).

Epitope blocking experiments were conducted to gain further insight into complex formation between Hsp70Nc and Hsp80Nc based on the premise that binding of specific antibody to the immobile protein, prior to interaction with the mobile partner, should block some of the interaction site(s), if the antibody and the mobile protein were to recognize the same epitope(s). To this end, Hsp70Nc and 80 were immobilized in separate wells and reacted with their cognate antibodies. Pretreatment with anti-Hsp70Nc-IgG of wells coated with Hsp70Nc resulted in a notable decrease in retention of Hsp80Nc. Likewise a marked reduction in binding of Hsp70Nc to Hsp80Nc-saturated wells, pre-treated with anti-Hsp80Nc IgG was recorded (Fig. 12B). Although the precise interacting surface regions of the two proteins can not be delineated by this technique, a direct physical interaction between the two was clearly indicated.

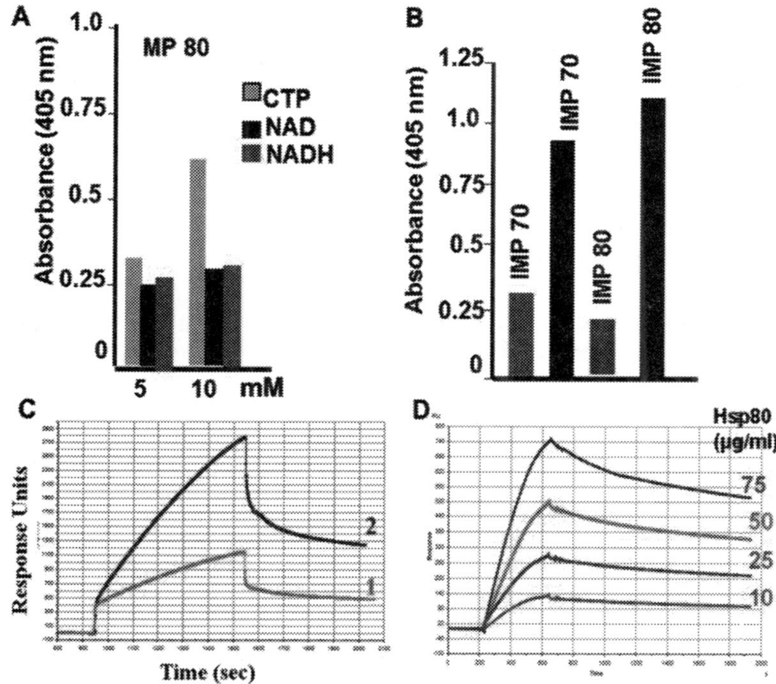


**Fig. 12**. Complex formation between Hsp70Nc and Hsp80Nc detected by ELISA assays (A, **B**) and Surface Plasmon resonance (SPR) measurements (**C, D**). One µg/well of Hsp70Nc was immobilized in ELISA plate wells and 1 µg/well Hsp80Nc was added. Bound mobile protein was detected with Hsp80Nc-specific IgG. **A,** effect of CTP, NAD and NADH on binding of mobile Hsp80Nc to immobile nHsp70. **B**, epitope blocking of Hsp70Nc and Hsp80Nc interaction by specific antibodies. One µg/well of the chaperone protein was immobilized (IMP) in the indicated wells. 1, 3 were then incubated with anti-Hsp70Nc and anti-Hsp80Nc, respectively and 2, 4 with buffer only. Interaction was detected by antibodies to the mobile protein. **C** and **D**, Interaction of Hsp70Nc and Hsp80Nc detected by SPR. Sensograms depicting effect of ATP on binding of Hsp80Nc in solution, to immobilized Hsp70Nc on the sensor chip. **C**, immobilization of 1267 RU of Hsp70Nc. **Sensogram 1**, Hsp80Nc (0.025mg/mL) + 5 mM ATP. **Sensogram 2**, Hsp80Nc (0.05 mg/mL) + 5 mM ATP. **D**, effect of increasing concentration of Hsp80Nc in solution. Samples applied: 0.01 (1), 0.025 (2), 0.05 (3) and 0.075 (4) mg/ml. All Sensograms were corrected for non-specific binding. Response units of bound Hsp80Nc are shown on the y axis and time (s) on the x axis.

### *3. Physical Interaction of Hsp70Nc and 80 in Real Time*

Surface Plasmon Resonance (SPR) was employed to seek confirmation of results obtained with affinity chromatography, chemical crosslinking and ELISA. It is a highly sensitive technique for detection of protein-protein interactions in real time and it provided a clear evidence of direct association of Hsp70Nc and Hsp80Nc. Briefly, the procedure entailed immobilization of ~1500 response units (RU)—1500 RU is equivalent to a ~1.5 ng/mm$^2$ surface protein concentration and of ~6 mg/ml in bulk protein—of Hsp70Nc and BSA (reference for non-specific binding), on separate flow cells, FC1 and FC2 respectively, of a sensor chip using amine-coupling chemistry. Appropriately diluted solutions of the analyte (Hsp80Nc) were then injected into the flow cells (flow path of FC1 to FC2) to permit it to bind to the immobile Hsp70Nc (ligand). SPR assays monitored the interaction of Hsp80Nc in solution with immobilized Hsp70Nc by recording the change in refractive index of the analyte, during its binding to the ligand in the initial phase and its dissociation during the terminal phase depicted by the sensograms in Fig. 12. Direct association between Hsp80Nc and Hsp70Nc is clearly evident; the protein samples had been prepared to ensure the absence of possible contaminating proteins. Using the software provided by the manufacturer (BIAcore), the data derived from the sensograms can be used to estimate apparent association/dissociation rate constants. Pre-treatment of Hsp80Nc with ATP resulted in a marked increase in its binding to immobile Hsp70Nc (Fig. 12 C, D). It is tempting to speculate that a greater interacting surface is available, in the ATP-bound form of Hsp80Nc, accounting for the increased binding to Hsp70Nc.

### *4. Limited Proteolysis*

Partial proteolysis has commonly been employed as a simple tool to study protein-protein interactions as well as to assess ligand-induced conformational changes in proteins. The experimental protocol comprised incubation of purified Hsp70Nc and Hsp80Nc—individually and in consort—with reaction mixtures containing predetermined amounts of trypsin. At regular intervals, aliquots were withdrawn and resolved by SDS-PAGE; protein blots prepared from duplicate gels were screened with

Hsp70Nc- and Hsp80Nc-specific IgG. Limited trypsin digestion of Hsp70Nc yielded four fragments of ~44, 42, 36, and 31 kDa, the major once being 44-kDa and 31-kDa corresponding to the N- and C-terminal domains of the typical Hsp70s. On the other hand, ATP-bound Hsp70Nc was digested at a faster rate yielding three additional fragments, as a consequence of exposure of more trypsin–sensitive sites.

Trysinolysis of free Hsp80Nc produced nine polypeptides ~75, 50, 45, 38, 38, 35, 31, 28, and 27 kDa; the effect of bound ATP was negligible. As revealed by western blot analysis, partial trypsin digestion of the Hsp70Nc:80 complex, on the other hand, generated 12 polypeptides, derived from both Hsp70Nc and Hsp80Nc. Significantly, two unique fragments each, derived from Hsp70Nc and Hsp80Nc—not seen previously in digests containing these chaperones alone—were also detected. These results indicate that both proteins undergo conformational changes resulting from complex formation, manifested as altered accessibility of trypsin-sensitive sites (6).

## Chaperoning activity of Hsp70Nc and Hsp80Nc

To evaluate the chaperoning activity of Hsp70Nc, Hsp80Nc and the hetero-oligomeric complex [Hsp70Nc:80] initially their ability to bind CMLA (reduced α-carboxymethylated lactalbumin), a classical substrate for molecular chaperones, was probed. This polypeptide lacks any secondary structure and exists in a permanently unfolded state. By means of ELISA we observed that both of these proteins, and their complex, were able to bind immobilized CMLA and appeared to be *bona fide* molecular chaperones (42). Their chaperoning activity was further verified by following the chaperone-mediated changes in the heat-induced denaturation profile of enzymes, reflective of aggregate formation by the target proteins.

### *Protection of enzymes from thermal denaturation in vitro*

Prevention of thermal aggregation of proteins is a facile, time-honoured method of evaluating suspected molecular chaperones. The ability of Hsp70Nc, Hsp80Nc and their complex to influence the denaturation of enzymes, on exposure to non-permissive temperatures, was analyzed by static light scattering measurements in the presence and absence of nucleotides. In these experiments, pre-heated assay mixtures containing the solvent (buffer + additives) and the chaperone protein (12.5 µg/mL) was equilibrated at the pre-determined denaturation temperature for 5-10 min in a thermostatted cuvette. When required, the nucleotides were included with the buffer and chaperone during the equilibration period. One to 2 µl of the enzyme solution was then added and light scattering was recorded while stirring at the specified temperature for 20 min, at an excitation and emission wavelength of 350 nm using the AMINCO-Bowman Series 2 Spectrophotofluorometer.

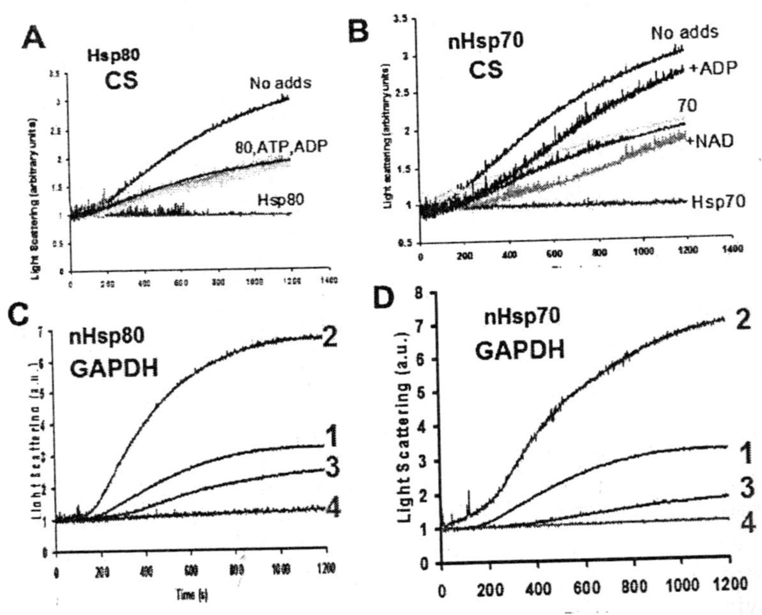


**Fig. 13**. Effect of molecular chaperones on thermally-induced aggregation of enzymes. Panels **A, B**: suppression of thermal aggregation of citrate synthase by Hsp70Nc and Hsp80Nc. Light scattering at 43°C was monitored over 20 min as described in the text. The nucleotides (0.25 mM) were included as indicated. Panels **C, D**, anti-chaperone activity of Hsp80Nc and Hsp70Nc vis-à-vis GAPDH. The reactions were carried out as described in the text, light scattering (at 44°C) was followed for 20 min. **C**, Reaction 1, enzyme alone; 2, enzyme and Hsp80Nc; 3, Enzyme+ Hsp80Nc + CTP; 4, Enzyme + Hsp80Nc + NAD. **D,** Hsp70Nc was used; nucleotides included were as in **C**.

It had been established in our previous work that Hsp70Nc and Hsp80Nc are completely stable at temperatures between 70 - 72°C for > 40 min. Both were found to elicit suppression of thermal aggregation of citrate synthase, a commonly used substrate for measurements of chaperone activity. Addition of ATP or ADP resulted in no additional protection over that provided by Hsp80Nc alone (Fig. 13A). However, addition of ADP appeared to counteract the protection provided by Hsp70Nc alone, while ATP had no observable effect (Fig. 13B).

It is well known that the major consequences of hyperthermal exposure comprise unfolding/aggregation of normal cellular proteins, an increase in reactive oxygen species, a partial shutdown of general metabolic pathways and a steep decline in the energy charge of the cell. To assess the importance of *Neurospora* molecular chaperones in relation to primary metabolism and the intracellular energy status, we employed static light scattering measurements to study their effect on temperature-dependent aggregation of some key enzymes of glycolysis/TCA cycle. A representative set of so-called housekeeping enzymes— Lactate dehydrogenase (LDH: 1.1.1.27), Glyceraldehyde phosphate dehy-drogenase (GAPDH: EC 1.2.1.12), Glutamate dehydrogenase (GDH), phosphoglycerate kinase (PGK: EC 2.7.2.3), Pyruvate kinase (PK: EC 2.7.1.40), Malate dehydrogenase (MDH: EC 1.1.1.32) and aldolase (EC 4.1.2.13)—was selected for this purpose. This group includes enzymes requiring energy charge-related nucleotides, ATP/ADP or NAD/NADH, for catalysis or allosteric regulation.

The thermal aggregation of GDH, PK, LDH and PGK was suppressed to varying degrees by both Hsp70Nc and 80. However, unexpectedly, that of GAPDH and MDH appeared to be markedly enhanced by both Hsp70Nc and Hsp80Nc (Fig. 13 C, D; Fig. 14). GAPDH is among one of the commonly employed generic substrates used for testing the activity of assorted chaperone proteins of eukaryotic and prokaryotic origin. A similar anti-chaperone-like activity has not been reported with the other Hsp70, Hsp90 or small-Hsp family members. This novel property of Hsp70Nc and Hsp80Nc, expressed vis-à-vis GAPDH and MDH, is probably not limited to these two enzymes. Examination of a wider list of enzymes may uncover similar interactions with many more proteins. The thermally-induced aggregation of these enzymes was also suppressed by NAD, ADP and ATP. This is not an altogether unexpected finding as the adenosine nucleotides are allosteric inhibitors of GAPDH, while NAD(H) is a coenzyme, binding to the catalytic site. Numerous examples are available in the classical biochemical literature of propagation of conformational changes in enzymes by binding of substrates, coenzymes and allosteric regulators, resulting in resistance to chaotropic agents. However, when energy-deficit conditions prevail—critically low intracellular levels of ATP (i.e. under extreme temperature and oxidative stress)—catalytic activity of such enzymes is likely to be severely compromised; the accumulation of inactive enzyme protein may be deleterious to the cell. Anti-chaperone activity, under such conditions, would prove beneficial by scavenging the surplus, catalytically defective protein which could be processed by the degradation machinery.

**Effect of Hsp70Nc and Hsp80Nc on thermal aggregation of enzymes**

| Enzyme | Hsp70Nc | Hsp80Nc |
|---|---|---|
| Citrate synthase (2-mer) | ↓ | ↓ |
| Pyruvate kinase (4-mer) | ↓ | ↓ |
| PGK (monomer) | ↓ | ↓ |
| GAPDH (4-mer) | ↑ | ↑ |
| LDH (4-mer) | ↓ | ↓ |
| MDH (2-mer) | ↑ | ↑ |

↓ Suppression ↑ Acceleration

**Fig. 14.** Summary of the effect of Hsp70Nc and Hsp80Nc on heat-induced aggregation of enzymes. Static light scattering was monitored for 20 min at the following temperatures (°C): Citrate synthase (43); Pyruvate kinase (51); GAPDH (44); LDH (43.5); GDH (48); MDH (43) and Aldolase (43).

Molecular chaperones can protect cellular proteins during extreme conditions in one or more of the following ways: (1) by binding to the native state and preventing unfolding/aggregation; (2) by actively promoting refolding of the unfolded state or (3) by holding the unfolded species temporarily in a "refoldable" state thereby preventing irreversible aggregation/denaturation. A plausible explanation for the acceleration of heat-induced aggregation of GAPDH and MDH by Hsp70Nc and Hsp80Nc is that these chaperones may bind to, and sequester a partially-unfolded intermediate ($U_I$) state of the enzyme, which aggregates immediately upon release from the chaperone. Alternatively, the chaperones fail to release the bound $U_I$. In this case, chaperone-bound $U_I$ may exhibit a strong tendency to form co-aggregates, containing the chaperones and the target protein.

GAPDH is a redox-sensitive enzyme of special interest due to the variety of functions attributed to it. Although traditionally viewed as a housekeeping enzyme, it is implicated in several, diverse pathways unrelated to glycolysis, including DNA replication, repair and transcriptional regulation. Furthermore, it has been shown to be associated with amyloid plaques and is thought to be an active participant in apoptosis, cancer, progression of Alzheimer's condition and other neurodegenerative diseases (59, 10).

## REFERENCES


1. Akerfelt M, Morimoto RI, and Sistonen L 2010 Heat shock factors: integrators of cell stress, development and life span, *Nature Reviews Molecular Cell Biology* **11** 545-555
2. Albanese V, Yam AY, Baughman J, Parnot C, and Frydman J 2006 Systems analyses reveal two chaperone networks with distinct functions in eukaryotic cells, *Cell* **124** 75-88
3. Ali MM, Roe SM, Vaughan CK, Meyer P, Panaretou B, Prodromou C, and Pearl LH 2006 Crystal structure of an Hsp90-nucleotide-p23/Sba1 closed chaperone complex, *Nature* **440** 1013-1017
4. Barends TR, Werbeck ND, and Reinstein J 2010 Disaggregases in 4 dimensions, *Current Opinion in Structural Biology* **20** 46-53
5. Borkovich KA, Alex LA, Yarden O, and Freitag M 2004 Lessons from the genome sequence of *Neurospora crassa*: tracing the path from genomic blueprint to multicellular organisms, *Microbiology and Molecular Biology Reviews* 1-108
6. Britton ME and Kapoor M 2002 The oligomeric state, complex formation, and chaperoning activity of Hsp70 and Hsp80 of *Neurospora crassa, Biochemistry and Cell Biology* **80** 797-809
7. Bron P, Giudice E, Rolland J-P, Buey RM, Barbier P, Diaz F, Peyrott V, Thomas D, and

Garnier C 2008 Apo-Hsp90 coexists in two open conformational states in solution, *Biology of the Cell* **100** 413-425

8. Bukau B, Weissman J, and Horwich A 2006 Molecular chaperones and protein quality control, *Cell* **125** 443-451
9. Calderwood SK, Khaleque MA, Sawyer DB, and Ciocca DR 2006 Heat shock proteins in cancer: chaperones of tumorigenesis, *Trends in Biochemical Sciences* **31** 164-172
10. Colell A, Ricci JE, Tait S, Milasta S, and Maurer U 2008 GAPDH and autophagy preserve survival after apoptotic cytochrome C release in the absence of caspase activation, *Cell* **129** 983-997
11. Davis R 2000 Neurospora: Contributions of a Model Organism, Oxford University Press. New York.
12. Dobson CM 2003 Protein folding and misfolding, *Nature* **426** 884-890
13. Driscoll R, Hudson A, and Jackson SP 2007 Yeast Rtt109 promotes genome stability by acetylating histone H3 on lysine 56, *Science* **315** 649-652
14. Duina AA, Chang HC, Marsh JA, Lindquist S, and Gaber RF 1996 *Science* **274** 1713-1715
15. Ellis RJ, van der Vies SM, and Hemmingsen SM 1989 The molecular chaperone concept, *Biochemical Society Symposia* **55** 145-153
16. Flaherty KM, DeLuca-Flaherty C, and McKay DB 1990 Three dimensional structure of the ATPase fragment of a 70-K heat shock cognate protein, *Nature* **346** 623-628
17. Freitag, DF, Ouimet PM, Girvitz TL, and Kapoor M 1997 Heat shock protein 80 of *Neurospora crassa*, a cytosolic molecular chaperone of the eukaryotic stress 90 family, interacts directly with heat shock protein 70, *Biochemistry* **36** 10221-10229
18. Frydman J 2001 Folding of newly translated proteins *in vivo*: the role of molecular chaperones. *Annual Review of Biochemistry* **70** 603-647
19. Galagan JE, Calvo SE, Borkovich KA, and Selker EU 2003 The genome sequence of the filamentous fungus *Neurospora crassa*, *Nature* **422** 859-868
20. Girvitz TL, Ouimet PM, and Kapoor M 2000 Heat shock protein 80 of *Neurospora crassa*: sequence analysis of the gene and expression during the asexual phase, *Canadian Journal of Microbiology* **46** 981-991
21. Grossman, AD, Erickson JW, and Gross CA 1984 The htpR gene product of *E. coli* is a sigma factor for heat-shock promoters, *Cell* **38** 383-390

22. Harris SF, Shiau AK, and Agard DA 2004 The crystal structure of the carboxy-terminal dimerization domain of htpG, the *Escherichia coli* Hsp90, reveal a potential substrate binding site, *Structure* **12** 1087-1097

23. Hartl FU and Hayer-Hartl M 2009 Converging concepts of protein folding *in vitro* and *in vivo*, *Nature Structural & Molecular Biology* **16** 574-581

24. Horovitz A and Williams KR 2005 Allosteric regulation of chaperonins, *Current Opinion in Structural Biology* **15** 646-651

25. Horwich AL, Farr GW, and Fenton WA 2006 GroEL-GroES-mediated protein folding, *Chemical Reviews* **106** 1917-1930

26. Jiang J, Prasad K, Lafer EM, and Sousa R 2005 Structural basis of interdomain communication in the Hsc70 chaperone, Molecular Cell **20** 513-524

27. Kadota Y, Shirasu K, and Guerois R 2010 NLR sensors meet at the SGT1-Hsp90 crossroad, *Trends in Biochemical Sciences* **36** 199-207

28. Kamal A, Thao L, Sensintaffar J, Zhang L, Boehm MF, Fritz LC, and Burrows FJ 2003 A high-affinity conformation of Hsp90 confers tumour selectivity on Hsp90 inhibitors, *Nature* **425** 407-410

29. Kampinga HH and Craig EA 2010 The Hsp70 chaperone machinery: J proteins as drivers of functional specificity, *Nature Reviews Molecular Cell Biology* **11** 579-592

30. Kapoor M 2009 A procedure for parallel purification of four stress-related proteins of *Neurospora* in their native state, *Fungal Genetics Reports* **56** 8-12

31. Kapoor M, Curle CA, and Runham C 1995 The *hsp*70 gene family of *Neurospora crassa*: cloning, sequence analysis, expression, and genetic mapping of the major stress-inducible member, *Journal of Bacteriology* **177** 212-221

32. Kapoor M, Sreenivasan GM, Goel N, and Lewis J 1990 Development of thermotolerance in *Neurospora crassa* by heat shock and other stresses eliciting peroxidase induction, *Journal of Bacteriology* **172** 2798-2801

33. Kriehuber T, Rattel T, Weinmaier T, Bepperling A, Haslbeck M, and Buchner J 2010 Independent evolution of the core domain and its flanking sequences in small heat shock proteins, *The FASEB Journal* **24** 3633-3642

34. Lai E, Teodoro T, and Volchuk A 2006 Endoplasmic reticulum stress: signaling the unfolded protein response, *Physiology* **22** 193-201

35. Lindquist S and Craig EA 1988 The heat shock proteins, *Annual Review of Genetics* **22** 15-29

36. Machwe A and Kapoor M 1993 Identification of the heat shock protein of *Neurospora crassa* corresponding to the stress-inducible peroxidase, *Biochemical and Biophysical Research Communications* **196** 692-698

37. Munch C and Bertolotti A 2010 Exposure of hydrophobic surface initiates aggregation of diverse ALS causing Superoxide dismutase1 mutations, *Journal of Molecular Biology* **399** 512-523

38. Murphy MP 2009 How mitochondria produce reactive oxygen species, *Biochemical Journal* **417** 1-13

39. Nover L, Scharf KD, and Neumann D 1989 Cytoplasmic heat shock granules are formed from precursor particles and are associated with a specific set of mRNAs, *Molecular and Cell Biology* **9** 1298-1302

40. Nystrom T 2005 Role of oxidative carbonylation in protein quality control and senescence, *The EMBO Journal* **24** 1311-1317

41. Osipiuk J, Gu M, Mihelic M, Orton K, Morimoto RI, and Joachimiak A X-ray crystal structure of C-terminal domain of human heat shock protein 70 kDa protein 1B,

42. Ouimet PM and Kapoor M 1998 Analysis of complex formation between Hsp80 and Hsp70, cytosolic molecular chaperones of *Neurospora crassa*, by enzyme-linked immunosorbent assays (ELISA), *Biochemistry and Cell Biology* **76** 97-106

43. Ouimet PM and Kapoor M 1999 Nucleotide binding and hydrolysis properties of *Neurospora crassa* cytosolic molecular chaperones, Hsp70 and Hsp80, heat inducible members of the eukaryotic stress-70 and -90 families, *Biochemistry and Cell Biology* **77** 89-99

44. Pearl LH and Prodromou C 2006 Structure and mechanism of the Hsp90 molecular chaperone machinery, *Annual Review of Biochemistry* **75** 271-294

45. Picard D 2011 Hsp90: facts and literature, http://www.picard.ch

46. Prodromou C, Roe SM, O'Brien R, Ladbury JE, Piper PW, and Pearl LH 1997 Identification and characterization of the ATP/ADP binding site in the Hsp90 molecular chaperone, *Cell* **90** 65-75

47. Richter K, Haslbeck M, and Buchner J 2010 The heat shock response: life on the verge of death, *Molecular Cell* **40** 253-266

48. Richter K, Moser S, Hagn F, Friedrich R, Hainzl O, Heller M, Schlee S, Kessler H, Reinstein J, and Buchner J 2006 Intrinsic inhibition of the Hsp90 ATPase activity, *The Journal of Biological Chemistry* **281** 11301-11311

49. Roychowdhury HS, Wong D, and Kapoor M 1992 *hsp80 of Neurospora crassa*: cDNA cloning, gene mapping, and studies of mRNA accumulation under stress, *Biochemistry and Cell Biology* **70** 1356-1367

50. Rüdinger S, Buchberger A, and Bukau B 1997 Interactions of Hsp70 chaperones with substrates, *Nature Structural and Molecular Biology* **4** 342-349

51. Rutherford SL and Lindquist S 1998 Hsp90 as a capacitor of phenotypic variation, *Nature* **396** 336-342

52. Schlecht R, Erbse AH, Bukau B, and Mayer MP 2011 Mechanics of Hsp70 chaperone enables differential interaction with client proteins, *Nature Structural and Molecular Biology* **18** 345-351

53. Schrader EK, Harstad KG, and Matouschek A 2009 Targeting proteins for degradation, *Nature Chemical Biology* **5** 815-822

54. Senczuk AM, Machwe A, and Kapoor M 2003 High constitutive peroxidase activity and constitutive thermotolerance in *Neurospora crassa*, *Mycoscience* **44** 129-137

55. Shetty NP, Jorgensen HJL, Jensen JD, Collinge DB, and Shetty HS 2008 Roles of reactive oxygen species in interaction between plants and pathogens, *European Journal of Plant Pathology* **121** 267-280

56. Shiau AK, Harris SF, Southworth DR, and Agard DA 2006 Structural analysis of *E. coli* hsp90 reveals dramatic nucleotide-dependent conformational rearrangements, *Cell* **127** 329-340

57. Shirasu K 2009 The Hsp90-SGT1 chaperone complex for NLR immune sensors, *Annual Review of Plant Biology* **60** 139-164

58. Shonhai A, Boshoff A, and Blatch GL 2007 The structural and functional diversity of Hsp70 proteins from *Plasmodium falciparum*, *Protein Science* **16** 1803-1818

59. Sirover MA 1999 New insights into an old protein: the functional diversity of mammalian glyceraldehyde-3-phosphate dehydrogenase, *Biochimica et Biophysica Acta* **1432** 159-184

60. Southworth DR and Agard DA 2008 Species-dependent ensembles of conserved conformational states define the Hsp90 chaperone ATPase

cycle, *Molecular Cell* **32** 631-640

61. Söti C, Pál C, Papp B, and Csermely P 2005 Molecular chaperones as regulatory elements of cellular networks, *Current Opinion in Cell Biology* **17** 210-215
62. Sreedhar AS and Csermely P 2004 Heat shock proteins in the regulation of apoptosis: new strategies in tumour therapy. A comprehensive review, *Pharmacology and Therapeutics* **101** 227-257
63. Stadtman ER and Levine RL 2000 Protein oxidation, *Annals of the New York Academy of Science* **899** 191-208
64. Stengel F, Baldwin AJ, Painter AJ, Jaya N, Basha E, Kay LE, Vierling E, Robinson CV, and Benesch JLP 2010 Quaternary dynamics and plasticity underlie small heat-shock protein chaperones function, *Proceedings of the National Academy of Science (USA)* **107** 2007-2012
65. Thirumalai D and Lorimer GH 2001 Chaperonin-mediated protein folding, *Annual Review of Biophysics and Biomolecular Structure* **30** 245-269
66. Toivola DM, Strnad P, Habtezion A, and Omary MB 2010 Intermediate filaments take the heat as stress proteins, *Trends in Cell Biology* **20** 79-91
67. Tsigelny IF and Nigam SK 2004 Complex dynamics of chaperone-protein interactions under cellular stress, *Cell Biochemistry and Biophysics* 40 263-276
68. Tyedmers J, Mogk A, and Bukau B 2010 Cellular strategies for controlling protein aggregation, *Nature Reviews Molecular Cell Biology* **11** 777-788
69. Vaughan CK, Piper PW, Pearl LH, and Prodromou C 2009 A common conformationally coupled mechanism for yeast and human cytoplasmic HSP90s, *The FEBS Journal* **276** 199-209
70. Wandinger SK, Richter K, and Buchner J 2008 The Hsp90 chaperone machinery, *The Journal of Biological Chemistry* **289** 18473-18477
71. Wang TF, Chang J, and Wang C 1993 Identification of the peptide binding domain of hsc70. 18-kilodalton fragment located immediately after ATPase domain is sufficient for high affinity binding, *Journal of Biological Chemistry* **268** 26049-26051
72. Welker S, Rudolph B, Frenzel E, Hagn F, Liebisch G, Schmitz G, Scheuring J, Kerth A, Blume A, and Weinkauf S 2010 Hsp12 is an intrinsically unstructured protein that folds upon membrane association and modulates membrane

function, *Molecular Cell* **39** 507-520

73. Wellen KE and Thompson CB 2010 Cellular metabolic stress: considering how cells respond to nutrient excess *Molecular Cell* **40** 323-332

74. Westerheide SD, Anckar J, Stevens SM Jr, Sistonen L, and Morimoto RI 2009 Stress-inducible regulation of the heat shock factor 1 by the deacetylase SIRT1, *Science* **323** 1016-1066

75. Winkler J, Seybert A, König L, Pruggnaller S, Haselmann U, Sourjik V, Weiss M, Frangakis AS, Mogk A, and Bukau B 2010 Quantitative and spatio-temporal features of protein aggregation in *Escherichia coli* and conseq-uences on protein quality control and cellular ageing, *The EMBO Journal* **29** 910-923

76. Yébenes H, Mesa P, Munoz IG, Montaya G, and Valpu-esta JM 2011 Chaperonins: two rings for folding, *Trends in Biochemical Sciences* **36** 424-432

77. Young JC, Agashe VR, Siegers K, and Hartl FU 2004 Pathways of chaperone-mediated protein folding in the cytosol, *Nature Reviews Molecular Cell Biology* **5** 781-791

78. Young JC and Hartl FU 2000 Polypeptide release by Hsp90 involves ATP hydrolysis and is enhanced by the co-chaperone p23, *The EMBO Journal* **19** 5930-5940.

*Rev. Plant Pathol. Vol. 6, 2014*
*Indian Society of Mycology and Plant Pathology*
*Scientific Publishers (India), Jodhpur pp. 59-91*

# HARNESSING THE POWER OF METABOLOMICS TO IMPROVE CROP TOLERANCE AGAINST STRESS

***Luis A.J. Mur[1, 4], Amanda J. Lloyd[1], Kapuganti J. Gupta [2], Usha Charkaborty[3] and Bishwanath Chakaborty[3]***

[1]Aberystwyth University, Institute of Environmental and Rural Science, Edward Llwyd Building, Aberystwyth, UK, SY23 3DA
[2] Department of Plant Physiology, University of Rostock , Albert Einstein Str-3, 18059 Rostock,Germany
[3] Department of Botany, University of North Bengal, Siliguri-734013, India email lum@aber.ac.uk

**KEY WORDS :** Metabolomics, MetMAS, FT-IR

## ABSTRACT

Metabolomics – simultaneously assessing large numbers of metabolites within a sample – is emerging as a powerful post-genomic approach. This review will introduce the field of plant metabolomics, its associated technologies and analytical approaches as well as the remaining challenges. It will outline how metabolomics is increasing the knowledge of plant responses to stress and likely tolerance mechanisms. We describe how metabolomics is now being developed to be in a position to (i) inform plant breeding strategies through the use of metabolite marker–assisted selection (MetMAS), (ii) acting as an early indicator of crop or environmental stress or (iii) to improve agricultural practices, in for example allowing the improvement of formulations of treatment with plant growth promoting microbes.

**Properly applied therefore, metabolomics stands ready to play an important role in securing plant productivity against the effects of adverse environmental changes.**

## WHAT IS METABOLOMICS?

The concept of metabolomics was first coined by (79) based on their work in yeast research and was subsequently applied to plants (30,31,32). These pioneering papers by Oliver Fiehn helped to establish the terminology associated with metabolomic approaches within the plant science community (Table 1). Metabolomics is progressive extension of omics concept but it deals with large scale small molecule information in living organisms (98). True metabolomic analyses seek to gain an overview of the entire metabolic complement of a organelle, cell, organ or organism under defined conditions. In addition parallel assessment of wide range of metabolites is possible. In the case of plant metabolomics this represents a uniquely difficult challenge due to the 100 - 200,000 different metabolites found within the plant kingdom (30). Even with the model plant Arabidopsis, with its very small genome and a tendency to low copy gene families (1), there are more 5,000 different low-molecular weight compounds. Thus, it is currently impossible to either quantitatively or and qualitatively detect all the metabolites within plants (28) due to their sheer numbers, chemical complexity (varying in molecular weight, polarity, solubility) as well as physical properties (for example, volatility) and huge differences in concentration (varying between pM and Mm) (24).

Metabolomic research encompasses at least three aspects. In one there is the employment of a wide range of technology platforms which aims to either rapidly provide comprehensive metabolite fingerprints (Table 1, Fig. 1), detect large numbers of metabolites or unambiguously confirm the identity of key metabolites (41). Metabolite fingerprinting commonly employs spectroscopic methods using electromagnetic (EM) radiation to energise metabolite bonds, causing the absorbance/redirection of the radiation or the transition of the sample atoms to a higher energy state. Fingerprinting methods include Nuclear Magnetic Resonance spectroscopy (NMR), Fourier Transform-Infra Red spectroscopy (FT-IR) and Raman spectroscopy.

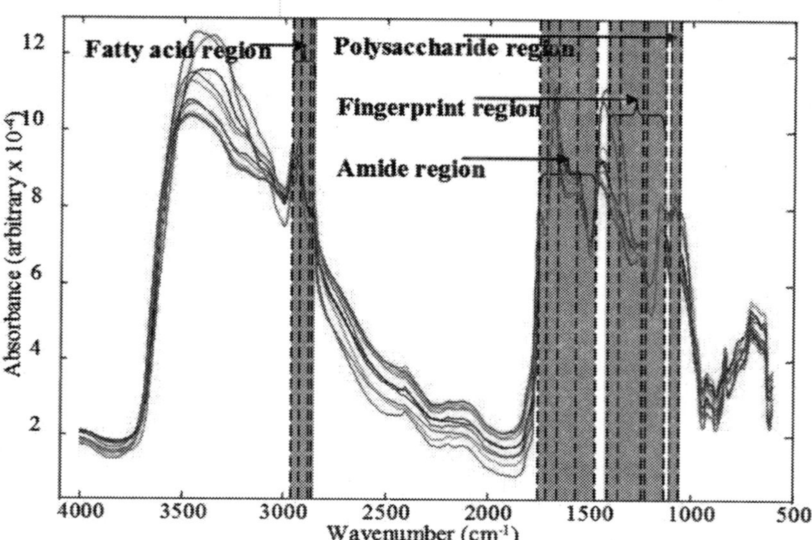


**Fig. 1.** Characteristic regions within Fourier Transform-Infra Red (FT-IR) spectra

FT-IR spectroscopy uses the mid-IR wavelengths (4000-600 $cm^{-1}$) which is broken down into sub-regions corresponding to general classes of compound (shaded in grey); fatty acids (2956-2850 $cm^{-1}$), amide region (1745-1454 $cm^{-1}$), fingerprint/mixed region (1400-1114 $cm^{-1}$) and polysaccharides 1085-1052 $cm^{-1}$). The IR absorbencies for the key functional groups that make up these regions are shown as dashed lines (Table 2).

**Table 1.** Definitions used in Metabolite analysis

| Term | Definition |
|---|---|
| Metabolomics | Non-biased identification and quantification of all metabolites in a biological system |
| Metabolite fingerprinting | High throughput, rapid, global screening of the metabolic composition of an organism or tissue, with the primary aim of sample classification and discrimination according to their origin or biological relevance, without identifying and quantifying individual metabolites. Often used as a forerunner to metabolic profiling. |
| Metabolite/ metabolic | Identification and quantification of a limited |

| | |
|---|---|
| profiling | number of pre-defined components, generally related to a specific metabolic pathway. |
| Target analysis | Qualitative and quantitative analysis of selected groups of metabolites by using optimized extraction and dedicated separation/detection techniques. |
| Metabolic footprinting | The analysis of the extra-cellular metabolome of the microbial cultures, which is a combination of spent growth medium and metabolites excreted. |
| Metabonomics | A non-plant term generally used to define the technology used to measure quantitatively the metabolic composition of body fluids following a response to pathophysiological stimuli or genetic modification. |

**Table 2.** Characteristic regions within Fourier Transform-Infra Red (FT-IR) spectra and absorbencies for key functional groups

| Region | Wavenumber ($cm^{-1}$) | Band assignment |
|---|---|---|
| Fatty acid region | 2956 | $CH_3$ asym. stretch |
| | 2920 | $CH_2$ asym. stretch |
| | 2870 | $CH_3$ sym. stretch |
| | 2850 | $CH_2$ sym. stretch |
| Amide region | 1745-1735 | >C=O stretch, ester, fatty acids |
| | 1705 | >C=O stretch, ester, carboxylic groups |
| | 1652-1648 | Amide I, (C=O) |
| | 1550-1548 | Amide II, N-H, C-N and proteins |
| | 1460-1454 | C-H bend, from $CH_2$ |
| Mixed region | 1400-1398 | C-O bend from carboxylate ions |
| | 1303 | Amide III (C-N) |
| | 1240 | P=O from phosphate |
| | 1222 | P=O |
| | 1114 | C-O-P, P-O-P |
| Polysaccharide region | 1085 | Ring vibrations |
| | 1052 | C-O, C-O-C from polysaccharides |

asym. = asymmetric; sym. = symmetric

For reasons of cost and efficiency our research employs FT-IR as a fingerprinting approach. FT-IR instruments measure the frequency and intensity bond absorption, thereby generating a unique biochemical spectrum/metabolic fingerprint of the sample. The mid-IR region (4000-600 $cm^{-1}$) is particularly useful in discerning chemical bond identity due to their distinctive vibrations (stretching or bending, [44]) at particular wave numbers with this region. Thus, different wave numbers used within the IR spectra indicate the types of bonds of some chemical classes within a sample. In particular, the fatty acid region (~2956 to 2850 $cm^{-1}$), the amide region (~1745 to 1454 $cm^{-1}$), the mixed or 'fingerprint' region consisting of a variety of weak features (1400 to 1114 $cm^{-1}$) and the polysaccharide region (~1085 to 1052 $cm^{-1}$). One drawback of FT-IR is the strong absorption of water in the mid-IR region and can interfere with the fingerprint. Thus, the sample must be dehydrated (for example by baking at 55 $^{o}$C for 30 min) prior to analysis.

FT-IR has been successfully applied as a metabolite fingerprinting technique in many areas of plant research. Due to the concentration of distinctive bond types, FT-IR has been most often used to analyse structural changes within the cell wall. Thus, FT-IR has been used to identify cell wall mutants (74), determine homogalacturonan content of the cell wall (25) assess variation in the xylem cell wall types in *Populus tremula* (42) or pectin distribution in response to cadmium in the root cell wall (27). FT-IR proved to be an effective screen for changed lipid content in *Chlamydomonas* mutants (49) and allowed the characterisation of pine pollen (113). FT-IR can also be used to probe the molecular structure of (for example) targeted proteins as exemplified by studies which examined variation in amino acid β-sheets in dehydration stress LEA7 protein (84) or oxidative variation in LOV domain with the cryptochrome 1 protein (58). In our research we mostly used FT-IR analysis test the suitability and reproducibility of plant systems prior to further targeted metabolite profiling (4).

In order to identify metabolites within a sample, scientists undertaking metabolomics usually follow one of two routes. One strategy employs nuclear magnetic resonance (NMR) and alternative is based on Mass Spectrometry (MS). It is a matter of person choice which is selected as whereas NMR allows a high degree of metabolite identification it is relatively insensitive compared to most forms of MS (97). NMR utilises nuclei with either odd atomic or mass number, such as those containing $^{1}H$, $^{13}C$, $^{15}N$ and $^{32}P$, which act as magnets that interact with an external magnetic field in a process termed nuclear spin. Nuclei are irradiated with EM radiation; and at specific frequencies different nuclei are in resonance, allowing promotion to high-energy spin states. The subsequent emission of radiation during the relaxation process is detected as the absorbance frequency. The NMR spectrum (called the chemical shift) depends on the effect of shielding by electrons orbiting the nucleus and can provide a unique signal for each chemically distinct nucleus within a molecule. The derived spectra may then be compared with a reference compound.

MS measures the mass-to-charge ($m/z$) ratio of low molecular weight charged particles. Ionisation is commonly by Electrospray Ionisation (ESI), Chemical Ionisation (CI), Photo-Ionisation (PI) or Matrix Assisted Laser Desorption Ionisation (MALDI). Mass spectrometers occur in many forms but in all the charged particle passes through a magnetic field resulting in a deflection the degree of which is proportional to the $m/z$ ratio. Mass spectrometer performance is often improved by prior chromatography (in "hyphenated" chromatography-MS techniques), such as Liquid Chromatography (LC) and Gas Chromatography (GC); or electrophoresis, such as Capillary Electrophoresis (CE), to separate the metabolites within a sample prior to introduction in the MS (97). Various other combinations of liquid chromatography (LC)-NMR, LC-solid-phase extraction (SPE)-NMR, and LC-MS-SPE-NMR have been developed (108)

Much early work in plant metabolomics used GC-MS to compare; for example, metabolite profiles in wild type and transgenic potato tubers (88,89,90) and mutant analysis in *Arabidopsis* 31, 101). GC separations tend to yield much greater chromatographic resolution than LC , but have the disadvantage of being limited to volatile compounds (alcohols, esters and monoterpenes) hence, samples must be derivatised to aid volatility (89). Multiple derivatised products can arise from the same metabolite can complicate the analysis of derived metabolomic data. Given these limitations the attractiveness of LC is immediately obvious. The effectiveness of LC-MS is directly dependent on the quality of the LC column used and many metabolomic groups use ultra-performance (UP) columns which use small chromatographic particles and higher mobile phase flow to provide higher resolution and greater biological information (22, 6). Given this, LC-based approaches are increasingly being exploited in metabolomics (for example, 57; 92). One drawback of hyphenated approaches is the loss of sensitivity so that low abundance metabolites may not be detected. Both Direct Injection (DI) and Flow Injection (FI) linked to MS detection devices, provides high-throughput analyses of relatively crude sample extracts; producing a single mass spectrum per sample representative of the metabolite composition and increasing MS sensitivity. We have used both DI and FI in own work (65, 4).

Following on from other post-genomic approaches; transcriptomics (12, 78) and proteomics (99,100), the Metabolomics Standards Initiative (MSI; http://msi-workgroups.sourceforge.net/) has started to formulate a minimum core set of necessary metabolomics metadata. Thus, MSI is moving to include growth environments, treatments, harvesting and storage condition as essential descriptors of metabolomic experiments (33,34). At the analytical level sampling and extraction methods should be clearly defined as these have a significant influence on the metabolite range that is detected (45). Crudely, metabolites with a high solubility (i.e. carbohydrates) are typically extracted with polar solvents such as methanol, methanol-water mixtures, or ethanol (96), whilst non-polar solvents such as chloroform, hexane or ethyl acetate recover such as membrane lipids and fatty acids (32). The success of strict metabolomic experimental approaches was recent demonstrated in some recent inter-laboratory comparisons of metabolite profiles (109,5).

Metabolite detection platforms must be married to robust data analysis strategies in order to derive statistically relevant results. In order to analyse the very large datasets that are typically derived from metabolomic experiments, computer-aided statistical and mathematical models; termed "chemometrics", are employed (41). Univariate methods – analysis of variance (ANOVA) and *t*-tests - can be used to identify to test for significantly modulated metabolites but interactions – a key feature of multivariate data sets – are very poorly revealed. Thus, most commonly used chemometric approaches are non-supervised and supervised multivariate methods. Non-supervised techniques require no information other than the original data and cluster the metabolite data into groups due to natural variation within the data. Hierarchical clustering analysis (HCA) is a non-supervised method involving progressive pair-wise comparison of sample variation and their ordering according to the relative similarity. Results can be viewed as dendrograms where branch lengths are related to group similarities. In Principal component analysis (PCA) reduces the dimensionality of multivariate data to “principal components” (PC) which progressively describes less of the variation within the data set (thus PC1 describes the greatest source of variation and PC2....$PC_n$ progressively less). PCA enables the visualisation of the natural variability within data in the form of 2D or 3D plots of different combinations of underlying variables, known as principal components (PC) (51). In supervised methods variables are attempted to be associated with given classes within the data set. These are particularly useful in metabolomic analyses as operator or technical errors are often the major source of variation. In Discriminant Function Analysis (DFA) analyses are based on reduced data from PCA and aims to maximise the variance between classes, while minimising variance within classes (39,41).

Clearly, there are many other chemometric approaches which could be followed when analysing metabolomic data and for further details the following papers are recommended (26,35,50) and particularly (50). Individual researchers must derive workflows based on available technology platforms and expertise. There are various software technologies and web based resources are available for peak analysis, analyzing and comparing the large scale metabolite data (102). The approach followed by our group is outlined in Figure 2 and illustrates our use of metabolite fingerprinting approaches and electrospray ionisation – mass spectrometry (ESI-MS).

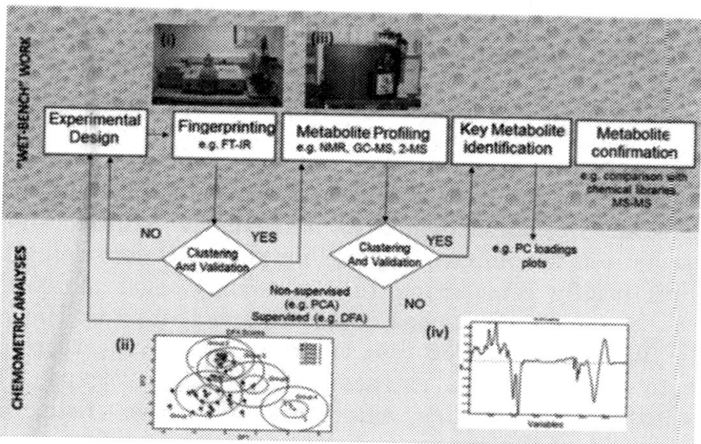


**Fig. 2.** An example metabolomic workflow

A schematic representation of steps undertaken in metabolomic based analyses to validate experimental set up, and to select profiling and chemometric analysis methods in order to confirm metabolite identification. An approach suggested from our experiences is presented. The variation which is a feature of biological experiments necessitates rigorous and statistically valid experimental design. The efficacy of the experimental design to inform the problem under investigation is assessed though a rapid and cost-effective metabolite fingerprinting approach involving such as Fourier Transform Infra Red Spectroscopy (FT-IR). (i) The Bruker Vertex FT-IR spectrometer is shown above the Fingerprinting box. Chemometric analysis involved non-supervised multivariate Principal Component Analysis (PCA) or supervised Discriminant Function Analysis (DFA). (ii) An example output from a DFA is shown where separation between FT-IR spectra for samples from five experimental groups is shown. The model is based on the used of 5 principal components (PCs) which explained 99.32% of the total variation. The circles present 90 and 95% confidence intervals for the samples associated with each group. Note that samples from Groups 1, 5 and 4 form distinctive groups. If the experimental approach is validated it is appropriate to proceed to more laborious metabolite profiling using mass spectrometry (MS) with any of a range of pre-separation techniques or direct infusion electrospray Mass Spectroscopy (ESI-MS). (iii) The Micromass LCT-MS used in our studies is shown. The results should again be validated by chemometric approaches. Key metabolites could be identified by either examination of PC-loading plots (iv) where those variable (here either infra-red absorbencies or mass-ions [$m/z$]) used by the respective algorithms to derive PCA or DFA models may be revealed. Shown is an example PC-loadings plot. Variables which lie close to the dotted line do not majorly contribute to the PCA/DFA models and corresponding variables plotted further from the dotted line contribute more. Based on the loadings, $m/z$ may be tentatively identified by comparison with chemical database and confirmed by tandem MS (MS-MS).

## Plant Stress responses: Important examples of metabolomic reprogramming

Early plant metabolomic analyses focused on establishing crop responses to herbicides and assessing the effects of genetic modification on crop species (10, 14). However, as will be outlined below, due to the role played by complex biochemical events, metabolomics approaches are particularly suitable for the assessment of the impact of stress on plants. Before doing this, it is work briefly considering this commonly used phrase "plant stress". Stress was first used in a biological context by Hans Selye (1907- 1982) in the 1936 (94) when describing the responses of mice to injection with extracts from various organs. Though described many decades ago and in another field, Selye's development of the general adaption syndrome offers lessons to plant scientists which can see stress as always linked to cellular damage. Selye proposed three phases in a general stress adaptation syndrome. The first stage is a perception phases ("alarm") where stressors – for infection or heat stress – are detected and elicit a short term adaptive response. If the stressor persists attempt are made to tolerant / resist its effects. Only in the last phase ("exhaustion") is gross physiology affected and, for example, crop yield is reduced. It is important to note that changes in genes, proteins and biochemistry usually linked with "stress" occur at each phase. It is thus unsurprising that "stress-associated" features such as the generation of reactive oxygen species (ROS) and reactive nitrogen species (RNS) are also involved in the regulation of stomatal function (77) or root hair formation (13).

Extensive characterization of plant responses to pathogens has revealed the dominant role of metabolite changes both as a consequence of disease and also of the exhibition of resistance. Preformed chemical defences, termed phytoanticipins, may be found at the plant surface or sequestered in vacuoles or organelles and released through a hydrolyzing enzyme after pest or pathogen challenge. Saponins represent one of the best classes of characterised phytoanticipins. These are isoprenoid based metabolites which structurally are classified as either triterpenoid, steroid or steroidal glycoalkaloid types (80). In oats (*Avena sativa*) the saponin avenacin complexes with sterols in the fungal pathogen membrane to form pores and thus affect viability (72).

If the passive defences are defeated, energy-requiring inducible defence mechanisms, involving interconnecting and multi-component signalling cascades are deployed; equivalent to Selye's "alarms". In the animal kingdom many of these alarms are peptide hormones but although examples of these types of signals are used in plants (62) small molecular weight metabolite signals have received much more attention. "Stress"-associated signals include salicylates (2-hydroxybenzoic acid and derivatives), ethylene, jasmonates (cyclopentanone derivatives from linolenic acid), abscisic acid (ABA) and now also hormones previously linked to plant development such as auxins (7). These may be localised to the site of infection or mobilised systemically to confer systemic resistance against subsequent attack.

These alarm signals are rather crude sub-categorised as regulating defence against pathogen groups following specific forms of pathogenesis. Thus, salicylates have been shown to mediate resistance against biotrophic pathogens - where the infected host cell remains viable for a considerable period during disease. In *Arabidopsis* the use of mutants demonstrated resistance against the oomycete *Peronospora parasitica* (downy mildew), the fungus *Erysiphe orontii* and strains of the bacteria *Pseudomonas syringae* was dependent on salicylate. Notably, resistance against the necrotrophic (where host cell death is an early feature of infection) fungus *Alternaria brassicicola* was not affected. Indeed, jasmonate and ethylene-dependent defence responses mediated resistance against necrotrophic fungi and bacteria including *A. brassicicola*, *Botrytis cinerea* and *Plectosphaerella cucumerina*.

The responses evoked by these signals may be related to Selye's adaptive phase where resistance/ tolerance to disease is actually effected and involves whole-scale reprogramming plant secondary metabolism. Classically, this involves production of phytoalexins at the infection site following from *de novo* gene synthesis where they are assumed to act as antimicrobials agents (46). In *Arabidopsis* the major phytoalexin is the indole-sulphur compound, camalexin (3-thiazol-2'-yl-indole), derived from L-tryptophan (105) and acts against *P. syringae* pathovars (36,105), fungal pathogens (103,29,23) and oomycetes (91). As camalexin, indole glucosinolates and auxin share a common intermediate, indole-3-acetaldoxime, the biosynthesis of each class is closely linked so that mutants are required to reveal the relative contribution of each to resistance (93).

Many other defence compounds are derived *L*-phenylalanine and have a basic $C_6C_3$ phenylpropane skeleton. A very large number of metabolites are derived from this skeleton. Hydroxycinnamate derivatives form the basis of lignin (monolignols) precursors, coumarins and hydroxycinnamate esters i.e. sinapate esters. Monolignols differ in the number methoxyl substitutions on their aromatic ring from zero (p-coumaryl-H unit), one (coniferyl-G-unit) and two methyl groups (sinapyl-S unit) (11). The monolignols do not accumulate to high levels within living plant cells but are immediately transported to the cell wall to be polymerised to form lignin polymer and represents a typical defence response against some fungal infections (69).

Flavonoids are another class of phenylpropanoid –derived metabolites consisting of two phenyl rings connected by a three-carbon bridge ($C_6C_3C_6$). This group may be subdivided into six major groups: chalcones, flavones, flavonols, flavandiols, anthocyanins and condensed tannins (polymers of anthocyanidins known as proanthocyanidin). Flavonols invariably occur naturally as glycosides (usually with D-glucose) and hydrolysis removes the sugar to form aglycones such as kaempferol, quercetin and myricetin. Flavonoids have a wide range of functions; including UV protection, pollinator and insect feeding attractants, feeding deterrents and anti-microbial compounds (48,104).

Adaptive responses are equally complex in response to abiotic stressors and often feature substantial changes in primary metabolism. Thermotolerance in plants has mostly clearly associated with the expression of heat shock proteins (hsp) which protect proteins through chaperone functions. Interestingly, some of the pathogen induced alarm hormones; salicylate, jasmonates and ethylene, are also involved in heat stress and can influence hsp expression (17,18). However, wider metabolomic changes are also vital to protect or modify cellular biochemistry in response to temperature effects on the structure, catalytic properties and function of enzymes (59). Such changes include the production of (i) osmolytes to reduce cellular dehydration, (ii) solutes that stabilize enzymes, membranes and other cellular components; and

(iii) chelating agents sequester metals and inorganic ions (59). Fructan accumulation in response to drought and temperature stress may be very important and following interaction of the polysaccharide with a phospholipid headgroup has led to increased stability of the membranes (64). Frost hardiness may also be dependent on carbohydrate changes (106) but again with some distinctive elements.

In the our opinion, Selye's exhaustive phase should only be of interest to plant scientists interested how cellular nutrients may be being mobilised out of the stressed area – possibly paralleling some senescence-like process. However, given the theme of this review, we suggest that this phase will offer few mechanisms through which plant tolerance could be improved. Indeed, we would recommend that the exhaustive phase be avoided in metabolomic studies as the processes involved with - for example – cell death would introduce a heavy bias in the data processing.

## NEW INSIGHTS FROM METABOLOMIC STUDIES

The previous section described metabolite changes as defined using targeted metabolite assays, so it is therefore unsurprising that the simultaneous profiling of large numbers of metabolites is yielding new insights. One of the earliest metabolomic studies into plant-pathogenic interactions was undertaken by our group where susceptible and resistant metabolomic changes in the model grass species *Brachypodium* challenged with the rice blast pathogen *Magnaporthe grisea* (4). Focusing on non-polar metabolite this revealed phosphatidic and phosphatidyl glycerol phospholipid processing during resistant responses which correlated with the production of jasmonate.

Many subsequent studies have focused on the interaction of well-characterised *Pseudomonas syringae* strains with its hosts. Although understandable, metabolomic investigations this using these model pathogens possess considerable problems as inoculations involve injecting leaves with high titre bacterial suspensions. Thus, it could prove difficult therefore so easily distinguish whether metabolites – particularly those associated with primary metabolism – originate from the host or pathogen (86). Our solution to this problems was to develop a "dual metabolomic" approach, were metabolite changes were determined in Arabidopsis cell culture inoculated with *Pseudomonas syringae* which allowed rapid separation of host and pathogen cells prior to their separate analysis (3).

The Allwood (3) study was a demonstration study based on FT-IR fingerprinting but recently, a more comprehensive *Pseudomonas syringae* analysis has examined actual metabolite changes leading to disease development (110). This demonstrated that there was a reduction in sucrose catabolism and thus an accumulation of sucrose. There was also a shift to indole metabolism but not accumulation of indole glucosinolates; instead indole carboxylic acid was incorporated into the cell wall. Another feature was the accumulation of γ-aminobutyric acid (GABA), glutamine, asparagine and phenylpropanoids. *Pseudomonas syringae* mutants in GABA transaminase are unable to break GABA down and have reduced virulence on plants (82). It seems likely therefore that pathogen induced GABA accumulation is an important feature of virulence.

Another recent study examined the responses of rice lines to *Xanthomonas oryzae* (*Xoo*) comparing disease (cv TP309) and resistant (isogenic cv TP309 *Xa21*) forming interactions and including a *Xoo rax* mutant which is required for avirulence gene (*Ax21*) production that is essential for the triggering of host defences (92). A common response to challenge with either bacterial strain was an increase in arginine which the authors tentatively associated with increased catabolism of pathogen-derived proteins. Resistance was associated with increases in ornithine, citrulline, tyrosine, phenylalanine, lysine as well as arginine but not proline indicating specific modulation of certain metabolic pathways in plant defence, perhaps specifically towards the production of anti-microbial alkaloids (92).

Considering plant responses to abiotic stress, one early metabolomic study employed GC-MS to obtain a metabolomic understanding of the response of Arabidopsis to heat (43°C and 44°C) and cold (-4 °C) (54). One very surprising outcome was that there was a considerable overlap between the induced metabolite changes which included salicylate (the "alarm" hormone), a range of amino acids and polyamines and such as maltose, galactinol, and raffinose. Most specific stress events where seen after cold treatment and included the elevation of glucose, proline, arginine, phenylalanine, fructose-6-phoshate and glucopyranose. In the case of heat shock, minor changes included those of uracil and quinic acid. In an interesting development, these authors could only partially relate metabolite changes with associated enzyme transcript abundance underlying the value of metabolomics in reflecting post-transcriptional and translation control mechanisms (54). A metabolomic study which compared transgenic *Arabidopsis* plants over expressing the C-repeat/dehydration responsive element-binding factor (CBF) 3 suggested that many of the cold-induce metabolite changes were regulated by CBF cold response (19). The increased concentrations of glucose and charged amino acids on chilling could suggest that they are involve in reducing cellular dehydration/ ice crystal formation and associated cellular damage. A dominant role for osmotically active metabolites in adaptive stress responses was also suggested by a study by Morsy (73) where a range of rice (*Oryza sativa*) genotypes were studies after chilling, salt and osmotic stress. Focusing on carbohydrates, they observed an association between stress tolerance in rice genotypes and the stress-responsive accumulation of galactose and raffinose. A requirement for increase carbohydrate metabolism may also reflect the considerable energetic demands required for tolerance as also suggested a metabolomic study of water and salt stressed grape which revealed much higher concentrations of glucose and malate (20). Interestingly, in line with a large number of other workers Morsy (73) also suggested that stress-tolerance was linked to a more effective ROS-scavenging system; presumably by preventing/ delaying entry into the "exhaustive" phase of stress responses. Attempts to focus on the effects of ROS on the metabolome have involved treating Arabidopsis cultures (9) or roots (63) with menadione which interferes with electron transport in organelles (21). These resulted in a substantial inhibition of the TCA cycle and amino acid metabolism, a dominant role for glycolysis and a flux of carbon into the pentose phosphate pathway which was consistent with a switch from anabolism to catabolism. Clearly, such observations were not consistent with those observed during responses to abiotic stress and further substantiate the hypothesis that high level ROS generation is a feature of entry into the exhaustive stress phase. This stated, in root tissues at least, it appears that metabolic activity recovered after oxidative inhibition (63).

Other authors focusing on salt tolerance in tomato used FT-IR to identify key changes in fruit flesh. Analyses employed PCA and DFA but a genetic algorthim (GA)-based approach proved to be most powerful (51). In GA analyses populations of data - called "chromosomes" – are re-arranged and tested for their appropriateness in explaining the observed experimental outcome(s). This process is repeated until the "fittest" solution is obtained. The parallel with biological evolution is self-evident. GA found that responses to salt lead to the production of saturated and unsaturated nitrile compounds, cyanide-containing compounds and other nitrogen-containing compounds. These authors suggested that the increased occurrence of cyanide or nitrile containing compounds was associated with detoxification of hydrogen cyanide a by-product of ethylene production which is known to occur during salt stress (71,52). A link with ethylene was further substantiated in a time-course experiment on salt-stressed *Arabidopsis* cell cultures (57) and revealed an early prominence of methylation cycle for the supply of methyl groups which is essential to ethylene biosynthesis (107). Other important activated pathways included lignin production, glycine betaine production, glycolysis and sucrose metabolism and ultimate suppression of the methylation cycle. Moving beyond Arabidopsis, when salt response of the Salt Cress (*Thellungiella halophile*), a close relative of *Arabidopsis* was examined and changes in other sugars, sugar alcohols, organic acids and phosphate were detected (38).

In a fascinating and impressive study of salt tolerance in Euphrates poplar trees (*Populus euphratica*) growing within the Negev desert, β-alanine, valine and proline were increased in osmotically stressed areas. β-Alanine is the precursor of β-alanine-betaine which has similar osmoprotective properties as glycine-betaine so that this may indicate a similar salt tolerance mechanism as revealed by the Kim *et al.*, (57) study. Interestingly however, there were very limited changes in the concentrations of carbohydrates and organic acids which have been found to be important in other studies (38,73). In another work where relevance to field conditions was uppermost in the author's minds the effects of combinational stress; drought and heat, was examined in *Arabidopsis* (87). These plants exhibited the expected accumulation in carbohydrates (for example, sucrose, maltose and glucose) but proline accumulation – a well-established drought marker (112) - was not observed under conditions of combinational stress. With such surprising observations more work is clearly needed to understand how signalling pathways interact (70).

## PUTTING METABOLOMICS TO WORK: APPLICATIONS TO AGRICULTURE

The aforementioned litany of metabolomic studies into plant stress responses is clearly impressive but a common feature of nearly all studies is the use of model plant systems grown under controlled conditions. Clearly, to have biological relevance, metabolomic changes in the field need to be deduced. Few such studies have been undertaken with perhaps the most notable being that of (14) who compared field grown wild type and genetically modified potatoes.

Following a strategy opposite to that of a strictly controlled MIAME/MIS- type experiment, we have recently tested how far natural variation could be incorporated into our metabolomic analyses and yet still reveal biologically relevant information (Fig. 3). Thus, we have sampled tea leaves infected with the tea blister fungus (*Exobasidium vexans*) from the Margaret Hope Tea garden in Darjeeling, India . The leaf symptoms were classified into healthy (or more precisely, asymptomatic) and three different stages of tea blister disease progression. For each stage of infection, 10 samples were taken. Initially, using FT-IR fingerprinting we assessed whether metabolite differences within the samples could be related to the symptom stages. Unsurprisingly, non-supervised PCA failed to distinguish between FT-IR fingerprints from each class (data not shown) but DFA succeeded in identifying IR wave numbers which allowed discrimination between healthy and early infection stage symptoms (S1 and S2) and particularly the last necrotic stage 3 (S3) which we consider corresponds to Selye's exhausted biological stress stage (Fig. 3). We have now identified those metabolites which significantly differed as Tea blister symptoms developed (76). This demonstration that metabolomics can be applied in field situations represents a "freeing of metabolomics from the lab" and now can be exploited to inform agricultural practice. Given the current state of technology, the authors suggest the following applications.

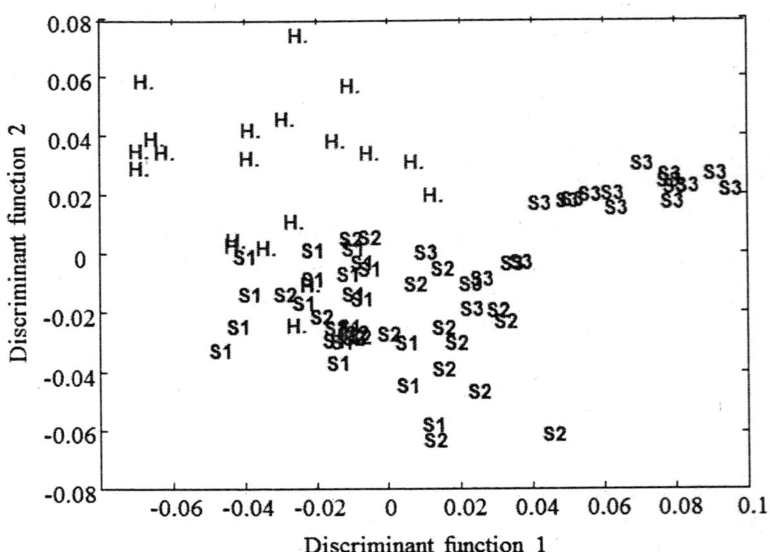


**Fig. 3:** FT-IR analysis of Tea Blister (*Exobasidium vexans*) infected tea (*Camellia sinensis*) leaves.
**(A)** Authors (left BNC, right LM, sampling tea blister exhibiting tea leaves from the Margaret's Hope Tea Gardens (North Kurseong, Darjeeling, India, 26.95059N, 88.279537E; Altitude 950-1830m). Inset: Tea blister exhibiting tea leaf (Bar = 2 cm). **(B)** Discriminant Function Analysis of Fourier Transform Infra Red (FT-IR) spectra from 10 samples from each of the following infection stages; H = Healthy (asymptomatic) ; S1 = stage 1; S2 = stage 2 and S3 = stage 3 leaves. The stages corresponds to S1; leaves showing anthocyanin accumulation; S2: leaves with blisters (as in Fig. 3A inset) and S3; necrotic, which suggested corresponds to the exhausted stage of biological stress as defined by (Selye, 1998). The model is based on 10 PC (principal components) which encompassed 99.23% of the total variance amongst the samples. Results from samples from healthy (asymptomatic) stages are coloured green whilst all results from diseased leaves are coloured red. Note, that samples from healthy leaves are mostly clearly separated from infected leaves in the DFA model.

## Metabolite markers for environmental stress

Perhaps the most immediate application of metabolomic approaches to agriculture involves the identification of early stress biomarkers. Predicted outcomes of climate change include a lack of climate predictability and oscillation between extreme weather events (81). Clearly, such climatic oscillations could give rise to crops which appear phenotypically normal during their early stages of development but in fact are in at "alarm" or "tolerance/resistance" stage of biological stress. The impact of such stress would undoubtedly lead to reduced yields and productivity (47). Metabolomic changes could therefore be used and an estimate of crop "health". Ideally, such biomarker assessments need not involve continuously sampling crop plants and analysis via fingerprinting or profiling based approaches and this may be avoided if the stress markers are volatile. Volatile organic compounds (VOC) - for example isoprenoids - are well-established indicators of stress (61) and sensitive cost-effective detection methods have been demonstrated for example, using electronic noses (60). Thus, coupling metabolomics to established VOC research is likely to lead to new targets for monitoring crop stress in the field.

## Using metabolomics to improve agricultural practice

If metabolomics could effectively be used to augment the "alarm" phase of crop plants, it could also aid to improve plant tolerance/ resistance. One technology driven solution is to combat increased environmental is to derive genetically modified (GM) crops. Clearly, by contributing to our understanding of plant biochemical processes, metabolomics can influence thinking on how crops could be altered to withstand environmental change. More prominently, the wide-ranging nature of metabolomic profiling has led metabolomics being employed to establish the "substantial equivalence" of a GM derivative to the unaltered parental line (14,85).

Field tolerance in crops can be improved through the use of such as plant-growth-promoting rhizobacteria (PGPR), which are widely used in India in agriculture (2, 8, 15, 95, 56, 66). Often PGPR act by inducing the production of defence hormones and plant developmental hormones so that besides increasing immunity, plant productivity is also improved (111, 68, 67). Metabolomics could be employed to perfect PGPR formulations so that defences are effectively induced with minimal impact on yield.

**Metabolite marker -assisted selection (MetMAS).**

Plant breeders are very familiar with the use of molecular markers linked to major genes controlling important traits; known as marker associated selection (MAS). A range of markers may be used; for example, single nucleotide polymorphisms (SNP) and microsatellites with which can be linked to traits in a segregating population (53). MAS has proven to be very effective is informing breeding programmes for traits that are based on small numbers of genes as there is often only a small number of DNA markers available to define genotypic variation. Thus, MAS is ideal for traits such as certain types of disease resistance, which are often dependent on single or only a very few genes. In practice, the more genetically complex a trait the more difficult it is be exploited using MAS. In particular genes making minor and yet significant contributions may be lost due to the effects of majorly acting genes or the large amount of data needed to accurately estimate their effects. This is now being remedied with the large genome wide SNP maps that are being derived for many crop species which, when coupled to high-power statistical approaches (37) will allow whole genome variation to be assessed and "genome selection" approaches to be applied to even wheat (83). Genomic selection represents very much a "genes-first" to plant breeding, but metabolomics could be used in a "traits-first" breeding tool. As indicted above, stress tolerance is often a multi-component and thus multi-genic trait. Metabolomics' trait- first approach could use multiple metabolite markers to follow key sergeant lines within a population and would guarantee the maintenance of desired trait. This latter is an especially important feature as it often difficult to determining the distinctiveness of individual lines by conventional morphological and field analysis, especially if the crop is an out-breeder and hence each variety is heterogeneous. Finally, it is important to emphasise that metabolomic approaches could be relatively inexpensive – particularly FT-IR fingerprinting.

Metabolite marker assisted selection requires a tight association between genetic variation and metabolomic patterns and this demonstrated by Aberystwyth (UK)- based scientists within a context relevant to plant breeding. (40) used FT-IR fingerprinting to distinguish between outbreeding *Lolium multiflorum, Festuca glaucescens* and *Festuca pratensi* species.

FT-IR placed the two most clearly related species belonging to the genus *Festuca* L closer together than to *Lolium* suggesting a correspondence between genetic and metabolite variation. A more significant challenge was offered in comparisons of the Lex86 and Dancer amenity grass lines which required 3 years of field trials before they could be identified as distinct varieties but these were easily distinguished by FT-IR. Wheat substitution and alien introgression lines are important tools in wheat genetics but require constant re-examination to confirm their cytological status. However, once again FT-IR readily distinguished plants lines lacking a specific chromosome.

The actual isolation of a gene based on data set which included metabolomic results was demonstrated by Mouille et al., (75). This group examined a Bay-0 x Shahdara recombinant inbred line population and identified, 33 QTL linked to various cell wall characteristics using classical genetic approach as well as FT-IR microspectroscopy. Based on assessments of cell wall arabinose-rhamnose ratios a gene accounting for 51 % variation within a major QTL was isolated by positional cloning. In a *tour de force* study, Chan et al., (16) have demonstrated the use of metabolomic data in genome selective approaches, again in *Arabidopsis*. This group compared >200,000 single nucleotide polymorphisms in wild populations of Arabidopsis and associations were attempted with genotypic variation in the accumulation of 327 metabolites. They demonstrated the co-selection of sections of the genome and linked metabolites, suggesting evolutionary selection acting at the level of the metabolome. These authors also noted that genotype-metabolite associations were heavily influenced by environmental variation which is likely to complicate metMAS approaches. This last caveat notwithstanding, technology, statistical approaches and expertise now seems to be in place to begin to exploit metMAS in crop breeding programmes.

Taking all of the above points together over the last decade, metabolomics has moved from a "Cinderella" post-genomic approach (43) to one which is providing some real insights. We now suggest that it is poised to make real impacts on agriculture and perhaps most potently, in crop stress tolerance.

**Acknowledgments**

Thanks to current and former members of the Mur Metabolomics team and collaborators for inspiring many of the points discussed in this review; namely J. Will Allwood, (AU-IBERS, UK now Manchester University, UK), Anushen Sivakuruman (AU-IBERS, UK), Royston Goodacre (Manchester University, UK) and the late Aileen Smith (AU-IBERS, UK). The Mur group is supported by grants from the BBSRC, European Union, Royal Society, Welsh Assembly and the Aberystwyth University Research Fund. This Indian-UK collaboration is supported by grants from the Royal Society of Chemistry (London UK).

## REFERENCES

1. Agi 2000 Analysis of the genome sequence of the flowering plant *Arabidopsis thaliana, Nature* **408** 796-815
2. Ahmad F, Ahmad I, and Khan M S 2008 Screening of free-living rhizospheric bacteria for their multiple plant growth promoting activities, *Microbiol Research* **163** 173-181
3. Allwood J W, Clarke A, Goodacre R, and Mur LA J 2010 Dual metabolomics: A novel approach to understanding plant-pathogen interactions, *Phytochemistry* **71** 590-597
4. Allwood JW, Ellis DI, Heald JK, Goodacre, R, and Mur LAJ 2006 Metabolomic approaches reveal that phosphatidic and phosphatidyl glycerol phospholipids are major discriminatory non-polar metabolites in responses by Brachypodium distachyon to challenge by *Magnaporthe grisea, Plant Journal* **46** 351-368
5. Allwood J W, Erban A, De Koning, S, Dunn WB, Luedemann A, Lommen A, Kay L, Loscher R, Kopka J, and Goodacre R 2009 Inter-laboratory reproducibility of fast gas chromatography-electron impact-time of flight mass spectrometry (GC-EI-TOF/MS) based plant metabolomics, *Metabolomics* **5** 479-496
6. Allwood JW and Goodacre R 2010 An introduction to liquid chromatography-mass spectrometry instrumentation applied in plant metabolomic analyses, *Phytochemical Analysis* **21** 33-47
7. Bari R and Jones J 2009 Role of plant hormones in plant

defence responses, *Plant Molecular Biology* **69** 473-488

8. Basnet M, Chakraborty B, and Chakraborty U 2009 Exploitation of Tea Rhizosphere Microorganism for Improvement of Plant Health Status, *Journal of Mycology and Plant Pathology* **39** 1-13

9. Baxter CJ, Redestig H, Schauer N, Repsilber D, Patil K R, Nielsen J, Selbig J, Liu J L, Fernie AR, and Sweetlove LJ 2007 The metabolic response of heterotrophic Arabidopsis cells to oxidative stress, *Plant Physiology* **143** 312-325

10. Beale M H, Ward JL, and Baker JM 2009 Establishing substantial equivalence: metabolomics, *Methods in Molecular Biology* **478** 289-303

11. Boerjan W, Ralph J, and Baucher M 2003 Lignin Biosynthesis, *Annual Review of Plant Biology* **54** 519-546

12. Brazma A, Hingamp P, Quackenbush J, Sherlock G, Spellman P, Stoeckert C, Aach J, Ansorge W, Ball CA, Causton HC, Gaasterland T, Glenisson P, Holstege FCP, Kim I F, Markowitz V, Matese JC, Parkinson H, Robinson A, Sarkans U, Schulze-Kremer, S, Stewart J, Taylor R, Vilo J, and Vingron M 2001 Minimum information about a microarray experiment (MIAME) - toward standards for microarray data, *Nature Genetics* **29** 365-371

13. Carol RJ and Dolan L 2006 The role of reactive oxygen species in cell growth: lessons from root hairs, *Journal of Experimental Botany* **57** 1829-1834

14. Catchpole GS, Beckmann M, Enot DP, Mondhe M, Zywicki B, Taylor J, Hardy N, Smith A, King, RD, Kell DB, Fiehn O, and Draper J 2005 Hierarchical metabolomics demonstrates substantial compositional similarity between genetically modified and conventional potato crops, *PNAS* USA **102** 14458-14462

15. Chakraborty U, Chakraborty BN, Basnet M, and Chakraborty AP 2009 Evaluation of *Ochrobactrum anthropi* TRS-2 and its talc based formulation for enhancement of growth of tea plants and management of brown root rot disease, *Journal of Applied Microbiology* **107** 625-634

16. Chan EK, Rowe HC, Hansen BG, and Kliebenstein DJ 2010 The complex genetic architecture of the metabolome, *PLoS Genetics* **6** 1001198

17. Clarke SM, Cristescu SM, Miersch O, Harren F J M, Wasternack C, and Mur LA J 2009 Jasmonates act with

salicylic acid to confer basal thermotolerance in *Arabidopsis thaliana*, *New Phytologist* **182** 175-187

18. Clarke SM, Mur LAJ, Wood JE, and Scott I M 2004 Salicylic acid dependent signaling promotes basal thermotolerance but is not essential for acquired thermotolerance in *Arabidopsis thaliana*, *Plant Journal* **38** 432-447

19. Cook D, Fowler S, Fiehn O, and Thomashow MF 2004. A prominent role for the CBF cold response pathway in configuring the low-temperature metabolome of *Arabidopsis*, *Proceedings of the National Academy of Sciences,USA* **101** 15243-15248

20. Cramer GR, Ergul A, Grimplet J, Tillett RL, Tattersall EAR, Bohlman, MC, Vincent D, Sonderegger J, Evans J, Osborne C, Quilici D, Schlauch KA, Schooley DA, and Cushman JC 2007 Water and salinity stress in grapevines: early and late changes in transcript and metabolite profiles, *Functional & Integrative Genomics* **7** 111-134

21. Criddle DN, Gillies S, Baumgartner-Wilson HK, Jaffar M, Chinje EC, Passmore S, Chvanov M, Barrow S, Gerasimenko OV, Tepikin AV, Sutton R, and Petersen OH 2006 Menadione-induced reactive oxygen species generation via redox cycling promotes apoptosis of murine pancreatic acinar cells, *The Journal of Biological Chemistry* **281** 40485-40492

22. De Villiers A, Lestremau F, Szucs R, Gelebart S, David F, and Sandra P 2006 Evaluation of ultra performance liquid chromatography. Part I. Possibilities and limitations, *Journal of Chromatography* **1127** 60-91

23. Denby KJ, Kumar P, and Kliebenstein DJ 2004 Identification of *Botrytis cinerea* susceptibility loci in *Arabidopsis thaliana*, *The Plant Journal* **38** 473-486

24. Dunn WB and Ellis DI 2005 Metabolomics: Current analytical platforms and methodologies, *TrAC Trends in Analytical Chemistry* **24** 285-294

25. Durand C, Vicre-Gibouin M, Follet-Gueye ML, Duponchel L, Moreau M, Lerouge P, and Driouich A 2009 The Organization Pattern of Root Border-Like Cells of Arabidopsis Is Dependent on Cell Wall Homogalacturonan, *Plant Physiology* **150** 1411-1421

26. Eliasson M, Rannar S, and Trygg J 2011 From Data Processing to Multivariate Validation - Essential Steps

in Extracting Interpretable Information from Metabolomics Data, *Current Pharmaceutical Biotechnology* **12** 996-1004

27. Fan L, Linker R, Gepstein S, Tanimoto E, Yamamoto R, and Neumann PM 2006 Progressive inhibition by water deficit of cell wall extensibility and growth along the elongation zone of maize roots is related to increased lignin metabolism and progressive stelar accumulation of wall phenolics, *Plant Physiology* **140** 603-612

28. Fernie AR, Trethewey RN, Krotzky AJ, and Willmitzer L 2004 Metabolite profiling: from diagnostics to systems biology, *Nature Reviews Molecular Cell Biology* **5** 763-769

29. Ferrari S, Plotnikova JM, De Lorenzo G, and Ausubel FM 2003 *Arabidopsis* local resistance to *Botrytis cinerea* involves salicylic acid and camalexin and requires EDS4 and PAD2, but not SID2, EDS5 or PAD4, *Plant Journal* **35** 193-205

30. Fiehn O 2002 Metabolomics - the link between genotypes and phenotypes, *Plant Molecular Biolology* **48** 155-171

31. Fiehn O, Kopka J, Dormann P, Altmann T, Trethewey RN, and Willmitzer L 2000a Metabolite profiling for plant functional genomics, *Nature Biotechnology* **18:** 1157-1161

32. Fiehn O, Kopka J, Trethewey RN, and Willmitzer L 2000b Identification of uncommon plant metabolites based on calculation of elemental compositions using gas chromatography and quadrupole mass spectrometry, *Analytical Chemistry* **72** 3573-3580

33. Fiehn O, Robertson D, Griffin J, Van Der Werf, M, Nikolau B, Morrison N, Sumner LW, Goodacre R, Hardy NW, Taylor C, Fostel J, Kristal B, Kaddurah-Daouk R, Mendes P, Van Ommen B, Lindon J C, and Sansone SA 2007a The metabolomics standards initiative (MSI), *Metabolomics* **3** 175-178

34. Fiehn O, Sumner LW, Rhee SY, Ward J, Dickerson J, Lange BM, Lane G, Roessner U, Last R, and d Nikolau B 2007b Minimum reporting standards for plant biology context information in metabolomic studies, *Metabolomics* **3** 195-201

35. Fukushima A, Kusano M, Redestig H, Arita M, and Saito K 2011 Metabolomic correlation-network modules in *Arabidopsis* based on a graph-clustering approach, *BMC Systems Biology* **5** 1-12

36. Glazebrook J and Ausubel FM 1994 Isolation of Phytoalexin-Deficient Mutants of *Arabidopsis-*

*thaliana* and Characterization of Their Interactions with Bacterial Pathogens, *Proceedings of the National Academy of Sciences ,USA* **91** 8955-8959

37. Goddard ME and Hayes BJ 2007 Genomic selection, *Journal of Animal Breeding and Genetics* **124** 323-30

38. Gong QQ, Li PH, Ma SS, Rupassara SI and Bohnert HJ 2005 Salinity stress adaptation competence in the extremophile *Thellungiella halophila* in comparison with its relative *Arabidopsis thaliana, Plant Journal* **44** 826-839

39. Goodacre R 2005 Making sense of the metabolome using evolutionary computation: seeing the wood with the trees, *Journal of Experimental Botany* **56** 245-254

40. Goodacre R, Roberts L, Ellis DI, Thorogood D, Reader SM, Ougham H, and King I 2007 From phenotype to genotype: whole tissue profiling for plant breeding, *Metabolomics* **3** 489-501

41. Goodacre R, Vaidyanathan S, Dunn WB, Harrigan GG, and Kell DB 2004 Metabolomics by numbers: acquiring and understanding global metabolite data, *Trends in Biotechnology* **22** 245-252

42. Gorzsas A, Stenlund H, Persson P, Trygg J, and Sundberg B 2011 Cell-specific chemotyping and multivariate imaging by combined FT-IR microspectroscopy and orthogonal projections to latent structures (OPLS) analysis reveals the chemical landscape of secondary xylem, *Plant Journal* **66** 903-914

43. Griffin JL 2006 The Cinderella story of metabolic profiling: does metabolomics get to go to the functional genomics ball? *Philosophical Transactions of the Royal Society B: Biological Science* **361** 147-161

44. Griffiths PR 1983 Fourier-Transform Infrared Spectrometry, *Science* **222** 297-302

45. Gullberg J, Jonsson P, Nordstrom A, Sjostrom M, and Moritz T 2004 Design of experiments: an efficient strategy to identify factors influencing extraction and derivatization of *Arabidopsis thaliana* samples in metabolomic studies with gas chromatography/mass spectrometry, *Analytical Biochemistry* **331** 283-95

46. Hagemeier J, Schneider B, Oldham NJ, and Hahlbrock K 2001 Accumulation of soluble and wall-bound indolic metabolites in *Arabidopsis thaliana* leaves infected with Virulent or avirulent *Pseudomonas syringae* pathovar tomato strains, *Proceedings of the National Academy of Sciences USA* **98** 753-758

47. Heil M and Baldwin IT 2002 Fitness costs of induced resistance: emerging experimental support for a slippery concept, *Trends in Plant Science* **7** 61-67

48. Iwashina T 2003 Flavonoid function and activity to plants and other organisms, *Biological Sciences in Space* **17** 24-44

49. James GO, Hocart CH, Hillier W, Chen H, Kordbacheh F, Price G D, and Djordjevic M A 2011 Fatty acid profiling of *Chlamydomonas reinhardtii* under nitrogen deprivation, *Bioresource Technology* **102** 3343-3351

50. Jansen JJ, Smit S, Hoefsloot HCJ, and Smilde AK 2010 The Photographer and the Greenhouse: How to Analyse Plant Metabolomics Data, *Phytochemstry Analysis* **21** 48-60

51. Johnson HE, Broadhurst D, Goodacre R, and Smith AR 2003 Metabolic fingerprinting of salt-stressed tomatoes, *Phytochemistry* **62** 919-928

52. Johnson HE, Gilbert RJ, Winson MK, Goodacre R, Smith AR, Rowland JJ, Hall M A, and Kell DB 2000 Explanatory Analysis of the Metabolome Using Genetic Programming of Simple, Interpretable Rules, Genetic Programming and Evolvable Machines **1** 243-258

53. Jones N, Ougham H, Thomas H, and Pasakinskiene I 2009 Markers and mapping revisited: finding your gene, New Phytology **183** 935-966

54. Kaplan F, Kopka J, Haskell DW, Zhao W, Schiller K C, Gatzke N, Sung DY, and Guy C L 2004 Exploring the temperature-stress metabolome of *Arabidopsis, Plant Physiology* **136** 4159-4168

55. Kaplan F, Kopka J, Sung DY, Zhao W, Popp M, Porat R, and Guy CL 2007 Transcript and metabolite profiling during cold acclimation of Arabidopsis reveals an intricate relationship of cold-regulated gene expression with modifications in metabolite content, *Plant Journal* **50** 967-981

56. Karthiba L, Saveetha K, Suresh S, Raguchander T, Saravanakumar D, and Samiyappan R 2010 PGPR and entomopathogenic fungus bioformulation for the synchronous management of leaf-folder pest and sheath blight disease of rice, *Pest Management Science* **66** 555-564

57. Kim JK, Bamba T, Harada K, Fukusaki E, and Kobayashi A 2007 Time-course metabolic profiling in Arabidopsis thaliana cell cultures after salt stress treatment, *Journal of Experimental Botany* **58** 415-424

58. Kottke T, Batschauer A, Ahmad M, and Heberle J 2006 Blue-light-induced changes in *Arabidopsis* cryptochrome 1 probed by FTIR difference spectroscopy, *Biochemistry* **45** 2472-2479

59. Kubien DS, Von Cammerer, S, Furbank RT, and Sage R F 2003 C-4 photosynthesis at low temperature. A study using transgenic plants with reduced amounts of Rubisco, *Plant Physiology* **132** 1577-1585

60. Laothawornkitkul J, Moore JP, Taylor J E, Possell M, Gibson TD, Hewitt CN, and Paul ND 2008 Discrimination of plant volatile signatures by an electronic nose: aA potential technology for plant pest and disease monitoring, *Environment Science Technology* **42** 8433-8439

61. Laothawornkitkul J, Taylor JE, Paul ND, and Hewitt CN 2009 Biogenic volatile organic compounds in the Earth system. *New Phytology* **183** 27-51

62. Lease KA and Walker JC 2006 The Arabidopsis unannotated secreted peptide database, a resource for plant peptidomics, *Plant Physiology* **142** 831-838

63. Lehmann M, Schwarzlander M, Obata T,Sirikantaramas S, Burow M, Olsen CE, Tohge T, Fricker MD, Moller BL, Fernie AR, Sweetlove LJ and Laxa M 2009 The Metabolic Response of *Arabidopsis* Roots to Oxidative Stress is Distinct from that of Heterotrophic Cells in Culture and Highlights a Complex Relationship between the Levels of Transcripts, Metabolites, and Flux, *Molecular Plant* **2** 390-406

64. Livingston DP, Hincha DK and Heyer AG 2009 Fructan and its relationship to abiotic stress tolerance in plants, *Cellular and Molecular Life Sciences* **66** 2007-2023

65. Lloyd AJ, Allwood JW, Winder C, Dunn W, Heald J, Cristescu S, Sivakumaranan A, Harren FJ M, Mulema J, Denby K, Goodacre R, Smith A and Mur LAJ 2011 Metabolomic approaches reveal that cell wall modifications play a major role in ethylene-mediated resistance against Botrytis cinerea, *Plant Journal* **66** 234-241

66. Loganathan M, Sible GV, Maruthasalam S, Saravanakumar D, Raguchander T, Sivakumar M, and Samiyappan R 2010 Trichoderma and chitin mixture based bioformulation for the management of head rot (*Sclerotinia sclerotiorum* (Lib.) deBary)-root-knot (*Meloidogyne incognita* Kofoid and White; Chitwood) complex diseases of cabbage, *Archives*

*of Phytopathology and Plant Protection* **43** 23 -29

67. Lucy M, Reed E, and Glick B R 2004 Applications of free living plant growth-promoting rhizobacteria, Antonie Van Leeuwenhoek Inter. *Journal of Genenral Molecular Microbiology* **86** 1-25
68. Lugtenberg B and Kamilova F 2009 Plant-Growth-Promoting Rhizobacteria, *Annual Review Microbiology* **63** 541-556
69. Mauch-Mani B and Slusarenko AJ 1996 Production of salicylic acid precursors is a major function of phenylalanine ammonia-lyase in the resistance of Arabidopsis to Peronospora parasitica, *Plant Cell* **8** 203-212
70. Mittler R 2006 Abiotic stress, the field environment and stress combination, *Trends in Plant Science* **11** 15-19
71. Mizrahi Y 1982 Effect of Salinity on Tomato Fruit Ripening, *Plant Physiology* **69** 966-970
72. Morrissey JP and Osbourn AE 1999 Fungal resistance to plant antibiotics as a mechanism of pathogenesis, *Microbiology and Molecular Biology Reviews* **63** 708-713
73. Morsy MR, Jouve L, Hausman JF, Hoffmann L, and Stewart JM 2007 Alteration of oxidative and carbohydrate metabolism under abiotic stress in two rice (*Oryza sativa* L.) genotypes contrasting in chilling tolerance, *Journal of Plant Physiology* **164** 157-167
74. Mouille G, Robin S, Lecomte M, Pagant S, and Hofte H 2003 Classification and identification of Arabidopsis cell wall mutants using Fourier-Transform InfraRed (FT-IR) microspectroscopy, *Plant Journal* **35** 393-404
75. Mouille G, Witucka-Wall H, Bruyant, MP, Loudet O, Pelletier S, Rihouey C, Lerouxel O, Lerouge P, Hofte H, and Pauly M 2006 Quantitative trait loci analysis of primary cell wall composition in Arabidopsis, *Plant Physiology* **141** 1035-1044
76. Neill SJ, Desikan R, Clarke A, Hurst RD, and Hancock JT 2002 Hydrogen peroxide and nitric oxide as signalling molecules in plants, *Journal of Experimental Botany* **53** 1237-1247
77. Oliver S 2003 On the MIAME standards and central repositories, of microarray, *Comparative and Functional Genomics* **4** 1-1
78. Oliver SG, Winson MK, Kell DB, and Baganz F 1998 Systematic functional analysis of the yeast genome, *Trends in Biotechnology* **16** 373-378
79. Osbourn AE 1996 Preformed antimicrobial compounds and

plant defense against fungal attack, *Plant Cell* **8** 1821-1831

80. Pachauri RK and Reisinger A 2007 Contribution of Working Groups I, II and III to the Fourth Assessment Report of the Intergovernmental Panel on Climate Change. Geneva, Switzerland.

81. Park DH, Mirabella R, Bronstein PA, Preston GM, Haring MA, Lim CK, Collmer A, and Schuurink RC 2010 Mutations in gamma-aminobutyric acid (GABA) transaminase genes in plants or *Pseudomonas syringae* reduce bacterial virulence, *Plant Journal* **64** 318-330

82. Paux E, Faure S, Choulet F, Roger D, Gauthier V, Martinant JP, Sourdille P, Balfourier F, Le Paslier MC, Chauveau A, Cakir M, Gandon B, and Feuillet C 2010 Insertion site-based polymorphism markers open new perspectives for genome saturation and marker-assisted selection in wheat, *Plant Biotechnology Journal* **8** 196-210

83. Popova AV, Hundertmark M, Seckler R, and Hincha DK 2011 Structural transitions in the intrinsically disordered plant dehydration stress protein LEA7 upon drying are modulated by the presence of membranes, *Biochimica et Biophysica Acta* **1808** 1879-87

84. Ren YF, Wang T, Peng YF, Xia B, and Qu LJ 2009 Distinguishing transgenic from non-transgenic *Arabidopsis* plants by H-1 NMR-based metabolic fingerprinting, *Journal of genetics and genomics* **36** 621-628

85. Rico A, Mccraw SL, and Preston GM 2011 The metabolic interface between Pseudomonas syringae and plant cells, *Current Opinion in Microbiology* **14** 31-38

86. Rizhsky L, Liang HJ, Shuman J, Shulaev V, Davletova S, and Mittler R 2004 When Defense pathways collide. The response of *Arabidopsis* to a combination of drought and heat stress, *Plant Physiology* **134** 1683-1696

87. Roessner U, Luedemann A, Brust D, Fiehn O, Linke T, Willmitzer L, and Fernie AR 2001a Metabolic profiling allows comprehensive phenotyping of genetically or environmentally modified plant systems, *Plant Cell* **13** 11-29

88. Roessner U, Wagne C, Kopka J, Trethewey RN, and Willmitzer L 2000 Simultaneous analysis of metabolites in potato tuber by gas chromatography-mass spectrometry, *Plant Journal* **23** 131-142

89. Roessner U, Willmitzer L, Fernie AR 2001b High-resolution metabolic phenotyping

of genetically and environmentally diverse potato tuber systems. Identification of phenocopies, *Plant Physiology* **127** 749-764

90. Roetschi A, Si-Ammour A, Belbahri L, Mauch F, and Mauch-Mani B 2001 Characterization of an *Arabidopsis-Phytophthora* pathosystem: resistance requires a functional PAD2 gene and is independent of salicylic acid, ethylene and jasmonic acid signalling, *Plant Journal* **28** 293-305

91. Sana TR, Fischer S, Wohlgemuth G, Katrekar A, Jung KH, Ronald P C, and Fiehn O 2010 Metabolomic and transcriptomic analysis of the rice response to the bacterial blight pathogen *Xanthomonas oryzae* pv. *Oryzae*, *Metabolomics* **6** 451-465

92. Schlaeppi K, Abou-Mansour E, Buchala A, and Mauch F 2010 Disease resistance of *Arabidopsis* to *Phytophthora brassicae* is established by the sequential action of indole glucosinolates and camalexin, *Plant Journal* **62** 840-851

93. Selye H 1998 A syndrome produced by diverse nocuous agents. 1936, *The Journal of Neuropsychiatry and Clinical Neurosciences* **10** 230-231

94. Siddiqui ZA and Futai K 2009 Biocontrol of *Meloidogyne incognita* on tomato using antagonistic fungi, plant-growth-promoting rhizobacteria and cattle manure, *Pest Manaement Science* **65** 943-948

95. Streeter JG and Strimbu CE 1998 Simultaneous extraction and derivatization of carbohydrates from green plant tissues for analysis by gas-liquid chromatography, *Analytical Biochemistry* **259** 253-257

96. Sumner LW, Mendes P, and Dixon RA 2003 Plant metabolomics: large-scale phytochemistry in the functional genomics era, *Phytochemistry* **62** 817-36

97. Sumner LW 2010 Recent advances in plant metabolomics and greener pastures, F1000 *Biology Reports* **2** 7

98. Taylor CE 2006 Minimum reporting requirements for proteomics: A MIAPE primer, *Proteomics* **6** 39-44

99. Taylor CF, Paton NW, Lilley KS, Binz PA, Julian RK, Jones A R, Zhu WM, Apweiler R, Aebersold R, Deutsch E W, Dunn MJ, Heck AJR, Leitner A, Macht M, Mann M, Martens L, Neubert TA, Patterson SD, Ping PP, Seymour SL, Souda P, Tsugita A, Vandekerckhove J, Vondriska TM, Whitelegge JP, Wilkins MR, Xenarios, I, Yates JR, and Hermjakob H 2007 The minimum information about a proteomics experiment

(MIAPE), *Nature Biotechnology* **25** 887-893

100. Taylor J, King RD, Altmann T, and Fiehn O 2002 Application of metabolomics to plant genotype discrimination using statistics and machine learning, *Bioinformatics* **18** S241-S248.

101. Tohge T and Fernie AR 2009 Web-based resources for mass-spectrometry-based metabolomics: a user's guide, *Phytochemistry* **70** 450–456

102. Thomma B P H J, Eggermont K, Tierens KFMJ, and Broekaert WF 1999 Requirement of functional ethylene-insensitive 2 gene for efficient resistance of *Arabidopsis* to infection by *Botrytis cinerea*, *Plant Physiology* **121** 1093-1101

103. Treutter D 2006 Significance of flavonoids in plant resistance: a review, *Environmental Chemistry Letters* **4** 147-157

104. Tsuji J, Jackson EP, Gage, DA, Hammerschmidt R, and Somerville SC 1992 Phytoalexin Accumulation in *Arabidopsis-Thaliana* During the Hypersensitive Reaction to *Pseudomonas-Syringae* Pv-*Syringae*, *Plant Physiology* **98** 1304-1309

105. Vagujfalvi A, Crosatti C, Galiba G, Dubcovsky J, and Cattivelli L 2000 Two loci on wheat chromosome 5A regulate the differential cold-dependent expression of the cor14b gene in frost-tolerant and frost-sensitive genotypes, *Molecular Genetics & Genomics* **263** 194-200

106. Wang KLC, Li H, and Ecker JR 2002 Ethylene biosynthesis and signaling networks, *Plant Cell* **14** S131-S151

107. Ward JL, Baker JM, and Beale MH 2007 Recent applications of NMR spectroscopy in plant metabolomics, *FEBS Journal* **274** 1126–1131

108. Ward JL, Baker JM, Miller SJ, Deborde C, Maucourt M, Biais B, Rolin D, Moing A, Moco S, Vervoort J, Lommen A, Schafer H, Humpfer E, and Beale MH 2010a An inter-laboratory comparison demonstrates that [H-1]-NMR metabolite fingerprinting is a robust technique for collaborative plant metabolomic data collection, *Metabolomics* **6** 263-273

109. Ward JL, Forcat S, Beckmann M, Bennett M, Miller SJ, Baker JM, Hawkins ND, Vermeer CP, Lu CA, Lin WC, Truman WM, Beale MH, Draper J, Mansfield JW, and Grant M 2010b The metabolic transition during disease following infection of *Arabidopsis thaliana* by *Pseudomonas syringae* pv. *tomato*, *Plant Journal* **63** 443-457

110. Yang J, Kloepper JW, and Ryu CM 2009 Rhizosphere

bacteria help plants tolerate abiotic stress, *Trends in Plant Science* **14** 1-4

111. Yoshiba Y, Kiyosue T, Nakashima K, Yamaguchishinozaki K, and Shinozaki K 1997 Regulation of levels of proline as an osmolyte in plants under water stress, *Plant Cell Physiology* **38** 1095-1102

112. Zimmermann B 2010 Characterization of pollen by vibrational spectroscopy, *Journal of applied spectroscopy* **64:** 1364-73.

*Rev. Plant Pathol. Vol. 6, 2014*
*Indian Society of Mycology and Plant Pathology*
*Scientific Publishers (India), Jodhpur pp. 93-126*

# CURRENT UNDERSTANDING ON PLANT *R*-GENES/PROTEINS AND MECHANISMS OF DEFENSE RESPONSES AGAINST BIOTIC STRESSES

***Soumitra Maiti*[1]*, Jolly Basak*[2] *and Amita Pal*[1*]**

[1]Division of Plant Biology, Bose Institute, Kolkata-700054, India.
[2]Centre for Biotechnology, Siksha Bhavana, Visva-Bharati University, Santiniketan-731235, India, *E-mail:amita@bic.boseinst.ernet.in

**KEY WORDS:** R-genes/proteins, Biotic stresses, NB-LRR class of proteins

## ABSTRACT

Plants have evolved sophisticated and complex mechanisms to trigger defense responses upon recognition of pathogen derived elicitor molecule/s which have been explained based on a two-branched innate immunity system. Plant defense system involves several components such as basal resistance, reactive oxygen species, hypersensitive reaction and programmed cell death, systemic acquired resistance and post transcriptional gene silencing. Resistance genes are broadly classified into eight classes of proteins the largest two classes of which encode the cytoplasmic NB-LRR class of proteins. Molecular genetic studies of closely related NB-LRR proteins with different recognition specificities indicate that the LRR domain is the primary determinant of R proteins that provides recognition specificity in the event of effector triggered immunity. Defense signalling may be either through salicylic acid or jasmonic acid and ethylene dependent pathways.

Pathogenesis related protein is a group of proteins may or may not be constitutively present in the host plant, but their synthesis is triggered when defense signals are transmitted from the R proteins after recognition of pathogen effecter/elicitor proteins. RNA silencing, an ancient cellular mechanism of defense against foreign nucleic acids that also functions in gene regulation is an important plant defense strategy.

## INTRODUCTION

A large number of plant-associated microbes are pathogens that impair plant growth and reproduction leading to disease symptom development; they depend on plant genetic machinery for their survival. Others are warded off by the plant immune system or defense mechanisms of plants. Plants have evolved sophisticated and complex mechanisms to trigger defense responses upon recognition of pathogen derived elicitor molecule/s to ward off the invading microbial pathogens, nematodes and insects (18, 121). Of late, the arsenals of plant defense system have been explained based on a two-branched innate immunity system. The first one recognizes and responds to molecules conserved in many classes of microbes, including non-pathogenic organisms and are referred as Microbe/Pathogen Associated Molecular Patterns (MAMPs/PAMPS). This forms the first line of defense in plants and known as MAMP/PAMP Triggered Immunity (MTI or PTI), which restricts further colonization of the pathogen. This branch of plant defense system uses transmembrane pattern recognition receptors (PRRs) which respond to slowly evolving MAMPS or PAMPs, like lipopolysaccharides and highly conserved bacterial proteins, e.g. flagellin and Ef-Tu (79, 130). Some pathogens, however, are able to bypass the MTI or PTI and that is where they encounter Effector Triggered Immunity (ETI), the second line of defense. These defense systems mainly function using the polymorphic NB-LRR protein products encoded by plant disease resistance (*R*) genes (18); named after characteristic nucleotide binding (NB) and leucine rich repeat (LRR) domains. *R* genes mainly encode receptors to avirulence (*Avr*) gene products expressed by the pathogen when it attacks the host plants (68).In case of compatible interaction, due to the unavailability of the plant R protein corresponding to the *Avr* protein of the invading pathogen, results in susceptibility; while in incompatible interaction the plant R proteins interact with the Avr proteins of the pathogen and thereby generating downstream signaling results in disease resistance. Compatible and incompatible host-pathogen interactions have been illustrated in the Figure 1.

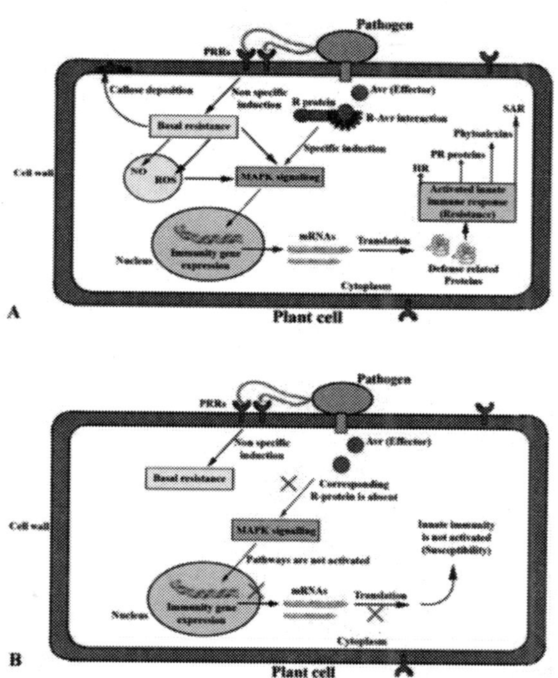


**Fig. 1:** Interaction between Plant R proteins and Pathogen Avr proteins: Pathogens attaches to the plant cell and express virulence proteins. These proteins enter into plant cells through appropriate receptors or via specialized mechanism. In the cytoplasm, target host proteins that control defense responses and transduce signals affect pathogen virulence. (A) The plant cell expresses an R protein which is capable of recognizing specific virulence protein. This recognition event activates a complex signal transduction network triggering defense responses. (B) The plant cell does not express an R protein that is capable of recognizing any virulence proteins; consequently, the plant cannot detect the pathogen virulence factor or effect or proteins efficiently and resulted in susceptibility.

The complicated relation between the plant resistance/ susceptibility and pathogen virulence was proposed as the 'zig-zag' model (43). They have quantitatively explained the zig-zag model that elucidates the interactions between tomato resistance proteins and *Fusarium oxysporium*'s effector proteins (Figure 2) in the following phases: In phase 1, tomato protein I-3 recognizes 'Avr3' to trigger MTI/PTI-triggered immunity. While, the fungus struggles to counteract host defense response and evolves a second effector 'Avr1' that suppresses I-3 mediated defenses. Thus, in phase 2, successful pathogens deliver effector protein/s that interfere with MTI/PTI, or enable pathogen nutrition and spread of infection using host machinery, resulting in effector-triggered susceptibility (ETS) at this phase R-Avr interactions are weak. In turn, the tomato evolves another R protein I-1 that recognizes 'Avr1' and activates the host defense against the fungus (36). Hence, phase 3 of the 'zig-zag' model depicts one effector is recognized by an NB-LRR protein, activating effector-triggered immunity (ETI); showing a much higher aptitude than PTI, like that of hypersensitive cell death.

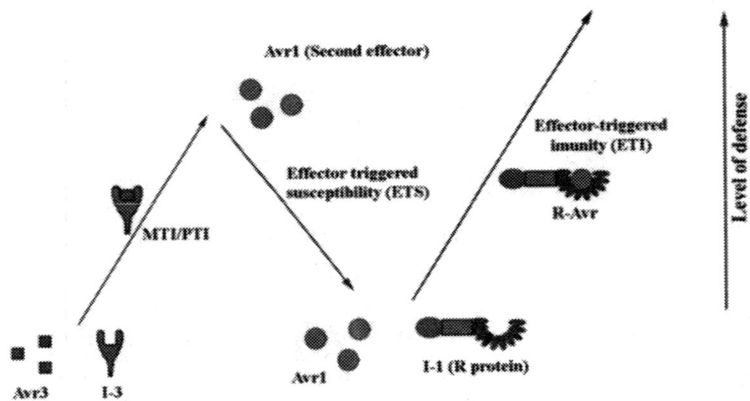


**Fig. 2:** Schematic representation of 'zig-zag' model as proposed by Jones and Dangl (2006) depicting complicated relation between the plant resistance/susceptibility and pathogen virulence factors. Detail explanation has been provided within the text.

## TYPES OF PLANT DEFENSE RESPONSE

Plants have historically been divided into two major categories (26) depending on their defense responses against various biotic stresses: non-host resistance and host-resistance.

### Non-host Resistance

Non-host resistance is a major arsenal of plants, where all genotypes within a plant species show either resistance or fail to be infected by a particular pathogen. It specifically signifies the state where genetic polymorphism for susceptibility against a particular pathogen has not been identified in the host taxa. Generally, most plant species are resistant to many plant viruses. Non-host resistance is durable and therefore is valuable for exploitation towards virus-resistant plants. However, it is not yet feasible to understand non-host resistance by classical genetics; and therefore remains unexplored. In case of viral pathogen this phenomenon could probably be related to an intrinsic lack of susceptibility than to active defense mechanisms. On the contrary, studies on related phenomena for other pathogens suggest that non-host resistance could be genetically complex, multigenic traits involving a number of distinct processes (34). Perhaps this complex phenomenon exploits both constitutive and inducible defense responses, in combination with a host physiology that is characteristically incompatible with pathogen mediated elicitors (32, 128). It has been shown that non-host resistance of *Arabidopsis* against wheat powdery mildew pathogen, *Blumeria graminis* f.sp. tritici (Bgt) is established via a series of defense responses which are deployed at specific steps during Bgt infection (128). It has been recently hypothesized that the relative contribution of NB-LRR- and PRR-triggered immunity to non-host resistance might have altered during phylogenetic divergence between host and nonhost. Likewise, alterations in pathogen host range, i.e. variations in pathogen effector repertoires leading to the extent of reproductive isolation leads to new pathotypes development (96).

### Host-Resistance

Host-resistance to plant pathogens has been more systematically investigated, because, unlike non-host resistance, it is genetically attainable. Perhaps the best characterized mechanism of plant host resistance is mediated by resistance (*R*) genes. Generally, this phenomenon is referred as host resistance, specific resistance, genotypic resistance and also as cultivar resistance. Genotypic resistance occurs when genetic polymorphisms for susceptibility is observed in some genotypes of the plant taxon, that show heritable resistance to a particular pathogen, while other genotypes in the same gene pool are susceptible (44). In resistant individuals, a pathogen may not multiply to that extent, but spread of the pathogen through the plant is demonstrably restricted relative to susceptible hosts; and disease symptoms generally are highly localized or are not evident at all.

## ARCHITECTURE OF PLANT DEFENSE SYSTEM

### *Basal Resistance*

Basal defense is a part of non-host resistance that remains active even in susceptible plants to limit pathogen growth. This first line defense of plant innate immune system is mediated by transmembrane PRRs. PRRs are often well conserved in structure and function and recognize well conserved MAMPs/PAMPs associated with broad classes of pathogens (79, 130). Responses induced by PRRs, often referred as MTI/PTI, are generally low-impact but sufficient to confer resistance against many bacterial, fungal and oomycetes pathogen (15). The basal defense response overlaps significantly with receptor-like plant resistance (R) protein mediated defense, but is comparatively slower and of low amplitude. However, host-adapted pathogens are able to interfere with and overcome MTI/PTI mechanisms through the deployment of so-called effector proteins, necessitating a second line of defense based on much more specific recognition of pathogens. Effector proteins are delivered to the host cytoplasm via the various secretion mechanisms of bacteria and eukaryotic pathogens and the main function of these proteins is to interfere with MTI/PTI signaling (15, 31).

### *Reactive Oxygen Species*

Infections of plants and animals are often accompanied by an oxidative burst which is characterized by the appearance of active oxygen species, e.g., hydrogen peroxide ($H_2O_2$), a hydroxide radical (·OH), and an anion radical ($O_2^-$). Reactive oxygen species (ROS) have long been recognized as signals in defense, most notably during the oxidative burst or bursts that occur very early in the hypersensitive reaction during a gene-for-gene response (33) or during recognition of MAMPS or PAMPS in basal resistance (63). The oxidative burst, which occurs in cells in the close vicinity of the infection site, is due mainly to the activation of NADPH oxidase associated with the plasma membrane. NADPH oxidase catalyses the formation of superoxide ($O_2^-$) anions that are readily converted to other ROS, including the perhydroxyl radical ($\cdot HO_2$) and hydrogen peroxide ($H_2O_2$), via non-enzymatic and enzymatic reactions (54). ROS is not only highly toxic molecules capable of localizing infection, but are also active participants in the signal system. ROS mediated signalling occurs by two different modes. Primarily, the oxidative burst activates $Ca^{2+}$ ion influx across the plasma membrane via cyclic nucleotide-gated channels, in addition to mobilization of $Ca^{2+}$ ions from intracellular resources (60,112). Another mode of alteration in cytoplasmic $Ca^{2+}$ ion concentrations is by triggering the activities of calcium-dependent protein kinases and that of highly complex mitogen-activated protein kinase (MAPK) cascades. It is now well established from studies on different plant species that MAPK cascades are implicated in the control of cell death, regulation of ROS production, pathogenesis related (*PR*) gene induction and establishing induced resistance in tissues within and beyond the infection site (86).

### *Hypersensitive Reaction and Programmed Cell Death*

An alternative line of plant defense is based on more specific recognition of pathogens or pathogen isolates. This type of recognition is mediated by the products of plant *R* genes and resulted in a much more reliable, high impact defense responses, often culminating in programmed cell death (PCD) known as the hypersensitive response (HR) (77). The hypersensitivity reaction constitutes a protective system which is activated in many higher plants in response to infection; the reaction is characterized by rapid necrotization of infected cells (87). The phenomenon of necrotization of plant cells is similar to that of apoptosis or PCD in mammals and insects, and plays an important role in limiting pathogens to the site of their penetration (19, 29, 30). Cell death in the HR involves breakdown and re-cycling of cellular materials, rather than their absolute destruction. HR is accompanied by a number of biochemical, physiological, and molecular processes, the appearance of active oxygen ($O_2^-$, $H_2O_2$, and $\cdot OH$), an increase in lipoxygenase activity, synthesis of phytoalexins, storage of lignin and also accumulation of pathogenesis related (PR) proteins (69). It has been illustrated that resistance is not always associated with hypersensitive death of plant cells, and it remains unclear whether the hypersensitive reaction is programmed and natural or whether it is only an accidental by product of inherent protective responses (41,100). However, hypersensitive reaction not only restricts the infection process to localize the pathogen by accumulating antibacterial, anti-viral substances, necrotizating infected cells etc., but also induces acquired resistance to repeated damage by the secondary pathogens and subsequently different defense responses are turned on.

## *Systemic Acquired Resistance*

PR proteins of plants are able to counter different types of pathogen; many of these are specific for specific pathogens or specific isolates/pathotypes and are implicated in multiple signaling pathways. These induced responses not only restrict pathogen infection locally, but also induce signals that enhance defense responses to pathogens in distal systemic tissues, a phenomenon known as systemic acquired resistance (SAR)(119). SAR responses are associated with the transcriptional up-regulation of a number of defense related genes, including those encoding PR proteins in both local and systemic tissues (23,78). Interestingly, SAR is durable and can last for several weeks. Salicylic acid (SA) plays a central role in the signal transduction pathway that results in SAR (4). During virus-induced HR lesion development, SA biosynthesis occurs initially in and around the developing lesions, but later throughout the entire plant (38,64,71,80,106)

## *Post Transcriptional Gene Silencing*

Post transcriptional gene silencing (PTGS) is a natural regulatory mechanism of eukaryotes acting against invasive nucleic acids or modulating endogenous gene expression (7,120). Silencing is initiated when double-stranded RNAs (dsRNAs) or hairpin RNAs (hpRNAs) re-processed into 21- to 24-nucleotide small interfering RNA (siRNA) or microRNA (miRNA) duplexes by dsRNA specific RNase III-type Dicer enzymes. These small RNAs (sRNAs) are then incorporated into the RNA-induced silencing complex (RISC) to guide the degradation or translation repression of complementary RNA (cRNA) targets (7,110). Unlike animals, higher plants have evolved diversified RNA silencing pathways to generate different sRNA classes with specialized functions (11), e.g. *Arabidopsis thaliana* encodes four Dicer-like (DCL) proteins (DCL1 to DCL4). Of these, DCL1 mainly processes imperfectly base-paired fold-back precursors to produce miRNAs (6); the other three DCL proteins are responsible for generating various endogenous siRNAs from perfect dsRNAs. In addition, plant DCL enzymes can also generate exogenous siRNAs that are derived from invading viruses or hpRNA-forming transgenes. siRNAs that mediate antiviral defense consist of primarily 21-nt and 22-nt species which are produced by DCL4 and DCL2, respectively (2,10,20,75,124). Specifically, the 21-nt siRNAs appear to form the first line of antiviral defense, whereas the DCL2-dependent 22-nt siRNAs serves as an alternative arsenal when DCL4 is inactivated (6,10,20). Functions of miRNAs and their involvement in developmental regulation and many other cellular processes in plants have been reviewed (6). However, the possibility of miRNA-guided RNA silencing in host defense against virus infection has been shown for the first time (57). Besides, it has been demonstrated that potyviruses are able to modulate gene expression of the host cell by interfering with miRNA-guided cleavage of multiple regulatory targets in plants (45). Such antiviral function of natural miRNAs has been demonstrated in animal systems (55) and later the plant miRNA pathway was also shown to be implicated in impairing the replication of an engineered potyvirus bearing a miRNA target (104).

## PLANT DISEASE RESISTANCE (*R*) GENES

Resistance genes are broadly classified into eight classes of proteins (Figure 3), the largest two classes of which encode the cytoplasmic NB-LRR class of proteins. The NB (Nucleotide Binding) domain aids in ADP/ATP binding and also in signal transduction pathways. The NB, also called NB-ARC domains, named after Nucleotide Binding domain and having high sequence homology with human APAF-1, Resistance Genes and *Caenorhabditis elegan*'s CED-4 gene (88) has a few conserved signature motifs and high homology with the animal and human innate immunity proteins (113). NB-ARC family of proteins are within the class of STAND (signal transducing ATPases with numerous domains) proteins (56). The NACHT sister family within this class comprised of the animal NLR (NACHT-LRR/NOD-LRR) innate immune receptors, where as the plant NB-LRR proteins comprise NB domains fused with LRR domains (89). STAND proteins are proposed to function as molecular switches, regulating cellular responses through nucleotide dependent conformational changes (56). In the NB-LRR protein, the LRR is towards the carboxyl terminus and contains irregularly spaced and conserved leucine residues. All the known LRRs are said to be involved in protein-protein interaction and in a wide range of processes like DNA repair, RNA processing, cell adhesion, signal transduction *etc*. The specificity of plant disease resistance genes lies in the LRR domains, which directly or indirectly binds to the ligand (47). NBS-LRR class of *R* genes are rapidly evolving and make up one of the largest and most variable gene families found in plants (16). Hence, in this review main emphasis on this class of proteins or protein coding genes has been given. Most of the isolated *R* genes are dominant traits and are involved in defense responses manifested by HR or SAR (66).

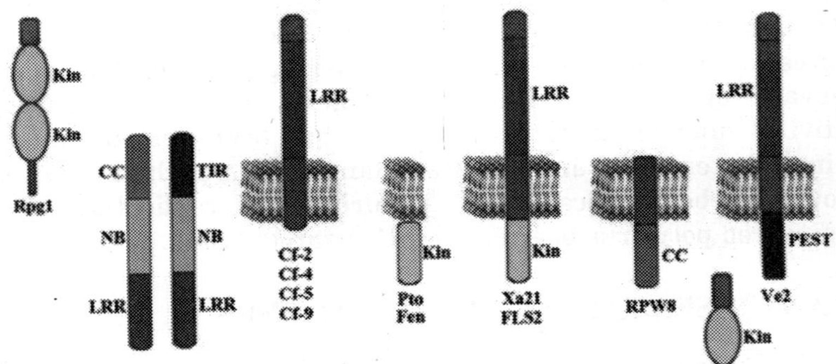


**Fig. 3:** Representation of the location and structure of the eight main classes of plant disease resistance proteins. Xa21 and Cf-X proteins carry trans-membrane domains and extracellular LRRs. The *RPW8* gene product with a putative signal anchor at the N terminus. The *Pto* gene encodes a cytoplasmic Ser/Thr kinase; the NB-LRR class, are presumably cytoplasmic, they could be membrane associated as well and anchored with distinct N-terminal domains. Barley Rpg1 contains two kinase domains; Tomato Ve2 protein comprised of an extracellular LRR anchored with a PEST domain.

The other six classes of *R* genes are structurally diverse. These are membrane bound and feature extracellular LRRs (47). *Cf* genes of tomato encode an extracytoplasmic LRR protein that resists the fungus *Cladosporium fulvium* (42). Another class of proteins consists of an extra-cytoplasmic leucine rich repeats anchored to a serine-threonine kinase domain. Whereas, *Pto* gene of tomato encodes a serine/threonine protein kinase domain without LRR that confers resistance to *Pseudomonas syringae* strains carrying AvrPto (18). *Xa21* from rice encodes extracellular LRR and intracellular kinase domain conferring resistance to *Xanthomonas oryzae* (105). Later, a novel disease resistance gene *RPW8* reported from *Arabidopsis* encodes a small membrane protein anchored with a putative coiled-coil domain at the N-terminus, without having any homology with other R-proteins (123). Unlike *Pto* gene, in barley *Rpg1* contains two kinase domains without LRR (13,35). Vel and Ve2 proteins contain putative extracellular LRRs, of which Ve2 comprised of a PEST domain for protein degradation and this short motif perhaps targets the pathogenic effector proteins for receptor mediated endocytosis (46).

## R-PROTEIN AND PATHOGEN RECOGNITION

Plant R proteins are the specificity determinants of plant immune response. NB-LRR class of R proteins show considerable similarity in their organization, in which three functional domains have been defined: an N-terminal domain similar to either a Toll/interleukin-1 receptor (TIR) domain or a coiled-coiled (CC) protein domain; a central nucleotide binding (NB) domain and a C-terminal LRR domain. Genetically, the LRRs of plant R proteins are determinant of pathogen specificity ( 24,90,101). There are two major subclasses of NB-LRR proteins that are distinguished by the domains present at their N-termini: TIR-NB-LRR (TNL or TIR) proteins have an N-terminal Toll and interleukin-1 receptor (TIR) like domains (39,122) and the others do not have TIR domain (non-TIR-NB-LRR) (72). Many non-TIR-NB-LRR proteins posses N-terminal α-helix endowed with coiled-coil (CC) motif and often referred to as CC-NB-LRR (CNL) proteins. The TIR and non-TIR bearing R-proteins are also distinct at the functional level due to their putative involvement in different signal transduction pathways (1). A number of plant disease resistance genes, conferring resistance against different viruses, encode NB-LRR proteins. Two different mode of actions proposed for pathogen recognition by NB-LRR proteins; one is the direct recognition of pathogen proteins (Avr) modeled as 'gene-for-gene' interaction (25) and the other is 'guard hypothesis' (18,114).

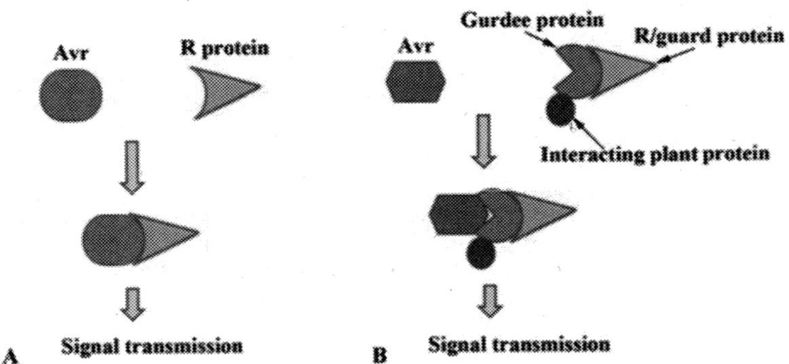


**Fig. 4:** Gene for gene hypothesis (A): Direct recognition of Avr protein by plant R protein induces defense signal. Guard hypothesis (B): Avr protein is recognized by a cellular protein (Guardee) which in turn activates R protein (Guard) and presence of other protein(s) in the complex.

## *Conserved Molecular Sensor within R Proteins: Direct Recognition of Pathogen Effector Proteins*

Molecular genetic studies of closely related NB-LRR proteins with different recognition specificities indicate that the LRR domain is the primary determinant of R proteins that provides recognition specificity in the event of effector triggered immunity (ETI) (24,90,101). Plant R proteins can recognize pathogen effector (Avr, avirulence protein) directly through physical association with the Avr like that of receptor-ligand interactions. The gene-for-gene hypothesis explained specific interactions between *Avr* gene loci and its corresponding *R* loci in the plant. He proposed that a specific plant *R* gene product recognizes and responds specifically to a pathogen effector molecule which he referred to avirulence (*Avr*) gene product. When corresponding *R* and *Avr* genes are present in respective host and pathogen, the consequence is disease resistance. On the contrary, when either is inactive or *R* is absent, plants show disease symptoms (25). The simplest interpretation of Flor's gene-for-gene hypothesis is that the *Avr*-gene dependant ligand binds directly to their cognate *R* gene product (Figure 4A) which then transmits signal and activates downstream cascade of pathogenesis related genes to induce various defense responses (27), subsequently pathogen growth is arrested.

### *Guard and Decoy Models: Indirect Recognition of Pathogen Effector Proteins*

Later it was hypothesized (114) that many plant R proteins may be activated by indirect recognition sensing the effector (elicitor) proteins of the pathogen. This is referred as "Guard Hypothesis" in which it is believed that the R proteins in the host plant are responsible for moderating one or more of the host defense proteins leading to the activation of enormous defense components against the different pathogens (9, 18). In this model, the interaction between the R protein (guard) and its cognate Avr determinant is mediated by a host cellular protein (guardee), which triggered innate immunity in the resistant plants. The pathogen causes structural modifications to the guardee (the target) after infestation and immediately detected by the guard by sensing the biochemical changes in the targeted protein caused by the pathogen Avr protein (Figure 4B). Any protein modification that can alter the quaternary structure of the guardee could result in detection of the pathogen. This activates the guard to initiate a signaling cascade that culminates in the resistance response (18,114).

The Guard Model was first suggested to explain the mechanism by which *Pseudomonas syringae* AvrPto is sensed independently by two tomato proteins Pto (serine threonine kinase) and Prf (NB-LRR). Prf protein was found to associate with the Pto protein and their association is presumably to keep Prf in an inactive state. It is activated upon Pto-AvrPto interaction which interrupts the inhibitory action of Pto and that reverts Prf in an active state to induce plant defense responses (76,114)

The most convincing evidence for the Guard hypothesis has been demonstrated by *Arabidopsis* RIN4 (RPM1 interacting 4), RPM1 (resistance to *P. syringae pv. maculicola 1*) and RPS2 (resistance to *P. syringae pv. tomato* 2) proteins. In this case, RIN4, the guardee, is a negative regulator of plant defense and found to be targeted by three *P. syringae* Avr proteins, Avr-Rpm1, AvrB and AvrRpt2, and guarded by two CC-NB-LRR proteins, RPM1 and RPS2. Guardee (RIN4) is modified in various ways, depending on the Avr that is associated with; and these modifications activate corresponding R proteins, RPM1 or RPS2 (guards) (5,62,63).. RIN4 is either phosphorylated in presence of AvrB and AvrRpm1 subsequently activates RPM1 (63); or degraded upon interaction with AvrRpt2 and thereby interferes with RPM1 function, but activates RPS2 (5,62).

Although several evidences put forwarded in the last decade, which could convincingly explain the interplay between pathogen effector molecules with guard and guardee proteins of the host plants to confer either resistance or susceptibility against the disease, yet, the main limitation of Guard hypothesis is guardee is always subjected to opposite selection forces, i.e., either promoting or opposing leading to differential disease reactions. Interpreting an evolutionary unstable situation, consequently justification of the hypothesis became contemptuous. To resolve such difference of opinion the 'Decoy Model' has been proposed (115). Decoys are imitators of pathogen effector targets or guardees, preferentially evolved for enhanced pathogen effector binding. These imitators compete with the guardees in effector binding, thus lowering the pathogenic load without impeding the pathogenic fitness irrespective of presence or absence of specific *R* genes (99,129).

## SWITCHING ON OR OFF OF PLANT DEFENSE SYSTEM MEDIATED BY NB-LRR

Biochemical and structural analyses of APAF-1 and CED-4 showed that the NB-ARC domain is perfectly suited as a molecular switch to regulate signaling pathways through conformational changes (48, 94). It has been shown that the interaction of the CC domain with the NB-ARC-LRR part of the potato R protein, Rx, is disrupted upon Avr perception, and this intra-molecular interaction requires a wild type P-loop (74). Similarly, it has also been shown that oligomerisation of the tobacco R protein, N, upon co-expression of the Avr protein requires an intact P-loop (70). The specific reduction of ATP hydrolysis in the auto-activating I-2 mutant R proteins of tomato was thought to result in the accumulation of the ATP-bound state and when this state exceeds a certain threshold, it would trigger defense signaling (108).

Furthermore, *in vitro* binding studies on I-2 indicated that the ADP-bound state is highly stable, which suggests that this state is probably the resting state, whereas the ATP-bound state is the activated form (108). The LRRs play an important negative role as well as a positive role in regulating NB-LRR protein activity. Negative regulation is suggested by several auto-activating mutations that were identified in this domain.

Based on all these available information, a model has been proposed that explains how NB-LRR proteins function as a molecular switch in plant defense (107). According to this model, in absence of a pathogen, a NB-LRR protein remains in its OFF-state (resting state), in which the LRR exerts its negative role by stabilizing the ADP-bound state. Presence of an Avr is recognized by the LRR domain, which induces a conformational change in the NB-ARC domain. This allows the exchange of ADP for ATP which subsequently triggers a second conformational change in the N-terminal effector domain, releasing its signaling potential. The ATPase activity of the protein attenuates the signalling response and returns the protein to its resting state. APAF-1 and many other STAND proteins were found to be oligomerised upon activation (56). Although it has been shown that Avr-dependent oligomerisation occurs in tobacco R protein (N) via a TIR–TIR interaction but it is still unclear whether oligomerisation is a common event for NBS-LRR protein activation or it occurs only in specific cases (70).

## DEFENSE SIGNALLING PATHWAYS

### *Salicylic acid (SA)-Dependent Pathway*

SA (2-hydroxybenzoic acid) plays a central role in the signal transduction pathway that results in SAR and it is required for localization of viral and other pathogens during the HR (4). SA is required for the expression of a cascade of pathogenesis-related (PR) proteins (78,116). The induction of SAR and, to some extent, maintenance of basal resistance to pathogen infection depends on SA-mediated signalling. SA can be detected in phloem sap (71, 92), consistent with a role in the spread of SAR induction to all parts of the plant. It has been demonstrated that transgenic plants silenced for the gene encoding a SA-binding protein (SABP2) were compromised in gene expression of PR proteins and SAR induction (84). SABP2 is an esterase that can convert methyl salicylic acid (Me-SA) into SA (53). It is a volatile chemical that can diffuse into the air and even induce SAR in nearby, non-infected plants (102). All these evidences suggest that Me-SA is an important long-distance signal in the induction of SAR following a HR.

### *The Jasmonic acid (JA) and Ethylene Dependent Pathway*

Jasmonic acid (JA) is an oxygenated fatty acid (oxylipin) that serves as a signal in response to certain insect pests, bacterial and fungal pathogens that generates resistance response in the host plants (93,109). Together with the gaseous plant hormone ethylene (ET), JA also regulates a systemic resistance pathway inducible by non-pathogenic microbes, 'induced systemic resistance' (ISR) leading to prime resistance in *A. thaliana* against fungus and bacterial pathogens (111). However, based on studies with *Turnip crinkle virus* (TCV), it emerges that ISR is not effective against virus infections (111).

ET is produced copiously during the HR and is a strong inducer of certain *PR* genes. ET-dependent signalling is indispensable for the maintenance of basal resistance to fungi and bacteria (28,116). However, ethylene does not appear to be essential for the induction of resistance to viruses, although since its biosynthesis from 1-aminocyclopropane-1-carboxylic acid results in the release of cyanide (103), it is conceivable that this may contribute to the induction of resistance to viruses via the mitochondrial signaling pathway. It is also possible that ethylene plays a role as a negative regulator of SA-mediated resistance to viruses. It has been proposed that this may explain a decrease in susceptibility to infection with CaMV observed by Milner and colleagues in the ethylene signaling mutant *etr1* in *A. thaliana* (58).

## STRUCTURE OF NB-LRR PROTEIN

NB-LRR class of proteins possesses at least three distinct domains joined by linker regions: a variable amino-terminal domain (TIR or CC), a central NB domain and C-terminal LRR (Figure 5A). The LRR region determines recognition specificity (41), whereas the amino-terminal domains are mainly involved in signalling (90). The NB forms part of a NB-ARC domain that presumably functions as a molecular switch (107). Crystal structure of any plant NB-LRR proteins has not yet been determined due to their presence in low abundance (65); on the other hand, crystal structures of NB-ARC domain of mammalian APAF-1 (94) and CED4 of nematodes (125) are, however, available. Three dimensional structural models of NB-ARC domains encoded in five diverse plant R genes were constructed (PDB codes: 1W71, 2B84, 2B85, 2FT4, 2FT5) based on crystal structure of APAF-1 (IZ6T) (14).

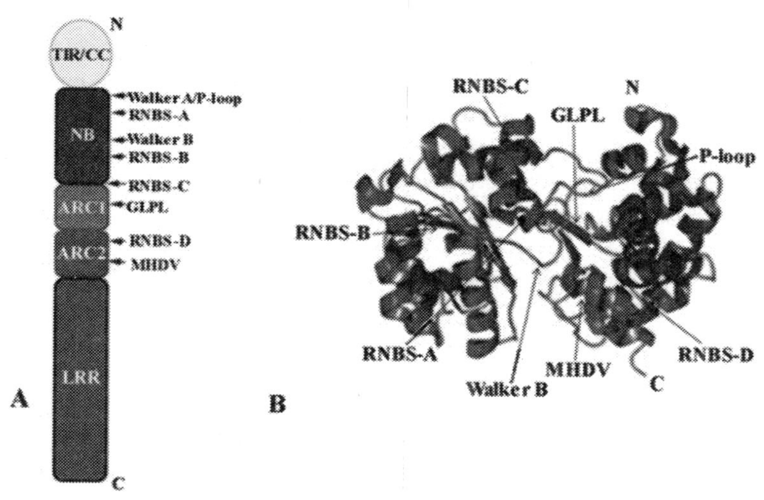


**Fig. 5:** A. Schematic representation of a typical NB–LRR protein. The (sub) domains are depicted as coloured boxes: CC/TIR domain (yellow), NB (green), ARC1 (blue), ARC2 (violet) subdomains and LRR domain (orange). B. NB domain of I2C-1 protein of *Lycopersicon esculentum*, PDB ID 2FT4 (Chattopadhyaya and Pal, 2008). Conserved motifs and N- and C-termini are marked.

### *NB-ARC Domains*

Structure and function of the NB-ARC domains of NB-LRR proteins are well characterized. Due to the distinct structural difference the ARC domain is divided in two regions, ARC1 and ARC2 (3, 90). The NB and ARC domains often referred to collectively as the NB-ARC comprise of a functional nucleotide-binding pocket capable of binding and hydrolyzing ATP (108). The

NB domain has few conserved signature motifs, these include the P-loop, kinase-2, and kinase-3/RNBS-B motifs of the NB domain, the GxP (GLPL) motif of the ARC1 domain as well as the MHDV motif of the ARC2 (Figure 5A, B); all of these were thought to contribute to the nucleotide binding and/or hydrolysis (72). NB domains of TNLs and CNLs can be distinguished by the sequences of three resistance nucleotide binding site (RNBS) motifs within them (RNBS-A, RNBS-C, and RNBS-D). Presence of eight new conserved motifs in the members of Fabaceae, in addition to the reported conserved motifs within the NB-ARC domains has been reorted (81). A critical analysis of these sequences revealed signatures of geminiviral coat proteins (GVCP) within this favoured allele, i.e., *R* gene or RGHs. The probable mechanism of integration of GVCP signatures within the host species of Geminivirus, that are absent within the NB-ARC domain of non-host species, was predicted and their biological significance has been analyzed (81).

### *Amino-Terminal Domains*

NB-ARC family of R protein has been split into two subfamilies, based on the presence or absence of a region homologous to the Toll protein of *Drosophila* and mammalian Interleukin-1-receptor (TIR) domain at the N-terminus (82,127). In contrast to the TIR, the amino acid residues at the N-terminal of NB-ARC domain of the non-TIR subfamily are organized into a putative coiled-coil (CC) structure, a repeated heptad sequence with interspersed hydrophobic amino acid residues, of which the leucine zipper is one example. It consists of two or more alpha helices that interact to form a supercoil and found in a variety of proteins involved in different biological processes also implicated in protein-protein interactions. However, many CC domains do not conform to coiled-coil prediction programs and there are few motifs within this domain that are broadly conserved with the exception of the EDVID motif which is present in most characterized CC domains (67, 91). The coiled-coil domain often extended by the DNA binding domain, such as BEAF/DREAF zinc finger domain or by a Solanaceous domain (59). CC/LZ domain has been implicated in the interaction of R proteins with molecules downstream of the signal transduction pathway (8).

Searches of plant EST databases suggest that monocots do not have TIR-NBS-LRR-like proteins (72, 87). The TIR and non-TIR bearing R-proteins are also distinct at the functional level due to their putative involvement in different signal transduction pathways (1). The N-terminal domains of NB-LRR proteins have conventionally been thought to act as a signalling domain due to similarities in domain structure between plant NB-LRR proteins and certain animal proteins involved in apoptosis and innate immunity (90).

### *LRR Domains*

Most of the plant R proteins contain a long C-terminal segment consisting of repeated leucine reach regions. The number of repeats in the LRR domains in TIR- and non-TIR bearing R-proteins of *Arabidopsis* ranges between 8 to 25. Usually, LRR domains are endowed with repeats of 25–30 amino acid motifs comprising of Leucine or other aliphatic residues at conserved positions and presence in a wide range of proteins and implicated in protein–protein interactions (50). The basic repeat motif corresponds to structural units containing alternately arranged ß-strand and parallel α-helix forming the concave face shaped like a horseshoe strands are more closely packed than the opposed helices, resulting in a characteristic curved structure (51). Most variations are clustered around ß-sheet region which appears to form a protein interaction surface.

The LRR domain emerged as principal determinant of recognition specificity. Most amino acid variation occurs within this domain is possibly due to the strong positive selection (73) and domain swaps between cognate R proteins with different recognition specificity illustrated positively that this region certainly controls recognition specificity (21, 90, 101). Further confirmation of the role of the LRR regions of this class of R proteins in Avr recognition comes from a work where it was shown that the LRR domain of the rice Pi-ta resistance protein interacts in a yeast two-hybrid assay with the corresponding Avr-Pita protein from *Magnaporthe grisea* (40).

The NB-LRR class of R-protins have been studied extensively for their structural organization, sequence evolution and genome distribution in various plant taxa. Phylogenetic analyses revealed that polymorphisms within the NBS-LRR domain were extremely low among ancient Angiosperms that diversified over a period of time independently in different taxonomic families. Current understanding based on the studies carried out during the last 25 years elucidates existence of polymorphisms of NB-LRR domain even within a plant genome (52).

## IDENTIFICATION AND CLASSIFICATION OF PR-PROTEINS

Pathogenesis related (PR) protein is a group of proteins may or may not be constitutively present in the host plant, but their synthesis is triggered when defense signals are transmitted from the R proteins after recognition of pathogen effecter/elicitor proteins (49, 116,117). PR proteins play cellular roles and contribute in an orchestrated manner towards defense response in the host plant. Some of these are found to be products of genes that are associated with SAR with provable antimicrobial activity; while others conferred some degree of protection against pathogen induced stresses when over-expressed in the transgenic plants (12, 117). PR-proteins have been defined as 'inducible defense-related proteins' and classified broadly in 17 families based on their primary structure, serological types, enzymatic as well as biological functions, although not every family is represented in every plant species (Table 1) (117). Additionally, PR-18 family is being considered that consists of proteins with antimicrobial activity, and occurring in *Helianthus annus* and *Amaranthus caudatus* (117). Many PR-proteins are extracellular and others have been localized in the plant cell vacuole. Most of the PR-proteins have antimicrobial activity supporting a role in plant defense. However, some PR-proteins of each family have no demonstrable antimicrobial activity (118). The defense function of chitinases (PR-3, PR-4, PR-8, and PR-11) and -1, 3-glucanases (PR-2) can be shown by their ability to degrade major cell-wall constituents of most filamentous fungi. PR-6 includes proteinase inhibitors which reduce the ability of the pathogen to obtain nutrition, complete its replication cycle, or generate secondary elicitors (97). PR-9 belongs to peroxidases and is involved in plant cell-wall reinforcement through increased lignification (85). PR-10s are ribonuclease-like proteins affecting the translation and replication of invading viral pathogens (83). PR-15, PR-16, and

PR-18 are enzymes that produce toxic hydrogen peroxide, which can also elicit additional plant defense responses ( 17, 22, 37). The defense function of thaumatin (THN)-like proteins (PR-5), defensins (PR-12), thionins (PR-13), and lipid-transfer proteins (LTPs; PR-14) are attributed to their ability to permeabilize pathogen membranes. Genome sequence data from many plant species are revealing that the PR protein families in plants are large. The PR-1, -6, -12, -13, and -14 families have been reviewed based on genome sequence data (97, 126).

**Table 1.** Classification of inducible defense related proteins following Van Loon and colleagues (2006)

| Family | Sources | Protein | Recognized targets |
|---|---|---|---|
| PR-1 | *Nicotiana tabacum,* and many other species | Unknown (antifungal) | *Phytophthora infestans* |
| PR-2 | *Hordeum vulgare, N. tabacum, Oryza sativa* | β-1,3-Glucanase (antifungal) | *Colletotrichum lagenarium, Fusarium solani, Rhizoctonia solani* |
| PR-3 | *N. tabacum, Phaseolus vulgaris, Triticum aestivum* | Chitinase (antifungal) | *Fusarium oxysporum, Physalospora piricola* |
| PR-4 | *H. vulgare, N. tabacum, Solanum tuberosum, Solanum lycopersicum, Zea mays* | Chitinase (antifungal) | *Alternaria radicina, Fusarium moniliforme, F. solani, Trichoderma viride* |
| PR-5 | *Cucurbita maxima, H. vulgare, N. tabacum, Zea mays* | Thaumatin-like (antifungal) | *Candida albicans, F. oxysporum, Neurospora crassa, P. infestans, Trichoderma reesei, Trich. viride* |
| PR-6 | *S. lycopersicum* | Proteinase inhibitor | Unknown |
| PR-7 | *S. lycopersicum* | Endoproteinase | Unknown |
| PR-8 | *Cucumis sativus* | Chitinase (antifungal) | *C. lagenarium* |
| PR-9 | *N. tabacum* | Lignin-forming peroxidase | Unknown |

| | | | |
|---|---|---|---|
| PR-10 | *Capsicum annuum, Lupinus albus, Oxalis tuberosa, Petroselinum crispum* | Ribonuclease-like (antifungal) | *Colletotrichum gloeosporioides, F. oxysporum, Phytophthora capsici, R. Solani* |
| PR-11 | *N. tabacum* | Chitinase (antifungal) | *A. radicina, Trichoderma viride* |
| PR-12 | *Raphanus sativus* | Defensin (antifungal) | *Aspergillus flavus, F. moniliforme* |
| PR-13 | *Arabidopsis thaliana* | Thionin (antifungal) | *F. solani* |
| PR-14 | *H. vulgare* | Lipid-transfer protein | Unknown |
| PR-15 | *H. vulgare* | Oxalate oxidase (antifungal) | *Blumeria graminis* |
| PR-16 | *H. vulgare* | Oxalate-oxidase-like | *B. graminis* |
| PR-17 | *H. vulgare, N. tabacum, T. aestivum* | Unknown (antifungal) | *B. graminis* |

## CONCLUSIONS

RNA silencing, an ancient cellular mechanism of defense against foreign nucleic acids that also functions in gene regulation is an important plant defense strategy. It plays an important role in the protection of eukaryotes against foreign molecules, such as, nucleic acids of viruses. In plants, post-transcriptional gene silencing controls numerous developmental processes and is required for innate immunity regulating virus accumulation. However, an in depth knowledge of plant immune function is still lacking, acquirement of such molecular in sight would enhance crop yield. A comprehensive understanding of the mechanisms by which infection fails during host-pathogen could be an important approach to achieve a breakthrough in the development of plants with durable broad-spectrum disease resistance.

### Acknowledgments

We are thankful to the Director, Bose Institute for providing us the lab facilities and for a Senior Research Fellowship to S.M., we also thank the Department of Biotechnology, India, for the financial assistance (Sanction no. BT/01/COE/06/03).

## REFERENCES

1. Aarts N, Metz M, Holub E, Staskawicz BJ, Daniels MJ, and Parker JE 1998 Different requirements for *EDS1* and *NDR1* by disease resistance genes define at least two *R* gene-mediated signaling pathways in *Arabidopsis*, *Proceedings of National Academy of Science* **95** 10306-10311

2. Akbergenov R, Si-Ammour A, Blevins T, Amin I, Kutter C, Vanderschuren H, Zhang P, Gruissem W, Meins F, Hohn Jr.T, and Pooggin MM 2006 Molecular characterization of geminivirus-derived small RNAs in different plant species, *Nucleic Acids Research* **34** 462-471

3. Albrecht M and Takken FL 2006 Update on the domain architectures of NLRs and R proteins, *Biochemical and Biophysical Research Communications* **339** 459-462

4. Alvarez ME 2000 Salicylic acid in the machinery of hypersensitive cell death and disease resistance, *Plant Molecular Biology* **44** 429-442

5. Axtell MJ and Staskawicz BJ 2003 Initiation of *RPS2*-specified disease resistance in *Arabidopsis* is coupled to the AvrRpt2-directed elimination of RIN4, *Cell* **112** 369-377

6. Bartel DP 2004 MicroRNAs: genomics, biogenesis, mechanism, and function, *Cell* **116** 281-297

7. Baulcombe D 2005 RNA silencing, *Trends in Biochemical Science* **30** 290-293

8. Bertioli DJ, Leal-Bertioli SCM, Lion MB, Santos VL, Pappas G, Cannon SB and Guimaraes PM 2003 A large scale analysis of resistance gene homologues in *Arachis*, *Molecular Genetics & Genomics* **270** 34-45

9. Bonas U and Lahaye T 2002 Plant disease resistance triggered by pathogen-derived molecules: refined models of specific recognition, *Current Opinion in Microbiology* **5** 44-50

10. Bouche N, Lauressergues D, Gasciolli V, and Vaucheret H 2006 An antagonistic function for *Arabidopsis* DCL2 in development and a new function for DCL4 in generating viral siRNAs, *The EMBO Journal* **25** 3347-3356

11. Brodersen P and Voinnet O 2006 The diversity of RNA silencing pathways in plants, *Trends in Genetics* **22** 268-280

12. Broekaert WF, Terras FRG, and Cammue BPA 2000 Induced and preformed antimicrobial proteins, in *Mechanisms of Resistance to Plant Diseases* (eds) AJ Slusarenko, RSS Fraser and

LC Van Loon, Dordrecht: Kluwer, 371–477 pages

13. Brueggeman R, Rostoks N, Kudrna D, Kilian A, Han F, Chen J, Druka A, Steffenson B, and Kleinhofs A 2002 The barley stem rust-resistance gene *Rpg1* is a novel disease-resistance gene with homology to receptor kinases, *Proceedings of National Academy of Science* USA, **99** 9328-9333
14. Chattopadhyaya R and Pal A 2008 Three-dimensional Models of NB-ARC Domains of Disease Resistance Proteins in Tomato, *Arabidopsis*, and Flax, *Journal of Biomolecular Structural Dynamics* **25** 357-372
15. Chisholm ST, Coaker G, Day B, and Staskawicz BJ 2006 Host-microbe interactions: Shaping the evolution of the plant immune response, *Cell* **124** 803-814
16. Clark RM, Schweikert G, Toomajian C, Ossowski S, Zeller G, and Shinn P 2007 Common sequence polymorphisms shaping genetic diversity in *Arabidopsis thaliana*, *Science* **317** 338-342
17. Custers JHHV, Harrison SJ, Sela-Buurlage MB,Van Deventer E, and LagewegW 2004 Isolation and characterization of a class of carbohydrate oxidases from higher plants, with a role in active defence, *Plant Journal* **39** 147–160
18. Dangl JL and Jones JDG 2001 Plant pathogens and integrated defence responses to infection, *Nature* **411** 826-833
19. Dangl JL, Dietrich RA, and Richberg MH 1996 Death Don't Have No Mercy: Cell Death Programs in Plant-Microbe Interactions, *Plant Cell* **10** 1793-1807
20. Deleris A, Gallego-Bartolome J, Bao J, Kasschau KD, Carrington JC, and Voinnet O 2006 Hierarchical action and inhibition of plant Dicer-like proteins in antiviral defense, *Science* **313** 68-71
21. Dodds PN, Lawrence GJ, and Ellis JG 2001 Six amino acid changes confined to the leucine rich repeat ß-strand/ß-turn motif determine the difference between the P and P2 rust resistance specificities in flax, *Plant Cell* **13** 163-178
22. Donaldson PA, Anderson T, Lane BG, Davidson AL, and Simmonds DH 2001 Soybean plants expressing an active oligomeric oxalate oxidase from the wheat gf-2.8 (germin) gene are resistant to the oxalate-secreting pathogen *Sclerotina sclerotiorum*, *Physiological and Molecular Plant Pathology* **59** 297–307

23. Durrant WE and Dong X 2004 Systemic acquired resistance, *Annual Reveiw of Phytopathology* **42** 185-209
24. Ellis JG, Dodds PN, and Lawrence GJ 2007 Flax rust resistance gene specificity is based on direct resistance-avirulence protein interactions, *Annual of Review of Phytopathology* **45** 289-306
25. Flor HH 1971 Current status of gene-for-gene concept, *Annual Review of Phytopathology* **9** 275-296
26. Fraser RSS 1990 The genetics of resistance to plant viruses, *Annual Review of Phytopathology* **28** 179-200
27. Gabriel DW and Rolfe BG 1990 Working models of specific recognition in plant-microbe interactions, *Annual Review of Phytopathology* **28** 365-391
28. Geraats BPJ, Bakker PAHM, Linthorst HJM, Hoekstra J, and van Loon LC 2007 The enhanced disease susceptibility phenotype of ethylene-insensitive tobacco cannot be counteracted by inducing resistance or application of bacterial antagonists, *Physiological and Molecular Plant Pathology* **70** 77-87
29. Glandorf DCM, Peters LGL, Van der Sluis I, and Bakker Greenberg JT 1997 Programmed Cell Death in Plant-Pathogen Interactions, *Annual Review of Plant Physiolpogy and Plant Molecular Biology* **48** 525-545
30. Greenberg JT, Guo A, Klessig DF, and Ausubel FM 1994 Programmed Cell Death in Plants: A Pathogen-Triggered Response Activated Coordinately with Multiple Defense Function Set, *Cell* **77(4)** 551-563
31. Guo M, Tian F, Wamboldt Y, and Alfano JR 2009 The majority of the type III effector inventory of *Pseudomonas syringae pv. tomato* DC3000 can suppress plant immunity, *Molecular Plant-Microbe Interactions* **22** 1069-1080
32. Heath MC 2000a Non-host resistance and non-specific plant defences, *Current Opinion in Plant Biology* **3** 315-319
33. Heath MC 2000b Hypersensitive response-related death, *Plant Molecular Biology* **44** 321-334
34. Holub EB and Cooper A 2004 Matrix, reinvention in plants: how genetics is unveiling secrets of non-host disease resistance, *Trends in Plant Science* **9** 211-214
35. Horvath H, Rostoks N, Brueggeman R, Brian SteffensonB, von Wettstein D, and Kleinhofs A 2003 Genetically engineered stem rust resistance in barley

using the *Rpg1* gene, *Proceedings of National Academy Science* USA, **100** 364–369

36. Houterman PM, Cornelissen BJC, and Rep M 2008 Suppression of Plant Resistance Gene-Based Immunity by a Fungal Effector, *Plos Pathogens* **4(5)** e1000061

37. Hu X, Bidney DL, Yalpani N, Duvick JP, and Crasta O 2003 Over expression of a Gene encoding hydrogen peroxide-generating oxalate oxidase evokes defense responses in sunflower, *Plant Physiology* **133** 170-181

38. Huang WE, Huang L, Preston G, Naylor M, Carr JP, Li Y, Singer AC, Whiteley AS, and Wang H 2006 Quantitative *in situ* assay of salicylic acid in tobacco leaves using a genetically modified biosensor strain of *Acinetobacter sp.* ADP1, *Plant Journal* **46** 1073-1083

39. Jebanathirajah J, Peri S, and Pandey A 2002 Toll and interleukin-1 receptor (TIR) domain-containing proteins in plants: a genomic perspective, *Trends in Plant Science* **7** 388-391

40. Jia Y, McAdams SA, Bryan GT, Hershey HP, and Valent B 2000 Direct interaction of resistance gene and avirulence gene products confers rice blast resistance, *The EMBO Journal* **19** 4004-4014

41. Jones DA and Jones JDG 1996 The Roles of Leucine-Rich Repeats in Plant Defenses, *Advances in Botanical Research* **24** 89-167

42. Jones DA, Thomas CM, Hammond-Kosack KE, Balint-Kurti PJ, and Jones JDG 1994 Isolation of the tomato *Cf-9* gene for resistance to *Cladosporium fulvum* by transposon tagging, *Science* **266** 789-793

43. Jones JDJ and Dangl JL 2006 The plant immune system, *Nature* **444** 323-329

44. Kang BC, Yea MI, and Jahn MM 2005 Genetics of plant virus resistance, *Annual Review of Phytopathology* **43** 581-621

45. Kasschau KD, Xie ZX, Allen E, Llave C, Chapman EJ,and Krizan KA 2003 P1/HC-Pro, a viral suppressor of RNA silencing, interferes with *Arabidopsis* development and miRNA function, *Development Cell* **4** 205-217

46. Kawchuk, LM, Hachey J, Lynch DR, Kulcsar F, van Rooijen G, Waterer DR, Robertson R, Kokko E, Byers R, Howard RJ, Fischer R, and Prüfer D 2001 Tomato *Ve* disease resistance genes encode cell surface-like receptors, *Proceedings of National Academy of Science* USA, **98** 6511–6515

47. Kedzierski, LJ, Montgomery JC, and Handman E 2004 Leucine-rich repeats in host-pathogen interactions, *Archivum Immunologiae et Therapia Experimentalis* **52** 104-112

48. Kim HE, Du F, Fang M, and Wang X 2005 Formation of apoptosome is initiated by cytochrome c-induced dATP hydrolysis and subsequent nucleotide exchange on APAF-1, *Proceedings of National Academy of Science* **102** 17545-17550

49. Kitajima S and Sato F 1999 Plant pathogenesis-related proteins: molecular mechanisms of gene expression and protein function, *Journal of Biochemistry* **125** 1-8

50. Kobe B and Deisenhofer J 1994 The leucine-rich repeat: a versatile binding motif, *Trends in Biochemoical Science* **19** 415-421

51. Kobe B and Kajava AV 2001 The leucine-rich repeat as a protein recognition motif, *Current Opinion in Structural Biology* **11** 725-732

52. Kohler A, Rinaldi C, Duplessis S, Baucher B, Geelen D, Duchaussoy F, Meyers B C, Boerjan W, and Martin F 2008 Genome-wide identification of NBS resistance genes in *Populus trichocarpa*, *Plant Molecular Biology* **66** 619-636

53. Kumar D, Gustafsson C, and Klessig DF 2006 Validation of RNAi silencing specificity using synthetic genes: salicylic acid-binding protein 2 is required for innate immunity in plants, *Plant Journal* **45** 863-868

54. Lamb C and Dixon RA 1997 The oxidative burst in plant disease resistance, *Annual Review of Plant Physioliology and Plant Molecular Biology* **48** 251-275

55. Lecellier CH, Dunoyer P, Arar K, Lehmann-Che J, Eyquem S, Himber C, Saib A, and Voinnet O 2005 A cellular microRNA mediates antiviral defense in human cells, *Science* **308** 557-560

56. Leipe DD, Koonin EV, and Aravind L 2004 STAND, a class of P-loop NTPases including animal and plant regulators of programmed cell death: multiple, complex domain architectures, unusual phyletic patterns, and evolution by horizontal gene transfer, *Journal of Molecular Biology* **343** 1-28

57. Llave C 2004 MicroRNAs: more than a role in plant development? *Molecular Plant Pathology* **5** 361-366

58. Love AJ, Laval V, Geri C, Laird J, Tomos AD, Hooks MA, and Milner JJ 2007 Components of *Arabidopsis* defense- and ethylene-signaling pathways regulate

susceptibility to cauliflower mosaic virus by restricting long-distance movement, *Molecular Plant Microbe Interactions* **20(6)** 659-670

59. Lukasik E and Takken FL 200 Standing strong, resistance proteins instigators of plant defence, *Current Opinion in Plant Biology* **12** 427-436
60. Ma W and Berkowitz GA 2007 The grateful dead: calcium and cell death in plant innate immunity, *Cell Microbiology* **9** 2571-2585
61. Mackey D and McFall AJ 2006 MAMPs and MIMPs: proposed classifications for inducers of innate immunity, *Molecular Microbiology* **61** 1365-1371
62. Mackey D, Belkhadir Y, Alonso JM, Ecker JR, and Dangl JL 2003 *Arabidopsis* RIN4 is a target of the type III virulence effector AvrRpt2 and modulates RPS2-mediated resistance, *Cell* **112** 379-389
63. Mackey D, Holt BF, Wiig A, and Dangl JL 2002 RIN4 interacts with *Pseudomonas syringae* type III effector molecules and is required for RPM1-mediated resistance in *Arabidopsis, Cell* **108** 743-754
64. Malamy J, Carr JP, Klessig DF, and Raskin I 1990 Salicylic acid: a likely endogenous signal in the resistance response of tobacco to viral infection *Science* **250** 1002-1004
65. Martin GB, Bogdanove AJ, and Sessa G 2003 Understanding the functions of plant disease resistance proteins, *Annual Review of Plant Biology* **54** 23-61
66. Maule AJ, Caranta C, and Boulton MI 2007 Sources of natural resistance to plant viruses: status and prospects, *Molecular Plant Pathology* **8** 223-231
67. Mazourek M, Cirulli ET, Collier SM, Landry LG, Kang BC, Quirin EA, Bradeen JM, Moffett P, and Jahn M 2009 The fractionated orthology of *Bs2* and *Rx/Gpa2* supports shared synteny of disease resistance in the *Solanaceae, Genetics* **182** 1351-1364
68. McDowell JM and Woffenden BJ 2003 Plant disease resistance genes: recent insights and potential applications, *Trends in Biotechnology* **21(4)** 178-183
69. Mehdy MC 1994 Active oxygen species in plant defense against pathogens, *Plant Physiology* **105** 467-472
70. Mestre P and Baulcombe DC 2006 Elicitor-mediated oligomerization of the tobacco N disease resistance protein, *Plant Cell* **18** 491-501
71. Métraux JP, Signer H, Ryals J, Ward E, Wyss-Benz M, Gaudin J, Raschdorf K,

Schmid E, Blum W, and Inverardi B 1990 Increase in salicylic acid at the onset of systemic acquired resistance in cucumber, *Science* **250** 1004-1006

72. Meyers BC, Dickerman AW, Michelmore RW, Sivaramakrishnan S, Sobral BW, and Young ND 1999 Plant disease resistance genes encode members of an ancient and diverse protein family within the nucleotide binding superfamily, *Plant* Journal **20** 317-332

73. Michelmore RW and Meyers BC 1998 Clusters of resistance genes in plants evolve by divergent selection and a birth-and-death process, *Genome Research* **8** 1113-1130

74. Moffett P, Farnham G, Peart J, and Baulcombe DC 2002 Interaction between domains of a plant NBS-LRR protein in disease resistance-related cell death, *The EMBO Journal* **21** 4511-4519

75. Moissiard G and Voinnet O 2006 RNA silencing of host transcripts by cauliflower mosaic virus requires coordinated action of the four *Arabidopsis* Dicer-like proteins, *Proceedings of National Academy of Science* 103 19593-19598

76. Mucyn TS, Clemente A, Andriotis VM, Balmuth AL, Oldroyd GE and Staskawicz BJ 2006 The tomato NBARC-LRR protein Prf interacts with Pto kinase *in vivo* to regulate specific plant immunity, *Plant Cell* **18** 2792-2806

77. Mur LA, Kenton P, Lloyd AJ, Ougham H, and Prats E 2008 The hypersensitive response; the centenary is upon us but how much do we know? *Journal of Experimental Botany* **59** 501-520

78. Murphy AM, Chivasa S, Singh DP, and Carr JP 1999 Salicylic acid induced resistance to viruses and other pathogens: a parting of the ways? *Trends in Plant Science* **4** 155-160

79. Nicaise V, Roux M, and Zipfel C 2009 Recent advances in PAMP-triggered immunity against bacteria: Pattern recognition receptors watch over and raise the alarm, *Plant Physiology* **150** 1638-1647

80. Nobuta K, Okrent RA, Stoutemyer M, Rodibaugh N, Kempema L, Wildermuth MC, and Innes RW 2007 The GH3 acyl adenylase family member PBS3 regulates salicylic acid-dependent defense responses in *Arabidopsis*, *Plant Physiology* **144** 1144-1156

81. Pal A, Chakrabarti A and Basak J 2007 New motifs within the NB-ARC domain of R proteins: Probable

mechanisms of integration of geminiviral signatures within the host species of Fabaceae family and implications in conferring disease resistance, *Journal of Theoretical Biology* **246** 564-573

82. Pan Q, Wendel J, and Fluhr R 2000 Divergent evolution of plant NBS-LRR resistance gene homologues in dicot and cereal genomes, *Journal of Molecular Evolution* **50** 203-13

83. Park CJ, Kim KJ, Shin R, Park JM, Shin YC, and Paek KH 2004 Pathogenesis-related protein 10 isolated from hot pepper functions as a ribonuclease in an antiviral pathway, *Plant* Journal **37** 186-98

84. Park SW, Kaimoyo E, Kumar D, Mosher S, and Klessig DF 2007 Methyl Salicylate Is a Critical Mobile Signal for Plant Systemic Acquired Resistance, *Science* **318** 113-116

85. Passardi F, Penel C, and Dunand C 2004 Performing the paradoxical: How plant peroxidases modify the cell wall, *Trends in Plant Sc*ience **9** 534-540

86. Pedley KF and Martin GB 2006 Role of mitogen-activated protein kinases in plant immunity, *Current Opinion in Plant Biology* **8** 541-547

87. Pennazio S 1995 The Hypersensitive Reaction of Higher Plants to Viruses: A Molecular Approach, *Microbiologica* **18(2)** 229-240

88. Peso L, Gonzalez VM, Inoharat N, Ellis RE, and Nunez G 2000 Disruption of the CED-9.CED-4 complex by EGL-1 is a critical step for programmed cell death in *Caenorhabditis elegans*, *Journal of Biological Chem*istry **275** 27205-27211

89. Rairdan G and Moffett P 2007 Brothers in arms? Common and contrasting themes in pathogen perception by plant NB-LRR and animal NACHT-LRR proteins, *Microbes and Infection* **9** 677-686

90. Rairdan GJ and Moffett P 2006 Distinct domains in the ARC region of the potato resistance protein Rx mediate LRR binding and inhibition of activation, *Plant Cell* **18** 2082-2093

91. Rairdan GJ, Collier SM, Sacco MA, Baldwin TT, Boettrich T, and Moffett P 2008 The coiled-coil and nucleotide binding domains of the potato rx disease resistance protein function in pathogen recognition and signaling, *Plant Cell* **20** 739-751

92. Rasmussen JB, Hammerschmidt R, and Zook MN 1991 Systemic induction of salicylic acid accumulation

in cucumber after inoculation with *Pseudomonas syringae* pv *syringae*, *Plant Physiology* **97** 1342-1347

93. Reymond P and Farmer EE 1998 Jasmonate and salicylate as global signals for defense gene expression, *Current Opiniopn in Plant Biology* **1** 404-411

94. Riedl SJ, Li W, Chao Y, Schwarzenbacher R, and Shi Y 2005 Structure of the apoptotic protease-activating factor 1 bound to ADP, *Nature* **434** 926-933

95. Rolland F, Moore B and Sheen J 2002 Sugar Sensing and Signaling in Plants, *Plant Cell* S185–S205

96. Schulze-Lefert P and Panstruga R 2011 A molecular evolutionary concept connecting nonhost resistance, pathogen host range, and pathogen speciation, *Trends in Plant Science* **16(3)** 117-125

97. Sels J, Mathys J, De Coninck BM, Cammue BP, and De Bolle MF 2008 Plant pathogenesis-related (PR) proteins: a focus on PR peptides, *Plant Physiology and Biochemistry* **46** 941-950

98. Seo PJ, Lee AK, Xiang F, and Park CM 2008 Molecular and functional profiling of Arabidopsis pathogenesis-related genes: Insights into their roles in salt response of seed germination, *Plant Cell Physiology* **49** 334-344

99. Shabab M, Shindo T, Gu C, Kaschani F, Pansuriya T, Chintha R, Harzenc A, Colby T, Kamoun S, and van der Hoorn ARL 2008 Fungal effector protein AVR2 targets diversifying defense related Cys proteases of tomato, *Plant Cell* **20** 1169-1183

100. Shamray SN 2003 Plant Resistance Genes: A Molecular and Genetic Organization, Function, and Evolution, *Zh Voshch Biologii* **64(3)** 195-214

101. Shen QH, Zhou F, Bieri S, Haizel T, Shirasu K, and Schulze-Lefert P 2003 Recognition specificity and *RAR1*/*SGT1* dependence in barley *Mla* disease resistance genes to the powdery mildew fungus, *Plant Cell* **15** 732-744

102. Shulaev V, Silverman P, and Raskin I 1997 Airborne signalling by methyl salicylate in plant pathogen resistance, *Nature* **385** 718-721

103. Siegien I and Bogatek R 2006 Cyanide action in plants - from toxic to regulatory, *Acta Physiologiae Plantarum* **15** 483-498

104. Simon-Mateo C and Garcia JA 2006 MicroRNA-guided processing impairs *Plum pox virus* replication, but the virus readily evolves to escape this silencing

mechanism, *Journal of Virology* **80** 2429-2436

105. Song WY, Wang GL, Chen LL, Kim HS, Pi LY, Holsten T, Gardner J, Wang B, Zhai WX, Zhu LH, Fauquet C, and Ronald P 1995 A receptor kinase-like protein encoded by the rice disease resistance gene, *Xa21*, *Science* **270** 1804-1806

106. Strawn MA, Marr SK, Inoue K, Inada N, Zubieta C, and Wildermuth MC 2007 *Arabidopsis* isochorismate synthase functional in pathogen-induced salicylate biosynthesis exhibits properties consistent with a role in diverse stress responses, *Journal of Biological Chemistry* **282** 5919-5933

107. Takken FLW, Albrecht M, and Tameling WIL 2006 Resistance proteins: molecular switches of plant defence, *Current Opinion in Plant Biology* **9** 383-390

108. Tameling WIL, Vossen JH, Albrecht M, Lengauer T, Berden JA, Haring MA, Cornelissen BJC, and Takken FLW 2006 Mutations in the NB-ARC domain of I-2 that impair ATP hydrolysis cause auto activation, *Plant Physiology* **140** 1233-1245

109. Thaler JS, Owen B, and Higgins VJ 2004 The role of the jasmonate response in plant susceptibility to diverse pathogens with a range of lifestyles, *Plant Physiology* **135** 530-538

110. Tomari Y and Zamore PD 2005 Perspective: machines for RNAi, *Genes and Development* **19** 517-529

111. Ton J, van Pelt JA, van Loon LC, and Pieterse CMJ 2002 Differential effectiveness of salicylate-dependent and jasmonate/ethylene-dependent induced resistance in *Arabidopsis*, *Molecular Plant Microbe Interactions* **15** 27–34

112. Torres MA and Dangl JL 2005 Functions of the respiratory burst oxidase in biotic interactions, abiotic stress and development, *Current Opinion in Plant Biology* **8** 397-403

113. van der Biezen EA and Jones JD 1998a Plant disease resistance proteins and the gene-for-gene concept, *Trends in Biochemical Sciences* **23** 454-456

114. van der Biezen EA and Jones JD 1998b The NB-ARC domain: A novel signalling motif shared by plant resistance gene products and regulators of cell death in animals, *Curr Biology* **8** 226-227

115. van der Hoorn ARL and Kamoun S 2008 From guard to decoy: A new model for perception of plant pathogen effectors, *Plant Cell* **20** 2009-2017

116. van Loon LC and van Strien EA 1999 The families of pathogenesis related proteins, their activities, and comparative analysis of PR-1 type proteins, *Physiological and Molecular Plant Pathology* **55** 85–97

117. van Loon LC, Rep M, and Pieterse CMJ 2006 Significance of inducible defense-related proteins in infected plants, *Annual Review of Phytopathology* **44** 135-162

118. Veronese P, Ruiz MT, Coca MA, Hernandez-Lopez A, and Lee H 2003 In defense against pathogens. Both plant sentinels and foot soldiers need to know the enemy, *Plant Physiology* **131** 1580–1590

119. Vlot AC, Klessig DF, and Park SW 2008 Systemic acquired resistance: the elusive signal(s), *Current Opinion in Plant Biology* **11** 436-442

120. Voinnet O 2005 Induction and suppression of RNA silencing: insights from viral infections, *Nature Reviews Genetics* **6** 206-220

121. Wang B, Wang Y, Wang Q, and Luo G 2004 Characterisation of an NBS-LRR resistance gene homologue from soybean, *Journal of Plant Physiology* **161** 815-822

122. Whitham S, Dinesh-Kumar SP, Choi D, Hehl R, Corr C, and Baker B 1994 The product of the tobacco mosaic virus resistance gene *N*: Similarity to toll and the interleukin-1 receptor, *Cell* **78** 1101-1115

123. Xiao SY, Ellwood S, Calis O, Patrick E, Li TX, Coleman M, and Turner JG 2001 Broad-spectrum mildew resistance in *Arabidopsis thaliana* mediated by RPW8, *Science* **291** 118-120

124. Xie Z, Johansen LK, Gustafson AM, Kasschau KD, Lellis AD, Zilberman D, Jacobsen SE, and Carrington JC 2004 Genetic and functional diversification of small RNA pathways in plants, *PLOS Biology* **2** 642-652

125. Yan N, Chai J, Lee ES, Gu L, Liu Q, He J, Wu JW, Kokel D, Li H, and Hao Q 2005 Structure of the CED-4–CED-9 complex provides insights into programmed cell death in *Caenorhabditis elegans*, *Nature* **437** 831-837

126. Yokoyama R and Nishitami K 2004 Genomic basis for cell-wall diversity in plants. A comparative approach to gene families in rice and *Arabidopsis*, *Plant Cell Physiology* **45** 1111-1121

127. Young ND 2000 The genetic architecture of resistance,

*Current Opinion in Plant Biology* **3** 285-290

128. Yun BY, Atkinson, HA, Gabroit C, Greenland A, Read N, Pallas JA, and Loake GJ 2003 Loss of actin cytoskeletal function and EDS1 activity, in combination, severely compromises non-host resistance in *Arabidopsis* against wheat powdery mildew, *Plant J*ournal **34** 768-777

129. Zhou JM and Chai J 2008 Plant pathogenic bacterial type III effectors subdue host responses, *Current Opinion in Microbiology* **11** 179-185

130. Zipfel C and Felix G 2005 Plants and animals: a different taste for microbes? *Current Opinion in Plant Biology* **8** 353-360.

*Rev. Plant Pathol. Vol. 6, 2014*
*Indian Society of Mycology and Plant Pathology*
*Scientific Publishers (India), Jodhpur pp. 127-154*

# HOST-PATHOGEN INTERACTION, DIAGNOSIS AND MANAGEMENT OF *TOBACCO MOSAIC VIRUS* AND *TOMATO MOSAIC VIRUS* INFECTING PEPPER AND TOMATO

***H.S. Prakash, A.C. Udayashankar and K.N. Madhusudhan***

Department of Studies in Biotechnology, University of Mysore, Manasagangotri, Mysore 570006, Karnataka, India. E-mail: hsp@appbot.uni-mysore.ac.in

**KEYWORDS:** Tobamovirus, molecular diagnosis, management, RT-PCR, movement protein, coat protein, multiplex PCR

## ABSTRACT

The viruses in the genus *Tobamovirus* are distributed worldwide, infecting solanaceous, cucurbitaceous, and cruciferous plants resulting in severe economic losses. Tobamoviruses are the major viral pathogens of tomato and bell pepper. The genus *Tobamovirus* currently consists of 22 definitive and one tentative species. Tobamoviruses can remain infective for many months on contaminated testae of seeds collected from infected mother plants, and be transmitted mechanically to young seedlings if handled during transplantation. Tobamovirus movement proteins play a determinant role in the establishment of infections in plants, allowing the local movement of viral RNA genome through plasmodesmata. Following initial infection , usually by mechanical inoculation, many plant viruses, such as TMV, spread from cell to cell through plasmodesmata by using the existing cellular pathways of plasmodesmal transport. Biological assays as well as serological detection and identification of species within the genus *Tobamovirus* can be time-consuming and might lead to ambiguous results. Procedures based on the polymerase chain reaction such as IC-RT-PCR, RT-PCR and multiplex PCR have been developed to circumvent these problems. The systemic and local lesion hosts of TMV/ToMV respond differentially in terms of defense-responsive enzymes and tissue-localization of hydrogen peroxide. Transient expression of RNAi constructs against TMV gene like replicase and host genes like *TOM* seem to be promising in evolving management strategy against tobamoviruses.

## INTRODUCTION

Viruses are unique disease causing entity representing fundamentally different form of life. Plant viruses cause serious economic losses in many major crops by reducing yield and quality. Viruses are among the most important and biologically intriguing group of plant pathogens (3).

The genus *Tobamovirus* include rigid rod shaped viruses with a positive sense single stranded RNA, which is encapsidated within helically arranged coat protein subunits. The genome is unipartite. The genome of tobamovirus is a single-stranded 5'-capped RNA encoding three non-structural proteins and the coat protein (CP). The 130K and 180K non-structural proteins are directly translated from the genomic RNA and are involved in viral RNA replication via negative-strand RNA. The other non-structural protein, with an approximate molecular mass of 30 kDa is necessary for viral cell-to-cell movement in plants and is referred to as the movement protein (MP). Both MP and CP are translated from respective subgenomic mRNAs (48). The gene at the 3' end of the genome codes for 18 kDa coat protein subunits.

Although these viruses affect field crops, they are more often a problem in greenhouse crops where plants are generally grown at a higher density and handled more frequently. Infection of many vegetables and ornamental plants with tobamoviruses results in economically important losses in all parts of the world.

Tomato (*Solanum lycopersicum* L.) and bell pepper (*Capsicum annuum* L.) are the two important members of the family Solanaceae, which are used as vegetables worldwide. Tomato is an important fruit vegetable cultivated all over the world. Tomato has found wide use in the preparation of jams, ketchups, and sauces. It is also used as a raw vegetable.

Bell pepper is a native of America. It was introduced to Africa and Asia by the Portuguese. Paprika, derived from *C. annuum* is used primarily in flavouring of garnishes, pickles, meats, ketchup, salads, sausages etc. Bell pepper has a pungent odour; pungency and colour varies with variety. Bell pepper also has medicinal values and has been used as a carminative, digestive irritant and stimulant.

Tobamovirus infecting tomato are Tobacco mosaic virus (TMV), Tomato mosaic virus (ToMV) and Tobacco mild green mosaic virus (TMGMV). At least six tobamoviruses infect peppers: Paprika mild mottle virus (PaMMV), Pepper mild mottle virus (PMMoV), Ribgrass mosaic virus (RMV), Tobacco mild green mosaic virus (TMGMV), TMV and ToMV (14).

## TAXONOMY AND NOMENCLATURE OF TOBAMOVIRUSES

Even after extensive research on TMV from past 100 years, still virus is prevalent worldwide and affects numerous plant species of several cultivated and uncultivated genera. ToMV was for many years considered to be a strain of TMV. However, because the two viruses can be distinguished by differences in host range, serological affinities, protein and nucleic acid sequences and compositions, ToMV is now recognized as a distinct virus. The genus *Tobamovirus* currently consists of 22 definitive and one tentative species (30). Overall nucleotide sequence difference is considered to characterize strains of the same species, although most of the sequenced species have considerably less than 90% sequence identity.

## HOST RANGE OF TMV

*Nicotiana tabacum* (tobacco), *Lycopersicon esculentum* (tomato), *Capsicum annuum* (bell pepper), *Solanum melongena* (aubergine), *Allium sativum* (garlic), *Beta vulgaris* (beetroot), *Phaseolus vulgaris* (kidney bean), *Vigna unguiculata* (cowpea), *Glycine max* (soyabean), Brassicaceae (cruciferous crops), *Apium graveolens* (celery) and *Solanum tuberosum* (potato) are reported as primary hosts of TMV. Other than primary hosts, TMV also infects Rosa (roses), Petunia, *Pyrus communis* (European pear), *Allium chinense* (rakkyo), *Malus pumila* (apple), Dahlia, *Vicia faba* (broad bean), Cymbidium, *Helianthus annuus* (sunflower), *Citrullus lanatus* (watermelon) plants as secondary hosts. *Cucumis sativus* (cucumber), *Petunia hybrida*. Solanum (nightshade), *Plantago* (Plantain), Physalis (Groundcherry), *Calendula officinalis* (Pot marigold), *Nicotiana glauca* (Tree tobacco) plants were recorded as wild hosts which were infected by TMV (16).

## HOST RANGE OF TOMV

ToMV also has a very wide experimental host range. It is reported to be transmissible to at least 127 other species in 23 families (29). Primary hosts of ToMV are *S. lycopersicum*, *C. annuum* and *C. frutescens*. The reported secondary hosts of ToMV are *Picea rubens* (red spruce), *N. tabacum*, *S. tuberosum*, *Cornus florida* (Flowering dogwood), *Syringa vulgaris* (lilac), *Gomphrena globosa* (Globe amaranth) and *Petunia* spp. (Petunia annual flower). ToMV infects *Chenopodium murale* (nettleleaf goosefoot) and *Physalis* (Groundcherry), which are recorded as wild hosts (16).

## SYMPTOMATOLOGY

In tobacco, TMV causes various kinds of symptoms, mainly depending on the presence or absence of *N*-gene. In susceptible tobacco, TMV causes burnt streak symptoms often occur on old leaves. Necrotic symptoms are the major characteristics of tobacco genotypes bearing the N-gene when TMV is inoculated on plants at the second to third leaf stage, or after keeping infected plants at temperatures above 30 °C for several hours during the day (50). In bell pepper, TMV produces progressive symptoms of veinal discoloration on young leaves 5-6 days after inoculation. This is followed by a mottling or mosaic pattern of light and dark-green areas (Fig. 1), and later by blistering and fern-shaped leaves. Abnormal growth and stunting, with distorted and dwarfed leaves, also occur. Plants infected at the late stages only develop symptoms on the apical leaves. Symptoms of blotch mosaic and distortion were rarely observed on flowers.

**Fig. 1.** Symptoms of TMV infection on pepper leaves

Symptoms in tomato produced by ToMV were greatly affected by various factors viz., temperature, day length, light intensity, age of plant when infected, virulence of infecting virus and cultivar. However, infected greenhouse-grown tomatoes in summer develop light and dark green leaf mottling and, sometimes, distortion of young leaves (Fig. 2). In winter, with short days, low light intensity and low temperatures (below 20 °C), plants are often severely stunted. The laminae of leaves are much reduced ("fern-leaf") and yield is greatly reduced. Some strains can cause conspicuous symptoms such as yellow mottling of leaves and fruit, so-called aucuba symptoms, and others so-called "single virus streak" in which stems, petioles, leaves and/or fruit develop necrotic streaks.

**Fig. 2.** Symptoms of ToMV infection on tomato leaves

## YIELD LOSSES

TMV causes heavy yield losses for tobacco, tomato and pepper worldwide. Early infection by TMV mainly destroys the young tobacco plants in both nurseries and in the field. A high percentage of plants infected at, or soon after transplanting can reduce yield by up to 50%. TMV causes up to 90% yield reduction in pepper fields in association with other viruses (28).

In India, up to 34% loss has been reported in tomato under experimental conditions (32). Approximately 20% of the yield loss in greenhouse grown tomatoes due to infection by ToMV. The yield of infected greenhouse or field grown susceptible crops can be reduced up to 25% (17).

## EPIDEMIOLOGY OF TOBAMOVIRUSES

TMV survives for months and years in stored or even manufactured tobacco. The contact between the roots, infected soil and by workers' contact during cultural operation serves as the main source of TMV infection in tobacco fields. After the initial establishment of infection, plant-to-plant spread can occur by leaf contact. Unlike most viruses, tobamoviruses are not transmitted by insects. All tobamoviruses (ToMV) are highly stable and naturally transmitted by mechanical means.

The virus (ToMV) can remain infective for many months on contaminated testae of seeds collected from infected mother plants, and be transmitted mechanically to young seedlings if handled during transplantation. ToMV occur in the endosperm of seeds, but not the embryo, collected from fruits that were set after plants were infected systemically; such seeds can act as center point of infection and a means by which the virus can be disseminated. ToMV spreads rapidly within crops from primary foci is largely attributable to mechanical transmission of virus by horticultural workers on contaminated hands, clothing and tools during routine horticultural operations. In greenhouse the virus can persist for many months on greenhouse structures and in debris of infected plants in soil from which it is difficult to eradicate. Healthy seedlings planted in contaminated soil can be infected through minor wounds caused by natural mechanical damage to roots (15).

The presence of ToMV has been confirmed in different parts of seed such as, external mucilage, testa and endosperm, but not the embryo of seeds collected from infected tomato fruits. The great variation in the extent of contamination was reported greatly in seeds of different fruits; approximately 50% of seed is often contaminated, but this can be up to 94% of seed (13).

## TRANSMISSION OF TOBAMOVIRUSES

Primary infection of TMV in tomato plantlets in seedbeds is caused by particles on seed. Seed transmission rates of 98, 18 and 7% have been found in tomato, pepper and aubergine, respectively (23). Seed transmission is frequent in pepper, and field disease outbreaks have been attributed to the use of contaminated seeds.

ToMV is transmitted mechanically from contaminated tomato seed if seedlings are handled during transplanting (13). ToMV has been recovered from debris and soil in a field where infected tomato had been grown the previous year.

Seed samples of tomato and bell pepper collected from various sources in India were contaminated with TMV and ToMV respectively. The concentration of virus in seed extract detected by ELISA was 0.18-0.56 in tomato and 0.20-0.52 in pepper. The rate of seed transmission for seed-borne ToMV and TMV was 1-13% and 1-10% respectively in transplanted seedlings (21). TMV and ToMV were infected at different growth stages of host. The infection was observed irrespective of growth stages and the virus established in the seeds. The concentration of virus in seeds was determined by local lesion assay. The concentration of virus was high in seeds when the plants were inoculated early (22).

## BIOLOGY OF TOBAMOVIRUSES

### Cell-to-cell movement

Minutes after infection virions disassemble, RNA synthesis begins, and new particles develop in the cytoplasm, spreading from cell to cell through plasmodesmata and in to other plant parts through the phloem.

Cell-to-cell movement of TMV is mediated by a 30-kDa Movement Protein (MP) encoded by TMV RNA (57). Tobamovirus movement proteins play a determinant role in the establishment of infections in plants, allowing the local movement of viral RNA genome through plasmodesmata. P30 has at least two functions, to cooperatively bind single-stranded nucleic acids and to increase plasmodesmatal permeability.

Following initial infection, usually by mechanical inoculation, many plant viruses, such as TMV, spread from cell to cell through plasmodesmata by using the existing cellular pathways of plasmodesmal transport. TMV encodes a specific cell-to-cell movement protein (MP), which mediates the transport of the viral genomic RNA through plasmodesmata. TMV-MP has been shown to bind TMV RNA in association with cytoskeleton and ER, targets to plasmodesmata within plant cell walls, increases plasmadesmatal permeability and undergo negative regulation by phosphorylation (11). To perform these functions, TMV MP most likely co-operates with different cellular factors, some of which, such as cell wall PMEs (18), microtubules (MTs) and actin filaments (11).

Utilizing affinity chromatography on an immobilized TMV MP, tobacco cell wall proteins that bind TMV MP have been isolated(18). Outcome of these experiments, showed that a major protein of about 38 kDa was detected by SDS-PAGE, later it was identified as PME. Along with this band, other, less abundant bands in the 50- to 60-kDa range were also detected. TMV-MP has been shown to interact with PME suggests that binding to the host cell factors containing signal sequences may represent a general strategy employed by TMV-MP to enter the ER secretory pathway; because ER strands span plasmodesmata, TMV MP may utilize its association with the ER for transport to plasmodesmata.

Models of movement protein of TMV, ToMV and host protein pectin methylesterase of tomato were constructed by homology modeling and active site for interaction was determined (41).

## DETECTION

### *Biological tests*

Species from the families Solanaceae and Chenopodiaceae are hypersensitive to TMV, and their response to the virus is used as a fast, simple and inexpensive diagnostic method. *Nicotiana glutinosa*, *N*-tobacco genotypes or *Datura stramonium* are routinely used as diagnostic hosts in plant pathology laboratories. Inoculated leaves develop local necrotic lesions, of 2-3 mm diameter, after 3-4 days. Necrotic lesions appear after 24 hours when temperatures are above 28°C for several hours in the day. On sensitive tobaccos and other *Nicotiana* species mosaic symptoms develop after 10-15 days. Biological tests are very efficient due to the high infectivity of TMV, long persistence in crude sap and the dilution end point limit, which is about 1/1,000,000.

ToMV is readily transmitted by mechanical inoculation of sap to herbaceous test plant species. The most useful species are *Chenopodium amaranticolor, C. murale* - chlorotic or necrotic local lesions followed by systemic mottling and pale green or necrotic flecks. The inoculated leaves of *D. stramonium, N. glutinosa, N. sylvestris* show necrotic lesions after ca 3 days, but no systemic infection unless plants are retained at 33°C or more. *Gomphrena globosa* show chlorotic and necrotic local lesions followed by systemic chlorotic mottling in infected leaves. In *Nicotiana rustica, N. tabacum* cv. White Burley, Dutch A selection - most ToMV isolates induce only necrotic local lesions, but a few strains also infect inoculated plants systemically.

TMV, unlike typical isolates of ToMV, induces both local and systemic infection in *D. stramonium, N. rustica* and *N. tabacum* cultivars White Burley and Dutch A. TMV induces systemic mosaic in *N. sylvestris* and local lesions in *N. glutinosa, N. benthamiana* and *N. xanthii.* ToMV induces local lesions in *N. glutinosa, N. sylvestris* and *N. xanthii.*

## *Immunological tests*

The symptoms of TMV and ToMV are not sufficiently distinctive for infection to be diagnosed with certainty by visual inspection. Detection and identification is therefore dependent on the use of one or more diagnostic methods.

The double immuno-diffusion test in agar gel, immunosorbent electron microscopy (ISEM) (33) and ELISA have been used successfully for detection of TMV (21). TMV can be identified in one day using the immuno-diffusion test in a petri dish. ELISA kits are commercially available for testing large number of samples.

As the virus is a good immunogen, high titred antisera are readily obtained by injecting rabbits with purified virus. Various formats of ELISA, immunosorbent electron microscopy, immuno-dot blotting and hybridization methods are now routine diagnostic procedures. Monoclonal antibodies have also been used to differentiate strains of ToMV.

## Nucleic acid based detection

Biological assays as well as serological detection and identification of species within the genus *Tobamovirus* can be time-consuming and might lead to ambiguous results. Especially for the tobamovirus, differentiation problems are caused by heterologous reactions of antisera detecting common epitopes as well as by antisera produced against virus mixtures (40). Moreover, cross reaction between ToMV and TMV antisera (36) limits serological differentiation of these two viruses by ELISA. To circumvent these problems researchers have developed procedures based on the polymerase chain reaction. The advent of PCR has resulted in further more variants for detection of viruses with more sensitivity and specificity. These methods not only allow the rapid detection but also help in differentiation of tobamoviruses.

A multiplex reverse-transcription polymerase chain reaction (RT-PCR) protocol has been described (19) for simultaneous detection and discrimination of ToMV, TMV, subgroups of CMV (subgroups I & II), including its satellite RNA, using 18S rRNA as an internal control in tomato (Table 2). Multiplex RT-PCR procedure were optimized so as to simultaneously amplify six fragments of 704, 593, 512, 421, 385, 255 bp, specific to CMV subgroup II, CMV subgroup I, ToMV, TMV, satellite RNA and 18S rRNA, respectively. The sensitivity of the multiplex RT-PCR method was 100 times higher than that of double-antibody sandwich-enzyme-linked immunosorbent assay (DAS-ELISA).

In another variant of PCR, a pair of specific primer designed based on TMV movement protein gene (Table 2), the 179-bp fragments for positive amplification was achieved (68) within 17 min by oscillatory-flow reverse transcription-polymerase chain reaction (RT-PCR) microfluidics based on the capillary reactor for detection of TMV. A detection limit of 0.01 $\mu$g cDNA was achieved at the flow rate of 70 $\mu$l min$^{-1}$.

Multiplex immunocapture (IC)-RT-PCR assay for the simultaneous detection and differentiation of TMV and ToMV was developed with primers designed based on movement protein gene (Table 1) of TMV and ToMV respectively (37) in spruce and pine extracts. When purified viruses were diluted in root or needle extracts of virus-free conifer seedlings, both IC-RT-PCR assays detected their respective target viruses at concentrations of 10–100 fg ml$^{-1}$ compared to ELISA detection sensitivities of 1 ng ml$^{-1}$. PCR distinguished more reliably between the two viruses than ELISA. When root extracts were seeded with both viruses simultaneously, the multiplex assay detected each virus at concentrations of 1–10 pg ml$^{-1}$. Multiplex PCR primers based on movement protein gene (61) and replicase gene (39) have been successfully employed for detection of TMV and ToMV simultaneously from infected leaves and seeds respectively (Fig. 3).

**Table 1.** List of specific primers for detection of TMV and ToMV by RT-PCR

| Virus | Gene | Primer (5'-------3') | Product | PCR | Reference |
|---|---|---|---|---|---|
| TMV | Movement protein | TMV1-GACCTGACAAAAATGGAGAAGATCT<br>TMV2-GAAAGCGGACAGAAACCCGCTG | 422 bp | IC-RT-PCR | Jacobi et al 1998 |
| ToMV | Movement protein | ToMV5-CTCCATCGTTCACACTCGTTACT<br>ToMV6-GATCTGTCAAAGTCTGAGAAACTTC | 508 bp | IC-RT-PCR | Jacobi et al 1998 |
| TMV | Movement protein | TMVf-TGAAAATGAATCATTGTCTG<br>TMVr-ACTCATCAACAACTTCTTCC | 454 bp | RT-PCR | Vinayarani et al 2011 |
| ToMV | Movement protein | ToMVf-TGAAAATGAATCATTGTCTA<br>ToMVr-CATCTTCAATCAAATTATC | 623 bp | RT-PCR | Vinayarani et al 2011 |
| TMV | Replicase gene | TMVF-CgACATCAgCCgATgCAgC<br>TMVR-ACCgTTTTCgAACCgAgACT | 880 bp | RT-PCR | Kumar et al 2011 |
| ToMV | Replicase gene | ToMVF-CgAgAggggCAACAAACAT<br>ToMVR-ACCTgTCTCCATCTCTTTgg | 318 bp | RT-PCR | Kumar et al 2011 |
| TMV | Movement protein | TMV-Fw-ATGGAAAGAGCCGACGAG<br>TMV-Rv-GAAAGCGGACAGAAACCC | 179 bp | Oscillatory-flow RT-PCR | Ying et al 2009 |
| TMV | Coat protein | TMV(F)-CGATGATGATTCGGAGGC<br>TMV(R)-GAGGTCCARACCAAMCCAG | 512 bp | RT-PCR | Chen et al 2011 |
| ToMV | Coat protein | ToMV(F)-CATCTGTATGGGCTGAC<br>ToMV(R)-GAGGTCCARACCAAMCCAG | 421 bp | RT-PCR | Chen et al 2011 |
| Tobamovirus group | Coat protein | Tob-Uni 1-ATTTAAGTGGASGGAAAAVCACT<br>Tob-Uni 2-GTTGTTGATGAGTTCRTGGA<br>TMVspec-CGGTCAGTGCCGAACAAGAA<br>TOMVspec-CGGAAGGCCTAAACCAAAAAG | 750-800 bp | Semi-nested RT-PCR and RT-PCR-RFLP | Letschert et al 2002 |

**Table 2.** Some virus inhibitory plant proteins possessing property of host ribosome inactivation

| Antiviral protein | Source | Virus-host system | Reference |
|---|---|---|---|
| Abrin | *Abrus precatorius* seeds | TMV/*N. glutinosa* | 53 |
| Gelonin | *Gelonium muliflorum*-Seeds | | |
| Modeccins | *Adenia digitata*- roots | | |
| Momordin | *Momordica charantia* seeds | | |
| Ricin | *Ricinus communis seeds* | | |
| Agrostins | *Agrostemma githago* seeds | TMV/*N. glutinosa* | 54 |
| Saporins | *Saponaria officinalis* leaves, roots, seeds | | |
| Bryodin | *Bryonia dioica* - leaves and roots | TMV/*N. glutinosa* | 55 |
| CA-SRI | *Clerodendrum aculeatum* leaves | TMV/*N.glutinosa*<br>TMV / *N. tabacum* SamsunNN | 60 |
| Dianthins | *Dianthus caryophyllus*- leaves | TMV/*N. glutinosa* | 56 |
| MAP | *Mirabilis jalapa* -roots | TMV/*N. tabacum* | 38 |
| PAP | *Phytolacca americana* Seeds, leaves, roots | TMV / *N. glutinosa,*<br>TMV / *P.vulgaris,*<br>TMV / *C. amaranticolor* | 7 |
| Yucca recurvifolia protein | *Yucca recurvifolia* leaves | TMV / *C. amaranticolor* | 35 |

A procedure involving RT-PCR using a single primer pair was developed for the detection of five tobamovirus species viz., TMV, ToMV, PMMV, TMGMV and CGMMV which are related serologically. For the genus-specific RT-PCR of tobamoviruses listed above, except CGMMV, the primer combination Tob-Uni 1 and Tob-Uni 2 was used. Species-specific RT-PCR was carried out with Tob-Uni 1 and a set of specific primers (Table 1). The forward primer Tob-Uni 1 was chosen within the 3′-non translated region and the reverse primer Tob-Uni 2 within the origin of assembly. The designed primers amplified TMV, ToMV, TMGMV and PMMV resulting in bands of around 750–800 bp. No PCR product was obtained when CGMMV was tested with primers Tob-Uni 1 and

Tob-Uni 2. However, CGMMV was detected with primer combination Tob-Uni 1 and $CGMMV_{spec}$ (40).

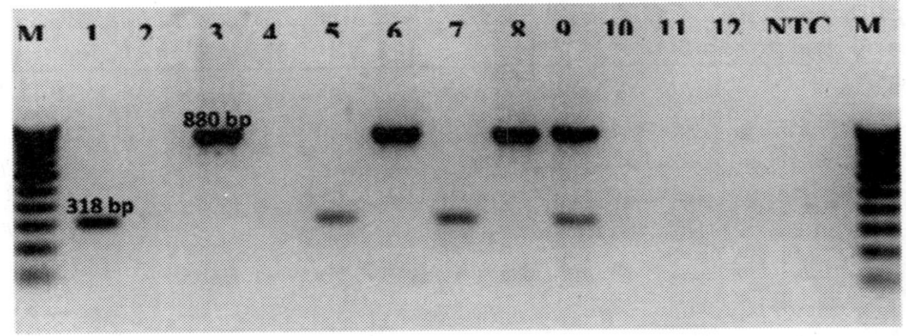


**Fig. 3.** Detection of TMV and ToMV by RT-PCR.
Amplified cDNA fragments (c. 880 and 318 bp) were analysed by electrophoresis on 2% agarose gel. M: 100 bp DNA ladder, lane 1-ToMV-infected total RNA with ToMV primers; lane 2-ToMV-infected total RNA with TMV primers; lane 3-TMV-infected total RNA with TMV primers; lane 4-TMV-infected total RNA with ToMV primers; lane 5-doubly infected total RNA with ToMV primers; lane 6-doubly infected total RNA with TMV primers; lane 7-ToMV-infected total RNA with multiplex primers; lane 8-TMV-infected total RNA with multiplex primers; lane 9-doubly infected total RNA with multiplex primers; lane 10-healthy control with ToMV primers; lane 11-healthy control with ToMV primers; lane 12-healthy control with multiplex primers; lane 13-non template control (NTC) (PCR reaction without cDNA).

## MANAGEMENT

### Seed Treatment

Seed treatment with chemicals is extensively being used for the eradication of seed-borne pathogens. Similarly, treatment of freshly harvested seed with dilute hydrochloric acid will inactivate virus contaminating seed coats. Heat treatment of dry seed for 3-5 days at 70°C is needed to inactivate ToMV inside the seeds without impairing seed germination. By soaking the seeds for 20 min in 10% (w/v) tri-sodium phosphate solution ToMV were removed from the external surface of seeds (12).

### *Host-Plant Resistance*

*N*-gene hypersensitivity was first recommended as a crop resistance strategy against TMV spread by Holmes (1938), who also bred the Samsun *NN* cultivar by transferring the *N* gene from *N. glutinosa* via the synthetic amphidiploid *N. digluta* (*N. glutinosa* (n=24) x *N. tabacum* (n=12)). The reaction localizes TMV and most tobamoviruses at the infection point. This type of resistance has proved durable (50).

*N*-gene resistance has proved practically valuable only in 'Burley' type varieties; in other types resistance is correlated with lower quality of the cured leaf (5). The introduction via biotechnology of the single *N*-gene could help overcome the correlated negative traits in the latter types (64).

Insertion of a TMV-RNA antisense construct in sensitive tobacco inhibits the gene encoding both 126 and 183 kDa replicase proteins (47). Transgenic tobacco plants expressing the gene for TMV replicase with an additional insertion in the middle of 183-kDa has been shown to be highly resistant to systemic infection by TMV and other tobamoviruses.

Plants transformed with a gene encoding a defective mutant of the TMV movement protein had delayed symptom appearance and reduced systemic spread of infection to upper leaves (24).

Three major genes (designated *Tm-1*, *Tm-2* and *Tm-2*$^2$) for resistance have been most used in tomato breeding programs. The earlier use of cultivars containing genes *Tm-1* and/or *Tm-2* resulted in the rapid emergence or increased incidence of strains which were able to overcome resistance (15). However, gene *Tm-2*$^2$ in homozygous cultivars, or together with genes Tm-1 and/or Tm-2, has been highly effective in controlling ToMV. Although the molecular basis for genetic resistance has yet to be determined, the genes probably inhibit virus multiplication and movement. Resistance to ToMV has also been reported in *C. annuum* (49).

### *Induction of resistance against tobamoviruses by using chemicals*

Salicylic acid (SA) is an important signal in plant defense (26). SA accumulates in plant tissues following a hypersensitive response (HR), a resistance response most often characterized by programmed host cell death at the point of attempted entry by an avirulent pathogen (34). The death of host cells at the site of invasion may not be enough to limit the spread of a pathogen, even in the case of an obligate intracellular parasite like a virus.

However, in many instances, the occurrence of the HR has a signaling role and lead to an enhanced in resistance to a broad spectrum of pathogens throughout the plant called systemic acquired resistance (SAR). In plants that have not undergone an encounter with a pathogen leading to a HR, an SAR-like state of enhanced disease resistance can be induced by treatment with solutions of SA or its synthetic functional analogs (26).

SA-induced resistance to TMV results from the inhibition of both virus replication and movement of the virus through the plasmodesmata linking adjacent cells. The inhibition of replication occurs in the meshophyll cells, while restriction of cell-to-cell movement occurs in the epidermal cells (45). The spread of a virus is inhibited if it induces HR, an effect that is dependent upon one or more mechanisms regulated by SA. Even in the cells undergoing death or already deed cells showed the viable particles of TMV at the periphery of HR lesions (44). Studies on mechanistic basis of SA-induced resistance to viruses revealed that the accumulation of TMV RNA, coat protein and viral RdRp activity was inhibited by SA in the inoculated leaves of *nn* genotype (i.e. TMV susceptible) tobacco (20). Inhibition of virus replication is not the sole anti-viral effect induced by SA. The inhibition of replication of viruses occurs in the mesophyll cells while restriction of cell-to-cell movement occurs in the epidermal cell layer (45).

Acibenzolar-S-methyl (ASM) pre-treatment reduced the concentration of ToMV and TMV in tomato and bell pepper seedlings respectively. The number of local lesions in indicator plant *N. glutinosa* was also reduced. The treatment evidenced higher expression of Alternative oxidase (AOX) and RdRp in the upper leaves of ASM-treated tomato plants challenge inoculated with ToMV. The concentration of virus was also lower in treated seedlings (43).A time course change in antioxidant enzymes, endogenous salicylic acid, endogenous hydrogen peroxide, lipid peroxidation and protein oxidation was studied in systemic and local lesion host of ToMV and TMV. The levels of peroxidase, salicylic acid, lipid peroxidation, protein oxidation, $H_2O_2$ increased and catalase levels decreased in local lesion host in comparison to systemic hosts. In local lesion host more number of cells showed $H_2O_2$ accumulation at 24 h post inoculation (Figs. 4-6).

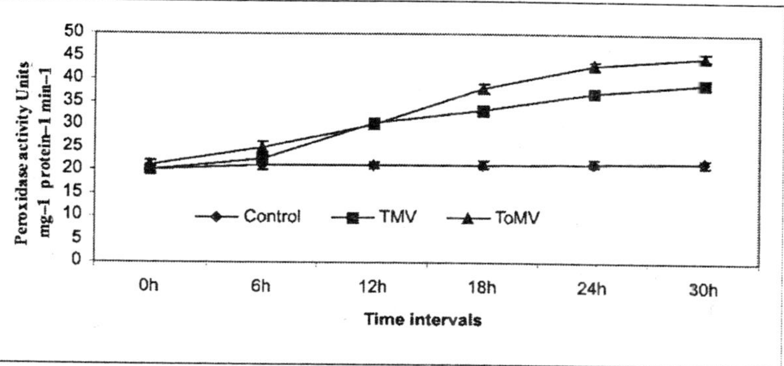


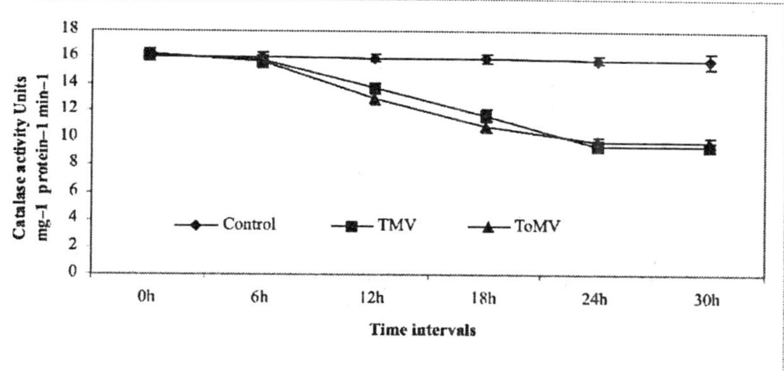


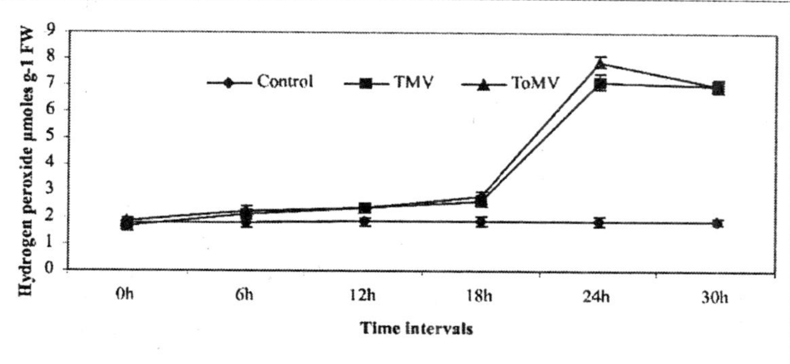


**Fig. 4.** Changes in peroxidase, catalase and hydrogen peroxide in *N. glutinosa* seedlings inoculated with TMV and ToMV. The data represents mean of three replicates ± SE.

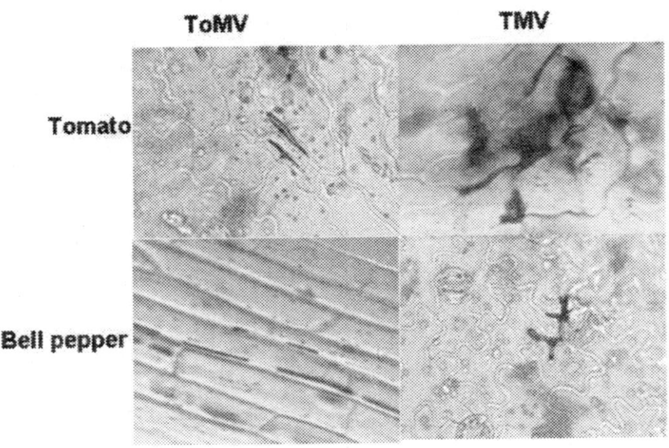


**Fig. 5.** Localization of hydrogen peroxide in tissues of tomato and bell pepper leaves on inoculation with TMV and ToMV at 24 hpi

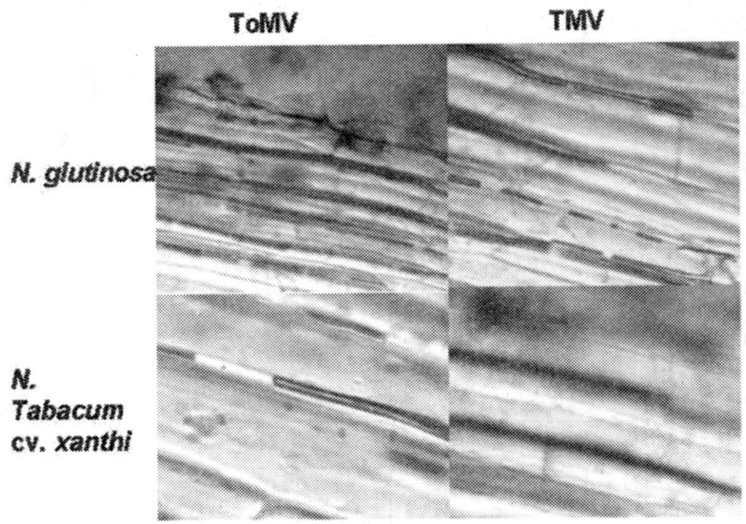


**Fig. 6.** Localization of hydrogen peroxide in tissues of *N. glutinosa* and *N. tabacum* cv. *xanthi* leaves on inoculation with TMV and ToMV at 24 hpi.

## MANAGEMENT OF TOBAMOVIRUSES BY USING PLANT EXTRACTS

Leaf extracts of *Alternanthera brasiliana, A. ficoidea, Bougainvillea spectabilis, Chenopodium amaranticolor, C. ambrosioides, Iresine herbstii, Mirabilis jalapa* and *Phytolacca thirsiflora* were very effective when applied to *Phaseolus vulgaris* plants before mechanical inoculation with TMV, inhibiting infection > 70% and even 100% in some cases (1). Mirabilis antiviral protein (MAP) (molecular weight of 24.2 KDa, pI of 9.8) isolated from roots inhibited the mechanical transmission of TMV, PVY and CGMMV and turnip mosaic virus on local lesion and systemic hosts (38).

A potent resistance inducing protein (Crip-31) from the leaves of *Clerodendrum inerme* which is highly stable, basic in nature, 31 kDa in molecular mass that induced a high degree of localized as well as systemic resistance against three plant viruses, ToMV, PVY and CMV and. The systemic inhibitory activity of the Crip-31 provided protection to whole plants within 40-60 min of its application (52).

The leaf extract of *Celosia cristata* was found to prevent lesion production by *Sunn hemp rosette virus* (SRV), TMV and PVX in several hosts (8). Studies on the partially purified inhibitor indicated that it might interfere with virus establishment in the host. Two glycoprotein virus inhibitors have been purified which could impart resistance to TMV and SRV on local lesion hosts (6)..

An extract from dried roots of mature plants of *Boerhaavia diffusa* or a glycoprotein isolated therefrom, when sprayed on leaves of test plants before inoculation, prevented infection by TMV in tobacco, CMV and TMV in tomato, CGMMV in melon, *Sunn hemp rosette virus* (SRV) in *Crotalaria juncea* and Gomphrena mosaic virus in *Gomphrena globosa* (4). Induction of systemic resistance by an antiviral protein from *Bougainvillea buttiana* in *N. glutinosa* and *Cyamopsis tetragonoloba* against TMV and SRV has been obsereved (46). Two antiviral proteins named CAP-I and CAP-II from the leaves of *Chenopodium album* have been isolated and purified (27). The protein was reported to inhibit TMV and SRV infections on their respective host plants. Both the proteins had a molecular weight of 24 KDa and a pI of 10.2. They differed in their amino acid composition and N-terminal sequence. The aqueous leaf extract of Guava, *Phyllanthus*, *Thuja* were effective in reducing the infection by ToMV. The acetone precipitated fraction of *Tridax*, *Thuja*, Guava and Tulsi were effective against TMV. The effective TLC solvent fraction of Guava against ToMV and *Thuja* against TMV were identified (25).

Plant extracts (10% w/v), viz. neem (*Azadirachta indica*), pongamia (*Pongamia glabra*), bougainvillea (*Bougainvillea spectabilis*), seed oils - neem oil (5%) and *Phyllanthus* oil (5%) were effective in reducing local lesion in *N. glutinosa* against tobamoviruses. Both spray and seed treatment reduced the concentration of tobamoviruses as evident from ELISA values. *B. spectabilis* was the most effective in reducing the concentration of tobamoviruses (42).

## MECHANISMS OF SYSTEMIC INDUCED RESISTANCE BY PLANT EXTRACTS

The mechanism of systemic induced resistance induced by plant extracts against plant viruses is still not elucidated clearly. So far, various workers working on SAR against plant viruses reported some mechanisms. Treatment with botanicals showed *de novo* synthesis of antiviral substances (59). Induction of viral inactivating agent (VIA) in treated plants have been demonstrated which completely inhibits viruses in hypersensitive hosts. Prior to production of VIA, a mobile inducing signal may be produced in treated leaves after these botanical resistance inducers binds with the host plant surface. The mobile signal produces VIA in the entire plant system. Some virus inhibitory plant proteins possessing property of host ribosome inactivation are listed in Table 2.

### *Transgenic Plants*

A new approach to the control of virus infection is based on plant immunological response: a reduction in TMV symptoms was obtained in tobacco plants producing antibodies against TMV after transformation with a gene from mice immunized with TMV (62). The transgenic methods made it possible to introduce new genes into existing commercial crop varieties. These genes are derived from the viruses themselves. Virus resistance in tobacco plants against TMV by utilizing the TMV coat protein gene has been demonstrated (51). Coat proteins from almost all plant virus groups have been expressed as transgenes in a variety of crops and model test plants. Non-conventional resistance to ToMV has been conferred on genetically engineered plants containing the virus coat protein gene (67), antisense viral RNA and a ribozyme (31).

### RNAi mediated resistance to tobamoviruses

RNA interference or RNAi, also known variously as RNA silencing, quelling or post-transcriptional gene silencing, is a sequence-specific RNA degradation process, affecting all highly homologous sequences in which foreign, overexpressed or aberrant RNA molecules are targeted for destruction in a sequence-specific manner (10). RNAi can contribute to resistance to virus infection and a variety of plant and animal viruses have been found to encode RNAi suppressor proteins (58). The possibility that a functional connection exists between SA-induced resistance to viruses and RNA silencing arose from the discovery that a tobacco gene encoding an RNA-dependent RNA polymerase (RdRp) was induced by SA (65). The significance of this is that host RdRps can play important roles in the induction of RNAi (2).

## FUTURE APPROACHES

### RNA silencing platforms in plants

There is an intrinsic mechanism, probably present in all multicellular eukaryotes, which converts double-stranded (ds) RNA into small (~21 nt) RNAs (by cleavage with endonucleases called Dicers) and uses them to direct sequence-specific degradation of cognate single-stranded RNAs. These small RNAs are termed short interfering (si) RNAs and the process is called RNA interference (RNAi). Curiously, most animals possess only one Dicer gene, whereas in plants: *Arabidopsis* has four and poplar and rice seem to have five and six, respectively. There are currently three methods of delivery: by virus infection, particle bombardment and *Agrobacterium* infection. Each method has advantages and disadvantages (63).

Exogenous sequences can be inserted into certain locations in a virus genome without destroying its infectivity and, when such modified genomes are used to infect plants, the foreign sequences also induce, and become the target of the host plant's RNAi response (9).. This ability of viruses to carry and induce RNAi against foreign sequences has been harnessed into a technology referred to as virus-induced gene silencing (VIGS).

TOM1 and TOM3 proteins are host proteins which associate with tonoplast membrane and interact with the helicase domain of tobamovirus-encoded replication proteins. These two proteins function to support tobamovirus multiplication from an early stage of infection after the uncoating of virions and when the invasion of viral RNA is not destined either to abortive infection or productive virus multiplication. TOM1 and TOM3 function in parallel to support the formation and maintenance of membrane-bound tobamovirus replication complex. TOM1 gene was first identified as a host gene which required for efficient multiplication of a tobamovirus in *Arabidopsis* (66). TOM1 and TOM3 like genes are present in other plant species like *Nicotiana* spp., tomato, melon and rice. Tobamovirus-resistant tobacco plants have been generated by RNA interference against *N. tabacum* TOM1 (NtTOM1) and TOM3 (NtTOM3) genes (3). Gene silencing against TOM1 and TOM3 genes resulted complete inhibition of the multiplication of tobamoviruses. We have also developed and demonstrated RNAi constructs against TMV replicase gene and TOM gene of pepper against TMV infection in pepper.

## REFERENCES


1. Alexandre MAV, Noronha AB and Vicente M 1987 Action of natural inhibitors on two virus diseases of bean, bean golden mosaic virus and tobacco mosaic virus legume strain, *Fitopatology Brasil* **12** 202-205
2. Ahlquist P 2002 RNA-dependent RNA polymerases, viruses and RNA silencing, *Science* **296** 1270-1273
3. *Asano M*, Satoh R, Mochizuki A, Tsuda S, Yamanaka T, Nishiguchi M, Hirai K, Meshi T, Naito S and Ishikawa M 2005 Tobamovirus-resistant tobacco generated by RNA interference directed against host genes, *FEBS Letters* **579** 4479-4484
4. Awasthi LP, Chowdhury B and Verma HN 1984 Prevention of plant virus disease by

*Boerhaavia diffusa* inhibitor, *International Journal of Tropical Plant Disease* **2** 41-44

5. Aycock MK and McKee CG 1995 Registration of 'MD 40' tobacco, *Crop Science* **35** 1207
6. Balasubramanyam A, Baranwal VK, Lodha ML, Varma A and Kapoor HC 2000 Purification and properties of growth stage-dependent antiviral proteins from the leaves of *Celosia cristata, Plant Science* **15** 413-421
7. Batelli MG and Stripe F 1995 Ribosome inactivating proteins from the plants In Antiviral Proteins in Higher plants, (eds) Borde DD and Zipf, CRC Press, USA, 39-64 pages
8. Baranwal VK and Verma HN 1992 Localized resistance against plant virus infection by leaf extract of *Celosia cristata, Plant Pathology* **41** 633-638
9. Baulcombe DC 1999 Fast forward genetics based on virus induced gene silencing, *Current Opinion of Plant Biology* **2** 109-113
10. Baulcombe DC 2001 Diced defence RNA silencing, *Nature* **409** 295-296
11. Boyko V, Van-der-Laak J, Ferralli J, Suslova E, Kwon MO and Heinlein M 2000 Cellular targets of functional and dysfunctional mutants of tobacco mosaic virus movement protein fused to green fluorescent protein, *Journal of Virology* **74** 11339-11346
12. Broadbent L 1965 The epidemiology of tomato mosaic VIII Virus infection through tomato roots, *Annals of Applied Biology* **55** 57-66
13. Broadbent L 1976 Epidemiology and control of tomato mosaic virus *Annual Review of Phytopathology* **14** 75-96
14. Brunt AA, Crabtree K, Dallwitz MJ, Gibbs AJ, Watson L and Zurcher EJ 1996 *Plant Viruses Online Descriptions and Lists from the VIDE Database* Wallingford, Oxfordshire, UK CABI Publishing
15. Brunt AA 1988 *Tomato mosaic virus The Plant Viruses* Vol II van-Regenmortel MHV, Fraenkel-Conrat H (Ed) Plenum Press New York USA, 181-204 pages
16. CABI 2001 Crop protection compendium Wallingford, UK, CABI Publishing
17. Candilo M, Faccioli G, Grassi G and Faeti V 1992 Effect of tomato mosaic virus (ToMV) on yield of machine-harvested processing tomatoes, *Phytopathology Mediterranea* **31** 32-36

18. Chen MH, Sheng J, Hind G, Handa AK and Citovsky V 2000 Interaction between the tobacco mosaic virus movement protein and host cell pectin methylesterases is required for viral cell-to-cell movement, *The EMBO Journal* **19** 913-920

19. Chen S, Gu H, Wang X, Chen J and Zhu W 2011 Multiplex RT-PCR detection of *Cucumber mosaic virus* subgroups and tobamoviruses infecting tomato using 18S rRNA as an internal control *Acta Biochimica et Biophysica Sinica* **43** 465-471

20. Chivasa S, Murphy AM, Naylor M and Carr JP 1997 Salicylic acid interferes with tobacco mosaic virus replication via a novel salicylhydroxamic acid-sensitive mechanism *Plant Cell* **9** 547-557

21. Chitra TR, Prakash HS, Albrechtsen SE, Shetty HS and Mathur SB 1999a Seed transmission of mosaic viruses in tomato and bell pepper, *Tropical Science* **39** 80-84

22. Chitra TR, Prakash HS, Albrechtsen SE, Shetty HS and Mathur SB 1999b *Infection of tomato and bell pepper by ToMV and TMV at different growth stages and establishment of virus in seeds, Journal of Plant Pathology* **81** 123-126

23. Cicek Y and Yorganci U 1991 Studies on the incidence of tobacco mosaic virus on certified seed of tomato, pepper and eggplant in Aegean region, *Journal Turkish Phytopathology* **20** 57-68

24. Cooper B, Lapidot M, Heick JA, Dodds JA and Beachy RN 1995 A defective movement protein of TMV in transgenic plants confers resistance to multiple viruses whereas the functional analog increases susceptibility, *Virology* **206** 307-313

25. Deepthi N, Madhusudhan KN, Uday Shankar AC, Kumar HB, Prakash HS and Shetty HS 2007 Effect of plant extracts and acetone precipitated proteins from six medicinal plants against tobamovirus infection, *International Journal of Virology* **3** 80-87

26. Dempsey DA, Shah J and Klessig DF 1999 Salicylic acid and disease resistance in plants, *Critical Review of Plant Science* **18** 547–575

27. Dutt S, Narwal S, Kapoor HC and Lodha ML 2003 Isolation and characterization of two protein isoforms with antiviral activity from *Cheropodium album* L leaves, *Journal of Plant Biochemistry and Biotechnology* **12** 117-122

28. Escudero J 1996 Survey of viruses affecting pepper

(*Capsicum annuum* L) in southern Puerto Rico *Journal of Agricultural of the University of Puerto Rico* **80** 77-80

29. Edwardson JR and Christie RG 1997 Viruses infecting peppers and other *Solanaceous* crops Vol 1 Monograph 18–1, University of Florida Agricultural Experiment Station
30. Fauquet CM, Mayo MA, Maniloff J, Desselberger U and Ball LA 2005 Virus taxonomy Classification and nomenclature of viruses Eighth report of the international committee on the taxonomy of viruses Elsevier Academic Press, Amsterdam, 1009-1013
31. Feyter R, Young M, Schroeder K, Dennis ES and Gerlach W 1996 A ribozyme and an antisense gene are equally effective in conferring resistance to tomato mosaic virus on transgenic tobacco *Mol Genetics and Gennomics* **3** 329-338
32. Giri BK and Mishra MD 1990 Effect of tomato mosaic causing viruses on pollen viability and yield of tomatoes, *Indian Phytopathology* **43** 487-490
33. Gupta VP, Garg ID and Naqvi QA 1994 Immunosorbent electron microscopic studies on a tobamovirus causing brinjal necrotic mosaic, *Acta Virology* **38** 43-45
34. Heath MC 2000 Hypersensitive response-related death *Plant Mol Biol* 44 321-334
35. Ito Y, Seki I, Tanfuji S and Hiramatsu A 1993 Inhibition of protein synthesis by antiviral protein from *Yucca recurvifolia* leaves, *Bioscience Biotechnollogy Biochemistry* **8** 518-519
36. Jacobi V and Castello JD 1991 Isolation of tomato mosaic virus from waters draining forest stands in New York State, *Phytopathology* **81** 1112-1117
37. Jacobi V, Bachand GD, Hamelin RC and Castello JD 1998 Development of a multiplex immunocapture RT-PCR assay for detection and differentiation of tomato and tobacco mosaic viruses, *Journal of Virological Methods* **74** 167-178
38. Kubo S, Ikeda J, Imaizumi S, Jakanami Y and Mikami Y 1990 A potent plant virus inhibitor found in *Mirabilis jalapa* L *Annals of Phytopathological Society of Japan* **56** 481-487
39. Kumar S, Udayashankar AC, Nayaka CS, Lund OS and Prakash HS 2011 Detection of *Tobacco mosaic* and *Tomato mosaic tobamovirus* in leaves and seeds of chilli and tomato by multiplex RT-PCR, *Letters of Applied Microbiology* **53** 359-363

40. Letschert B, Adam G, Lesemann DE, Willingmann P and Heinze C 2002 Detection and differentiation of serologically cross-reacting tobamoviruses of economical importance by RT-PCR & RT-PCR-RFLP, *Virological Methods* **106** 1-10

41. Madhusudhan KN, Siva-kumar D and Prakash HS 2010 Pectin Methylesterase Estimation in Plant-Tobamovirus Interaction and Comparative Modelling, Active Site Prediction and Docking Between the Proteins Involved in the Movement of Tobamoviruses, *Journal of Computational Intelligence in Bioinformatics* **3** 163-174

42. Madhusudhan KN, Vinayarani G, Deepak SA, Niranjana SR, Prakash HS, Singh GP, Sinha AK and Prasad BC 2011 Antiviral Activity of Plant Extracts and other Inducers against Tobamoviruses Infection in Bell Pepper and Tomato Plants, *International Journal of Plant Pathology* **2** 35-42

43. Madhusudhan KN, Srikanta BM, Shylaja MD, Prakash HS and Shetty HS 2008 Changes in antioxidant enzymes, hydrogen peroxide, salicylic acid and oxidative stress in compatible and incompatible host-tobamovirus interaction, *Journal of Plant Interaction* **4** 157-166

44. Marathe R, Anandalakshmi R, Liu Y and Kumar SP 2002 The tobacco mosaic virus resistance gene, *New Moleccular Plant Pathology* **3**167-172

45. Murphy AM and Carr JP 2002 Salicylic acid has cell-specific effects on tobacco mosaic virus replication and cell-to-cell movement, *Plant Physiology* **128** 552-563

46. Narwal S, Balasubramanyam A, Sadhna P, Kapoor H and Lodha ML 2000 A systemic resistance inducing antiviral protein with N-glucosidase activity from *Bougainvillea xbuttiana* leaves, *Indian Journal of Experimental Biology* **39** 600-603

47. Nelson A, Roth DA and Johnson JD 1993 Tobacco mosaic virus infection of transgenic *Nicotiana tabacum* plants is inhibited by antisense constructs directed at the 5' region of viral RNA, *Gene* **127** 227-232

48. Nishikiori M, Dohi K, Mori M, Meshi T, Naito S and Ishikawa M 2006 Membrane-bound tomato mosaic virus replication proteins participate in RNA synthesis and are associated with host proteins in a pattern distinct from those that are not membrane bound, *Journal of Virology* **80** 8459-*8468*

49. Patagas KG, Schuerger AS and Wetter C 1989

Management of tomato mosaic virus in hydroponically grown pepper, *Plant Disease* **73** 570-573

50. Piccirillo P and Piro F 1996 La resistenza a tobamovirus in *Nicotiana tabacum* L *Informaton of Fitopatology* **11** 26-33
51. Powell-Abel P, Nelson RS, De B, Hoffmann N, Rogers SG, Fraley RT and Beachy RN 1986 Delay of disease development in transgenic plants that express the tobacco mosaic virus coat protein gene, *Science* **232** 738-743
52. Praveen S, Tripathi S and Varma A 2001 Isolation and characterization of an inducer protein (Crip-31) from *Clerodendrum inerme* leaves responsible for induction of systemic resistance against viruses, *Plant Science* **161** 453-459
53. Stevens WA, Spurdon C, Onyon LJ and Stirpe F 1981 Effect of inhibitors of protein synthesis from plants on tobacco mosaic virus infection, *Experientia* **37** 257-259
54. Stripe F, Gaseri-Campani A, Barbieri L, Falasca AI, Abdondaza A and Stevens WA 1983 Ribosome inactivating protein from the seeds of *Saponaria officinalis L, Agrostemma githag L* and *Asparagus officinalsi* L and from the latex of *Hura crepitans L, Journal of Biochemistry* **216** 617-625
55. Stripe F, Barbier LMG, Falasca AL, Abdondanza A, Lorenzoni E and Stevens EA 1986 Bryodin, a ribosome inactivating protein from the roots of *Bryoniae dioica* L *Journal of Biochemistry* **240** 659-665
56. Stripe F, Williams DG, Onyon LJ, Legg RF and Stevens WA 1981 Dianthins, ribosome inactivating proteins with antiviral properties from *Dianthus caryophyllus* L (Carnation), *Journal of Biochemistry* **195** 399-405
57. Tzfira T, Rhee Y, Chen MH, Kunik T and Citovsky V 2000 *Nucleic acid transport in plant–microbe interactions the molecules that walk through the walls Annual Review of Microbiology* **54** 187-*219*
58. Vance V and Vaucheret H 2001 RNA silencing in plants-defense and counter defense, *Science* **292** 2277-2280
59. Verma HN and Awasthi LP 1979 Antiviral activity of *Boerhaavia diffusa* root extract and the physical properties of the virus inhibitor, *Canadian Journal of Botany* **57** 926-932
60. Verma HN, Srivastava S, Varsha and Kumar D 1996 Induction of systemic resistance in plants against viruses by a basic protein from *Clerodendrum acule-*

*atum* leaves, *Phytopathology* **86** 485-492

61. Vinayarani G, Madhusudhan KN, Deepak SA, Niranjana SR and Prakash HS 2011 Detection of Mixed Infection of Tobamoviruses in Tomato and Bell Pepper by using RT-PCR and Duplex RT-PCR *International Journal of Plant Pathology,* **2** 89-95
62. Voss A, Niersbach M, Hain R, Hirsch HJ, Liao YC, Kreuzaler F and Fischer R 1995 Reduced virus infectivity in *N tabacum* secreting a TMV-specific full-size antibody, *Molecular Breeding* **1** 39-50
63. Watson JM, Fusaro AF, Wang M and Waterhouse PM 2005 RNA silencing platforms in plants, *FEBS Letters* **579** 5982–5987
64. Whitham S, Kumar DSP, Choi D, Hehl R, Corr C and Baker B 1994 The product of the tobacco mosaic virus resistance gene N similarity to Toll and the interleukin-1 receptor, *Cell* **78** 1101-1115
65. Xie ZX, Fan BF, Chen CH and Chen ZX 2001 An important role of an inducible RNA-dependent RNA polymerase in plant antiviral defense, *Proceeding of National Academy of Science* **98** 6516-6521
66. Yamanaka T, Ohta T, Takahashi M, Meshi T, Schmidt R, Dean C, Naito S and Ishikawa M 2000 *TOM1*, an *Arabidopsis* gene required for efficient multiplication of a tobamovirus, encodes a putative transmembrane protein *Proceeding of National Academy of Science USA* **97** 10107–10112
67. Yeh SH, Pon JC, Yeh SD and Pon JC 1996 Coat protein mediated resistance to tomato mosaic virus in systemic and local lesion hosts, *Plant Protection Bulletin* **5** 15-27
68. Ying HW, Sun CZ and Yuan YL 2009 Rapid Detection of Tobacco Mosaic Virus from Crude Samples on an Oscillatory-flow Reverse Transcription-Polymerase Chain Reaction Microfluidics, *Chinese Journal of Analytical Chemistry* **37** 1286-1290.

*Rev. Plant Pathol. Vol. 6, 2014*
*Indian Society of Mycology and Plant Pathology*
*Scientific Publishers (India), Jodhpur pp. 155-188*

# TREHALOSE, AN ANTI-STRESS DISACCHARIDE PROTECTING LIFE-FORMS IN ADVERSE CONDITIONS

***Sumita Ghosh[1], Bipasha Basistha[1] and Anil K Ghosh[2]***

1 Department of Botany, Presidency University, Kolkata
2 Drug Development Diagnostics and Biotechnology Department, Indian Institute of Chemical Biology, Kolkata

**KEY WORDS:** Trehalose, Non-reducing disaccharide, Chemical synthesis

## ABSTRACT

Trehalose, a non-reducing disaccharide, plays an immense role in stress protection among different organisms and consequently its metabolism is extremely an important event. Presence of trehalose is observed in a wide range of organisms, starting from microbes to lower and higher plant groups, and also in animals, excepting mammals. Trehalose possesses physical and/or chemical properties that make it unique and different from other sugars. This imparts it the characteristic to play an important role in defence mechanism, stabilizing proteins and biological membranes from a large number of stresses, including oxidative stress, hydrostatic pressure, desiccation, osmotic stress, and exposure to toxic chemicals. It also plays a role in controlling primarily reducing different abiotic and biotic stress related diseases. Trehalose, though not produced in a mammalian system has found extensive use as therapeutics. It has been used in large amount for the preservation and protection of biological materials like vaccines in hot climate and cellular blood transfusion products.

Biotechnological trials have been already done for the production of stress tolerant crops, by incorporating genes for trehalose synthesis in stress susceptible plants.

## INTRODUCTION

The disaccharide trehalose involved in stress response, can be found in animals, plants, and microorganisms, except in mammals [11, 97, 106, 111, 114]. In animal kingdom, trehalose is prevalent in shrimp, and also in insects, including grasshoppers, locusts, butterflies, and bees, in which the blood-sugar trehalose serves as source of energy [13]. In plants, the presence of trehalose is found in Moonwort (*Selaginella*) [11], *Botrychium lunaria*, and sea algae [7]. Spores and macrocysts of *Dictyostelium mucoroides* contain trehalose, which is about 7% of its total dry weight [17]. For years, trehalose was thought to be restricted to resurrection plants, such as *Myrothamnus flabellifolius* [32] or *Selaginella lepidophylla* because, in these plants, trehalose reaches an easily detectable level upto 10 mg $g^{-1}$ fresh weight. Initially in vascular plants as a whole, trehalose had been found rarely and had been identified only in the ripening fruits of several members of Apiaceae. Later trehalose was detected in the model plant *Arabidopsis* (*A. thaliana*) using validamycin A, an inhibitor of the trehalose degrading enzyme, trehalase. More recently, trehalose has been detected in crops, such as rice (*Oryza sativa*) [8] and tobacco (*Nicotiana tabacum*), at concentrations of approximately $10\mu g\ g^{-1}$ fresh weight. Within the fungus family, it is prevalent in some mushrooms such as shiitake (*Lentinula edodes*), maitake (*Grifola frondosa*), nameko (*Pholiota nameko*), and Judas's ear (*Auricularia auricula-judae*), which contains 1% to 17% percent of trehalose in dry weight form, thus it is also referred to as mushroom sugar. Trehalose is also found in microorganisms such as baker's yeast and wine yeast, and it is metabolized by a number of bacteria, including *Streptococcus mutans*, the common oral bacterium responsible for plaque. This widespread distribution of trehalose suggests that it plays an important role in the biological world. The larvae of sleeping chironomid (*Polypedilum vanderplanki*) and artemia (sea monkeys, brine shrimp) are able to withstand dehydration because they store trehalose within their cells. Even within the plant kingdom, *Selaginella* , which grows in desert and mountainous areas, appears to be crinkled and dried in drought condition, but turns green and revives after a rain, which is a function of trehalose. It is also seen that the reason behind the dried shiitake mushrooms springing back into shape and swells in water because they contain trehalose. The two prevalent theories to explain how trehalose works within the organism in the state of cryptobiosis are "the vitrification theory", a state that prevents ice formation, or "the water displacement theory", where water is replaced by trehalose. It is also possible that a combination of both the theories may be responsible for the trehalose activity. Trehalose is the major carbohydrate energy storage molecule used by insects for flight. One possible reason for this is that the double glycosidic linkage of trehalose, when acted upon by an insect trehalase, releases two molecules of glucose, which is required for the rapid energy requirements of flight. This is doubles the efficiency of glucose release from the storage polymer starch, for which cleavage of one glycosidic linkage releases only one glucose molecule.

## CHEMICAL NATURE OF TREHALOSE

Trehalose is a disaccharide composed of two molecules of glucose joined by a 1-1 alpha bond [Fig: 1]. Trehalose is quite an oddity to most biochemists. Among disaccharides, trehalose has particular properties as shown in Table 1. Both the reducing ends of the molecule are involved in forming the glycosidic bond. This bonding makes trehalose very resistant to acid hydrolysis, and therefore is stable in solution at high temperatures, even under acidic conditions. Due to the absence of reducing ends in the chemical structure, trehalose is highly resistant to heat, pH and Millard's reaction. The name trehalose is given to the D-glucosyl D-glucosides and originates from the preparation of α, α-trehalose, an important sugar forming the "trehala manna", or oval shell of insects belonging to the family Curculionides or Rhynchophores [13]. Hence trehalose is commonly referred to as "insect sugar". Among three possible anomers of trehalose, viz. α, α 1,1-, α, β 1,1-, and β, β 1,1-, only the α, α-trehalose has been isolated from and also biosynthesized in living organisms. α, β trehalose is known as neotrehalose and β, β trehalose is called isotrehalose and they are only available in the synthesized form. It (α, α-trehalose) was probably first isolated from the ergot of rye, in 1832 by Wiggers

[13]. Since then, it has been reported in almost all organisms, except in mammals.

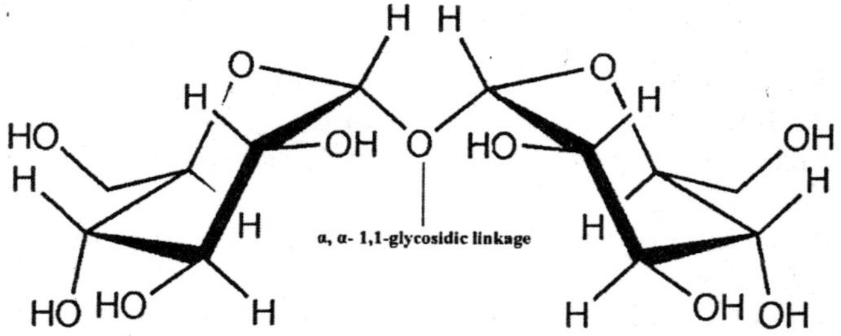

**Fig. 1.** Structure of trehalose [Courtesy: Wikipedia]

**Table 1.** Properties of trehalose

| | |
|---|---|
| Melting point | a) Dihydrate 97.0°C<br>b) Anhydride 210.5°C |
| Heat of fusion | a) Dihydrate 57.8kJ $mol^{-1}$<br>b) Anhydride 53.4 kJ $mol^{-1}$ |
| Solubility | 68.9 g/100 g H2O at 20°C |
| Relative sweetness | 45% of sucrose |
| Digestibility | digested and absorbed by the small intestine |
| pH stability of solution | >99% (pH 3.5–10, at 100°C for 24 h) |
| Heat stability of solution | >99% (at 120°C for 90 min) |

Trehalose has about 45% the sweetness of sucrose. It is less soluble than sucrose, except at high temperatures (>80°C). Trehalose forms a rhomboidal crystal as the dihydrate, and has 90% of the calorific content of sucrose in that form. Anhydrous forms of trehalose readily regain moisture to form the dihydrate.

## TREHALOSE METABOLISM

The discovery of trehalose metabolism has been one of the most exciting developments in plant metabolism and plant science in recent past. Metabolism and signaling of this sugar is an area of emerging significance. Mutant and transgenic organisms which have the capability of trehalose synthesis, display wide ranging and unprecedented phenotypes for withstanding stress. A consensus is emerging that trehalose-6-phosphate plays a central role in the co-ordination of metabolism with development. Molecular physiology and genomics have provided a glimpse of trehalose biology that had not been possible with conventional techniques, largely because the products of the synthetic pathway, trehalose-6-phosphate and trehalose are produced in trace amount and difficult to measure in most plants. In many ways it has similarities with sucrose-6-phosphate and sucrose synthesis in plants [Fig. 2]. In spite of the fact that its biosynthesis is similar to that of sucrose, its evolutionary origin is probably more ancient because of its presence in all kingdoms. Biosynthesis of trehalose in higher plants only occurs in the trehalose-6-phosphate synthase (TPS) - trehalose-6-phosphate phosphatase (TPP) pathway [Fig. 3], where TPS activity is encoded by the gene OtsA and TPP activity is encoded by OtsB gene. Degradation of trehalose is done by the enzyme trehalase. It breaks down the disaccharide trehalose to form two glucose residues for further use. This process has been found in all organisms that synthesize trehalose, even when distinct forms of trehalase co-exist, for example, in yeast.

Five different trehalose biosynthetic pathways exist among the wide range of organisms where trehalose is found to be present [Fig: 3].

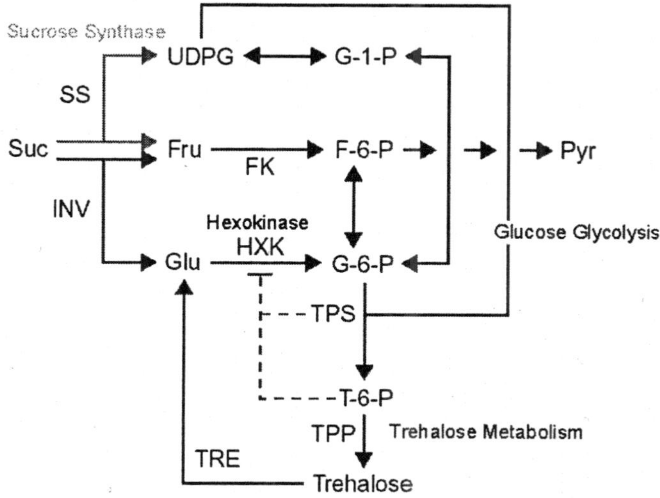


**Fig. 2.** Sugar Flux through Glycolysis & Trehalose Metabolism in Yeast

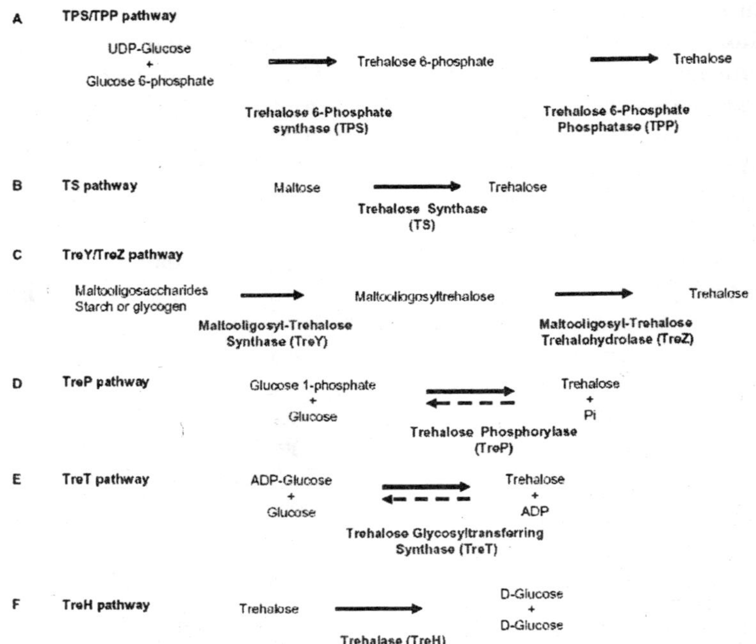


**Fig. 3.** Different pathways for trehalose biosynthesis. [Courtesy: Insights on the evolution of trehalose biosynthesis by Nelson Avonce and *et al.*, 2006]

The first pathway, i.e., the TPS-TPP pathway [Fig: 3, A] was discovered about 50 years ago. It is the most widely distributed and it has been reported in eubacteria, archaea, fungi, insects and plants. According to this pathway the building blocks of trehalose are UDP-Glucose (UDPG) and Glucose-6-phosphate (G6P). The enzyme trehalose-6-phosphate synthase (TPS) joins one molecule of glucose from UDPG with G6P to form a single molecule of Trehalose-6-phosphate (T6P). Subsequently, the resulting molecule T6P is dephosphorylated into trehalose by the enzyme trehalose-6-phosphate phosphatase (TPP) [Fig: 4] [36], although unspecific phosphatases are also able to dephosphorylate T6P. This final product of biosynthesis, trehalose is hydrolyzed by the trehalase enzyme which can be classified into two categories according to their pH optima, namely, cytosolic neutral trehalase (NT) and vacuolar acidic trehalase (AT). A balance between the synthesizing and hydrolyzing enzymes controls intracellular trehalose levels [108]. Post-translational modifications of protein molecules are necessary for regulation of enzyme activities. Besides protein phosphorylation- dephosphorylation, protein methylation is also a key post-translational event. S-adenosyl-L-methionine (AdoMet), the universal methyl group donor and Oxidized Adenosine (AdOx), the methylation inhibitor shows effect on overall trehalose metabolism. AdoMet is reported as a positive controller for the two biosynthesizing enzymes while it plays a crucial role in negatively controlling the NT activity. Recently few reports strongly suggest the enhancement of trehalose production using AdoMet as an additive [2, 9, 12, 18, 99].

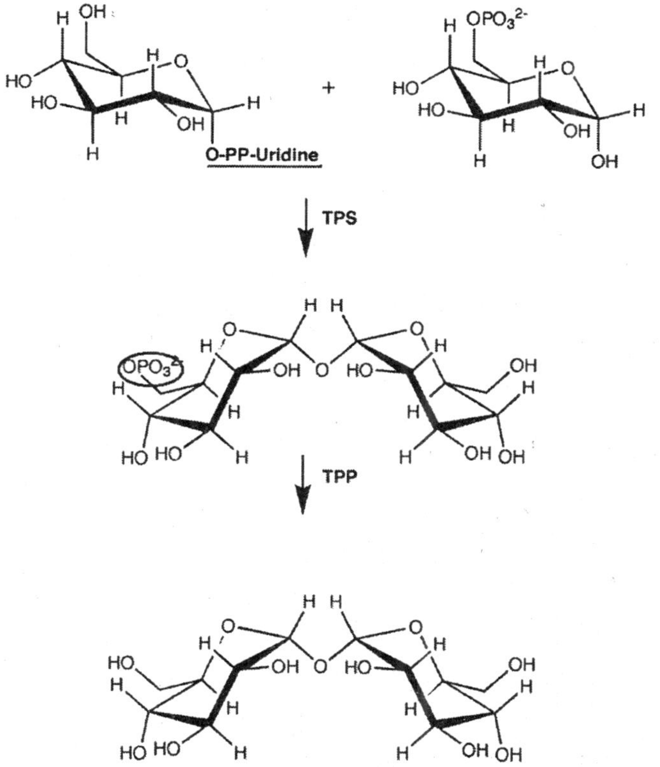

**Fig. 4.** Trehalose biosynthesis pathways: One glucose molecule is transferred from UDPG to G-6-P to form T6P by trehalose-6-phosphate synthase (TPS). T6P is next dephosphorylated to form trehalose by trehalose-6-phosphate phosphatase (TPP)

In the second biosynthetic pathway [Fig: 3, B], the enzyme Trehalose synthase (TS) isomerizes the α1-α1 bond, forming trehalose. This enzyme was first reported in *Pimelobacter* sp. and orthologs of this protein have been found in other eubacteria.

The third pathway [Fig: 3, C] involves the conversion of maltodextrines [i.e., malto-oligosaccharides, glycogen and starch] to trehalose. This pathway was reported in thermophilic archaea of the genus *Sulfobolus*. These organisms synthesize trehalose in two enzymatic steps catalyzed by maltooligosyl trehalose synthase (TreY), coded by the treY gene; which promotes the transglycosylation of the last glucose moiety at the reduced end of maltodextrines from a α1-α4 to a α1-α1 bond leading to malto-oligosyltrehalose, which contains a trehalose moiety at the end of the polymer. Next, maltooligosyl trehalose trehalohydrolase (Tre Z), coded by the tre Z gene, catalyzes the hydrolytic release of trehalose.

In the fourth pathway [Fig: 3, D], trehalose phosphorylase (Tre P), present in some fungi, catalyzes the reversible hydrolysis of trehalose in the presence of inorganic phosphate. The transfer of a glucose molecule to a phosphate generates glucose-1-phosphate and releases the other glucose residue. There is uncertainty about the participation of the Tre P enzyme in the synthesis or degradation of trehalose, since the biosynthetic reaction has only been shown in vitro.

The newer biosynthetic pathway [Fig: 3, E] for trehalose was found in the hyper-thermophilic archeon *Thermococcus litoralis* and involves the trehalose glucosyl transforming synthase (Tre T), which catalyzes the reversible formation of trehalose from ADP-Glucose and Glucose. It can also use UDPG and GDPG, although it is less efficient with these substrates. The Tre T enzyme transfers the glucose moiety from ADPG and joins it at position 1 of another glucose molecule to form trehalose. Trehalose is degraded [Fig: 3, F] by the enzyme trehalase (Tre H) into two glucose molecules. In general, the fate of ingested or parenterally administered trehalose corresponds to that of glucose, since trehalose is rapidly hydrolyzed to glucose by the enzyme trehalase. Trehalase is found in humans and most animals at the brush border of the intestinal mucosa, as well as in the kidney, liver, and blood plasma [28, 30, 39, 50, 62, 79, 90, 117]. Trehalase activity has been found in the small intestine of humans, mice, rats, guinea-pigs, rabbits, pigs, and baboons. No trehalase activity however, is found in the small intestine of cats [50].

## CHEMICAL SYNTHESIS OF TREHALOSE

α, α -Trehalose has been synthesized chemically using the ethylene oxide addition reaction between 2,3,4,5-tetra- O-acetyl-D-glucose and 3,4,6-tri-O-acetyl-1,2-anhydro-D- glucose [63]. This same series of reactions also gives rise to one of the other trehalose anomers, specifically, α,ß-trehalose, also referred to as neotrehalose. Neotrehalose has also been synthesized using the Koenigs-Knorr reaction [47]. On the other hand, this anomer has not been isolated from any living organisms, although it was identified in koji extract [72]. The other anomer of trehalose, ß, ß-trehalose, or isotrehalose, also has not been isolated from any living organisms, but it was found in starch hydrolysates [94]; it also has been synthesized chemically using the Koenigs-Knorr reaction, as well as by a dehydration reaction [13]. Trehalose also can be produced chemically by an acid reversion of glucose [104]. On the other hand, α, α -trehalose is the only anomer of trehalose that has been shown to be biosynthesized in many different types of organisms.

## ROLE OF TREHALOSE IN STRESS PROTECTION

The definition of stress in physics is any force applied to an object; whereas in biology, stress is any change in external factors or environmental conditions such as freeze, chill, heat, drought, flood, salt, pest and air pollution etc, which exert a disadvantageous effect on organism that might reduce or adversely change their growth or development. Resistance to stress is achieved through adaptation, avoidance and tolerance. Adaptation is permanent resistance to stress in morphology and structure, and also in physiology and biochemistry under long term stress condition. Stress disrupts metabolic homeostasis and requires an adjustment of metabolic pathways in a process that is usually referred to as acclimation. Several different phases are thought to be involved in acclimation. In the initial stages, the change in environmental condition is sensed by the organism which activates a network of signaling pathways. In later phases, the signal transduction pathways activated in the first phase trigger the production of different proteins and compounds that restore or achieve a new state of homeostasis. From the standpoint of metabolomics, at least three different types of compounds are important for these processes: (1) compounds involved in the acclimation process such as antioxidants or osmoprotectants (eg. trehalose); (2) by-products of stress that appear in cells because of the disruption of normal homeostasis by the alteration(s) in growth conditions; and (3) signal transduction molecules involved in mediating the acclimation response.

Initially it was thought that trehalose served as a carbohydrate reservoir like glycogen [33]. Trehalose accumulates intracellularly during periods of starvation, desiccation and after exposure to mild heat stress in various organisms, where it serves as a potent membrane stabilizer. According to the water replacement theory, trehalose replaces water by establishing hydrogen bonds with membranes or macromolecules during dehydration or freezing. In another process of glass formation, trehalose not only crystallizes but can also solidify into a glassy state when completely dehydrated. It is assumed, the glassy state of trehalose prevents denaturation of biomolecules at the time of water starvation, which ensures their activity to be retained in due course of rehydration [20]. Trehalose in its role as a crucial defense mechanism compound, stabilizes proteins and biological membranes from a large number of stresses, including temperature [27, 31, 110], oxidative stress [10], hydrostatic pressure [56], desiccation [27], osmotic stress [15, 26, 48, 54, 58, 84], and exposure to toxic chemicals [11, 108]. Trehalose is also important for the control of glucose influx during the cellular response to adverse condition [54]. Besides acting as a storehouse of glucose for energy and/ or synthesis of cellular components, it performs a large number of other functions as well, some of which are protective in nature. In some organisms, like yeast and plants, it may also serve as a signaling molecule to direct or control certain metabolic pathways or even to affect growth [57]. Trehalose also stabilizes small molecules like the anti-ageing compound, S-adenosyl-L-methionine [74].

## Protection from dehydration: [8, 16, 22, 33, 42, 46, 51, 63, 76, 85, 93, 114]

Although water is obviously necessary for life, some organisms are able to survive almost complete dehydration, for instance, even when 99% of their water content is removed. This includes common organisms such as plants, yeast cells, and fungal spores, but also microscopic animals such as nematodes, rotifers, and the cysts of brine shrimp [64]. Some of these dried organisms may remain in this state, known as anhydrobiosis, for decades under unfavorable conditions until water becomes available. When this happens, these organisms swell and resume the active state. Recent studies on these organisms have demonstrated mechanisms that allow them to survive dehydration; understanding some of these mechanisms have enabled researchers to develop new methods for the preservation of biological materials that are normally sensitive to drying [20, 22, 23, 24].

Anhydrobiotic organisms mentioned above generally contain high concentrations of trehalose (and sometimes other disaccharides and oligosaccharides). Thus it was shown that when the nematode *Aphelenchus avenae* was slowly dehydrated, it converted as much as 20% of its dry weight into trehalose [22]. The ability of this organism and others to survive in the absence of water has shown a strong correlation with the synthesis of trehalose [21, 22, 113]. Log-phase cultures of yeast have low concentrations of trehalose and are quite susceptible to dehydration, but as they enter the stationary phase of growth the levels of trehalose increase, along with their ability to survive dehydration [42]. This ability to survive in the presence of trehalose is independent of the growth phase of the cells because log-phase cells subjected to heat shock rapidly synthesize trehalose and also acquire the ability to survive dehydration [51]. These results are also applicable to a variety of other organisms ranging from brine shrimp [64] to the resurrection plant. This use of trehalose to enable cells to survive dehydration (and other stresses) may be an ancient adaptation because even Archaebacteria have been found to accumulate trehalose in response to stress [77]. Interestingly, in plants the disaccharide sucrose plays a similar role to that of trehalose in yeast and nematodes [4]. According to Leopold (1986), trehalose is preferred over sucrose by most organisms because it has fewer tendencies to form crystals than does sucrose.

The two primary stresses that are proposed to destabilize lipid bi-layers during dehydration are fusion and lipid phase transitions. Studies by laser light scattering and other techniques demonstrate that trehalose and other sugars inhibit fusion between the vesicles during drying, but the inhibition of fusion alone is not sufficient to preserve the dry vesicles. Thus trehalose is also necessary to prevent phase transitions [23]. The evidence suggests that trehalose depresses the phase transition temperature of the dry lipids, which maintains them in the liquid crystalline phase in the absence of water [21]. A large body of evidence indicates that the stabilizing effect of trehalose is due to its structure and stereochemistry. X-ray diffraction studies show that trehalose fits well between the polar head groups with multiple sites of interaction and suggests that the strong stabilizing effects of trehalose are related to its stereochemistry, which provides the most favorable fit with the polar head groups [93].

Trehalose also preserves labile proteins during drying. For example, phosphofructokinase is a tetramer that irreversibly dissociates to inactive dimers during the drying process. Although many compounds that stabilize proteins can prevent this dissociation when excess water is still present, only disaccharides are effective in stabilizing this protein during extreme drying. As in the case of membranes, trehalose probably interacts directly with the dry protein by hydrogen bonding between its hydroxyl groups and polar residues in the protein [16]. It is not clear why this interaction results in stabilization, but the fact that trehalose is a nonreducing sugar means that it cannot undergo the typical browning reaction between the aldehyde group on reducing sugars and amino groups on proteins. The browning reaction usually leads to denaturation of proteins.

An increased resistance to withstand drought conditions has been conferred on several different plants and also animal cells by increasing their levels of intracellular trehalose. For example, transgenic tobacco plants were engineered to produce substantial amounts of trehalose by introducing the *E. coli* otsA and otsB genes for trehalose synthesis. The introduction of these genes had a pronounced effect on plant morphology and growth performance under drought conditions. Thus the transgenic plants showed considerably improved growth [85]. The same results were obtained with human fibroblasts that had the otsA and otsB genes inserted and expressed. These cells could be maintained in the dry state for up to 5 days, as compared with controls cells that were very sensitive to drying [46].

**Protection against hydrostatic pressure: [41]**

Effect of hydrostatic pressure on the morphology and ultrastructure of wild-type and trehalose synthase mutant cells of *S. cerevisiae* were investigated by transmission electron microscopy. Pressure was seen to induce several morphological changes in both wild-type and tps1 cells, of which the latter showed greater structural alterations, as it was 1000 fold more barosensitive than the wild type strain.

**Role in cryo-protection: [21, 31, 38, 55, 60, 61, 65]**

Trehalose has been reported to protect living cells from cold stress as well. The addition of trehalose to cells has been found to protect them against freezing. Exogenous trehalose enhances viability of bacteria and yeast during freezing. Cryo-preservation of intact yeast cells is important not only for scientific research, but also as an industrial tool. The increasing demand for frozen yeasts with conserved desirable metabolic features could be regarded as an example of this fact. Mild cold shock treatment was seen to enhance cell tolerance to low temperature, just like mild heat shock treatment, only difference being that while heat treatment was able to increase the accumulation of intracellular trehalose whereas, during cold shock exposure, the intracellular amount of trehalose remained unaltered. This protection is thought to be due to retardation of membrane fluidity. The undesirable effects of freezing and freeze-drying on intact yeast cells stored for a month can be reduced utilizing simple procedures such as a pre-exposure at 5–10°C for 3 h (or less) and the addition of trehalose during the low-temperature process.

Mammalian cells transfected with genes for trehalose synthesis were able to survive much longer in the frozen state[i]. A mutant *E. coli* strain unable to produce trehalose has been shown to die much faster than the wild type at 4 °C. Transformation of the mutant with the otsA/otsB genes, responsible for trehalose synthesis, restored trehalose content and cell viability at 4 °C. Researchers have demonstrated that a cold shock from 37 °C to 16 °C induced up to 8 fold increase of trehalose in wild type *E. coli* cells. This trehalose enhanced cell viability when temperature was further decreased to 4 °C. Before the increased accumulation of trehalose, levels of mRNA encoding OtsA/OtsB increased markedly. The mechanism by which trehalose mediates tolerance to freezing is not clear, but presumably involves a stabilization of certain cell proteins and/or lipid membranes.

**Protection against heat: [5, 6, 45, 52, 76, 81, 100, 101, 102, 112]**

In yeast, stimuli that trigger the heat shock response also cause the accumulation of trehalose. In fact at least two subunits of the trehalose-6-P synthase complex of *S. cerevesiae* are actively synthesized during heat shock, and physiological concentrations of trehalose (up to 0.5 M) were found to protect enzymes of yeast and other organisms from heat inactivation in vitro. Trehalose also reduced the heat-induced formation of protein aggregates. Trehalose was as good or better as a protein stabilizer than any of a number of compatible solutes, including polyols, sugars, and amino acids [26]. Yeast mutants were prepared that were defective in genes coding for key enzymes in trehalose metabolism (TPS1, TPS2), and these mutants showed an inability to accumulate trehalose on mild heat shock and were much less resistant to heating than was the wild-type organism. These various studies strongly implicate trehalose as playing a key role in thermo-tolerance and also indicate that the enzymes that synthesize trehalose are induced in response to the stress in order to increase the levels of trehalose.

An important in vivo and in vitro study showed that trehalose protects cells from heat by stabilizing proteins at high temperatures. Using two different temperature-sensitive reporter proteins, these investigators showed that enzymes are better able to retain activity during heat shock in cells that are producing trehalose [100]. These studies showed an additional and important role of trehalose, that is, the ability to suppress aggregation of proteins that have already been denatured. Based on these studies, these researchers also explained why trehalose must be degraded rapidly after the heat shock, that is, if the unfolded luciferase, one of their reporter proteins, is removed from or diluted out of the cell, and it can be refolded by molecular chaperones. On the other hand, if the trehalose concentration remains high, it interferes with the refolding process, and the protein is not renatured by the chaperone [100]. Thus, it may be important to have active trehalase present once the heat stress is alleviated.

## Protection from damage by oxygen radicals: [3, 34]

Another role for trehalose is in protecting cells against oxygen radicals. Exposure of *S. cerevisiae* to a mild heat shock or to a proteosome inhibitor induced trehalose accumulation (as indicated) and also markedly increased the viability of cells on exposure to a free radical–generating system ($H_2O_2$/iron). However, when the cells were returned to the normal growth temperature, both the trehalose content and the resistance to oxygen stress decreased rapidly and returned to the wild-type level. A mutant cell line defective in trehalose synthesis was much more sensitive to oxygen killing than was the wild type, but adding trehalose to the medium enhanced the resistance of these cells to $H_2O_2$. The effect of oxygen radicals on these cells was to damage amino acids in cellular proteins and the presence of high concentrations of trehalose in the cells prevented this damage. The trehalose-defective mutant showed a much higher content of damaged proteins even after only a brief exposure to oxygen stress. The suggestion is that trehalose acts as a free radical scavenger. In these studies, mannitol and galactose also protected but less so than trehalose, whereas sucrose was ineffective. This lack of protection by sucrose may have to do with its inability to quench oxygen radicals or be taken up by cells [11].

As indicated earlier in this review, trehalose and sucrose are both nonreducing disaccharides that may have considerable similarity in synthesis and function. Both are stored in the cytosol of cells, and both may be present in their respective cells in high concentrations depending on various environmental conditions. Another commonality is that when either of these oligosaccharides are present in high concentration, the cells become quite resistant to a variety of stress conditions, including heat, dehydration, oxygen stress, and so on. In plants, oligosaccharides of the raffinose series (Gal$\alpha$1-6sucrose and higher) can accumulate in cells to a level of 15% of the dry weight, and these plants may have considerable stress resistance. This striking correlation cannot be ignored and needs further investigation to determine whether raffinose and stachyose and other sucrose oligosaccharides are in fact produced as a protection against different stress conditions. The same may be true of the trehalose oligosaccharides that have been isolated from the cytosol of *M. smegmatis*. Although their concentrations in these cells were fairly low, that could be because those cells were not stressed. It will be important to examine the levels of these various trehalose analogs after cells have been exposed to stress.

### Protection from cold: [60]

A mutant strain of *E. coli* that was unable to produce trehalose died much faster that did the wild type at 4°C. However, transformation of this mutant with otsA/otsB genes restored the ability to synthesize trehalose and also cell viability in the cold [60]. Additional studies showed that downshifting cells from 37°C to 16°C caused an eightfold increase in trehalose levels and a marked increase in mRNA levels for otsA and otsB. The authors speculate that because a number of proteins denature and precipitate in the cold where the hydrophobic effects are relatively weak, it is possible that trehalose also prevents the denaturation and aggregation of specific proteins at cold temperatures. Trehalose may also stabilize cell membranes whose fluidity decreases during temperature downshift. Thus, exogenous trehalose has been shown to protect a variety of organisms against freezing, with maximum protection seen when trehalose is present on both sides of the membrane.

## APPLICATION OF TREHALOSE IN ENHANCEMENT OF STRESS TOLERANCE AND DISEASE CONTROL

Trehalose or its precursor trehalose-6-phosphate are involved in providing protection from various types of stress to the organism producing them as can be observed by some interesting examples listed below. Induction of defense mechanism in plants against pathogens by application/ use of trehalose has also been carried out in several cases. Trehalose confers resistance by activating plant defense responses. Trehalose though not synthesized in mammals, has found extensive use in therapeutic fields due to its anti stress properties.

Genetically engineered (transgenic) crop plants start the desired trehalose metabolism genes have been tested for their stress tolerance. Field trials proved that the crops in most cases showed improved growth under stressed condition, drought tolerance, enhanced rate of photosynthesis (Table: 2), less photo-oxidative damage and favorable mineral balance under salt, drought and low temperature stress.

**Table 2.** Biotechnological trials for production of stress tolerant crops [courtesy: voh.chem.ucla.edu/vohtar/spring04/.../ **trehalose** together1.ppt]

| Origin | Gene used | Promoter | Target plant | Prominent effects |
|---|---|---|---|---|
| *E. coli* | *otsA, otsB* | Constitutive (CaMV35S) | Tobacco | Improved growth under stress conditions, morphological alterations |
| Yeast | *TPS* | Tissue-specific rbcS | Tobacco | More trehalose levels |
| Yeast | *TPS1* | Constitutive (CaMV35S) | Tobacco | Stunted growth, lancet-shaped leaves, reduced sucrose content and improved drought tolerance |
| *E. coli* | *TPS* | Tuber specific (Patatin) | Potato | No trehalose levels detected |
| *E. coli* and *yeast* | *TPS* | Constitutive (CaMV35S) | Tobacco | Enhanced rate of photosynthesis (TPS) |
| | *TPP* | | | Reduced photosynthesis (TPP) |
| *E. coli* | *otsA, otsB* | Tissue specific (rbcS) and stress dependant (abscisic acid inducible) | Rice | Sustained plant growth, less photo-oxidative damage, favorable mineral balance (under salt, drought and low temperature stress) and more trehalose, Increased stress tolerance |

Reignault et al. (2001) [89], showed a reduction in the degree of powdery mildew infection of wheat leaves after treatments with trehalose. Wheat (*Triticum aestivum*) cv. Sideral plants grown in phytotrons were inoculated with *Blumeria graminis* f.sp. *tritici*. In addition to degree of infection, the effect of trehalose solution was further investigated using light and fluorescence microscopy and enzyme assays. Infection in wheat leaves was reduced by 50 and 95% with trehalose solution (15 g $l^{-1}$) following a single spraying and three sprayings, respectively; in a detached leaf assay, trehalose was effective at concentrations as low as 0.01 g $l^{-1}$. Trehalose did not inhibit conidial germination and differentiation of appressoria (*in vitro* or on the leaf epidermis), but enhanced papilla deposition in epidermal cells. Trehalose also enhanced phenylalanine ammonia-lyase (PAL) and peroxidase (PO) activities, both markers of plant defense responses. However, the level of three cinnamyl alcohol dehydrogenase (CAD) activities (conyferyl, *p*-coumaryl and sinapyl alcohol dehydrogenase) was unchanged. Trehalose treatment of wheat confers resistance to *B. graminis* infection by activating plant defense responses (e.g. papilla deposition, PAL and PO activities).

Studies on the pathobiology of *Cryptococcus neoformans* identified the presence of a functioning trehalose pathway during infection and suggested its importance for *C. neoformans* survival in the host. Both *TPS1* and *TPS2* are required for high-temperature (37°C) growth and glycolysis but the block at *TPS2* results in the apparent toxic accumulation of T6P, which makes this enzyme a fungicidal target. Sorbitol suppresses the growth defect in the *tps1* and *tps2* mutants at 37°C, which supports the hypothesis that these sugars (trehalose and sorbitol) act primarily as stress protectants for proteins and membranes during exposure to high temperatures in *C. neoformans* [37].

To analyze how trehalose and trehalase respond to salt stress in root nodules of legumes, tissue expression of *Medicago truncatula* trehalase gene (MTTRE1) and the expression of MTTRE1 under salt stress were analyzed by real-time quantitative reverse transcription-PCR method. MTTRE1 expression is induced in nodules compared with leaves and roots, indicating a transcriptional regulation of trehalase in the presence of the microsymbiont. Under salt stress conditions, trehalase activity is down regulated at the transcriptional level, allowing trehalose accumulation [71].

Bacteria of the genus *Azospirillum* increase the grain yield of several grass crops. Transformed bacteria with a plasmid harboring a trehalose biosynthesis gene-fusion from *Saccharomyces cerevisiae* were able to grow up to 0.5 M NaCl and to accumulate trehalose, whereas wild-type *A. brasilense* did not tolerate osmotic stress or accumulate significant levels of the disaccharide. Moreover, 85% of maize plants inoculated with transformed *A. brasilense* survived drought stress, in contrast with only 55% of plants inoculated with the wild-type strain. A 73% increase in biomass of maize plants inoculated with transformed *A. brasilense* compared with inoculation with the wild-type strain was found. In addition, there was a significant increase of leaf and root length in maize plants inoculated with transformed *A. brasilense* [92].

In February 2003, a severe disease with bacterial leaf necrosis symptoms was observed on container-grown melon (*Cucumis melo* cv. Nun) seedlings at a nursery in Adana, located in eastern Mediterranean region of Turkey. A fluorescent, opaque, semi fluid, gram-negative bacterium was isolated from diseased tissues onto King's medium B (King et al., 1954). Fatty acid analysis identified the strains as *Pseudomonas viridiflava*. The bacterial strains did not produce levan type colonies on sucrose nutrient agar, but were able to induce hypersensitive reaction on tobacco leaves. The strains were capable of producing acid which is needed for its growth, from sorbitol, fructose, glucose, L (+) arabinose, D (+) xylose, D (-) mannitol, but not from sucrose, trehalose, maltose, melibiose, lactose and D (-) arabinose. [116]

The heterothallic ascomycete *Magnaporthe grisea* causes rice blast disease of cultivated rice. The potentially devastating effects of this disease are determined by the ability of the fungus to mechanically penetrate the plant cuticle and epidermis. The force necessary for cuticle penetration is due to the enormous hydrostatic turgor generated within the appressorium as a result of the high concentrations of compatible solutes, notably glycerol, which accumulate in appressoria. Glycogen, lipid and trehalose are the most likely potential precursors for glycerol biosynthesis in *M. grisea* as they represent the most abundant storage products in spores of the fungus. Here the role of trehalose synthesis and metabolism in the infection cycle of *Magnaporthe grisea* was studied thoroughly. Trehalose is synthesized by the trehalose 6 phosphate synthase complex (T6PS), and it was shown that *TPS1*, encoding the catalytic sub-unit of T6PS, is required for pathogenicity of *M. grisea* and appressorium turgor generation (Foster *et al.,* 2003). However, the subsequent predicted breakdown of trehalose, which would be required for glycerol synthesis, is dispensable for appressorium turgor generation, and, therefore, it seems more likely that trehalose accumulation contributes to appressorium function, either due to trehalose acting in a protective capacity, or by accumulating as a solute to contribute to turgor.

*Spiroplasma citri* is a plant-pathogenic mollicute, phylogenetically related to Gram-positive bacteria. *Spiroplasma* cells are restricted to the phloem sieve elements and are transmitted by leafhopper vectors. Recent research has allowed depicting a unique scenario in *S. citri* pathogenicity, where sugar metabolism plays a major role. In vitro *S. citri* uses fructose, glucose and trehalose, which are imported through the Phosphoenolpyruvate phosphoTransferase System (PTS). When both fructose and glucose are present, fructose is used preferentially. The fructose-PTS permease is made of one single polypeptide IIABC sup(Fru), whereas the glucose permease is split into two polypeptides IICB sup(Glc) and IIA sup(Glc), encoded by two separated genes, ptsG and crr. The glucose IICB sup(Glc) and trehalose IIBC sup(Tre) permeases function with a single IIA sup(Glc) domain, enabling the *Spiroplasma* to rapidly adapt from glucose to trehalose and vice versa. Upregulation of expression of the PTS permease genes by the relevant sugar also illustrates the adaptive capacity of the *Spiroplasma*. *S. citri* mutants unable to import glucose and trehalose are still highly pathogenic [90].

Improving stress tolerance and yield in crops are major goals for agriculture. A new strategy is applied to increase drought tolerance and yield in legumes by overexpressing trehalose-6-phosphate synthase in the symbiotic bacterium *Rhizobium etli*. *Phaseolus vulgaris* (common beans) plants inoculated with *R. etli* overexpressing trehalose-6-phosphate synthase gene had more nodules with increased nitrogenase activity and higher biomass compared with plants inoculated with wild-type *R. etli*. In contrast, plants inoculated with an *R. etli* mutant in trehalose-6-phosphate synthase gene had fewer nodules and less nitrogenase activity and biomass. Macroarray analysis of 7,200 expressed sequence tags from nodules of plants inoculated with the strain overexpressing trehalose-6-phosphate synthase gene revealed upregulation of genes involved in stress tolerance and carbon and nitrogen metabolism, suggesting a signaling mechanism for trehalose. As a result, trehalose metabolism in rhizobia is key for signaling plant growth, yield, and adaptation to abiotic stress, and its manipulation has a major agronomical impact on leguminous plants [88].

The genome of the foliar pathogen *Pseudomonas syringae* pv. *tomato* strain DC3000 contains genes for two trehalose biosynthetic pathways, treS and treYZ, and lacks genes for the more common OtsAB pathway. Deletion of either the treS (PSPTO_2760-2762) or treY/treZ (PSPTO_3125-3134) locus eliminated trehalose accumulation and reduced bacterial growth under hyper osmotic conditions. In evaluating the role of trehalose in *P. syringae* fitness on leaves, it is found that a double deletion mutant lacking these loci exhibited poorer survival than the wild type on tomato leaves over a 2-week period in a growth chamber [15].

Trehalose is a non-reducing disaccharide found at high concentrations in *Aspergillus* conidia required for its viability and rapidly degraded upon induction of conidial germination. Trehalose-degradation is concomitant with the accumulation of a glycerol pool in the conidia [31]. Recent work in yeast indicates that trehalose also promotes survival under conditions of extreme heat, by enabling proteins to retain their native conformation at elevated temperatures and suppressing the aggregation of denatured proteins. The latter property, however, seems to impair the recovery of cells from heat shock if they fail to degrade trehalose after the stress has been removed. These multiple effects of trehalose on protein stability and folding suggest a host of promising applications [102].

Studies on stress protection indicate that trehalose protects *Drosophila* and mammalian cells from hypoxic and anoxic injury. The mechanism of this protection is probably related to a decrease in protein denaturation through protein–trehalose interactions [87].

Recent studies have shown that fine structural modifications of *Mycobacterium tuberculosis* cell envelope lipids mediate host cell immune activation during infection. One such alteration in lipid structure is cis-cyclopropane modification of the mycolic acids on trehalose dimycolate (TDM) mediated by proximal cyclopropane synthase of α mycolates (*pcaA*), a proinflammatory lipid modification during early infection [108]. Yamaguchi et al., 2006 [118] showed mitochondria frozen with trehalose retain a number of biological functions and preserve outer membrane integrity.

Not only in stressed conditions, but in some cases trehalose is found to be produced in a normal state too. Excised mycorrhizae of spruce trees, grouped according to state of development and integrity, were tested for their ability to produce fungal and plant specific, soluble carbohydrates from glucose. Trehalose was the principal sugar formed. A strong correlation was found between the rate of trehalose synthesis and the proportion of intact, turgid, young and vigorous appearing mycorrhizae [79].

As stated earlier, no report is available on trehalose synthesis by mammals. But to combat some inevitable diseases related to stress in them, trehalose is used extensively and successfully by pharmacists in therapeutic fields. A few of this aspect is discussed below [10, 20, 21, 23, 24, 25, 26, 27, 29, 38, 39, 40, 50, 70, 74, 83, 96, 102, 103, 104, 110, 118].

Trehalose has become an important target for biotechnologists, where it is produced for different pharmaceutical needs like a non-toxic cryoprotectant of vaccines and organs for surgical transplants and in cosmetic products such as lip sticks. It has been used in large amounts for the preservation and protection of biological materials. It stabilizes bioactive soluble proteins such as monoclonal antibodies and enzymes for medical use. It increases viability of embryos during freeze-drying. It is used for preserving cellular blood transfusion products and to extend the shelf life of platelets, and cord blood. It also increases viability of transplant cells and tissue during cryo-preservation and decreases host immune response. Cord blood (CB) and fetal liver (FL) cells are two alternative sources of human hematopoietic stem cells. Among multiple factors that are responsible for the cryo-damage of cells, membrane leakage and oxygen free-radical generation have been shown to contribute substantially towards the process. Trehalose, when used in combination with 10 % dimethyl sulfoxide (DMSO), affords better cryo-protection of these cells as evidenced by significantly increased colony formation as compared to 10 % DMSO alone. The cryo-protection afforded by trehalose persists at least for 1.5 years. This increased cryo-protective effect has been reported both at -196 °C and at -80 °C storage temperatures.

The ability of trehalose to inhibit lipid and protein misfolding has been used as the focus of study in several neurodegenerative diseases which are associated with the misfolding of disease-specific proteins. These include Alzheimer's disease (AD) - an amyloid proteinopathy, Huntington's disease (HD) - an expanded polyglutamine proteinopathy, and oculopharyngeal muscular dystrophy (OPMD) - an expanded polyalanine proteinopathy. Benefit of trehalose was first shown in Huntington's model system. It has also been shown that trehalose could protect corneal epithelial cells in culture from death by desiccation. The sugar has been considered as a component of eye drops for dry eyes syndrome.

Although trehalose absorption in humans has not been well studied, a small fraction (0.5 %) is likely to be absorbed by passive diffusion. In mammalian cell culture, trehalose is moved from the extracellular to intracellular compartment via a fluid phase endocytotic mechanism, and is dependent on extracellular concentration of the sugar. Tanaka et al. (2004) [104] have postulated that low concentration may be sufficient to exert effect as trehalose is resistant to hydrolysis. In cell culture, intracellular trehalose concentration does not change for extended periods of time after it has once been introduced. As such, its benefits are expected to persist for relatively extended times.

**Role in Alzheimer's disease**

Sporadic Alzheimer's disease is pathologically characterized by aggregation of small beta-amyloid (Aβ) peptides into distinctive amyloid plaques and neurofibrillary tangles. Like other proteinopathy neurodegenerative diseases, Alzheimer's progression is also correlated with increasing load of aggregated amyloid; though toxicity is increasingly linked to the formation of oligomeric forms of Aβ peptides. Although Aβ40 is the most prominent peptide (10:1), Aβ42 is more toxic.

Liu et al. (2005) [70] have reported effects of trehalose on aggregation *in vitro*, and neurotoxicity in a neuron cell system, using solutions of Aβ peptide either singly or in combination. Co-incubation with trehalose inhibited aggregation of the Aβ40 peptide in a dose-dependent manner. High concentration of trehalose completely inhibited peptide aggregation. Trehalose also inhibited aggregation of Aβ42 in a dose-dependent manner, but was significantly less efficient than the corresponding Aβ40. High dose of trehalose prevented only 50 % aggregation of this more toxic peptide. The authors summarized that trehalose has potential as part of a therapeutic 'cocktail' strategy for treating Alzheimer's because of its effect on Aβ40, the predominant form of Aβ peptide in amyloid plaques.

### Role in Huntington's disease

Huntington's disease is an autosomal dominant neurodegenerative disease. The disease includes cognitive impairment, involuntary choreiform movements, and psychiatric manifestations as its symptoms. Huntington's is characterized by an expanded CAG repeat within the first exon of the Huntington gene. The mutant protein generated has an expanded polyglutamine (polyGN) tract. The pathologic hallmark of this and other polyGN diseases is the formation of aggregates, containing misfolded mutant protein in both cytoplasm and nucleus of affected cells. Tanaka et al. demonstrated that trehalose inhibits polyglutamine-mediated protein aggregation of a model polyGN protein *in vitro* solution, and that it decreases aggregate formation and prolongs viability in a model cell culture. They also showed that trehalose ameliorated motor symptoms, decreased aggregate number and size, and prolonged life by 20% in the R6/2 transgenic mouse model of Huntington's. The authors showed that trehalose interacted with the expanded polyglutamine tract of the proteins, stabilizing the protein structure, and inhibiting protein aggregation at the initial stage of aggregate formation. They found that extracellular trehalose application decreased intracellular aggregates and improved cell viability in a dose-dependent fashion.

## FUTURE PROSPECT

Correlation of trehalose biosynthesis with increased tolerance to adverse environmental conditions in different micro-organisms is still being explored. Research on post-translational modification of trehalose metabolism has been initiated for the enhancement of trehalose production. In practice, the role of trehalose in disease management and its utilization in various therapeutic and pharmaceutical fields needs to be further studied.

Inspite of extensive studies on trehalose and its metabolism in plant system, there still exist lacunae between the research findings and field trials of agricultural crops. Exploitation of anti-stress effect of trehalose in transgenic plants requires further understanding of this aspect. Biotechnological applications have already been done for the production of stress tolerant crops. Increased trehalose production will lead to the enhanced capability of crop plants to withstand different abiotic and biotic stresses leading to higher yield through cultivation of crops in those fields which were previously un-cultivable due to stress intolerance. Increasing food production will fulfill the goal to feed the growing world population which is at present one of the major concerns.

## ACKNOWLEDGEMENT

Authors acknowledge UGC, for financial support for their work on trehalose metabolism.

## REFERENCES

1. Adams RP, Kendall E, and Kartha KK 1990 Comparison of free sugars in growing and desiccated plants of *Selaginella lepidophylla, Biochemistry System Ecology* **18** 107-110
2. Aletta JM, Cimato TR, Ettinger MJ 1998 Protein methylation a signal event in post-translational modification, *Trends in Biochemistry Science* **23** 89-91
3. Alvarez-Peral FJ, Zaragoza O, Pedren Y, Arguelles JC 2002 Protective role of trehalose during severe oxidative stress caused by hydrogen peroxide and the adaptive oxidative stress response in *Candida albicans, Microbiology* **148** 2599-2606
4. Anandarajah K and Mckersie BD 1990 Manipulating the desiccation tolerance and vigor of dry somatic embryos of *Medicago sativa* L with sucrose, heat shock and abscissic acid, *Plant Cell Reports* **9** 451-455
5. Arguelles JC 1994 Heat-shock response in a yeast tps1 mutant deficient in trehalose synthesis, *FEBS Letters* **350** 266-270
6. Arguelles JC 1997 Thermotolerance and trehalose accumulation induced by heat shock in yeast cells of *Candida albicans FEMS Microbiology Letters* **146** 65-71
7. Augier J 1954 The biochemistry of a North American algae, *Tuomeya fluviatilis, Comptes Rendus* **239** 87-89
8. Bajaj S, Mohanty A 2005 Recent advances in rice biotechnology – towards genetically superior transgenic rice *Plant Biotechnology Journal* **3** 275-307
9. Basistha B, Ghosh AK and Ghosh S 2010 Regulation of Trehalose metabolism by protein methylation in a

mutant strain of *Saccharomyces cerevisiae, NBU Journal of Plant Sciences* **4** 39-45

10. Beattie GM, Crowe JH, Lopez AD, Cirulli V, Ricordi C, Hayek A 1997 Trehalose a cryoprotectant that enhances recovery and preserves function of human pancreatic islets after long-term storage *Diabetes* **46** 519-523
11. Benaroudj N, Lee DH, Goldberg AL 2001 Trehalose accumulation during cellular stress protects cells and cellular proteins from damage by oxygen radicals, *The Journal of Biological Chemistry* **276** 24261-24267
12. Bhattacharyya S, Chaudhuri P, Basu A, Sengupta S, Ghosh AK 2005 Regulation of trehalose metabolism by AdOx and AdoMet in *Saccharomyces cerevisiae, Indian Journal of Experimental Biology* **43** 360-368
13. Birch GG 1963 Trehaloses In Adv Carbohyd Chem, Wolform ML (ed) Acad Press New York and London, **18** 201-225
14. Bredereck H, Hoschele G and Ruck K 1953 Synthesis of non-reducing disaccharides, *Chemische Berichte* **86** 1277-1286
15. Brian C Freeman, Chiliang Chen and Gwyn A Beattie 2010 Identification of the trehalose biosynthetic loci of *Pseudomonas syringae* and their contribution to fitness in the phyllosphere, *Environmental Microbiology*, **12** (6) 1486–1497
16. Boos W, Ehmann U, Forki W, Klein M, Rimmele M, Postma P 1990 Trehalose transport and metabolism in *Escherichia coli, Journal of Bacteriology* **172** 3450-3461
17. Carpenter JF, Crowe JF 1989 An infrared spectroscopic study of the interactions of carbohydrates with dried proteins, *Biochemistry* **28** 3916-3922
18. Chaudhuri P, Basu A, Sengupta S, Lahiri S, Dutta T, Ghosh AK 2009 Studies on substrate specificity and activity regulating factors of trehalose-6-phosphate synthase of *Saccharomyces cerevisiae Biochimica et Biophysica Acta* **1790** 368-374
19. Clegg JS, Filosa MF 1961 Trehalose in the cellular slime mold, *Dictyostelium mucoroides, Nature* **192** 1077-1078
20. Colaco C, Sen S, Thangavelu M, Pinder S, Roser B 1992 Extraordinary stability of enzymes dried in trehalose simplified molecular biology, *Biotechnology* **10** 1007-1011
21. Colaco C, Smith CJS, Sen S, Roser DH, Newman Y, Ring S, Roser BJ 1994 Chemistry of protein stabilization by trehalose In Formulation and Delivery of Proteins and

Peptides, Cleland, JL Langer R (eds) American Chemical Society, Washington DC; **567** 222-240

22. Crowe JH, Carpenter JF, Crowe LM 1998 The role of vitrification in anhydrobiosis *Annual Review of Physiology* 60 73-103
23. Crowe JH, Crowe LM, Chapman D 1984 Preservation of membranes in anhydrobiotic organisms, *Science* **223** 701-703
24. Crowe JH, Crowe LM 1990 Lyotropic effects of water on phospholipids In Water Science Reviews, Franks F (ed) Cambridge Univ Press England 1-23
25. Crowe JH, Tablin F, Wolkers WF, Gousset K, Tsvetkova NM, Ricker J 2003 Stabilization of membranes in human platelets freeze-dried with trehalose, *Chem Phys Lipids* **122** 41-52
26. Davies JE, Sarkar S, Rubinsztein DC 2006 Trehalose reduces aggregate formation and delays pathology in a transgenic mouse model of oculopharyngeal muscular dystrophy, *Hum Mol Genet* **15** 23-31
27. de Castro AG, Tunnacliffe A 2000 Intracellular trehalose improves osmotolerance but not desiccation tolerance in mammalian cells, *FEBS Letters,* **487** 199-202
28. De Virgilio C, Hottiger CT, Dominiguez J, Boller T, Wiemken A 1994 The role of trehalose synthesis for the acquisition of thermotolerance in yeast I Genetic evidence that trehalose is a thermoprotectant *Europian Journal Biochemistry* **219** 179-186
29. Demelier JF, Labat J and Courtois JE 1975 [Influence of parenteral injection of trehalose on mammals with distinct trehalose equipment *CR Acad Sc (Paris) Serie D,* **280** 669–672 (in French)
30. dEnfert C 1997 Fungal spore germination insights from the molecular genetics of *Aspergillus nidulans* and *Neurospora crassa, Fungal Genetic Biology* 21, 163–172
31. dEnfert C, Bonini BM, Zapella PDA, Fontaine T, da Silva AM, Terenzi HF 1999 Neutral trehalases catalyse intracellular trehalose breakdown in the filamentous fungi *Aspergillus nidulans* and *Neurospora crassa, Molecular Microbiology* **32** 471–484
32. Diniz-Mendes L, Bernardes E, de Araujo PS, Panek AD, Paschoalin VM 1999 Preservation of frozen yeast cells by trehalose *Biotechnol Bioengg* **65** 572-578
33. Drennan PM, Smith MT, Goldsworthy D, van Staden J 1993 The occurrence of trehalose in the leaves of the desiccation-tolerant angiosp-

erm *Myrothamnus flabellifolius* Welw *J Plant Physiol* **142** 493-496

34. Elbein AD, Mitchell M 1973 Levels of glycogen and trehalose in *Mycobacterium smegmatis*, and the purification and properties of the glycogen synthase *J Bacteriol* **113** 863-873
35. Elbein AD, Pan YT, Pastuszak I, Carroll D 2003 New insights on trehalose a multifunctional molecule *Glycobiol* **13** 17R-27R
36. Elbein AD 1974 The metabolism of α, α-trehalose In Adv Carbohydr Chem Biochem, Tipson RS, Horton D (eds) Acad Press New York; **30** 227-256
37. Elizabeth Wills Petzold, Uwe Himmelreich, Eleftherios Mylonakis, Thomas Rude, Dena Toffaletti, Gary M Cox, Jackie L Miller, and John R Perfect, (2006), Characterization and Regulation of the Trehalose Synthesis Pathway and Its Importance in the Pathogenicity of *Cryptococcus neoformans*, Infect Immun; **74(10)** 5877–5887
38. Erdag G, Eroglu A, Morgan J, Toner M 2002 Cryopreservation of fetal skin is improved by extracellular trehalose Cryobiol **44** 218-228
39. Eroglu A, Russo MJ, Bieganski R, Fowler A, Cheley S, Bayley H, Toner M 2000 Intracellular trehalose improves the survival of cryopreserved mammalian cells *Nat Biotechnol* **18** 163-167
40. Eze, LC (1989) Plasma trehalase activity and diabetes mellitus *Biochem Gen,* **27,** 487–495
41. Fairbairn D 1958 Glucose, trehalose and glycogen in *Porrocaecum decipiens* larvae *Nature* **181** 1593-1594
42. Fernandes PMB, Farinal M, Kurtenbach E 2001 Effect of hydrostatic pressure on the morphology and ultrastructure of wild-type and trehalose synthase mutant cells of *Saccharomyces cerevisiae* Lett Appl Microbiol **32** 42-46
43. Gadd GM, Chalmers K, Reed RH 1987 The role of trehalose in dehydration resistance of *Saccharomyces cerevisiae FEMS Microbiol Letters* **48** 249-254
44. Gancedo C, Flores CL 2004 The importance of a functional biosynthetic pathway for the life of yeasts and fungi *FEMS Yeast Res* **4** 351-359
45. Goddijn OJM, van Dun K 1999 Trehalose metabolism in plants *Trends in Plant Science* **4** 315-319
46. Gross C, Watson K 1998 De novo protein synthesis is essential for thermotolerance acquisition in a *Saccharomyces cerevisiae* trehalose synthase mutant *Biochem Mol Biol Int* **45** 663-671

47. Guo N, Puhlev I, Brown DR, Mansbridge J, Levine F (2000) Trehalose expression confers desiccation tolerance on human cells *Nat Biotechnol* **18** 168-171

48. Helferich and Weis, B Helferich and K Weis, Zur (1956) Synthese von Glucosiden und nicht-reduzierenden Disacchariden *Chem Ber* **89** 314–315

49. Hershkovitz N, Oren A, Cohen Y 1991 Accumulation of trehalose and sucrose in cyanobacteria exposed to matric water stress *Appl Environ Microbiol* **57** 645-648

50. Higashiyama T 2002 Novel functions and applications of trehalose *Pure Appl Chem* **74** 1263-1269

51. Hore, P & Messer, M (1968) Studies on disaccharidase activities of the small intestine of the domestic cat and other carnivorous mammals *Comp Biochem Physiol*, **24**, 717–725

52. Hottiger T, Boller T, Wiemken A 1987 Rapid changes of heat and desiccation tolerance correlated with changes of trehalose content in *Saccharomyces cerevisiae* cells subjected to temperature shifts FEBS Lett ers **220** 113-115

53. Hottiger T, De Virgilio C, Bell W, Boller T, Wiemken A 1992 The 70-kilodalton heat-shock proteins of the SSA subfamily negatively modulate heat-shock-induced accumulation of trehalose and promote recovery from heat stress in the yeast, *Saccharomyces cerevisiae Europian Jounal of Biochemistry* **210** 125-132

54. Hottiger T, De Virgilio C, Hall MN, Boller T, Wiemken A 1994 The role of trehalose synthesis for the acquisition of thermotolerance in yeast II Physiological concentrations of trehalose increase the thermal stability of proteins *in vitro Europian Jounal of Biochemistry* **219** 187-193

55. Hounsa CG, Brandt EV, Thevelein J, Hohmann S, Prior BA (1998) Role of trehalose in survival of *Saccharomyces cerevisiae* under osmotic stress *Microbiology* **144** 671–680

56. Israeli E, Shaffer BT, Lighthart B 1993 Protection of freeze-dried Escherichia coli by trehalose upon exposure to environmental conditions Cryobiol **30** 519-523

57. Iwahashi H, Nwaka S, Obuchi K 2000 Evidence for contribution of neutral trehalase in barotolerance of *Saccharomyces cerevisiae Applied Environmental Microbiology* **66** 5182-5185

58. Jules M, Guillou V, Francois J, Parrou JL 2004 Two distinct pathways for trehalose assimilation in the yeast *Saccharomyces cerevisiae*

*Applied Environmental Microbiology* **70** 2771-2778

59. Kaasen I, Falkenberg P, Sryrvold OB, Strom AR 1992 Molecular cloning and physical mapping of the otsBA genes upon exposure, which encode the osmoregulatory trehalose pathway of *Escherichia coli, J Bacteriol* **174** 889-898
60. Kandler O, Hopf H 1980 Occurrence, metabolism, and function of oligosaccharides In The Biochemistry of Plants, Preiss J (ed) Acad Press, New York; 221-270
61. Kandror O, DeLeon A, Goldberg AL 2002 Trehalose synthesis is induced upon exposure of *Escherichia coli* to cold and is essential for viability at low temperatures *Proc Natl Acad Sci* **99** 9727-9732
62. Kaul SC, Obuchi K, Komatsu Y 1992 Cold shock response of yeast cells induction of 33 kDa protein and protection against freezing injury *Cell Mol Biol* **38** 553-559
63. Labat-Robert, J 1982 Trehalases In Lee, CK & Lindley, MG, eds, Developments in Food Carbohydrate, Vol 3, Disaccharidases, London Applied Science Publishers, 84–106
64. Lemieux RU and Bauer HF, (1954),"A Chemical Synthesis of D-Trehalose" *Can J Chem*, **32** 340-344
65. Leopold AC 1986 Membranes, metabolism, and dry organisms Cornell Univ Press New York; 1-377
66. Leslie SB, Teter SA, Crowe LM, Crowe JH 1994 Trehalose lowers membrane phase transitions in dry yeast cells *Biochimical Biophysics Acta* 11927-13
67. Lillie SH, Pringle JR 1980 Reserve carbohydrate metabolism in *Saccharomyces cerevisiae* responses to nutrient limitation *J Bacteriol* **143** 1384-1394
68. Limaye LS, Kale VP 2001 Cryopreservation of human hematopoietic cells with membrane stabilizers and bioantioxidants as additives in the conventional freezing medium *J Hematother Stem Cell Res* **10** 709-718
69. Lindberg B 1955 Studies on the chemistry of lichens Investigation of a *Dermatocarpon* and some *Roccella* species *Acta Chem Scand* **9** 917-919
70. Liu R, Barkhordarian H, Emadi S, Park CB, Sierks MR 2005 Trehalose differentially inhibits aggregation and neurotoxicity of beta-amyloid 40 and 42 *Neurobiol Dis* **20** 74-81
71. Lopez, Miguel, Tejera, A Noel, Iribarne, Carmen, Lluch, Carmen, Herrera-Cervera, A Jose 2008 Trehalose and trehalase in root nodules of *Medicago truncatula* and

*Phaseolus vulgaris* in response to salt stress

72. Martin MC, Diaz LA, Manzanal MB, and Hardisson C 1986 Role of trehalose in the spores of *Streptomyces FEMS Microbiology Letters* **35** 49-54
73. Matsuda K 1956 Studies on the unfermentable disaccharides in koji extract IV Isolation of a,a- and a,b-trehalose Tohoku *Journal of Agriculture Reserach,* **6** 271-277
74. Matsuo T 2001 Trehalose protects corneal epithelial cells from death by drying Br *J Ophthalmol* **85** 610-612
75. Morana A, Stuso P, Colonna G, Lamberti M, Cratenì M, and De Rosa M 2002 Stabilization of S-adenosyl-L-methionine promoted by trehalose *Biochim Biophys Acta* **1573** 105-108
76. Muller J, Aeschbacher RA, Wingler A, Boller T, Wiemken A 2001 Trehalose and trehalase in *Arabidopsis, Plant Physiology* **125** 1086-1093
77. Neves MJ, Francois J 1992 On the mechanism by which a heat shock induces trehalose accumulation in *Saccharomyces cerevisiae, Biochemistry Journal* **288** 859-864
78. Nicolaus B, Gambacorta A, Basso AL, Riccio R, DeRosa M, and Grant WD 1988 Trehalose in *Archaebacteria* systems, *Applied Microbiology* **10** 215-217
79. Niederer M, Pankow W, and Wiemken A 1989 Trehalose synthesis in mycorrhiza of Norway spruce an indicator of vitality, *European Journal of Forest Pathology* **19 (1)** 14–20
80. Niederst, C & Dauça, M (1985) Activities of renal hydrolases during pre- and postnatal development of mice. *Can J Physiol Pharmacol*, 63, 731–734 (in French)
81. Nwaka S, Holzer H 1998 Molecular biology of trehalose and trehalases in the yeast, *Saccharomyces cerevisia, Prog Nucleic Acid Res Mol Biol* **58** 197-237
82. Odumeru JA, D'Amore T, Russell I, Stewart GG 1993 Alterations in fatty-acid composition and trehalose concentration of Saccharomyces brewing strains in response to heat and ethanol shock *J Ind Microbiol* **11** 113-119
83. Oliver AE, Jamil K, Crowe JH, and Tablin F 2004 Loading human mesenchymal stem cells with trehalose by fluid-phase endocytosis *Cell Preservation Technology* 235-49
84. Panek AD, Mattoon JR 1977 Regulation of energy metabolism in *Saccharomyces cerevisiae* Relationships between catabolite repression, trehalose synthesis, and mitochondrial development *Arch Biochem Biophys* **183** 306-316

85. Paschinger H 1977 DCCD induced sodium uptake by Anacystis nidulans, *Archives of Microbiology* **113** 285-291

86. Pilon-Smits EAH, Terry N, Sears T, Kim H, Zayed A, Hwang S, van Dun K, Voogd E, Verwoerd TC, Krutwagen RWHH and Goddijn OJM 1998 Trehalose-producing transgenic tobacco plants show improved growth performance under drought stress, *Journal of Plant Physiology* **152** 525-532

87. Chen Q and Haddad GG 2004 Role of trehalose phosphate synthase and trehalose during hypoxia from flies to mammals, *Journal of Experimental Biology* **207** 3125-3129

88. Ramón Suárez, Arnoldo Wong, Mario Ramírez, Aarón Barraza, María del Carmen Orozco, Miguel A Cevallos, Miguel Lara, Georgina Hernández, and Gabriel Iturriaga 2008 Improvement of Drought Tolerance and Grain Yield in Common Bean by Over expressing Trehalose-6-Phosphate Synthase in Rhizobia, *The American Phytopathological Society* **21(7)** 958 - 966

89. Reignault PH, Cogan A, Muchembled J, Lounes-Hadj Sahraoui A, Durand R, and Sancholle M 2001 trehalose induces resistance to powdery mildew in wheat, *New Phytologist* **149(3)** 519–529

90. Renaudin J 2006 Sugar metabolism and pathogenicity of *Spiroplasma citri* [Citrus species] *Journal of Plant Pathology* **88(2)** 129-139

91. Riby J, Sunshine S and Kretchmer N 1990 Renal trehalase Function and development *Comp Biochem Physiol* **95A** 95–99

92. Rodríguez-Salazar J, Suárez R, Caballero-Mellado J, and Iturriaga G 2009 Trehalose accumulation in *Azospirillum brasilense* improves drought tolerance and biomass in maize plants, *FEMS Microbiology Letters* **296(1)**52-9

93. Roser B 1991 Trehalose, a new approach to premium dried foods *Trends Food Sci Technol* **2** 166-169

94. Rudolph BR, Chandrasekhar I, Gaber BP, and Nagumo M 1990 Molecular modelling of saccharide-lipid interactions *Chem Phys Lipids* **53** 243-261

95. Sato, A and Aso K 1957 Kojibiose (2-O alpha-D-glucopyranosyl-D-glucose) Isolation and structure-isolation from hydrol, *Nature* **180** 984-985

96. Satpathya GR, Töröka Z, Balia R, Dwyrea DM, Little E, Walker NJ, Tablina F, Crowe JH, and Tsvetkova NM 2004 Loading red blood cells with trehalose a step towards biostabilization, *Cryobiol* **49** 123-136

97. Schiraldi C, Di Lernia I, De and Rosa M 2002 Trehalose production exploiting novel

approaches, *Trends Biotechnol* **20** 420-425

98. Sengupta S, Lahiri S, Banerjee S, Basistha B, and Ghosh AK 2011 Arginine mediated purification of trehalose-6-phosphate synthase (TPS) from *Candida utilis* Its characterization and regulation doi101016/jbbagen 201106025

99. Sengupta S, Chaudhuri P, Lahiri S, Dutta T, Banerjee S, Majhi R, and Ghosh AK, 2010 Possible regulation of trehalose metabolism by methylation in Saccharomyces cerevisiae, *J Cell Physiol* **226** 158–164

100. Shimakata T, Minatagawa Y 2000 Essential role of trehalose in the synthesis and subsequent metabolism of corynomycolic acid in *Corynebacterium matruchotii, Arch Biochem Biophys* **380** 331-338

101. Simola M, Hanninen AL, Stranius SM, and Makarow M 2000 Trehalose is required for conformational repair of heat-denatured proteins in the yeast endoplasmic reticulum but not for maintenance of membrane traffic functions after severe heat stress, *Mol Microbiol* **37** 42-53

102. Singer MA, Lindquist S 1998 Multiple effects of trehalose on protein folding *in vitro* and *in vivo, Mol Cell* **1** 639-648

103. Suzuki T, Boediono A, Takagi M, Saha S, Sumantri C 1996 Fertilization and development of frozen-thawed germinal vesicle bovine oocytes by a one-step dilution method *in vitro, Cryobiol* **33** 515-524

104. Tanaka M, Machida Y, Niu S, Ikeda T, Jana NR, Doi H, Kurosawa M, Nekooki M, Nukina N 2004 Trehalose alleviates polyglutamine-mediated pathology in a mouse model of Huntington disease, *Nat Med* **10** 148-154

105. Thompson, A, Anno, K, Wolfrom, ML, and Inatome, M (1954) Acid reversion products from D-glucose, *J Am Chem Soc,* **76** 1309-1311

106. Trevelyan WE, Harrison JS 1956 Studies on yeast metabolism The trehalose content of baker's yeast during anaerobic fermentation, *Biochem J* **62** 177-182

107. van Elburg RM, Uil JJ, Kokke FT, Mulder AM, van de Broek WG, Mulder CJ, Heymans HS 1995 Repeatability of the sugar-absorption test, using lactulose and mannitol, for measuring intestinal permeability for sugars *J Pediatr Gastroen-terol Nutr* **20** 184-188

108. Vivek Rao, Feng Gao, Bing Chen, William R Jacobs and Michael S Glickman (2006), Trans -cyclopropanation of mycolic acids on trehalose dimycolate suppresses *Mycobacterium tuberculosis* – induced inflammation and virulence, *J Clin Invest;* **116(6)**1660–1667

109. Voit EO 2003 Biochemical and genomic regulation of the trehalose cycle in yeast

review of observations and canonical model analysis *J Theor Biol* **22** 355-78

110. Walsh DM, Klyubin I, Shankar GM, Townsend M, Fadeeva JV, Betts V, Podlisny MB, Cleary JP, Ashe KH, Rowan MJ, Selkoe DJ 2005 The role of cell-derived oligomers of Abeta in Alzheimer's disease and avenues for therapeutic intervention *Biochem Soc Trans* **33** 1087-1090
111. Werner-Washburne M, Braun E, Johnston GC, Singer RA 1993 Stationary phase in the yeast *Saccharomyces cerevisiae Microbiol Rev* **57** 383-401
112. Wiemken A 1990 Trehalose in yeast, stress protectant rather than reserve carbohydrate *Anton van Leeuwen* **58** 209-217
113. Winkler K, Kienle I, Burgert M, Wagner JC, Holzer H 1991 Metabolic regulation of the trehalose content of vegetative yeast, *FEBS Letters* **291** 269-272
114. Womersley C 1981 Biochemical and physiological aspects of anhydrobiosis, *Comp Biochem Physiol* **70B** 669-678
115. Wyatt GR, Kalf GF 1957 The chemistry of insect hemolymph Trehalose and other carbohydrates, *J Gen Physiol* **40** 833-846
116. Y Aysan, M Mirik, A Ala, F Sahin and O Cinar 2003 First report of *Pseudomonas viridiflava* on melon in Turkey, New Disease Reports, **7** 22
117. Yamaguchi R, Andreyev A, Murphy A N, Perkins G A, Ellisman M H and Newmeyer DD 2006 Mitochondria frozen with trehalose retain a number of biological functions and preserve outer membrane integrity, Edited by JC Martinou, *Cell Death and Differentiation*, **14** 616–624
118. Yoshida, K 1993 Serum trehalase activity in patients with rheumatoid arthritis *Clin Chim Acta* **215** 123–124
119. Zan LS, Hu JH, Zhao XL,Li QW, Jiang ZL, Li YK and Li X, 2010, Effects of trehalose supplementation on semen quality and oxidative stress variables in frozen thawed bovine semen, *J Anim Sci,* **88-5**, 1657-1662
120. Zhang XB, Li K, Yau KH, Tsang KS, Fok TF, Li CK, Lee SM, Yuen PM 2003 Trehalose ameliorates the cryopreservation of cord blood in a preclinical system and increases the recovery of CFUs, long-term culture-initiating cells, and no obese diabetic-SCID repopulating cells *Transfusion* **43** 265-272.

*Rev. Plant Pathol. Vol. 6, 2014*
*Indian Society of Mycology and Plant Pathology*
*Scientific Publishers (India), Jodhpur pp. 189-213*

# MOLECULAR UNDERSTANDING OF RED ROT RESISTANCE IN SUGARCANE THROUGH GENOMICS AND PROTEOMICS

***R. Viswanathan*, A. Ramesh Sundar, P. Malathi, V. Ganesh Kumar, M. Sathyabhama, M. Muthumeena and P. R. Rahul.***

Sugarcane Breeding Institute, ICAR, Coimbatore 641007, India
*E-mail: rasaviswanathan@yahoo.co.in

**KEYWORDS:** Sugarcane, red rot, *Colletotrichum falcatum*, resistance, phytoalexins, differential display, RACE-PCR, subtractive libraries, proteomics

## Abstract

Red rot of sugarcane caused by *Colletotrichum falcatum* Went is one of the devastating diseases of sugarcane causing significant loss to sugarcane production in the country. Complex polyploidy and lack of information on inheritance to red rot in sugarcane make breeding for red rot resistance more difficult. Studies were initiated at this institute on identifying red rot resistance mechanism in sugarcane. Studies conducted during the previous decades indicated possible role of oxidative enzymes, pathogenesis-related (PR) proteins and 3-deoxyanthocyanidin phytoalexins in governing red rot resistance. Recent studies involving semi-quantitative RT-PCR post pathogen challenge from sugarcane cultivars varying in red rot resistance, revealed differential accumulation of transcripts of the flavanoid biosynthetic pathway like coumarate-4-hydroxylase, chalcone synthase, chalcone reductase, flavanoid 3`-5` hydroxylase and flavanoid glycosyl transferase and this transcript analysis, further confirmed the role of sugarcane phytoalexins in red rot resistance. Similarly, the role of PR- proteins like chitinase and β-1,3-glucanase was established at the transcript level. Detailed molecular analysis applying differential display (DD)-RT-PCR identified expression of more number of differentially expressed transcripts during the host-pathogen interaction. Full length sequences of many potential transcripts were identified and are being functionally annotated. Also to identify specific proteins involved in red rot resistance, proteomics approach has been attempted by optimizing sample preparation from stalk tissues, 2-D electrophoresis (2-DE), down stream processing of identified spots and bioinformatics. Several resistance-associated proteins spots were identified and they are being analyzed critically for their functional relevance.

## INTRODUCTION

Sugarcane (*Saccharum* spp. hybrids) is an important commercial crop grown in the Indian sub-continent and other countries. Sugarcane meets ~60% of the global requirement of sweeteners. India is the second largest producer of sugarcane in the world and the crop is cultivated in almost all the states occupying an area of 4.2 million ha with an annual production of ~300 million tonnes of canes. Sugarcane production in various regions of the country is affected by different biotic and abiotic stresses. Among the biotic constraints, red rot - a fungal disease caused by *Colletotrichum falcatum* Went (Perfect state: *Glomerella tucumanensis* (Speg) Arx Muller) is the major constraint affecting cane production in most of the sugarcane growing areas in Indian sub-continent, Thailand, Australia, Brazil, Fiji, USA etc. The disease occurs in 77 countries representing all the sugarcane growing continents worldwide(18,42). At present, the disease is prevalent in most of the sugarcane growing states in the country at varying intensities. The disease was first recorded in Java during 1893 and in India during 1901. Series of disease epidemics in the sub-tropical India had devastated cane cultivation in the entire Gangetic plains and Punjab till 1960s. Later the disease spread to parts of tropical India and caused severe damage to the crop resulting in the elimination of many commercial varieties, through disease epidemics which occurred in different regions (34,42). Severe epiphytotics on cv CoC 671- the wonder cane of tropical region, during 1980s and 1990s have caused extensive damage to the cane industry in Peninsular India. Likewise many important commercial varieties BO 3, BO 10, BO 11, BO 17, BO 54; Co 312; Co 419, Co 453, Co 658, Co 785, Co 997, Co 1148, Co 1158, Co 6304, Co 7805, CoC 671, CoC 92061, CoJ 64, CoS 8436 etc were lost due to the disease (42). The repeated red rot epidemics brought heavy damage to the growers and millers in India.

In the early decades of the last century, breeding programmes in India were aimed at identifying varieties for subtropical India, to replace poor yielding Indian sugarcane *Saccharum barberi*. The breeding and selection process gave emphasis to adaptability, yield and quality improvement and resistance to red rot utilizing *S. officinarum*, *S. barberi* and the wild species *S. spontaneum*. The result was the release of outstanding varieties from SBI, which were high yielding and tolerant to the major disease - red rot. But recurrent outbreaks of red rot in epidemic forms resulted in the replacement of varieties which succumbed to this disease. Changing fungal race flora is considered as another major factor responsible for the breakdown of host resistance resulting in shorter life span of new varieties (25). In the absence of clear-cut information on inheritance of red rot resistance, exact mechanism governing red rot resistance in sugarcane remains yet to be understood clearly. However, two kinds of resistance *viz.*, the structural or mechanical that is static and the physiological, which is dynamic, forms part of the plant defence against the pathogen. Only limited information is available on this area of work in sugarcane but for the work done at Sugarcane Breeding Institute (SBI), Coimbatore. This chapter summarizes red rot resistance with more emphasis on the recent findings on physiological resistance.

## BIOCHEMICAL RESISTANCE

Different biochemical parameters that impart resistance in a particular genotype have been reported. Red rot resistant varieties in response to *C. falcatum* infection develop gum deposits that block the spread of the advancing hyphae by filling up the intercellular spaces. This process takes place in advance of infection and seals off further spread of the pathogen in adjoining tissues (20). Gum formation may also take place in susceptible varieties, but to a lesser extent and usually after the tissue has been invaded which was termed as 'hypersensitive gummy reaction'. Later studies have reported on the role of phenolic compounds in red rot resistance. Higher quantities of phenols have been reported in resistant and moderately resistant varieties (16,44). However, further studies revealed that there was no correlation between total phenolic content and degree of resistance to red rot (3,17). Level of total phenols increased in resistant varieties after infection and it maintained, while in susceptible varieties, the level of phenolic content dropped after an initial increase. Number of host enzymes like peroxidase (POX), polyphenol oxidase (PPO), phenylalanine ammonia lyase (PAL) and tyrosine ammonia-lyase (TAL) were reported to be associated with resistance in many sugarcane varieties (5,13,14,21,22)

## PHYTOALEXINS

The reddish compound released in cells and intercellular spaces near the invading hyphae in the infected cane stalks is referred to as red rot pigment (RRP). The RRPs formed in response to *C. falcatum* infection in sugarcane inhibited conidial germination *in vitro* and slowed down the mycelial growth (3,29). Fractionation of the RRPs showed presence of several compounds and some of them are phenolics. Red rot pigments (RRPs) have been detected between 24 and 48h in the infected susceptible host (3) and it was found that conidial germination *in vitro* takes place much before the pigment accumulation. Hence, the pigment may not be effective in stopping the initial germination. The RRPs from resistant cane tissues showed seven compounds, while in susceptible cultivar CoC 671 has shown only four of them in thin layer chromatography (29). The loss of three pigment fractions in the susceptible cultivar after pathogen infection indicated that the invading pathogen in the stalks might have suppressed these resistance-contributing fractions.

In the chromatographic analysis, qualitative differences were not found between the resistant and susceptible varieties, before or after infection (3). Studies conducted at Sugarcane Breeding Institute confirmed the differential patterns of pigment accumulation between resistant and susceptible genotypes in response to pathogen infection. It was found that in resistant variety pigment localized at the site of inoculation at a concentration of 1200 μg/g tissue, with little spread of pigment into the surrounding tissue, whereas susceptible showed a diffuse pigmentation throughout the pith with a lower concentration of 600 μg/g tissue. Studies conducted at the SBI also very clearly indicated that pigment accumulation concentrated in higher quantities around the site of inoculation in resistant varieties, whereas in the susceptible tissues the concentration was too low and in a diffused pattern. These studies are the firm evidences to show that a resistant mechanism operating in sugarcane against *C. falcatum*. Luteolinidin as the major component in RRP was identified and it appeared 24 to 48 hours after infection in cane tissues. Other components identified were luteolin, chlorogenic acid and a glucoside of luteolinidin. Fractionation of the pigments by high performance liquid chromatography (HPLC) revealed the presence of 3-deoxyanthocyanidin compounds, which were identified as luteolinidin, apigeninidin and caffeic acid ester of 5-0-apigeninidin (27,28). In resistant interaction, these compounds present at much higher concentration, whereas in susceptible or compatible reaction they were completely absent or present in very low concentration. Among the three compounds, luteolinidin level was the highest and present in all the varieties tested. Apigeninidin was also present in all the varieties but in low concentration.

Further studies revealed that susceptible variety accumulated reasonable amount of these phytoalexins, when toxin was used for inoculation, but in response to pathogen inoculation the same variety failed to synthesize phytoalexins probably due to their degradation by the pathogen. However, here also resistant variety recorded multifold phytoalexins synthesis than in the susceptible variety (28). The resistant variety may resist pathogen mediated phytoalexin degradation, thereby arresting the pathogen growth in the canes after pathogen inoculation. This was confirmed by inoculation of varieties varying in red rot resistance. Here induction of phytoalexins was recorded in both varieties. These findings clearly demonstrated that the phytoalexins were accumulated only in incompatible host pathogen interactions and compatible interactions had no such phytoalexins or with trace quantities. These three compounds were identified as phytoalexins in sorghum and *C. sublineolum* interaction, a host-pathogen interaction close to sugarcane and *C. falcatum* (19).

Subsequent studies with a set of host differentials proved that there is accumulation of anthocyanin compounds in incompatible interactions and no such induction/accumulation in compatible interactions (30,41). It has been conclusively proved that 3-deoxyanthocyanidin compounds act as sugarcane phytoalexins (6) and probably govern resistance in sugarcane to red rot. HPLC analysis with phytoalexin standards revealed differential accumulation of luteolinidin and apigeninidin phytoalexins in the cv Co 93009 resistant to red rot as compared to traces of compounds in the susceptible cv CoC 671.

The classical incompatible interaction of pathotypes from Co 1148 and Co 7717 on the host cultivars is very much demonstrated earlier at SBI, Coimbatore (7). This was further elaborated by quantifying phytoalexins in such interactions. The variety Co 1148 accumulates 3-deoxyanthocyanidins in incompatible interaction with Cf7717, which results in resistant reaction. However, the same variety fails to synthesize phytoalexins in case of matching pathotype Cf1148 infection. Similar behaviour of Co 7717 to Cf1148 and Cf7717 pathotype interactions was proved. Information on phytoalexin induction in susceptible varieties by phytotoxin also amply demonstrates the role of phytoalexin in red rot resistance in sugarcane. Though previous studies have been unsuccessful in pinpointing biochemical basis of disease resistance, these studies at SBI have shown the possible role of phytoalexins in red rot resistance.

## PATHOGENESIS RELATED (PR) PROTEINS

Accumulation of PR-proteins is an important phenomenon of plant defence responses upon infection by pathogens. Many of the PR-proteins have been purified and the purified proteins exhibited strong antifungal activity against many fungal pathogens under *in vitro* conditions. Induction of high levels of these proteins in host plants reduced disease development in many crop plants. Specific involvement of many proteins in disease resistance has resulted in isolation and cloning of these PR-proteins and subsequently, developing transgenic plants expressing PR1a, PR-2 (β-1,3-glucanase), PR3 (chitinases) and PR5 (thaumatin-like proteins) in various crop plants (23). Such transgenic plants expressing foreign genes showed enhanced resistance to fungal pathogens. Detailed studies on PR-proteins in red rot resistance showed that constitutive activities of chitinase and β-1,3-glucanase were higher in disease resistant varieties as compared to susceptible varieties. Upon pathogen inoculation, the resistant variety accumulated higher hydrolytic enzymes as compared to susceptible cultivars (34). This information suggests a possible role of these enzymes in red rot resistance.

Further studies were conducted to identify the PR-proteins using Western blot technique in a set of resistant and susceptible varieties differing in resistance to the disease (26). The red rot resistant cv Co 93009 showed differential induction of four chitinase proteins with molecular mass ranging from 34-39 kDa after pathogen inoculation in leaf tissues. The intensity of these proteins increased with time from 6 to 42 h after inoculation (Fig. 3). In susceptible variety CoC 671, induction of a 35-kDa chitinase protein was recorded. In stalk tissues, induction of a 35-kDa chitinase protein was recorded 24 h after inoculation in resistant variety whereas, in susceptible plants such induction was delayed. Similarly, early induction of 43 and 37.5kDa TLPs by 24 h was observed in response to pathogen inoculation in the resistant variety, whereas in the susceptible variety such induction was less intense and could be seen 9 days post inoculation. These studies gave a clear indication that the PR-proteins may form part of the important defense mechanisms operating in sugarcane against red rot as in other host-pathogen interactions.

## MECHANISM OF INDUCED SYSTEMIC RESISTANCE IN SUGARCANE

Earlier studies have clearly established the induction of systemic resistance in sugarcane against red rot by fluorescent pseudomonads (37,40). These studies gave evidence of enhanced resistance in disease susceptible cultivars like CoC 671, CoC 90063, CoC 92061 etc under controlled and field conditions. Studies were conducted in detail on the mechanism of induced resistance mediated by strains of *Pseudomonas* against the disease. Involvement of different PR-proteins such as β-1,3-glucanases, chitinases and thaumatin-like proteins (TLPs) was found to be associated with *Pseudomonas*-mediated induced resistance (31,35) Chitinases purified from systemically protected stalk tissues of sugarcane against *C. falcatum* showed strong anti-fungal activities (31). It has also been clearly demonstrated that disease susceptible sugarcane plants following treatment with *Pseudomonas* sp. is able to restrict disease development to a level equivalent to moderately resistant varieties and many PR-proteins are involved in that ISR (Fig. 4). In addition, enzymes of phenyl-propanoid pathway and oxidative pathway were also found to be involved in ISR (38). Characterization of *Pseudomonas* strains revealed that production of different metabolites/antibiotics such as salicylic acid, auxins, siderophores, pyocyanine, pyoluteorin and 2,4-diacetyl phloroglucinol and hydrolytic enzyme chitinase contribute to suppression of *C. falcatum*, induced resistance and growth promotion in sugarcane (36,39,43).

## MOLECULAR BASIS OF RED ROT RESISTANCE IN SUGARCANE

Plants under are constant threat to infection by pathogens armed with a diverge array of effector molecules to colonize their host. Plants have in turn evolved a sophisticated detection and response systems that decipher pathogen signals and induce appropriate defence. Recent genetic analysis involving plant mutants defective in resistance response to the invading pathogens has revealed a number of distinct, but interconnecting, signalling networks that are under both positive and negative regulatory control. These pathways operate atleast partly through the action of small signalling molecules such as salicylate, jasmonate and ethylene. The interplay of signals probably allows the plant to fine-tune defense responses in both local and systemic tissue. AT SBI, we have taken up both genomic and proteomic studies to understand the host-pathogen interaction of sugarcane-*C. falcatum* at molecular level. In genomics, we have employed reverse transcriptase - polymerase chain reaction (RT-PCR), differential display RT-PCR, reverse northern and rapid amplification of cDNA ends (RACE) to identify and characterize the resistance/defense genes involved in red rot resistance.

### Molecular interactions of sugarcane-*C. falcatum*

Plants are able to recognize fungal pathogens by their secreted products, referred to as 'elicitors'. These elicitor molecules are signal molecules and elicit synthesis of phytoalexins, phenolics, lignin, PR-proteins, hydroxyl-proline rich glycoproteins and other defense-related metabolites in host plant and thus mediates pathogen recognition. Defence responses are induced upon perception of either specific or non-specific elicitors (1). Specific elicitors are products of pathogen avirulence genes and are hypothesized to be recognized by the products of corresponding pathogen race-specific resistant genes. Detailed studies were carried out to purify and characterize elicitors from *C. falcatum* and to study their role in recognition of *C. falcatum* by sugarcane.

A high molecular weight elicitor was isolated from the mycelial walls of *C. falcatum* and partially purified by gel filtration. The elicitor is characterized as a glycoprotein and the activity of elicitor resides in the carbohydrate moiety. The partially purified elicitor induced the accumulation of phenolics and the activities of phenylalanine ammonia-lyase (PAL) and peroxidase (POX) in both sugarcane leaves and suspension-cultured cells. Sugarcane cells in culture responded to *C. falcatum* elicitor in a manner similar to sugarcane leaves (9). Similarly, induction of hydrogen peroxide ($H_2O_2$), reactive oxygen species ($O_2^-$), lipoxygenase, lipid peroxidation, super oxide dismutase (SOD) and catalase was also observed in cell suspension cultures of sugarcane primed with the fungal elicitor. A rapid outburst and a spurt in the generation of active oxygen species especially $H_2O_2$ was observed indicating an early molecular event in recognition of the pathogen by the host cells. However, higher levels of the suppressor enzymes viz., catalase and superoxide dismutase were found to be maintained throughout in the cells of sugarcane suspension cultures without any elicitor treatment (10). When elicitor isolated from *C. falcatum* was compared with *C. lindemuthianum* - a non-pathogen elicitor, differential induction of POX isoforms in suspension-cultured cells of sugarcane cv. CoC 671 was found (11).

For studying the defense gene activation in sugarcane, the intact plant-pathogen system may not be ideal because the time course of pathogen infection along with the concomitant accumulation of host defense could not be monitored precisely. Cell cultures and elicitors instead of whole plants and live pathogens respectively are found to be suitable models to study the defense gene activation in bean, rice and in sugarcane because of their high degree of reproducibility and rapid experimental cycles. Further, the homogenous cell suspension is uniformly exposed to the elicitor preparation and hence the response of cells is relatively uniform. Preliminary results of the study conducted at SBI has provided insight into the mechanisms regulating the pathogen recognition at the interface which facilitates further elaboration of inducible defense response against *C. falcatum* in sugarcane (11,12). These studies confirmed that the elicitor molecules from *C. falcatum* are responsible for specific recognition of the pathogen by the host and resistance is determined by the rapidity of the downward signalling of defence pathway. Probably variation in initiation of signalling process between the resistant and susceptible genotype determines the pathogen colonization and disease development *vis a vis* host resistance.

## GENOMIC STUDIES IN DEFENCE/RESISTANCE GENE EXPRESSION

To understand genome complexity of sugarcane, a large scale expressed sequence tag (EST) programme known as 'SUCEST' was taken up recently in Brazil. More than 2,60,000 cDNA (complimentary DNA) clones were partially sequenced from 26 standard cDNA Libraries generated from different sugarcane tissues. They annotated 43,141 assembled sequences and found 50% of the putative identified sugarcane genes coding for protein metabolism, cellular communication/signal transduction, bioenergetics and stress responses. Eighty SUCEST sugarcane assembled sequences (SASs) encoded proteins (24) with clear similarity to the nucleotide-binding adaptor shared by APAF-1, R proteins, and CED-4 (NB-ARC) domain was found which is characteristic of one of the major classes of disease resistance genes (R genes). The database contained more than 200 SASs encoding WRKY transcription factor domains, which have been implicated in the defense gene regulation in plants. They also found other genes related to defense responses like chitinases, β-1,3-glucanases, chalcone syntheses, chalcone isomerases, isoflavone reductases, hydroxylproline rich glycoproteins, proline-rich proteins, catalases, superoxide dismutases etc. have been putative orthologs in sugarcane, which indicates a high conservation of defense strategies among plants. The wealth of information generated in the SUCEST database promises exciting prospects for the scientists involved in sugarcane improvement and other crops. Many transcripts including the disease resistance and defense responses-related ESTs will be a basic resource for the understanding of the biology of this complex polyploidy plant. This information may facilitate in identifying gene(s) involved in disease resistance in sugarcane.

We have carried out detailed studies on identifying specific transcripts induced during host-pathogen interaction to identify candidate genes involved in red rot resistance. After pathogen inoculation, total RNA was isolated from sugarcane stalk tissues and reverse transcribed to cDNA for expression analysis. The level of expression and the interval at which a particular gene expression etc, were analyzed using the custom made primers. The following transcripts R30, chitinase, metallothionein, receptor protein kinase (RPK) and reversibly glycosylated protein (RGP) were differentially expressed in resistant and susceptible varieties and these clones shared sequence similarity with disease resistance genes in other crops (33). The results also showed an early induction of defense/resistant gene(s) in the resistant variety and in case of susceptible variety, the induction was delayed significantly (Fig. 5). In further studies, they employed differential display (DD)-RT-PCR a powerful technique to identify more number of transcripts involved in pathogen recognition, signal transduction and plant defense in sugarcane. Differential display technique revealed ~450 transcripts to be differentially expressed upon pathogen inoculation in sugarcane. About 202 transcripts were selected for their down-regulation and 243 transcripts were selected for their up-regulation upon pathogen inoculation. After homology search, the expressed sequence tags (ESTs) with a match in the databases (both characterized and uncharacterized) were categorized into eight groups primarily based on putative function. The total up-regulated and down-regulated transcripts were 63% and 37% respectively, indicating a higher percentage of transcripts that seems to be induced in response to pathogen inoculation. Among the known proteins, the signal transduction group represented the highest percentage of up-regulated sequences, followed by protein synthesis and storage, general metabolism, transport, defense, cell structure/growth/division, transcription/post transcription and bioenergetics. The important transcripts identified include 14-3-3-like protein, Senescence-associated protein DH, xylanase inhibitor protein 1 precursor, Putative chitinase, Leucine Rich Repeat family protein, F-box domain containing protein, UMP/CMP kinase-a and Putative hydroxyproline-rich glycoprotein. Similarly, DD-RT-PCR was conducted with sugarcane cell lines treated with an elicitor molecule identified from red rot pathogen and similar transcripts were found to be differentially regulated upon elicitor treatment. Further northern and reverse northern blot analyses using radioactive α [$^{32}$P] dCTP confirmed differential expression of potential DD transcripts. This approach would be more useful to identify the potential transcripts during molecular interaction between the host and the pathogen.

## SUBTRACTIVE LIBRARIES

DD-RT-PCR studies clearly established differential expression of defence related transcripts in cane tissues as well as suspension cells in response to red rot fungus infection and Cf-elicitor treatment, respectively. Further studies were carried out to construct cDNA libraries in the resistant cv Co 93009, 12h and 72h after pathogen infection. In suppression subtractive hybridization (SSH) only the differentially expressed transcripts in two populations of comparison are subtracted, amplified and cloned. In the study, RNA population from the resistant cv Co 93009 was used as tester and the susceptible cv CoC 671, was used as the driver. Also injured and uninjured controls were maintained. A total of 700 subtracted transcripts encoding different functional categories were obtained and they gave new information concerning disease resistance in sugarcane. The early (12 hr) and late (72 hr) response transcripts were grouped into several functional categories as signal transduction, defense, energy metabolism, carbohydrate metabolism, cell wall strengthening transcripts and transcription factors. The early response library transcripts identified include the following: lipoxygenase, involved in signal transduction, peroxidise a defence response transcript and structural motif transcripts helix loop helix protein and transducin WD-40 repeat protein. Further, a group of cytokine members like thylakoid formation protein and actin related proteins were also induced. Apart from this, acyl carrier protein involved in fatty acid metabolism was also induced in the resistant cultivar. The late response transcripts include the following: signal transduction transcripts of ACC oxidase and ACC synthase, involved in ethylene biosynthesis pathway and serine/threonine kinase, which signals defense response in plants. The well-documented defense responsive, antifungal genes like chitinase and ß-glucanase were found to be expressed in the resistant cultivar. Adenosine kinase involved in viral defense was also expressed in the resistant cultivar. Transcripts associated with energy metabolism identified include mitochondrial alternative oxidase, uncoupling protein and mitochondrial alternative oxidase. The differentially expressed transcripts involved in carbohydrate metabolism were sucrose transporter, xylosyl transferase and 6- phospho fructo kinase. This is a new information on sugar metabolism and disease resistance in sugarcane. Cell wall strengthening gene transcripts included pectin methyl transferase precursor, xyloglucan endo-trans glycosylase and cellulose synthase. In the study, transcripts of transcription factors like leucine zipper, MYB, ethylene responsive and LIM involved in lignin biosynthesis were also identified as differentially expressed. It was noted that the transcripts involved in protein structure folding and fatty acid metabolism were induced in the early response where as, transcripts associated with sugar metabolism and more of signal transduction were observed in the late response. We have generated detailed information on differential expression of transcripts responsive to red rot infection in sugarcane which are new to sugarcane and further studies are in progress to cluster the transcripts with their functional groups and to validate their functions. Eighty five clusters of expressed gene tags (ESTs) that preferentially express upon *C. falcatum* infection have been identified, which were previously unreported (4). By real time RT-PCR profiling of selected EST clusters, they identified several sugarcane clusters that show differential expression in response to biotic and abiotic stress conditions. In addition to six resistant gene analogues, (RGAs), a full-length R-gene (TIR-NBS-LRR) termed *SNLR* have been isolated (45) from NCo 376, a smut resistant variety in mainland China and its expression profile have been charcterized by real-time RT-PCR,in response to treatments with *Sporisorium scitaminea* causing smut, salicylic acid and $H_2O_2$. Further studies in this area of work will be more rewarding and would lead to identifying several candidate genes involved in disease resistance in sugarcane.

## PHYTOALEXIN PATHWAY TRANSCRIPTS

Since we have established the role of 3-deoxyanthocyanidin phytoalexins in red rot resistance, the genes involved in the flavanoid biosynthetic pathway like coumarate-4-hydroxylase (C4H), chalcone synthase (CHS), chalcone reductase (CHR), flavanoid 3`-5` hydroxylase (F3'5'H) and flavanoid glycosyl transferase (FGT) were further studied to establish relation between transcript accumulation and differential accumulation of phytoalexins in the resistant and susceptible cultivars. The pathogen inoculation enhanced their expression of C4H especially in resistant variety, which may have positive correlation with their ability to synthesize various downstream compounds like phenolics and flavonoids as reported earlier from various biochemical studies. Further along the phenypropanoid pathway after the conversion of 4-coumaryl CoA to flavone, the flavone and isoflavonoid pathway is initiated. The CHS and CHR are key enzymes of phenyl-propanoid pathway diverting the substrate, naringenin chalcone to the flavanoid and isoflavanoid branches of the phenyl-propanoid pathway that synthesizes the precursor of a large number of secondary metabolites, including pro-anthocyanidins, anthocyanins, flavones, flavonols and isoflavanoid-phytoalexins among others (46). Maximum induction of CHS was found in resistant variety, whereas in susceptible variety this induction was minimal. The F3'5'H transcript was found to be induced upon pathogen inoculation in the leaves of both the varieties and the flavonoid 3'5'-hydroxylase enzyme functions at an important branch point between flavonol and anthocyanin synthesis, as is evident from studies in petunia (*Petunia hybrida*), and potato (*Solanum tuberosum*). Induction of F3'5'H leads to the synthesis of flavanoids, flavanols, flavanones and phenolics. Reports of plant molecular responses to elicitor or pathogenic infections have pinpointed increase in activity of several genes of the phenyl-propanoid pathway leading to the synthesis of phenyl-propanoid metabolites, lignin and flavanoids. The accumulation of phytoalexins in sorghum leaves and mesocotyl tissue after infection with *Colletotrichum sublineolum* (causing anthracnose) was established (2). After penetration of the fungus into the host cell, it was restricted within infected regions by 72 h and an intense reddish pigmentation was observed by 36 h in the resistant cultivar, whereas in the susceptible variety the phytoalexin appears only by 48 h with proliferation of the pathogen. The HPLC profile of the phytoalexin from the resistant variety showed presence of luteolinidin, 5-methoxyluteolinidin, apigeninidin and the caffeic acid ester (CAE) of arabinosyl 5-O-apigeninidin, while susceptible variety showed presence of only two fractions apigeninidin and CAE of arabinosyl 5-*O*-apigeninidin. Their temporal studies revealed the accumulation of CHS and PR-10 to occur by the 24 h in the resistant variety, while the same occurred only by the 36 h in the susceptible cultivar. Also the level of PR-10 transcript accumulation was lower in the susceptible cultivar than that in the resistant cultivar. Timely expression of the specific transcript at the right place in sufficient quantity is a pre-requisite in the development of resistance in a host system. Our studies have clearly shown the differential pattern of spatial accumulation of transcripts of phenylpropanoid pathway in resistant and susceptible varieties after inoculation of *C. falcatum* in sugarcane.

## RAPID AMPLIFICATION OF CDNA ENDS (RACE)

The potential defense related transcripts were selected for RACE analysis to clone the full length cDNAs. Compatible primers were designed at the 3' side using the available partial sequence information and used in the RACE protocol. Using RNA ligase mediated - RACE, we have isolated full length sequences of the following four genes *viz.* 14-3-3 like protein, chitinase, xylanase inhibitor and basal antifungal peptide. After full length isolation of 14-3-3 like protein by RACE-PCR, it has been characterized using bioinformatics tools. The CDD search in the NCBI database revealed the presence of conserved 14-3-3 superfamily domain, 5'UTR (1-86bp), ORF (87-857 bp) and 3'UTR (858-1094 bp) in the full length sequence of 1094 bp. The 14-3-3 homologues are known to mediate signal transduction by binding to phosphoserine-containing proteins. They are involved in growth factor signalling and also interact with MEK kinases.

## CHARACTERIZATION OF SUGARCANE CHITINASE

Differential expression of sugarcane chitinase in stalk tissues during pathogenesis of *C. falcatum* was established by RT-PCR studies in a set of resistant and susceptible cultivars and the result was confirmed by reverse northern analysis. In further studies full length of the gene was isolated and bioinformatic analysis was done to identify its functional domains and to predict the three dimensional structure of full length chitinase sequence. Translated sequences revealed the typical characteristics of family 19 glycosyl hydrolase, class I/IV chitinase starting with a signal peptide and ending with a signature domain. Phylogenic study grouped sugarcane chitinase in class IV, based on major deletions in catalytic domain. The close structural template 2DKV (Rice class-I chitinase) was successfully used for the prediction of sugarcane chitinase 3D model. The chitinase gene was cloned in an expression vector and studies are in progress to purify the expressed protein and to assess its biological activities.

### Transcriptomic analysis of SAR and *C. falcatum* – responsive transcription factors in sugarcane

Two types of stalk tissue samples viz. systemic acquired resistance (SAR) – responsive tissues of disease susceptible (cv CoC 671) and resistant (Co 93009) and susceptible (cv CoC 671) were selected to perform transcriptomic analysis of SAR mediated resistance and pathogen responsive transcription factors (TFs) involved in sugarcane. Among the 41 TFs screened for differential regulation, 15 of them were found to be up-regulated upon SAR inducer priming and 24 TFs were up-regulated upon pathogen challenge. Upon comparison of SAR inducer priming and incompatible interaction, 8 TFs were found to be differentially up-regulated in both the cases. Genes representing WRKY family, MYB class of TFs, TLP, NAC and bZIP TFs were screened for their differential response against SAR-priming. Upon SAR inducer treatments, WRKY 40 and 44 were induced at earlier time interval (6 hpi) in BTH and *C. falcatum* elicitor treated samples and also in pathogen challenged samples, but the level of induction was high at a later interval (72 hpi) due to BTH priming. Up-regulation of MYB 78 and 83 were observed in response to BTH-priming. Upon comparison of Co 93009 (incompatible) and CoC 671 (compatible) response to pathogen challenge, most of TLP genes (TLP K and TLP L) were induced at earlier intervals between 0 and 24 hpi in Co 93009; however in CoC 671, these genes were induced only at later intervals between 24 and 48 hpi,, indicating that these TFs might be involved in activating defense early against pathogen attack in in-compatible interaction and SAR response. Similarly NAC C, D and E were induced at earlier intervals between 6 and 12 hpi in BTH primed tissues, but there was no such induction observed in *Cf* elicitor treated samples, thus indicating that TF regulated gene expression involves multiple signals forming a network of gene activation cascade. Variation in the expression pattern of individual bZIPs at different time intervals was observed. In inoculated control, there was an inconsistent regulation of these genes, induced between 6 and 72 hpi. The overall result indicated the possible role of few of these TFs in regulating SAR induced defense in sugarcane.

**Molecular basis of systemic acquired resistance (SAR) mediated resistance**

A consistent control of red rot was observed from the treatment of BTH, SA and Cf-elicitor in two different cvs, CoC 671 and CoC 92061 of sugarcane which are highly susceptible to the disease. This study not only explored the potential of SAR-elicitors but also the modes of pre-treatment such as sett treatment, foliar spraying and drenching in marcotted plants and timing of their application in sugarcane. Nevertheless marcotting was found to be effective; it is impractical to apply this method under field conditions. BTH (250 µM), SA (250 µM) and Cf-elicitor (60 µg glc equivalent) have proven to be effective inducers of systemic resistance in sugarcane against red rot. The results of the *in vitro* bioassay and antifungal assays revealed pathogen growth inhibition by metabolites/hydrolytic enzymes from the primed tissues indicating a possible involvement of SAR induced defence-related enzymes in restricting the pathogen growth (15). Possibly antifungal assay of marcotted stalk tissues was the first study to show the direct antifungal effect of primed tissues under *in vitro* conditions against the target pathogen.

The mechanism of induction of systemic resistance conferred by these SAR inducers (BTH, SA and Cf-elicitor) was studied through biochemical studies of defense related metabolites/ enzymes as well as isoenzyme and Western blot analyses of PR-proteins. It was observed that the induced resistance against red rot was correlated with increased activities of peroxidise, polyphenol oxidase, phenylalanine ammonia-lyase, tyrosine ammonia-lyase, super oxide dismutase, lipoxigenase, $H_2O_2$ generation, accumulation of phenolics and coumarins. Many of the PR proteins viz., PR1-C, PR-2, PR-3, PR-5 were also found to be induced in SAR primed sugarcane stalk and leaf tissues. Strong correlation was observed between increased activities of PR proteins in the leaf tissues and reduction of disease incidence. Microtitre plate assay gave further evidence for the induction of systemic resistance from the treatment with chemical activators in sugarcane.

BTH, SA and Cf-elicitor induced changes in the expression of key enzymes involved in the phenylpropanoid pathway was analyzed by reverse transcription (RT)-PCR approach. The results evidenced that induced resistance in sugarcane is largely correlated with the phenylpropanoid pathway enzymes which lead to the synthesis of coumarins, lignins, cell wall bound phenolics, flavanoids, flavanols, flavanones, isoflavanones and isoflavanols. RT-PCR work with the genomic scanning of the induced transcripts for R-genes and other defense-related genes revealed that SAR mediated resistance in sugarcane is largely correlated with R-genes, DIR, BAF and defense related enzymes which are involved in plant defense mechanism and pathogen degradation processes. Future work underlies on elucidating specific signalling pathways and identification of key molecules that could be modulated for inducing systemic resistance in sugarcane. Such approaches hold promise in future to achieve a tangible red rot management strategy in sugarcane and to understand disease resistance mechanism in sugarcane.

## DEFENCE PROTEOMICS STUDIES IN SUGARCANE

The term proteome refers to the complete set of proteins that are specified by the genome, and analogous to genomics, proteomics describes the study and characterization of this complete set of proteins present in a cell, organ or organism at a given time. Genome-level studies (genomics) reveal or suggest what could theoretically happen, whereas the proteome-level investigation (proteomics) provides insights into the actual players involved in mediating specific cellular processes. In addition, the study of proteins introduces another level of complexity at the level of the post-translational modification (PTM) and the biological relevance of such modifications. These changes in PTM during the growth and development of organisms (including plants) or in response to stress (including disease) cannot be deduced from studies investigating genome sequences and/or transcript abundance. Such changes can only be deciphered through proteomics and it is a powerful tool in understanding which proteins are present in particular tissue under given conditions and it is one of the fastest growing areas of biological research. In addition to the enzymes, transport and regulatory proteins, many proteins contribute in cascades of reactions leading to the metabolites involved in disease resistance. This makes the proteome an essential target for studying metabolic pathways.

Information on the clearly identified proteins during host-pathogen interactions and defense signaling in sugarcane is lacking to understand disease resistance mechanism(s). Hence we have standardized a protocol for extraction of proteins for 2-DE from the rigid, fibrous sugarcane stalk tissues to identify specific proteins involved in resistance to red rot in sugarcane tissues for the proteome analysis. Among the five methods tested, 2D cleanup-phenol method was found to be the most suitable for producing high number of good quality spots and reproducibility. Thirty protein spots commonly present in three methods were selected and subjected to *eLD*-IT-TOF-MS/MS analysis and a reference map has been established for sugarcane stalk tissue proteome. This is the first study on sugarcane stalk proteome analysis which possibly will show a new light on unexplored areas of sugarcane proteome research (8).

After establishing stalk specific proteome in sugarcane, detailed studies were conducted on SAR proteomics using BTH as an abiotic elicitor in susceptible cv CoC 671 and defence proteomics in resistant cv Co 93009. The SAR induced proteins were identified by 2-DGE, wherein four protein extraction methods were employed to identify the method specific protein induction in SAR primed tissues. After 2-DGE, the gels were silver stained and differentially expressed spots were picked, digested with trypsin and subjected to MS/MS analysis. The results of peptide mass fingerprinting was analyzed using MASCOT protein identification software. Comparative 2-D gels analysis identified 19 highly confident differential protein spots in BTH treatment over control. The differentially expressed spots were identified as COMT, profilin 4, Cu/Zn SOD, UDP glucose

DHase, isocitrate lyase, GAPDH, Os05g0501300 protein, IN2-1 protein, Nucleoside diphosphate kinase 1, unknown protein, PAL, Quinone reductase 2, Glutathione transferase 9 etc. Detailed defence proteomics in sugarcane revealed that the number of protein spots was found to be higher in the resistant cultivar cv Co 93009 after 12 h of pathogen challenge, whereas the inoculated susceptible cv CoC 671 had the lowest number of protein spots. More than 250 protein spots that were detected in stalk tissues by proteomic analysis showed reproducible abundance within replications. Approximately 50 protein spots were additionally induced in the resistant cultivar upon pathogen inoculation, whereas ~ 24 proteins have disappeared in susceptible cultivar. These studies have established that proteome analysis has the potential to provide significant insights into the molecular events that occur during sugarcane–*C. falcatum* interactions and this is the first attempt to standardize proteome analysis and to identify proteins involved in red rot resistance in sugarcane (32). Further studies using peptide mass finger printing would characterize the up/down regulated proteins and their role(s) in red rot resistance will be established.

## CONCLUSION

Overall, the studies conducted at SBI revealed definite role of 3-deoxyanthocyanidin phytoalexins and certain PR-proteins in red rot resistance. Although they are not the primary determinants of disease resistance they have been well established as potential antifungal weapons used by the plants to arrest the invading pathogen. Ongoing studies in our lab on identifying molecular basis of red rot resistance would reveal clear understanding of the molecular communication between the host and the pathogen that ultimately decides disease resistance. The newly identified transcripts will be good candidates for in-depth analysis to elucidate *C. falcatum*-responsive pathways in sugarcane and facilitate genetic manipulation to tailor this crop for tolerance to red rot and other diseases. Also disease resistance genes identified in our studies that showed differential regulation can provide preferred targets for breeding or to engineer durable disease resistance in sugarcane. Creating subtractive libraries to identify more resistance associated genes are in progress and detailed analyses of gene functions of the newly identified genes would help better understanding of the host resistance mechanisms underlying the response of sugarcane plant to different biotic and abiotic stresses.

## ACKNOWLEDGEMENTS

The authors are grateful to Dr N. Vijayan Nair, The Director of the Institute for the constant encouragement and suggestions. We are thankful to DBT, New Delhi, ICAR (AP-Cess fund on SAR mediated resistance, Outreach Project-Leaf spots) and DST, New Delhi for the financial support.

## REFERENCES

1. Bent AF, Mackey D 2007 Elicitors, Effectors, and R Genes The new paradigm and a lifetime supply of questions, *Annual Review of Phytopathology* **45** 399–436
2. Clive LS, De verdier K and Nicholson NI 1999 Accumulation of 3-deoxyanthocyanidin phytoalexins and resistance to *Colletotrichum sublineolum* in *sorghum, Physiology of Molecular Plant Pathology* **55** 263-273
3. Godshall MA, Lonergan TA 1987 The effect of sugarcane extracts on the growth of the pathogenic fungus *Colletotrichum falcatum Physiology of Molecular Plant Pathology* **30** 19-42
4. Gupta V, Raghuvanshi S, Gupta A, Saini N, Gaur A, Khan MS, Gupta RS, Singh J, Duttamajumder SK, Srivastava S, Suman A, Khurana JP, Kapoor R, and Tyagi AK 2009 The water-deficit stress-and red-rot-related genes in sugarcane *Functional Integrated Genomics* **10** 207-214
5. Madan KV, Soni N, Nigam M, Solomon S, and Agnihotri VP 1991 Enzyme activity and cane genotype resistant to red rot, *Sugar Cane* 2 6-8
6. Malathi P, Viswanathan R, Padmanaban P, Mohanraj D, Ganesh Kumar V, Salin and KP 2008 Differential accumulation of 3-deoxy anthocyanidin phytoalexins in sugarcane varieties varying in red rot resistance in response to *Colletotrichum falcatum* infection, *Sugar Technology* **10** 154-157
7. Padmanaban P, Mohanraj D, Viswanathan R, Rao MM, Prakasam N, Jothi R, and Alexander KC 1996 Differential interaction of sugarcane clones to pathotypes of *Colletotrichum falcatum* Went, *Sugar Cane*, **4** 16-20

8. Ramesh Sundar A, Nagarathinam S, Ganesh Kumar V, Rahul PR, Raveendran M, Malathi P, Agrawal GK, Rakwal R and Viswanathan R 2010 Sugarcane proteomics Establishment of a protein extraction method for 2-DE in stalk tissues and initiation of sugarcane proteome reference map, *Electrophoresis*, **31** 1959–1974

9. Ramesh Sundar A, Velazhahan R, Vidhyasekaran P 2002a A glycoprotein elicitor isolated from *Colletotrichum falcatum* induces defense mechanisms in sugarcane leaves and suspension-cultured cells *J Plant Dis Protect*, 109 601-611

10. Ramesh Sundar A, Velazhahan R, Viswanathan R and Vidhyasekharan P 2002b Induction of active oxygen species (AOS), lipoxygenase and lipid peroxidation in suspension-cultured sugarcane cells by a glycoprotein elicitor from *Colletotrichum falcatum, Jounal of Plant Disease Protection* **109** 441-451

11. Ramesh Sundar A, Vidhyasekaran P 2003a Differential induction of phenylpropanoid metabolites in suspension-cultured cells of sugarcane by fungal elicitors, *Acta Phytopathology Entomology Hungarica* **38** 29-42

12. Ramesh Sundar A, Vidhyasekaran P 2003b Induction of defense-related biochemical changes by elicitors of red rot pathogen and a non-pathogen in sugarcane cell culture, *Indian Phytopathology*, **56** 255-261

13. Ramesh Sundar A, Viswanathan R, Mohanraj D and Padmanaban P 1998 Role of oxidative enzymes in sugarcane and *Colletotrichum falcatum* Went interaction, *Acta Phytopathology Entomology Hungarica* **33** 297-304

14. Ramesh Sundar A, Viswanathan R, Mohanraj D and Padmanaban P 1999 Studies on possible use of pathogen toxin as a molecular marker for red rot resistance in sugarcane, *Acta Phytopathology Entomology Hungarica* **34** 211-217

15. Ramesh Sundar A, Viswanathan R and Nagarathinam S 2009 Induction of systemic acquired resistance (SAR) using synthetic signal molecules against *Colletotrichum falcatum* in *Saccharum officinarum, Sugar Technology* **11** 274-281

16. Rao KC, Krishnamoorthy TN, Lalitha E and Rajalakshmi C 1968 Phenols in relation to resistance of sugarcane varieties to red rot, *Current Science* **37** 532

17. Singh K, Singh RP and Agnihotri VP 1976 Phenolics

in relation to sugarcane resistance against red rot disease *Sugarcane, Pathologist's News* **15/16** 37-41

18. Singh RP, Lal S 2000 Red rot In A Guide to Sugarcane Diseases Ed P Rott, RA Bailey, J C Comstock, B J Croft, AS Saumtally, CIRAD/ ISSCT, Montpellier, France, 153-158 pages
19. Snyder BA, Leite B, Hipskind J, Butler LG and Nicholson RL 1991 Accumulation of sorghum phytoalexins induced by *Colletotrichum graminicola* at the infection site, *Physiological and Molecular Plant Pathology* **39** 463-470
20. Srinivasan KV, Bhat NR 1961 Red rot of sugarcane – criteria for grading resistance, *Jounal of Indian Botanical Society* **40** 566-577
21. Srinivasan KV 1969 Physiology of disease resistance in sugarcane with particular reference to red rot, *Proceedings of Indian Academic Science, Sect B* **59** 120-132
22. Thirupathiraja C, Viswanathan R and Padmanaban P 2004 Time course of peroxidase accumulation in sugarcane cultivars in response to *Colletotrichum falcatum* infection *Sugar Technology*, **6 (1&2)** 47-52
23. Van Loon LC, Rep M and Pieterse CMJ 2006 Significance of inducible defense-related proteins in infected plants *Annual Review of Phytopathology*, **44** 135–62
24. Vettore AL, da Silva FR and Kemper EL 2003 Analysis and functional annotation of an expressed sequence tag collection for tropical crop sugarcane, *Genome Research*, **13** 2725–2735
25. Viswanathan R, Malathi P and Padmanaban P 2003a Variation in sugarcane red rot pathogen *Colletotrichum falcatum* Went In *Frontiers of Fungal Diversity in India* (Eds G P Rao, C Manoharachari, D J Bhat, R C Rajak and T N Lakhanpal), International Book Distributing Co, Lucknow, 639-667 pages
26. Viswanathan R, Malathi P, Ramesh Sundar A, Aarthi S, Premkumari SM and Padmanaban P 2005 Differential induction of chitinases and thaumatin-like proteins in sugarcane in response to infection by *Colletotrichum falcatum* causing red rot disease *Jounal of Plant Disease Protection* **112** 537-542
27. Viswanathan R, Mohanraj D, Padmanaban P and Alexander KC 1996a Accumulation of 3-deoxyanthocyanidin phytoalexins luteolinidin and apigeninidin in sugarcane in relation to red rot disease *Indian Phytopathology* **49** 174-175
28. Viswanathan R, Mohanraj D, Padmanaban P and

Alexander KC 1996b Synthesis of phytoalexins in sugarcane in response to infection by *Colletotrichum falcatum* Went *Acta Phytopathology Entomology Hungarica* **31** 229-237

29. Viswanathan R, Mohanraj D and Padmanaban P 1996c Role of red rot pigments in relation to red rot resistance in sugarcane *Indian Jounal of Sugarcane Technology*, **11** 151-154
30. Viswanathan R, Mohanraj D and Padmanaban P 2000a Possible involvement of anthocyanin compounds in resistance of sugarcane red rot, *Indian Phytopathology*, **53** 311-313
31. Viswanathan R, Nandakumar R, Samiyappan R 2003b Involvement of pathogenesis-related proteins in *Pseudoonas* spp induced systemic resistance against *Colletotrchum falcatum* in sugarcane, *Journal of Plant Disease Protection*, **110** 524-537
32. Viswanathan R, Ramesh Sundar A, Malathi P and Ganesh Kumar V 2010 Application of proteomics research in sugarcane, *SBI News* **30(3)** 4-7
33. Viswanathan R, Ramesh Sundar A, Malathi P, Rahul PR, Ganesh Kumar V, Banumathy R, Prathima PT, Raveendran M, Kumar KK and Balasubramanian P 2009a Interaction between sugarcane and *Colletotrichum falcatum* causing red rot Understanding disease resistance at transcription level, *Sugar Technology* **11** 45-50
34. Viswanathan R, Samiyappan R 1999 Red rot disease in sugarcane a major constraint for the Indian Sugar Industry, *Sugar Cane* 5 9-15
35. Viswanathan R, Samiyappan R 2001a Role of chitinases in *Pseudomonas* spp induced systemic resistance against *Colletotrichum falcatum* in sugarcane, *Indian Phytopathol* **54** 418-423
36. Viswanathan R, Samiyappan R 2001b Antifungal activity of chitinases produced by some fluorescent pseudomonads against *Colletotrichum falcatum* Went causing red rot disease in sugarcane, *Microbiological Research* **155** 309-314
37. Viswanathan R, Samiyappan R 2002a Induced systemic resistance by fluorescent pseudomonads against red rot disease of sugarcane caused by *Colletotrichum falcatum, Crop Protect* **21** 1-10
38. Viswanathan R, Samiyappan R 2002b Role of oxidative enzymes in the plant growth promoting rhizobacteria (PGPR) mediated induced systemic resistance against *Colletotrichum falcatum* in sugarcane, *Journal Plant Disease Protect* **109** 88-100

39. Viswanathan R, Samiyappan R 2006 *Pseudomonas* spp colonization in sugarcane rhiosphere reduces titre of *Colletrichum falcatum* Went-causing red rot disease of sugarcane, *Archives of Phytopathlogy and Plant Protect* **39** 39-44

40. Viswanathan R, Samiyappan R 2008 Bio-formulation of fluorescent *Pseudomonas* spp induces systemic resistance against red rot disease and enhances commercial sugar yield in sugarcane *Archives of Phytopathology and Plant Protection* **41** 377 – 388

41. Viswanathan R 2001 Accumulation of anthocyanin compounds in sugarcane tissues as a factor responsible for red rot resistance, *Madras Agricultural Journal* **88** 571-576

42. Viswanathan R 2010 *Plant Disease Red Rot of Sugarcane,* Anmol Publishers, New Delhi, India, 306 page

43. Viswanathan R, Samiyappan R 2004 Secondary metabolites production by *Pseudomonas* spp strains antagonistic to *Colletotrichum falcatum* causing red rot disease in sugarcane *Acta Phytopathology Entomology Hungarica*, **39** (1-3) 28-37

44. Wilson KI, Srivastava DN 1970 Phenolic content of sugarcane in relation to red rot disease, *Experientia* 16 10-74

45. You-Xiong Q, Jian-Wei L, Ji-Sen Z, Miao-Hong R, Li-Ping X and Mu-Qing Z 2007 Molecular cloning and characterisation of a non-TIR-NBS-LRR type disease resistance gene analogue from sugarcane, *Proceeding of International Society of Sugar Cane Technology* **26** 927-931

46. Zabala G, Zou J, Tuteja J, Gonzalez DO, Clough SJ and Vodkin LO 2006 Transcriptome changes in the phenylpropanoid pathway of *Glycine max* in response to *Pseudomonas syringe* infection, *BMC Plant Pathology* **6** 1-28.

*Rev. Plant Pathol. Vol. 6, 2014*
*Indian Society of Mycology and Plant Pathology*
*Scientific Publishers (India), Jodhpur pp. 215-268*

# PLANT GROWTH PROMOTING RHIZOBACTERIA : DIVERSITY, MECHANISMS OF ACTION AND PERSPECTIVES IN AGRICULTURE

***U. Chakraborty, B.N. Chakraborty, A.P. Chakraborty, K. Sunar and P.L. Dey***

Immuno-Phytopathology Laboratory, Department of Botany, University of North Bengal, Siliguri- 734013, West Bengal

**KEY WORDS :** PGPR, *Bacillus, Pseudomonas, Paenibacillus,* Biocontrol

## ABSTRACT

Among the large group of microorganisms in the soil or in the rhizosphere, plant growth promoting rhizobacteria (PGPR) are a common group that can actively colonize plant roots and increase plant growth through different mechanisms. These do not belong to a single group, but have great diversity, and some of the common PGPR which have been identified so far belong to genera such as *Bacillus, Pseudomonas, Paenibacillus* and many others. Though a large number of rhizospheric bacteria have been isolated and identified as potential PGPR, this is just the tip of the iceberg. More than 95% bacteria still remain uncluturable and need techniques other than usual isolation if they are to be unravelled. An alternative approach, is the analysis of genes within environmental samples. These genes may be functional genes, i.e. those coding for proteins performing particular metabolic reactions of relevance to ecosystem processes. However, most applications have analysed genes encoding the small subunit (SSU) of ribosomal RNA. PGPR act through different mechanisms such as phosphate solubilization, phytohormone production, nitrogen fixation, ACC deaminase production or indirectly through disease suppression leading to crop improvement. Biocontrol may again be through direct inhibition of pathogen in soil or induction of systemic resistance in the host. However, in spite of successful demonstrations of the use of PGPR in laboratory and green house conditions, their use in large scale agriculture is still limited. Hence for maximum benefits, such microorganisms need to be used in combination with low doses of chemical fertilizers and/or pesticides in integrated management. On the other hand, while introducing microorganisms into the field, it is also essential to know how they will interact with the resident microbial community. The selection of the appropriate strains for the enhanced efficiency of PGPR under different conditions is of significance.

## INTRODUCTION

The rhizosphere can be defined as any volume of soil specifically influenced by plant roots and/or in association with roots and hairs, and plant-produced material (138, 35). This space includes soil bound by plant roots, often extending a few mm from the root surface and can include the plant root epidermal layer (35). Plant exudates in the rhizosphere, such as amino acids and sugars, provide a rich source of energy and nutrients for bacteria, resulting in bacterial populations greater in this area than outside the rhizosphere. Most rhizosphere organisms occur within 50 µm of root surface and populations within 10 µm of root surface may reach $10^9$–$10^{12}$ microbial cells g $^{-1}$ soil. Despite large numbers of bacteria in the rhizosphere, only 7–15% of the total root surface is generally occupied by microbial cells (168). The functions of soil biota are central to decomposition processes and nutrient cycling. Therefore, major microbial activity is confined to the 'hot-spot', i.e. aggregates with accumulated organic matter, rhizosphere (134, 168). The region around the root, the rhizosphere, is relatively rich in nutrients, due to the loss of as much as 40% of plant photosynthates from the roots (133). Consequently, the rhizosphere supports large and active microbial populations capable of exerting beneficial, neutral, or detrimental effects on plant growth. The importance of rhizosphere microbial populations for maintenance of root health, nutrient uptake, and tolerance of environmental stress is now recognized (34). These beneficial microorganisms can be a significant component of management practices to achieve the attainable yield, which has been defined as crop yield limited only by the natural physical environment of the crop and its innate genetic potential (52). It is a critical resource not only for agricultural production and food security but also towards maintenance of most life processes. Soil microorganisms play an important role in soil processes that determine plant productivity. For successful functioning of introduced microbial bioinoculants and their influence on soil health, exhaustive efforts have been made to explore soil microbial diversity of indigenous community, their distribution and behaviour in soil habitats (75).

Plant growth promoting rhizobacteria (PGPR) are a common group of bacteria that can actively colonize plant roots and increase plant growth. They are beneficial soil bacteria, which may facilitate plant growth and development both directly and indirectly (85). Direct stimulation may include providing plants with fixed nitrogen, phytohormones, iron that has been sequestered by bacterial siderophores, and soluble phosphate, while indirect stimulation of plant growth includes preventing phytopathogens (biocontrol) and thus, promote plant growth and development (82). PGPR perform some of these functions through specific enzymes, which provoke physiological changes in plants at molecular level. However, in field conditions, the above traits may not be sufficient to account for the observed growth promotion. The biochemical or physiological changes induced in the host that are activated by the PGPR also lead to plant growth promotion and develop resistance capacity in the host against pathogens. The prospect of manipulating crop rhizosphere microbial populations by inoculation of beneficial bacteria to increase plant growth has shown considerable promise in laboratory and greenhouse studies, but responses have been variable in the field (34). The potential environmental benefits of this approach, leading to a reduction in the use of agricultural chemicals and the fit with sustainable management practices, are driving this technology. Recent progress in our understanding of the biological interactions that occur in the rhizosphere and of the practical requirements for inoculant formulation and delivery should increase the technology's reliability in the field and facilitate its commercial development.

## DIVERSITY OF PGPR

PGPR are associated with plant roots, and either directly or indirectly stimulate plant growth. Among Rhizobacteria there is a gradient of root proximity and intimacy as follows: (i) bacteria living in the soil near roots, utilizing metabolites leaked from roots as C and N sources, (ii) bacteria colonizing the rhizoplane (root surface), (iii) bacteria residing in root tissue, inhabiting spaces between cortical cells and lastly (iv) bacteria living inside cells in specialized root structures, or nodules, which generally fall into two groups, the legume associated rhizobia and the woody plant associated *Frankia* species (88). Rhizobacteria that establish inside plant roots, forming more intimate associations, are endophytes. These include a wide range of soil bacteria forming less formal associations than the rhizobia–legume symbiosis; endophytes may stimulate plant growth, directly or indirectly (119) and include the rhizobia. Given the semantic overlap, Hallmann et al. (97) proposed the following definition for endophytic bacteria: 'those bacteria that can be isolated from surface-disinfected plant tissue or extracted from within the plant, and that do not visibly harm the plant'. In general, a greater proportion of endophytes are PGPR than is the case for bacteria inhabiting the rhizoplane or rhizosphere (156).

To aid in this conceptualization, Gray and Smith (88) proposed two simple terms : intracellular PGPR—(iPGPR) bacteria residing inside plant cells, producing nodules and being localized inside those specialized structures, and extracellular PGPR (ePGPR) - those bacteria living outside plant cells and not producing nodules, but enhancing plant growth through production of signal compounds that directly stimulate plant growth, improve plant disease resistance, or improve mobilization of soil nutrients. ePGPR can be subdivided into three types, based on degree of association with plant roots: those living near, but not in contact with, the roots; those colonizing the root surface; and those living in the spaces between cells of the root cortex (Fig. 1).

Since the first studies on PGPR in the 1950's, many hundreds of candidate PGPR strains have been screened and evaluated in laboratory, greenhouse and field studies across the world. Today PGPR are commonly used in developing countries, and inoculants are used on millions of hectares of land (239). Nevertheless, implementation of this biotechnology has been hindered by the lack of consistency and variation in responses that are obtained in field trials from site to site, year to year, or for different crops (123).

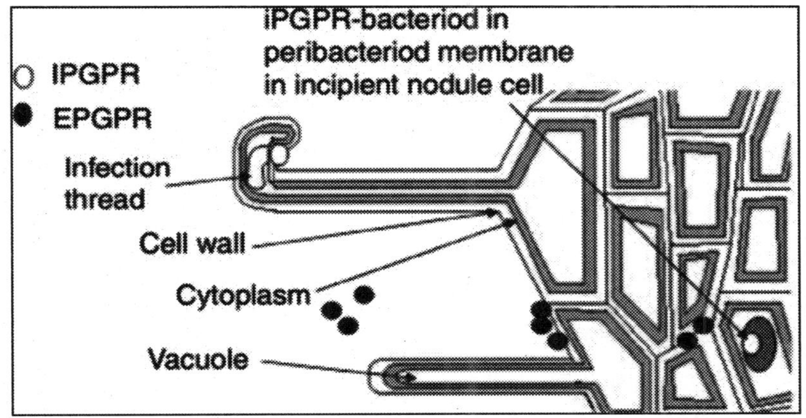


**Fig. 1**. Degree of bacterial associations among plant roots: iPGPR producing nodules and inhabiting the plant root (in this case (i) on the surface of a root hair during infection and (ii) in the plant cytoplasm); and ePGPR ((i) living near, but not in contact with roots; (ii) colonizing the root surface; (iii) living in spaces between cells of the root cortex).Adopted from Gray and Smith, 2005.

A large number of soil bacteria from various groups have now been identified and recorded in laboratories across the world as having the ability to promote growth under green house and field conditions. Those in the genera *Rhizobium, Bradyrhizobium, Sinorhizobium, Azorhizobium, Mesorhizobium and Allorhizobium,* belonging to the family Rhizobiaceae, invade plant root systems and form root nodule (141). Collectively they are often referred to as rhizobia. These iPGPR are mostly Gram-negative and rod-shaped, with a lower proportion being Gram-positive rods, cocci and pleomorphic forms. The primary mechanism by which rhizobia increase plant growth is $N_2$ fixation. ePGPR do not form nodules, but increase plant growth through a variety of mechanisms; these include genera such as *Bacillus, Pseudomonas, Erwinia, Caulobacter, Serratia, Arthrobacter, Micrococcus, Flavobacterium, Chromobacterum, Agrobacterium, Hyphomycrobium* and free-living nitrogen-fixing bacteria (74, 169, 47). A list of some of the reported PGPR is provided below in Table 1.

**Table 1.** List of Some plant growth promoting rhizobacteria along with the plants in which growth improvement has been recorded

| Microorganisms | Plant | References |
|---|---|---|
| *Bacillus subtilis* | *Arabidopsis thaliana*<br>*Pelargonium graveolens* | 118,150 |
| *Bacillus amyloliquefaciens* | *Arabidopsis thaliana* | 118 |
| *Bacillus pumilus* | *Camellia sinensis*<br>*Arabidopsis thaliana* | 47,118 |
| *B. pasteurii* | *Arabidopsis thaliana* | 118 |
| *Bacillus megaterium* | *Camellia sinensis*<br>*Hordeum vulgare* | 44,70 |
| *Bacillus licheniformis* | *Hordeum vulgare* | 70 |
| *Bacillus simplex* | *Lycopersicon esculentum* | 100 |
| *Bacillus cereus* | *Lycopersicon esculentum* | 100 |
| *Pseudomonas fluorescens* | *Arabidopsis thaliana* | 118 |
| *Pseudomonas putida* | *Hordeum vulgare* | 100 |
| *Serratia marcescens* | *Arabidopsis thaliana*<br>*Camellia sinensis* | 118, 46 |
| *Ochrobactrum anthropi* | *Camellia sinensis* | 45 |
| *Paenibacillus polymyxa* | *Arabidopsis thaliana*<br>*Hordeum vulgare* | 118, 70 |
| *Rhodobacter capsulatus* | *Hordeum vulgare* | 70 |
| *Azotobacter sp.*<br>*Pseudomonas sp.* | *Triticum aestivum* | 229 |
| *Pseudomonas fluorescens* | *Arachis hypogaea* | 64 |
| *Paenibacillus polymyxa E681* | *Sesamum indicum* | 185 |
| *Mesorhizobium loti MP6* | *Brassica campestris* | 48 |
| *Paenibacillus lentimorbus GBR158* | *Solanum lycopersicum* | 202 |
| *Paenibacillus polymyxa GBR-462* | *Capsicum sp* | 117 |
| *Azotobacter sp. WPR-51* | *Triticum aestivum* | 72 |
| *Serratia proteamaculans 1-102* | *Glycine max* | 12 |
| *Bradyrhizobium japonicum*<br>*Azotobacter chroococum* | *Glycine max* | 27 |

| Microorganisms | Plant | References |
|---|---|---|
| *Bacillus coagulans*<br>*Pseudomonas fluorescens* | *Santalum sp.* | 140 |
| *Bacillus megaterium C4* | *Zea mays*<br>*Oryza sativa* | 131 |
| *Azospirillum lipoferum CRT1* | *Zea mays* | 57 |
| *Bacillus megaterium strain B388* | *Pinus sp.* | 212 |
| *Bacillus megaterium* | *Lactuca sativa* | 143 |
| *Bacillus cereus (NAS4-3)*<br>*Serratia marcescens (KBS9-R)* | *Sorghum bicolor* | 107 |
| *Pseudomonas fluorescens* | *Solanum lycopersicum*<br>*Lactuca sativa*<br>*Brassica rapa* | 122 |
| *Pseudomonas fluorescens*<br>*Pseudomonas putida* | *Triticum aestivum*<br>*Brassica napus* | 148 |
| *Pseudomonas fluorescens (AURB15)*<br>*Burkholderia cepacia (AURB34)*<br>*Bacillus coagulans (AURB56)*<br>*Bacillus subtilis (AURB65)* | *Eragrostis tef* | 232 |
| *Rhizobium leguminosarum bv. trifolii SN10* | *Trifolium alexandrium*<br>*Oryza sativa* | 25 |
| *Enterobacter cancerogenus MSA2* | *Jatropha cucas* | 111 |
| *Citrobacter sp., Shigella sp., Klebsiella sp., Bacillus circulans* | *Halophila ovalis*<br>*Halodule pinifolia* | 81 |
| *Serratia marcescens NBR11213* | *Piper bittle* | 125 |
| *Pseudomonas chlororaphis DF190 and PA23, Bacillus cereus DFE4, Bacillus amyloliquefaciens DFE16* | *Brassica napus* | 174 |
| *Pseudomonas plecoglossicida, Brevibacterium antiquum*<br>*Bacillus altitudinis* | *Oryza sativa* | 87 |

## MOLECULAR DIVERSITY

Though a large number of rhizospheric bacteria have been isolated and identified as potential PGPR, this is just the tip of the iceberg. Morethan 95% bacteria still remain uncluturable and need techniques other than usual isolation if they are to be unravelled. An alternative approach, is the analysis of genes within environmental samples. These genes may be functional genes, i.e. those coding for proteins performing particular metabolic reactions of relevance to ecosystem processes. However, most applications have analysed genes encoding the small subunit (SSU) of ribosomal RNA. Analysis of 16S rRNA genes is now widely used for analysis of bacterial populations, and analysis of 18S rRNA genes and internal transcribed spacer (ITS) regions is increasingly being used to analyse fungal populations. Ribosomal rRNA genes are ideal for this purpose in that they possess regions with sequences conserved between all bacteria or fungi, facilitating alignment of sequences when making comparisons, while other regions exhibit different degrees of variation, enabling distinction between different groups. These differences provide the basis for a phylogenetic taxonomy and enable quantification of evolutionary differences between different groups. Discrimination of bacteria, using 16S rRNA gene sequences, is greater than that that for fungi, using 18S rRNA sequences, but finer scale information may be obtained by analysis of ITS regions.

The presence of regions of rDNA sequence with different degrees of conservation enables the identification of sequences that are common to all bacteria or fungi, or to specific phylogenetic groups, sometimes to the level of species. These sequences may then be used to design primers for the specific amplification, using the polymerase chain reaction (PCR), of rRNA genes belonging to particular groups or to design specific probes for these groups. These primer sequences provide the basis for analysis of species in natural populations. Two approaches may be adopted, the first based on PCR amplification of rRNA genes and the second involving *in situ* detection of rRNA within cells.

The era of molecular microbial ecology has uncovered only a part of novel microbiota, most of which is based on rRNA and rDNA analysis . The molecular methods used globally for diversity assessment of different cropping systems include, phospholipid fatty acid (PLFA) analysis (147, 49) terminal-restriction fragment length polymorphism (T-RFLP) (68), single-strand conformation polymorphism (SSCP) (193, 240, 192), and denaturing/ temperature gradient gel electrophoresis (DGGE/TGGE).

The quantitative description of microbial communities in terms of gene expression of particular function is now possible through the development of DNA microarray technology and its applications in the study of microbial community structure of agro/natural ecosystem(26–30). In conjunction with DNA microarray, direct RNA-based analysis of community dynamics to measure the functionality of environmental microbial populations without PCR amplification has been developed and it is equally applicable to direct detection and characterization of 16S rRNA of microbial species, and analysis of environmental samples (89, 40). To understand the dynamics of community life on a broader scale, metagenomics (study of collective genome of an ecosystem) provide insights of functional information through genomic sequences and expression of traits.

Another promising molecular tool in the study of PGPR is the use of potential marker genes for PGPR functions which include those encoding enzymes for antibiotic production, hydrogen-cyanide, ethylene destruction and auxin promotion. HCN genes are broadly distributed among many 2,4-DAPG producing pseudomonad strains (93). PCR primers for a conserved sequence in the *hcnAB* genes have been shown to be specific for detection of HCN producing pseudomonads from a world-wide collection of isolates (206). Similarly, primers are available to target production of 2,4-DAPG (24). Typically these primers have been used to test isolates that are cultivated from soil, but can be used with soil with the caveat that gene products that are produced. The PCR products should be further analyzed by DNA melting and size check (570 bp fragment) and sequencing to determine the primers have not amplified other genes. An advantage with these primers is that the forward primer can be combined with a GC clamp for analysis of the PCR products by DGGE to assess the genotypic diversity. Another target gene for quantification is the *accA* gene that encodes aminocyclopropane carboxylic acid deaminase. The *accA* gene is broadly distributed among a wide range of Gram negative bacteria including PGPR pseudomonads, and is common in many Gram-positive bacteria, rhizobia, and fungi (84). Here again, caution must be used during interpretation of changes in *accA* gene copy numbers, as some bacteria have sequences that are highly similar, but they code instead for enzymes with other functions such as serine deamination. Using PCR methods to detect and quantify PGPR target genes in fallow soils, our experience has been that cell numbers of PGPR and copy numbers of PGPR relevant genes may fall below detection limits, such that the populations are better estimated by culture of a host plant in the soil and baiting of the populations on to the roots where they can readily be quantified in their working habitat. An overview of the methods that are recommended to study the diversity and function of bacterial inhabitants of the mycosphere are given in Fig. 2.

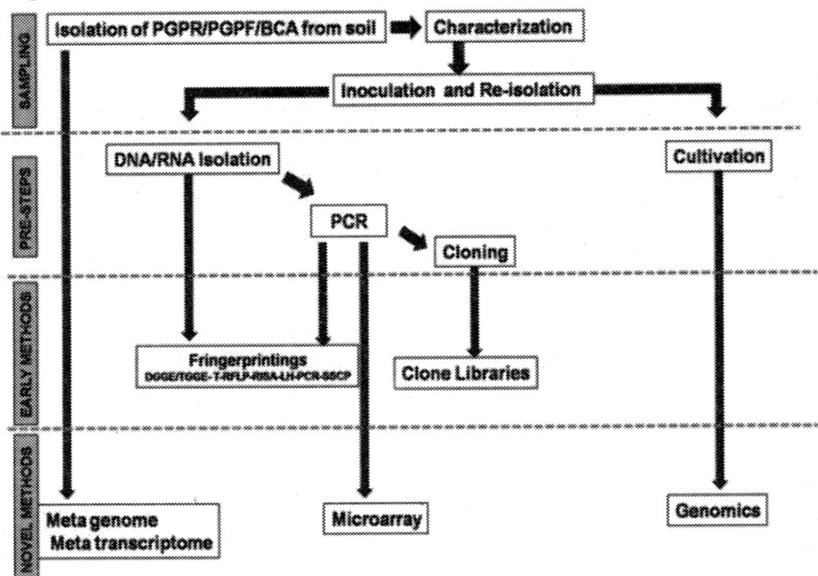


**Fig. 2.** An overview of methods that are recommended to study the diversity and function of bacterial inhabitants of the mycosphere. The combination of cultivation-independent with cultivation-dependent methods is predicted to enable great advances in our understanding of the system (217).

## MECHANISMS OF ACTION

There has been considerable confusion over the precise effects of PGPR, which confounds scientific studies aimed at quantifying their contribution to plant growth. This is largely due to poor understanding of the interactions between PGPR and their plant hosts and the resident microflora, as well as a paucity of information on how environmental factors influence processes that contribute to plant growth promotion (227).

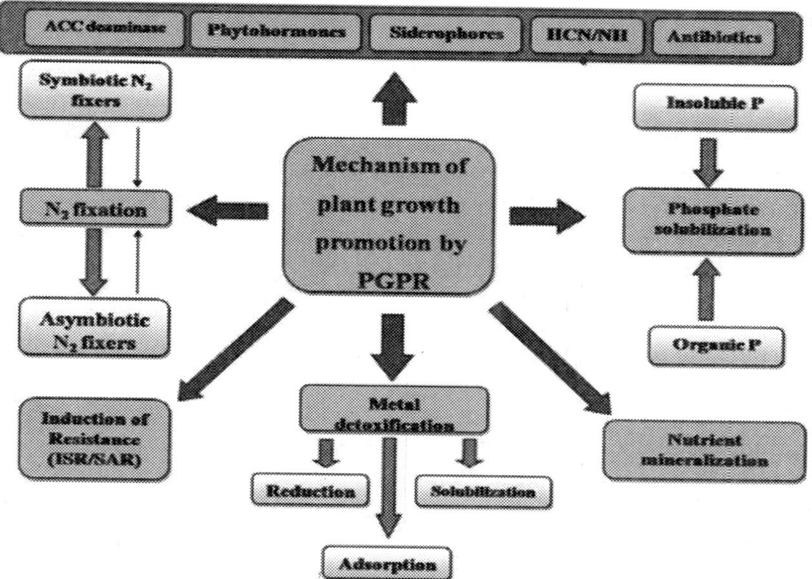


**Fig. 3.** Probable mechanisms of action for plant growth promotion by PGPR

Two paradigms that have emerged so far from the study of PGPR is that many of the best strains are multifunctional, and secondly, that PGPR traits are commonly distributed among many different species and genera of microorganisms, many of which are indigenous members of the soil microbial community. Many PGPR simultaneously solubilize phosphorus, produce auxins that stimulate root growth, and produce antibiotics and siderophores that may function in suppression of root disease. Other traits that may contribute to plant growth promotion include production of substances that induce systemic resistance or enzymes degrading hydrogen cyanide or ethylene and reactive oxygen species that are produced by plants during environmental stress. Altogether any and all of the cumulative effects of PGPR that influence root growth rates, root system architecture, root hair formation and longevity, will indirectly affect the ability to acquire water and nutrients and to tolerate root loss to disease (Fig. 3). Deciphering which mechanisms are most important and how to manage the soil microflora to obtain expression of these traits is so the remaining great challenge for consistent PGPR use in agricultural systems.

### *Phosphate solubilizing bacteria*

Phosphorus (P) is a major essential macronutrients for biological growth and development. P in soils is immobilized or becomes less soluble either by absorption, chemical precipitation, or both. A survey of Indian soils revealed that 98% of these need phosphorus fertilization either in the form of chemical or biological fertilizer. Although P content in an average soil is 0.05%, only 0.1% of the total P present is available to the plants because of its chemical fixation and low solubility. Application of chemical phosphatic fertilizers is practised though a majority of the soil P reaction products are only sparingly soluble. Under such conditions, microorganisms offer a biological rescue system capable of *solubilizing* the insoluble inorganic P of soil and make it available to the plants (208). Phosphate solubilizing micro-organisms (PSM) include largely bacteria and fungi, which can grow in media containing tricalcium, iron and aluminium phosphate, hydroxyapatite, bonemeal, rock phosphate and similar insoluble phosphate compounds as the sole phosphate source. Such microbes not only assimilate P but a large portion of soluble phosphate is released in quantities in excess of their own requirement. A large number of bacteria have been reported to have the ability of phosphate solubilization both *in vivo* and *in vitro*. *Bacillus* sp. are reported to be most common among phosphate solubilizers, followed by *Pseudomonas* sp. The reported bacilli include, *B. brevis*, *B. cereus*, *B. circulans*, *B. firmus*, *B. licheniformis*, *B. megaterium*, *B. mesentricus*, *B. mycoides*, *B. polymyxa*, *B. pumilis*, *B. pulvifaciens* and *B. subtilis* from the rhizosphere of legumes, cereals (rice and maize), arecanut palm, oat, jute chilli and tea ( 14, 224, 189, 7, 76, 172, 26, 120, 47). *Pseudomonas striata*, *P. cissicola*, *P. fluorescens*, *P. pinophillum*, *P. putida*, *P. syringae*, *P. aeruginosa*, *P. putrefaciens* and *P. stutzeri* have been isolated from rhizosphere of *Brassica*, chickpea, maize, soybean and other crops, desert soils and Antarctica lake(91,158,159). In addition, *Escherichia freundii, E. intermedia, Serratia phosphaticum* and species of *Achromobacter, Brevibacterium*, *Corynebacterium*, *Erwinia, Micrococus, Sarcina* and *Xanthomonas* are active in solubilizing insoluble phosphates.

Besides, Chakraborty et al. (45,46) reported phosphate solubilization by microorganisms isolated from tea such as *Serratia marcescens* and *Ochrobactrum anthropi*, in addition to *Bacillus megaterium* and *Bacillus pumilus* .

Goldstein et al. (86) demonstrated that an efficient mineral phosphate solubilizing phenotype in Gram-negative bacteria resulted from extracellular oxidation of glucose to gluconic acid via the quinoprotein glucose dehydrogenase. A unique bacterial population isolated from the roots of *Helianthus annus jaegeri* growing at the edge of an alkaline dry lake in the Mojave Desert in Israel showed no mineral phosphate solubilizing activity and no gluconic acid production. Addition of a concentrated solution containing material washed from the roots to these bacteria in culture however resulted in production of high levels of gluconic acid. This suggested that signalling between bacteria and plant root regulated expression of direct oxidation pathway in this bacterium. The resultant acidification of the rhizosphere played a key role in nutrient availability and/or other ecophysiological parameters essential for the survival of this desert plant. The establishment and performance of PSM is however affected severely under stressed conditions such as high salt, pH and temperature prevalent in degraded ecosystems represented by alkaline soils with tendency to fix phosphorus. In a screening of 4800 bacterial isolates from the root-free soil, rhizosphere and rhizoplane of *P. juliflora* growing in alkaline soils, 857 morphotypes solubilized phosphate in agar. The incidence of PSB was highest in the rhizoplane, followed by rhizosphere and root-free soil.

On the other hand, organic P can constitute between 30 and 50% of the total P of the soil, a high proportion of it corresponding to phytate (214). In this context, there are bacteria capable of producing phytase enzymes for the mineralization of phytates (130, 114). To date, there are only few studies reporting rhizobacteria capable of mineralizing the phytate. Among the phytase producing rhizobacteria, species belonging to *Bacillus*, *Burkholderia*, *Enterobacter*, *Pseudomonas*, *Serratia* and *Staphylococcus* genera are the most common culturable bacteria (179, 106, 198). Many of these bacteria are remarkably efficient. Richardson and Hadobas (179) isolated *Pseudomonas* spp. that utilized phytate from different soils in Australia. The isolated strains exhibited a high phytase activity, releasing over 80% of the P content in the phytate. In a later study utilizing plants with a limited capacity to obtain the P from phytate, Richardson *et al.* (178) observed that the ability of pasture plants to acquire P from phytate was enhanced followed by inoculation with the specified *Pseudomonas* sp. strains. Similarly, Unno et al. (216), isolated diverse bacteria with the ability to utilize phytate from the rhizosphere from white lupin (*Lupinus albus*). Almost all the isolates were classified as members of the *Burkholderia* genus and some of them significantly promoted the growth of the lupin. Jorquera *et al.* (113) isolated P solubilizing bacteria from the rhizospheres of five cultivated plants (*Lolium perenne*, *Trifolium repens*, *Triticum aestivum*, *Avena sativa*, *Lupinus luteus*), which presented more than one mechanism for utilizing insoluble forms of phosphorus. Moreover, all strains showed the capacity to produce P hydrolases. The major limitation today for use of these organisms is the lack of consistent effects in mobilizing P under field conditions. This is likely due to competition with the native microflora and environmental factors that either limit the population size or activity of the PGPR (227). It is now clear from many studies that evaluation and ranking of P solubilizing bacteria under laboratory conditions do not necessarily correspond to the efficacy of the PGPR for enhancing plant P uptake under field conditions (178, 176). Finally, it is likely that many so-called PGPR have dual action effects that are mediated by direct solubilization of inorganic P, mineralization of organic P, and stimulatory effects on plant root growth or mycorrhizae formation.

### *Nitrogen fixation*

Non-symbiotic nitrogen fixation is known to be of great agronomic significance. The main limitation to non-symbiotic nitrogen fixation is the availability of carbon and energy source for the energy intensive nitrogen fixation process. This limitation can be compensated by moving closer to or inside the plants, viz. in diazotrophs present in rhizosphere, rhizoplane or those growing endophytically. Some important nonsymbiotic nitrogen-fixing bacteria include, *Achromobacter*, *Acetobacter*, *Alcaligenes*, *Arthrobacter*, *Azospirillum*, *Azotobacter*, *Azomonas*, *Bacillus*, *Beijerinckia*, *Clostridium*, *Corynebacterium*, *Derxia*, *Enterobacter*, *Herbaspirillum*, *Klebsiella*, *Pseudomonas*, *Rhodospirillum*, *Rhodopseudomonas* and *Xanthobacter* (190) Various crops in India have been inoculated with diazotrophs particularly *Azotobacter* and *Azospirillum* (209, 210). Application of *Azotobacter* and *Azospirillum* has been reported to improve yields of both annual and perennial grasses Saikia and Bezbaruah (188) reported increased seed germination of *Cicer arietinum, Phaseolus mungo, Vigna catjung* and *Zea mays*. Members of the genus *Azospirillum* fix nitrogen under microaerophilic conditions, and are frequently associated with root and rhizosphere of a large number of agriculturally important crops and cereals. Due to their frequent occurrence in the rhizosphere these are known as associative diazotrophs. Azospirilla proliferate in the rhizosphere of numerous plant species and the genus *Azospirillum* now contains seven species – *A. brasilense, A. lipoferum, A. amazonense, A. haloprae-ferens, A. irakense, A. dobereinerae* and *A. largimobile*.

*Acetobacter diazotropicus* (family Acetobacteriaceae), isolated from roots and stems of sugarcane, was first reported as an N2-fixing bacterium from Brazil, and subsequently from Australia, India, Mexico, Uruguay, Canada and Cuba (208). Isolation of this bacterium from most tissues of sugarcane, and its absence from the soils of sugarcane fields suggested these to be systemic endophytes. The occurrence of this organism has been reported in sugar-rich plants like *Pennisetum purpureum* and sweet potato (161). This bacterium successfully colonizes sugarcane varieties in India where the chemical N fertilization is completely avoided for at least two successive years and replaced by organic manures. *Acetobacter* has gained importance as an inoculant for sugarcane (152). The family Acetobacteriaceae includes genera, *Acetobacter, Gluconobacter*, *Gluconoacetobacter* and *Acidomonas*. Based on 16S rRNA sequence analysis, the name *Acetobacter diazotrophicus* has been changed to *Gluconoacetobacter diazotrophicus* (233)

## *Phytohormone production*

Ability of most PGPR to produce phytohormones such as IAA has been considered to be of the most important traits for plant growth promotion. Many important plant-microbial interactions center on the production of auxins, IAA being the main plant auxin. IAA is responsible for the division, expansion and differentiation of plant cells and tissues and stimulates root elongation. A list of some of the phytohormones released by PGPR have been presented in Table 2.

**Table 2**. Phytohormones released by PGPR

| Plant Growth regulators | PGPR | References |
|---|---|---|
| IAA | *Rhizobium leguminosarum*<br>*Azotobacter sp.*<br>*Enterobacter sp.*<br>*Pseudomonas fluorescens*<br>*Mesorhizobium loti* MP6<br>*Bacillus sp.*<br>*Paenibacillus sp.*<br>*Bacillus* RC23 | 28, 237,149,64,48,21,71 |
| Gibberellin | *Acetobacter diazotropicus*<br>*Bacillus pumilus* | 18,92 |
| Cytokinin | *Paenibacillus polymyxa*<br>*Pseudomonas fluorescens*<br>*Bacillus licheniformis*<br>*Bacillus subtilis BC1* | 211, 61,105 |
| ACC deaminase | *Pseudomonas putida*<br>*Pseudomonas cepacia*<br>*Pseudomonas putida* biovar B<br>*Pseudomonas aeruginosa*<br>*Serratia proteamaculans* M35 | 145, 43,180,238 |

The ability to synthesize IAA has been detected in many rhizobacteria as well as in pathogenic, symbiotic and free living bacterial species (53, 213).At present, auxin synthesizing rhizobacteria are the most well-studied phytohormone producers (213, 203). These rhizobacteria synthesize IAA from tryptophan by different pathways, although it can also be synthesized via tryptophan-independent pathways, though in lower quantities (203). Among PGPR species, *Azospirillum* is one of the best studied IAA producers (65). Other IAA producing bacteria belonging to *Aeromonas* (96), *Azotobacter* (4), *Bacillus* ( 108, 207, 46), *Burkholderia* (96), *Enterobacter* (199), *Pseudomonas* (98) and *Rhizobium* (80) genera have been isolated from different rhizosphere soils. Inoculation with IAA producing PGPR has been used to stimulate seed germination, to accelerate root growth and modify the architecture of the root system, and to increase the root biomass.

Besides IAA, other phytohormones involved in PGPR action are cytokinins ( 157) and $GA_3$ (154). Plants and plant associated microorganisms have been found to contain over 30 growth promoting compounds of the cytokinin group. Cytokinins are produced by bacteria such as *Azospirillum* and *Pseudomonas* spp. (78). Some PGPR strains were reported to produce cytokinins, such as *Rhizobium leguminosarum*, *Paenibacillus polymyxa*, and *Pseudomonas fluorescens* (155, 211, 61, 23, 225).

A number of authors have reported that PGPR also produce gibberellins (GAs). The most widely recognized gibberellin is GA3 (gibberellic acid); the most active GA in plants is GA1, which is primarily responsible for stem elongation (58). Dobbelaere *et al.* (66) reported that over 89 GAs are known to date. PGPR such as *Rhizobium phaseoli*, *Azospirillum lipoferum*, *Azotospirillum brasilense*, *Acetobacter diazotropicus*, *Herbospirillum seropedicae*, *Bacillus licheniformis*, *B. pumilus*, *Bacillus cereus* MJ-1, *Bacillus macroides* CJ-29 were reported to produce GAs (33, 110, 18, 92, 112).

### *ACC deaminase production for ethylene regulation*

Another mechanism of plant growth promotion involving ethylene has been proposed by (39). Ethylene is a gaseous plant growth hormone produced endogenously by almost all plants. It is also produced in soil through a variety of biotic and abiotic mechanisms, and plays a key role in inducing diverse physiological changes in plants at cellular level.Bacterial 1-aminocyclopropane-1-carboxylate (ACC) deaminase plays a well-understood role in the regulation of a plant hormone, ethylene and thus, growth and development of plants are modified (9,84). This trait in inhibiting ethylene production by bacteria is, in most cases, correlated to their ability of plant growth promotion. It is extensively studied in numerous species of plant growth promoting bacteria like *Agrobacterium genomovars* and *Azospirillum lipoferum* (29), *Alcaligenes* and *Bacillus* (20), *Burkholderia* (29, 160, 194), *Enterobacter* (162), *Methylobacterium fujisawaense* (137), *Pseudomonas* (20, 29, 104), *Ralstonia solanacearum* (29), *Rhizobium* (136; 215), *Rhodococcus* (204), and *Sinorhizobium meliloti* (19) and *Variovorax paradoxus* (20). The mechanism of action of the PGPR secreting ACC deaminase is illustrated in Fig.4.

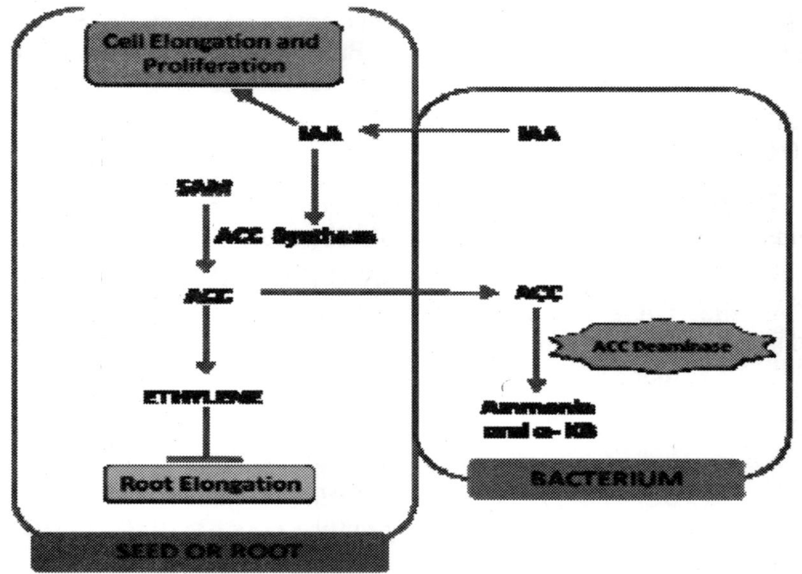


**Fig 4.** The ACC deaminase metabolizes the root's ACC into $\alpha$-ketobutyrate and ammonia and checks the production of ethylene which otherwise inhibits plant growth (103).

The plants treated with bacteria containing ACC-deaminase may have relatively extensive root growth due to less ethylene (38,195) and can better resist various stresses (38, 187 ).Various studies have demonstrated that plants treated with PGPR bacteria that produce ACC deaminase have increased their resistance to environmental stress. Grinchko and Glick (90) inoculated tomato seeds with the ACC deaminase expressing bacteria *Enterobacter cloacae* and *Pseudomonas putida* and registered an increase in plant resistance on 55 days of age to 9 consecutive days of flooding. Ghosh et al. (79) found ACC deaminase activity in three *Bacillus* species (*Bacillus circulans* DUC1, *Bacillus firmus* DUC2 and *Bacillus globisporus* DUC3), which stimulatedroot elongation of *Brassica campestri* plants.

Mayak et al. (146) evaluated tomato plants inoculated with the bacterium *Achromobacter piechaudii* under water and saline stress conditions. The authors reported a significant increase in fresh and dry weight of inoculated plants. In soils with a high copper content, Reed and Glick (175) reported an increase in dry matter content of the root and the air part in raps seeds inoculated with the ACC deaminase producing bacterium *Pseudomonas asplenii*.

## PLANT DISEASE MANAGEMENT THROUGH BIOCONTROL OF PATHOGENS

Phytopathogenic microorganisms have a great impact on crop yields and can significantly reduce plant performance and crop quality. Use of chemical pesticides for disease control has been the most common practice over the years, but this strategy has led to increased concerns over environmental contamination and has resulted in the so-called pesticide treadmill in which pathogens develop resistance to individual chemical controls over time, needing a constant development of new pesticides (73). In this context, rhizobacteria that can provide biocontrol of disease or insect pests (biopesticides) are considered an alternative to chemical pesticides (236). A large number of mechanisms are involved in biocontrol and can involve direct antagonism via production of antibiotics, siderophores, HCN, hydrolytic enzymes (chitinases, proteases, lipases, etc.), or indirect mechanisms in which the biocontrol organisms act as a probiotic by competing with the pathogen for a niche (infection and nutrient sites). Biocontrol can also be mediated by activation of the acquired systemic resistance (SAR), induced systemic resistance (ISR) responses in plants, and by modification of hormonal levels (34, 219) in the plant tissues.

### *Antibiotic production*

The production of antibiotics is considered one of the most powerful and studied biocontrol mechanisms for combating phytopathogens. Antibiotics constitute a wide and heterogeneous group of low molecular weight chemical organic compounds that are produced by a wide variety of microorganisms (171). Under laboratory conditions many different types of antibiotics produced by PGPR have shown to be effective against phytopathogenic agents (34). The antibiotics produced by PGPR include: butyrolactones, zwittermycin A, kanosamine, oligomycin A, oomycin A, phenazine-1-carboxylic acid, pyoluteorin, pyrrolnitrin, viscosinamide, xanthobaccin, and 2,4-diacetyl phloroglucinol (2,4- DAPG) (231, 73). Production of DAPG has been considered to be of great importance in several cases and can be produced by various strains of *Pseudomonas*, one of the most common bacterial species of the rhizosphere (177). The 2, 4-DAPG has a wide spectrum of properties in that it is antifungal (132, 177), antibacterial (223) and antihelmintic (54). In soils, it suppresses the growth of the wheat pathogenic fungus *Gaeumannomyces graminis* var. *tritici*, Raaijmakers et al. (170) reported a production of 0.62 ng 2,4-DAPG per 105-107 CFU $g^{-1}$ root by *P. fluorescens*, strain Q2-87.

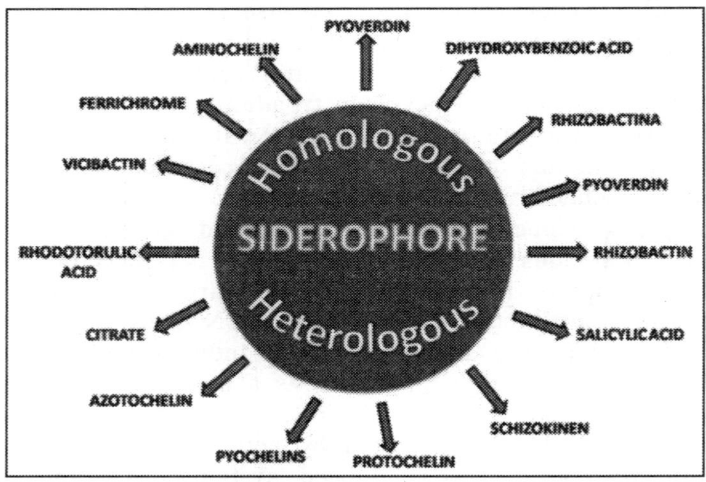


**Fig 5.** Some of the known siderophores which may be homologous or heterologous

### *Hydrogen cyanide (HCN) production*

Apart from the production of 2,4-DAPG, some rhizobacteria are capable of producing HCN (177). HCN is a volatile, secondary metabolite that suppresses the development of microorganisms and that also affects negatively the growth and development of plants (200). HCN is a powerful inhibitor of many metal enzymes, especially copper containing cytochrome C oxidases. HCN is formed from glycine through the action of HCN synthetase enzyme, which is associated with the plasma membrane of certain rhizobacteria (30). To date many different bacterial genera have shown to be capable of producing HCN, including species of *Alcaligenes, Aeromonas, Bacillus, Pseudomonas* and *Rhizobium* (63,4). HCN production is a common trait within the group of

*Pseudomonas* present in the rhizosphere, with some studies showing that about 50% of pseudomonads isolated from potato and wheat rhizosphere are able to produce HCN *in vitro* (13, 191).

Various studies attribute a disease protective effect to HCN, e.g. in the suppression of "root-knot" and black rot in tomato and tobacco root caused by the nematodes *Meloidogyne javanica* and *Thielaviopsis basicota*, respectively (228, 200). However, there are investigations reporting harmful effects on plants, inhibition of energy metabolism of potato root cells (13), and reduced root growth in lettuce (8). Likewise, HCN produced by *Pseudomonas* in the rhizosphere inhibits the primary growth of roots in *Arabidopsis* due to the suppression of an auxin responsive gene (182).

### *Siderophore-production*

Siderophores are low molecular weight compounds that are produced and utilized by bacteria and fungi as iron (Fe) chelating agents. These compounds are produced by various types of bacteria in response to iron deficiency which normally occurs in neutral to alkaline pH soils, due to low iron solubility at elevated pH (196).

Iron is essential for cellular growth and metabolism, such that Fe acquisition through siderophore production plays an essential role in determining the competitive fitness of bacteria to colonize plant roots and to compete for iron with other microorganisms in the rhizosphere (55, 56). Siderophore producing PGPR can prevent the proliferation of pathogenic microorganisms by sequestering $Fe^{3+}$ in the area around the root (201). Fe depletion in the rhizosphere does not affect the plant, as the low Fe concentrations occur at microsites of high microbial activity during establishment of the pathogen. Many plants can use various bacterial siderophores as iron sources, although the total concentrations are probably too low to contribute substantially to plant iron uptake. Plants also utilize their own mechanisms to acquire iron; dicots via a root membrane reductase protein that converts insoluble $Fe^{3+}$ into the more soluble $Fe^{2+}$ ion, or in the case of monocots by production of phytosiderophores (56). Various studies have isolated siderophore producing bacteria belonging to the *Pseudomonas* (32), *Serratia* (121, 46), *Streptomyces* (121) *Bacillus* sp. (44) and *Ochrobactrum anthropi* (45) which could promote growth of crop plants and suppress disease. However, the growth promotion that occurred may be due to other mechanisms or combinations of mechanisms that increase nutrient availability, suppress pathogens, or affect root growth via hormone production.

One of the most important functions of siderophores in mediating competitive interactions between microorganisms occurs in certain disease suppressive soils where fungal pathogens such as *Fusarium* and *Pythium* are inhibited by pseudomonads. Early research showed that production of large amounts of desferri-siderophore by pseudomonads in pure culture may sequester all available iron to the extent that they suppress growth of fungal pathogens that are cultivated on the same plate. Based on this observation, it was initially proposed that siderophores play a key role in antibiosis of plant pathogenic fungi. However, two decades of research on this topic have now provided detailed insight into the mechanisms of disease suppression. Some studies have provided direct evidence that siderophores have a direct role in the suppression of plant disease causing fungi (67). In most cases, however, siderophores appear to be more important for contributing to the rhizosphere competence of disease suppressive bacteria than for causing iron deprivation to the pathogenic fungi. Effective strains of plant disease suppressive bacteria that suppress pathogenic fungi may use a variety of mechanisms in combination, including production of hydrogen cyanide, antibiotics, and lytic enzymes (36; 158). Another mechanism by which bacteria can enhance disease suppression is by induction of systemic acquired resistance. This is stimulated by salicylic acid, which has the ability to chelate iron and is induced by iron stress. The pseudobactin produced by WCS374 may also induce the SAR response (126, 128). The discovery of this secondary effect of salicyclic acid and siderophores on the SAR response has led to reconsideration of the possible mode of action of siderophores for enhancing plant disease suppression. Siderophore defective mutants of these disease suppressive bacteria sometimes have reduced ability to suppress fungal pathogens due to their impaired ability to acquire iron. However, rhizosphere competence of disease suppressive bacteria can be enhanced by use of heterologous siderophores that are produced by other bacteria in the plant rhizosphere (56).

**Table 3:** Siderophore classes and representative siderophore types produced by soil microorganisms (102).

| Bacterial siderophores | Types | Siderophore producing microbes |
|---|---|---|
| Phenol catacholates | enterobactins | *Escherichia coli* |
| | | *Enterobacteriaceae* |
| | linear catechols | *Agrobacterium* |
| | mycobactins | *Mycobacteria spp.* |
| | pyochelin | *Pseudomonas aeruginosa* |
| | tropolone | *P. plantarii* |
| Hydroxamates | schizokinen | *Bacillus megaterium* |
| Citrate type | aerobactin | *Aerobacter* |
| | | *Enterobacteriaceae* |
| Ferrioxamines | arthrobactin | *Arthrobacter spp.* |
| | ferrioxamine | *Actinomyces spp.* |
| | | *Streptomyces spp.* |
| Cyclic Di- | | *Nocardia spp.* |
| Cyclic Mono | | *Arthrobacter spp.* |
| | alcaligin | Alcaligenes sp. |
| | cepabactin | *P.cepacia* |
| Carboxylate | rhizobactin | *Rhizobium meliloti* |
| Pyoverdines | pyoverdine | *Pseudomonas sp.* |
| | Pseudobactin | *Pseudomonas sp.* St B10 |
| **Fungal siderophores** | | |
| Rhodotorulic acid | Rhodotorulic acid | *Rhodotorula* |
| | | *Cryptococcus* |
| | | *Ustilago* |
| | dimerum acid | *Verticillium sp.* |
| | | *Fusarium sp.* |
| | | *Gaeumannomyces* |
| | coprogen | *Penicillium* |
| | | *Neurospora* |
| Ferrichromes | ferrichrome | *Aspergillus* |
| | | *Penicillium* |
| Fusarinines | fusarinnie | *Fusarium* |
| | | *Aspergillus* |
| | | *Penicillium* |

## INDUCTION OF SYSTEMIC RESISTANCE IN HOST

Biopriming plants with some PGPR can also provide systemic resistance against a broad spectrum of plant pathogens. Diseases of fungal, bacterial, and viral origin, and in some instances even damage caused by insects and nematodes, can be reduced after application of PGPR (115, 167, 173, 184,186). Certain bacteria trigger a phenomenon known as ISR (Induced systemic resistance) phenotypically similar to systemic acquired resistance (SAR). SAR develops when plants successfully activate their defense mechanism in response to primary infection by a pathogen, notably when the latter induces a hypersensitive reaction through which it becomes limited in a local necrotic lesion of brown, desiccated tissue. Similar to SAR, ISR is effective against different types of pathogens but differs from SAR in that the inducing PGPB does not cause visible symptoms on the host plant (218). PGPR-elicited ISR was first observed on carnation (*Dianthus caryophillus*) with reduced susceptibility to wilt caused by *Fusarium* sp. (220) and on cucumber (*Cucumis sativus*) with reduced susceptibility to foliar disease caused by *Colletotrichum orbiculare* (230). Manifestation of ISR is dependent on the combination of host plant and bacterial strain (116). There are several reports of ISR triggered by PGPR such as- *P. fluorescens* EP1 against red rot caused by *Colletotrichum falcatum* on sugarcane (226), *Burkholderia phytofirmans* PsJN against *Botrytis cinerea* on grapevine (5,6) and *Verticllium dahliae* on tomato (197), *P. denitrificans* 1-15 and *P. putida* 5-48 against *Ceratocystis fagacearum* on oak (37), *P. fluorescens* 63-28 against *F. oxysporum* f. sp. *radicis-lycopersici* on tomato (153) and *Pythium ultimum* and *F. oxysporum* f. sp. *pisi* on pea roots (22), *Bacillus pumilus* SE34 against *F. oxysporum* f. sp. *pisi* on pea roots (13) and *F. oxysporum* f. sp. *vasinfectum* on cotton roots (51), *P.fluorescencs* and *B.subtilis* against *Macrophomina phaseolina* in mulberry (77). Triggering of ISR in tea induced by *B.megaterium, Ochrobactrum anthropi* and *Serratia marcescens* against root rot pathogens such as *Fomes lamoensis* and *Poria hypolaterita* have been reported by (44,45,46).

Several bacterial traits (i.e., flagellation and production of siderophores and lipopolysaccharides) have been proposed to trigger ISR (101, 128, 127, 218, 221), but there is no compelling evidence for an overall ISR signal produced by bacteria (94). It has recently been reported that volatile organic compounds may play a key role in this process (167, 186,184). For example, volatiles secreted by *B. subtilis* GBO3 and *B. amyloliquefaciens* IN937a were able to activate an ISR pathway in *Arabidopsis* seedlings challenged with the soft-rot pathogen *Erwinia carotovora* subsp. *carotovora* (183). A major distinction often drawn between ISR and SAR is the dependence of the latter on the accumulation of salicylic acid (SA) (163). Some PGPR do trigger an SA-dependent signaling pathway by producing nanogram amounts of SA in the rhizosphere (60, 62). However, the majority of PGPR that activate ISR appear to do so via a SA-independent pathway involving jasmonate and ethylene signals (60, 166). ISR is associated with an increase in sensitivity to these hormones rather than an increase in their production, which might lead to the activation of a partially different set of defense genes (165, 99).

Besides, investigations on mechanisms of biological control by PGPR have also revealed that several strains protect the plants from pathogen attack by strengthening the epidermal and cortical cell walls with deposition ofnewly formed barriers beyond infection sites including callose, lignin and phenolics (22, 135) and by activating defense genes encoding chitinase, PO, PPO and PAL (135, 50) and enzymes which are involved in the synthesis of phytoalexins (144, 77).

## Multiple mechanisms of action

The notion of multiple mechanisms emerged from early studies on *Azospirillum,* when the results of the field inoculation experiments failed to demonstrate that N-fixation was the main mechanism by which plant growth was stimulated (17, 16). Failing N fixation as an explanation, the additive hypothesis was then proposed to describe the effect of *Azospirillum* on plant growth (16). The additive hypothesis proposes the possibility of multiple mechanisms that function simultaneously or sequentially (15). In the case of *Azospirillum*, N fixation has largely been discounted and primary plant growth promotion mechanism is now attributed to several other functions including phytohormones production (65, 139), ACC deaminase activity (129) and hydrolytic enzyme production (151). Today, it is increasingly recognized that many PGPR strain likely function by more than one mechanism (59). Vassilev et al. (222) reviewed the potential of phosphorus solubilizing microorganisms that can provide simultaneously phytopathogen biocontrol, and also affect plant growth via production of siderophores, hydrolytical enzymes and IAA. Avis et al. (11) classified PGPR bacteria into two groups based on the main action mechanisms by which they are known. These two groups are: (i) microorganisms with direct plant growth promoting mechanisms (e.g. phytohormone production, phosphorus solubilization, etc.), and (ii) microorganisms which indirectly promote plant growth and productivity through biocontrol of phytopathogens (e.g. production of siderophores, antibiotics, HCN, etc.). Despite this arbitrary classification, microorganisms of both groups can simultaneously contain secondary mechanisms.

## APPLICATIONS IN AGRICULTURE

In spite of successful demonstrations of the use of PGPR in laboratory and green house conditions, their use in large scale agriculture is still limited. Hence for maximum benefits, such microorganisms need to be used in combination with low doses of chemical fertilizers and/or pesticides in integrated management. On the other hand, while introducing microorganisms into the field, it is also essential to know how they will interact with the resident microbial community. The selection of the appropriate strains for the enhanced efficiency of PGPR under different conditions is of significance. (148). Different strains of PGPR can enhance the availability of different nutrients such as nitrogen or phosphorous. *Rhizobium* spp. and *Azospirillum* spp. can fix N and hence are important in combination with N fertilization. In other words, at the time of inoculating plants with *Rhizobium* or/and *Azospirillum* it must be exactly indicated that if the bacteria alone is able to supply the host plant with adequate amount of N or N fertilization must also be utilized to provide the plant with N requirement.

Soil P solubilizing microbes including AM fungi and bacteria can enhance P solubility by producing enzymes such as phosphatase. The synergistic interactions between AM fungi and PGPR including N-fixing *Rhizobium* can enhance the efficiency of soil microbes. Especially, in tripartite symbiosis AM hypha in the soil can result in the movement of soil bacteria through different parts of the soil (31, 10). This is of importance for the production of biofertilizers, because the right combination of soil microbes can enhance the biofertilizer efficiency (3, 234).

An interesting and important point about the combined use of soil microbes and chemical or/and organic fertilization is to determine their appropriate rates. This is of environmental and economical significance. There are several important key issues about this: 1) the microbial potential for providing nutrients under certain conditions for plant utilization, 2) plant species, 3) soil properties, and 4) climate properties. Soil microbial potential for enhanced nutrient availability differs under different conditions. Such abilities must be tested under different conditions to determine their capacity for providing nutrients for plant use. Plant species are also of importance as there are high variations among different plant species for their nutrient requirements. Soil properties can affect both plant and microbial growth (148). PGPR efficiency to increase plant growth is dependent on the level of nutrients in the soil. According to (42, 2,1, 235) increased level of NPK fertilization decreased the efficiency of *Pseudomonas* spp. They attributed such observations to the production of stress level ethylene under low levels of NPK, whose prerequisite is catabolized by ACC deaminase to $NH_3$ and α-ketobutyrate. In addition root growth and efficiency is of importance under nutrient deficiency and higher root growth results in higher nutrient uptake and hence plant growth. Accordingly, PGPR, with higher efficiency under nutrient efficient conditions, are able to increase plant growth by enhancing root growth and hence nutrient uptake (83, 109, 235).

Determining the efficiency of chemical and biological fertilization is of significance as the contribution and hence the optimum rate of each would be clearly indicated. Canbolat et al. (41) examined the effects of chemical and biological fertilization on plant growth and nutrient uptake and indicated that chemical fertilization resulted in higher NPK uptake by plants. However, on the other hand, Elcoka et al. (69) indicated that the efficiency of biological fertilization is a matter of the microbial combination as double and triple inoculation with PGPR may significantly enhance the bio fertilizer efficiency.

However, more research is necessary to clearly indicate the contributing role of chemical and biological fertilization in providing necessary nutrients for plant growth and yield production.

According to the research, conducted so far, use of chemical fertilization is necessary as biological fertilization has not yet been proven to be good enough for completely providing plant nutrient requirement. However, what matters about chemical fertilization is to enhance its efficiency environmentally and economically using biofertilizers. It has been indicated that about less than 50% of chemical fertilizers is absorbed by plant and the rest would not be accessible by plant as it is subjected to leaching, run-off and emission from the soil (1,3). Hence, use of biological fertilizers as supplementary fertilization to chemical fertilization is necessary with the above mentioned advantages. Accordingly, the right and proper application of chemical and biological fertilization is much dependent on realizing the interactions between soil, plant and microorganisms. Soil microbes are a big help to plant and the environment as they own some abilities that collectively enhance plant growth. Among such abilities enhanced nutrient uptake by plant is also of importance; in the presence of soil microbes, plant absorb higher amounts of nutrients and less risk of environmental pollution is likely.

Since no microbe can live in isolation in the soil, PGPR efficacy is dependent on establishing an effective population density of active cells in plant rhizosphere. Though this is a simple principle, it has proved to be difficult to establish dose response effects in which the degree of plant growth promotion or disease suppression can be directly correlated with size of the PGPR population. In general, bacterial suspensions of PGPR are prepared at densities of $10^8$ to $10^9$ CFU $ml^{-1}$ for root dipping and soil inoculation. After inoculation at these high densities, the cell numbers will undergo a rapid decline depending on whether or not the soil has been sterilized (142). In nonsterile soils where there is competition with the resident flora and predation by protozoa and nematodes, bacterial populations will decline rapidly by orders of magnitude per week until the population reaches equilibrium with its environment. This likely accounts for differences that are observed in lab and greenhouse studies where soils are sterilized, versus in the field where results of PGPR inoculation are much more inconsistent.

Inoculation efficacy depends on the rhizosphere competence of the bacteria for the particular host plant. To maintain effective cell densities under field conditions, it is often necessary to reinoculate at intervals during the production period but this is limited by the high cost of inoculum production, difficulty in maintaining viable cells for long periods of time in storage, and cost for delivery and incorporation of inoculants into the field. So far, the most commercially successful inoculants have been gram positive spore forming bacteria, which can persist in storage from months to years, and that can withstand starvation, temperature, moisture and other environmental stresses better than nonspore forming bacteria.

Methods for inoculation with either gram negative or gram positive PGPR bacteria require the use of a carrier to deliver the inoculum into the soil and allow mixing of the cells in the soil profile. This can involve low cost carriers such as peat, calcined clay, or powdered corn cobs that are mixed with the bacterial suspensions and dried. Alginate microbeads are also used and provide many advantages by incorporating the cells into a protected matrix that undergoes decomposition in the soil and slowly releases the bacteria. Lastly, bacteria can be introduced into the irrigation water via on-site fermentation equipment that automatically cultures the bacteria and pumps them into the irrigation water at desired intervals.

PGPR population densities are typically much higher in the plant rhizosphere than in the bulk soil. However, correlating population density to activity is a great challenge. The rhizosphere is very heterogenous with respect to nutrient availability. Mature roots are typically colonized by bacteria at densities of $10^8$ to $10^9$ CFU $g^{-1}$ in the mature root zones. A number of approaches are commonly used to quantify PGPR, including measurements of cell densities based on 16S rRNA gene copy numbers or plating on agar and CFU enumeration. Other approaches employ PCR methods to quantify the copy number of a particular functional gene, or expression of relevant mRNA for genes encoding PGPR traits. A common problem in much of the research on PGPR has been the failure to monitor the cell density of the introduced bacteria over time to confirm that inoculation was effective. In such cases, it is not possible to determine whether PGPR are responsible for the observed effects or to explain variations in efficacy of the inoculants that may be caused by management or environmental factors.

Mathematical modeling of the behavior of PGPR soil inoculants has been used to predict how various environmental factors affect the survival and activity of PGPR soil inoculants

(205). Supporting much experimental work, the model by Strigul and Kravchenko illustrates that survival and growth of newly introduced bacteria are strongly limited by competition for organic substrates with the resident microflora. PGPR are predicted to be the most effective in soils with low organic matter or stressed soils where growth of the indigenous population is restricted.

To date, there is very limited knowledge of how specific inoculants interact with resident microbial populations (95). Nonetheless, many resident bacteria, possibly including noncultu-rable bacteria, will carry genes encoding common PGPR functions. In the case of quorum regulated genes, such as those for antibiotic and siderophore production, there is a broad intra and inter-specific communication level between different bacterial popula-tions. Thus, it may result in either positive and negative feedback on quorum sensor mediated behavior (164). Prior experiments examining the effects of wheat inoculation with 2,4- DAPG producing pseudomonads have shown that there are broad interaction levels not only with various genotypes of resident fluorescent pseudomonads, but with populations of diverse bacterial species including *Arthrobacter, Chyrseobacterium, Flavobacteria*, and other species that are significantly enriched in the presence of 2,4-DAPG producers (124). Similarly, a study by Roesti *et al.* (181) showed striking shifts in rhizobacterial community structures following inoculation with various combina-tions of PGPR pseudomonads. A larger question is whether inoculation can result in shifts in community structure that increase plant growth promotion and disease suppression functions of the resident community. This question will only be answered once molecular tools are available to detect and quantify all the PGPR relevant phenotypes in the microbial community. As most bacteria in the rhizosphere are still uncultured, this will require a metagenomics approach to identify the genes. Quantitative PCR arrays or DNA microarrays will also provide valuable tools for examining the response patterns of microbial communities to soil inoculants.

## FUTURE PROSPECTS

Thus, it is clear that in spite of great promise shown by rhizo-bacteria over the last few decades as biofertilizers and bioprotec-tors, in most cases, results still remain confined to laboratories and glass houses. There still exists a wide gap in results obtained under controlled conditions and in the field. This is mostly due to the rhizospheric conditions in which the bacteria are introduced in the field and the environmental stresses they would be subjected to. More and more research is going on to unravel the mechanisms of action of these bacteria and it is becoming increasingly apparent that most PGPR can promote plant growth by several mechanisms, but most studies currently focus on individual mechanisms and have not been able yet to sort out the relative contributions of different processes that are responsible for plant growth promotion. Screening strategies for selecting the best strains will require more comprehensive knowledge of the traits required for rhizosphere competence, and studies on the ecology of introduced PGPR with the resident PGPR and other microbial species in the plant rhizosphere (227). While inoculation is now viewed as a means to enhance plant growth, the effects of various management practices or soil amendments on PGPR activity of indigenous bacteria remain unknown. The use of PGPR inoculants in agriculture is already proceeding, and offers many opportunities to improve plant nutrition, crop yields, and disease management, while improving sustainability by reducing the need for chemical inputs. Nevertheless, as our understanding of the ecology of these bacteria improves, it should be possible to obtain a more informed explanation of the mechanisms that are involved in plant growth promotion and identify situations in which bio-augmentation with soil inoculants may be useful for increasing crop yields. Another area which still requires considerable attention is the means of application of these bacteria into the soil, taking into account suitable carriers, shelf life and sustainability of the rhizobacteria.

## REFERENCES


1. Adesemoye A, Kloepper J 2009a Plant–microbes interactions in enhanced fertilizer-use efficiency, *Applied Microbiology and Biotechnology* **85** 1–12
2. Adesemoye AO, Torbert HA, Kloepper JW 2008 Enhanced plant nutrient use efficiency with PGPR and AMF in an integrated nutrient management system, *Canadian Journal of Microbiology* **54** 876–886
3. Adesemoye AO, Torbert HA, Kloepper JW 2009b Plant growth promoting rhizobacteria allow reduced appli-

cation rates of chemical fertilizers, *Microbial Ecology* **58** 921-929

4. Ahmad F, Ahmad I, Khan MS 2008 Screening of free-living rhizospheric bacteria for their multiple plant growth promoting activities, *Microbiology Research* **163** 173-181
5. Ait Barka E, Belarbi A, Hachet C, Nowak J, Audran JC 2000 Enhancement of *in vitro* growth and resistance to gray mould of *Vitis vinifera* cocultured with plant growth-promoting rhizobacteria, *FEMS Microbiology Letters* **186** 91-95
6. Ait Barka E, Gognies S, Nowak J, Audran JC, Belarbi A 2002 Inhibitory effect of endophyte bacteria on *Botrytis cinerea* and its influence to promote the grapevine growth, *Biological Control* **24** 135-142
7. Ali ME, Massoud AM, El-Xhander IAI 1989 Effect of different isolates of PSB on soil pH and available soil P In *Proceedings of the Conference Agricultural Development and Research*, Ain Shams University, Cairo, Egypt
8. Alström S, Burns RG 1989 Cyanide production by rhizobacteria as a possible mechanism of plant growth inhibition, *Biology and Fertility of Soils* **7** 232-238
9. Arshad M, Frankenberger WT 2002 Ethylene agricultural sources and applications Kluwer Academic / Plenum Publishers, New York
10. Artursson V, Finlay R, Jansson J 2006 Interactions between arbuscular mycorrhizal fungi and bacteria and their potential for stimulating plant growth, *Environmental Microbiology* **8** 1–10
11. Avis TJ, Gravel V, Antoun H, Tweddell RJ 2008 Multifaceted beneficial effects of rhizosphere microorganisms on plant health and productivity, *Soil Biology and Biochemistry* **40** 1733-1740
12. Bai Y, Souleimanov A, Smith DL 2002 An inducible activator produced by a *Serratia proteamaculans* strain and its soybean growth promoting activity under green house conditions, *Journal of Experimental Botany* **53** 1495-1502
13. Bakker AW, Schippers B 1987 Microbial cyanide production in the rhizosphere in relation to potato yield reduction and *Pseudomonas spp* mediated plant growth stimulation, *Soil Biology and Biochemistry* **19** 249-256
14. Banik S, Dey B K 1981 Solubilization of inorganic phosphate and production of organic acids by microorganisms isolated in sucrose tricalcium phosphate agar plates, *Zbl Bakteriology* **136** 478-486

15. Bashan Y, de-Bashan LE 2010 How the plant growth-promoting bacterium *Azospirillum* promotes plant growth – a critical assessment, *Advance in Agronomy* **108** 77-136

16. Bashan Y, Levanony H 1990. Current status of *Azospirillum* inoculation technology *Azospirillum* as a challenge for agriculture, *Canadian Journal of Microbiology* **36** 591-608

17. BashanY, Singh M, Levanony H 1989 Contribution of *Azospirillum brasilense* Cd to growth of tomato seedlings is not through nitrogen fixation, *Canadian Journal of Microbiology* **67** 2429-2434

18. Bastian F, Cohen A, Piccoli P, Luna V, Baraldi R, Bottini R 1998 Production of indole-3-acetic acid and gibberellins A1 and A3 by *Acetobacter diazotrophicus* and *Herbaspirillum seropedicae* in chemically defined media, *Plant Growth Regulation* 247–11

19. Belimov AA, Hontzeas N, Safronova VI, Demchinskaya SV, Piluzza G, Bullitta S , Glick BR 2005 Cadmium-tolerant plant growth-promoting bacteria associated with the roots of Indian mustard (*Brassica juncea* L Czern), *Soil Biology and Biochemistry* **37** 241-250

20. Belimov AA, Safronova VI, Sergeyeva TA, Egorova TN, Matveyeva VA, Tsyganov VE, Borisov AY, Tikhonovich IA, Kluge C, Preisfeld A, Dietz KJ, Stepanok VV 2001 Characterization of plant growth promoting rhizobacteria isolated from polluted soils and containing 1-amino-cyclopropane-1-carboxylate deaminase, *Canadian Journal of Microbiology* **47** 242-252

21. Beneduzi A, Peres D, Vargas LK, Bodanese-Zanettini MH, Passaglia LMP 2008 Evaluation of genetic diversity and plant growth promoting activities of nitrogen-fixing Bacilli isolated from rice fields in South Brazil, *Applied Soil Ecology* **39** 311–320

22. Benhamou N, Belanger RR, Paulitz TC 1996 Pre-inoculation of Ri T-DNA-transformed pea roots with *Pseudomonas fluorescens* inhibits colonization by *Pythium ultimum* Trow an ultrastructural and cytochemical study, *Planta* **199** 105-117

23. Bent E, Tuzun S, Chanway CP, Enebak S 2001 Alterations in plant growth and in root hormone levels of lodge ploe pines inoculated with rhizobacteria *Canadian Journal of Microbiology* **47** 793-800

24. Bergsma-Vlami M, Prins ME, Staats M, Raaijmakers JM 2005 Assessment of genotypic diversity of antibiotic-producing *Pseudomonas* species in

the rhizosphere by denaturing gradient gel electrophoresis, *Applied Environmental Microbiology* **71** 993-1003

25. Bhattacharjee RB, Journad P, Chaintreuil C, Dreyfus B, Singh A, Mukhopadhyay SN 2012 Indole acetic acid and ACC deaminase-producing *Rhizobium leguminosarum* bv *trifolii* SN10 promote rice growth, and in the process undergo colonization and chemotaxis, *Biology and Fertility of Soils* **48** 173–182
26. Bhattacharya P, Ghosh TK, Jain RK1998 Evaluation of native phosphate solubilizing microorganisms from Vidarbha soils, *Journal Maharashtra Agricultral University Publication* **22** 252-253
27. Bhattarai HD, Prasad BN 2003 Effect of dual inoculation of *Bradyrhizobium japonicum* and *Azotobacter chroococum Indian Journal of Micobiology* **43**(2) 139-140
28. Biswas JC, Ladha JK, Dazzo FB, Yanni YG, Rolf BG 2000 Rhizobial inoculation influences seedling vigor and yield of rice, *Agronomy Journal* **90** 880–886
29. Blaha D, Combaret CP, Mirza MS, Loccoz YM2006 Phylogeny of the 1 aminocyc-lopropane-1-carboxylic acid deaminaseencoding gene acdS in phytobeneficial and pathogenic Proteobacteria and relation with strain biogeography *FEMS Microbiology Ecology* **56** 455-470
30. Blumer C, Haas D 2000 Mechanism, regulation, and ecological role of bacterial-cyanide biosynthesis *Archives Microbiology* **173** 170-177
31. Bonfante P 2003 Plants, mycorrhizal fungi, and endobacteria a dialog among cells and genomes *Biology Bulletin* **204** 215–220
32. Boopathi E, Rao KS 1999 A siderophore from *Pseudomonas putida* type A1 structural and biological characterization *BBA-Protein Structure M* 14330-40
33. Bottini R, Fulchieri M, Pearce D, Pharis RP 1989 Identification of gibberellins A1, A3, and iso- A3 in cultures of *Azospirillum lipoferum Plant Physiology* 9045–9047
34. Bowen GD, Rovira AD 1999 The rhizosphere and its management to improve plant growth *Advances in Agronomy* 661-102
35. Bringhurst RM, Cardon ZG, Gage DJ 2001 Galactosides in the rhizosphere utilization by *Sinorhizobium meliloti* and development of a biosensor *Proceedings of the National Academy of Sciences* 98 4540-4545
36. Brisbane PG, Harris JR, Moen, R 1989 Inhibition of fungi from wheat roots by *Pseudomonas fluorescens* 2-79

and fungicides *Soil Biology Biochemistry* **21** 1019-1026

37. Brooks DS, Gonzalez CF, Apple DN, Filer DH 1994 Evaluation of endophytic bacteria as potential biological control agents for oak wilt, *Biological Control* 4 373–381
38. Burd GI, Dixon DG, Glick BR 2000 Plant growth-promoting bacteria that decrease heavy metal toxicity in plants *Canadian Journal of Microbiology* 4623-245
39. Burdman S, Jurkevitch E, Okon Y 2000 Recent advances in the use of plant growth promoting rhizobacteria (PGPR) in agriculture In *Microbial Interactions in Agriculture and Forestry* Subba Rao N S and Dommergues Y R Eds Science Publishers, Enfield, USA Vol II 229-250 pages
40. Burgmann H, Widmer F, Sigler WV, Zeyer J 2003 mRNA extraction and reverse transcription-PCR protocol for detection of *nif* H gene expression by *Azotobacter vinelandii* in soil *Applied Environmental Microbiology* **69** 1928-1935
41. Canbolat MY, Bilen S, Cakmakci R, Sahin F, Aydin A 2006 Effect of plant growth-promoting bacteria and soil compaction on barley seedling growth, nutrient uptake, soil properties and rhizosphere microflora *Biology and Fertlity of Soils* **42** 350– 357
42. Carlier E, Rovera M, Rossi A, Jaume S, Rosas 2008 Improvement of growth, under field conditions, of wheat inoculated with *Pseudomonas chlororaphis* subsp *aurantiaca* SR1, *World Journal Microbiology and Biotechnology* **24** 2653–2658
43. Cattelan AJ, Hartel PG, Fuhrmann JJ 1999 Screening for plant growth rhizobacteria to promote early soybean growth, *Soil Science Society American Journal* **63** 1670–1680
44. Chakraborty U, Basnet M, Chakraborty BN 2006 Plant growth promotion and induction of resistance in *Camellia sinensis* by *Bacillus megaterium, Journal Basic Microbiology* **3** 186-195
45. Chakraborty U, Chakraborty BN, Basnet M, Chakraborty AP 2009 Evaluation of *Ochrobactrum anthropi* TRS-2 and its talc based formulation for enhancement of growth of tea plants and management of brown root rot disease, *Journal Applied Microbiology* **107** 625-634
46. Chakraborty U, Chakraborty BN, Chakraborty AP 2010 Influence of *Serratia marcescens* TRS-1 on growth promotion and induction of resistance in *Camellia sinensis* against *Fomes lamaoensis*

*Journal of Plant Interaction* **5** 261-272

47. Chakraborty U, Chakraborty BN, Tongden C, Roy Chowdhury P, Basnet M 2004 Assessment of tea rhizosphere microorganisms as plant growth promoters *Tea* **25**(2) 38-47

48. Chandra S, Choure K, Dubey RC, Maheshwari DK 2007 Rhizosphere competent *Mesorhizobium loti* MP6 induces root hair curling, inhibits *Sclerotinia sclerotiorum* and enhances growth of Indian mustard (*Brassica campestris*) *Brazilian Journal of Microbiology* **38** 124–130

49. Chelius MK, Triplett EW 2001 The diversity of archaea and bacteria in association with the roots of *Zea mays* L, *Microbiology Ecology* **41** 252–263

50. Chen C, Belanger RR, Benhamou N, Paullitz TC 2000 Defense enzymes indu-ced in cucumber roots by trea-tment with plant-growth pro-moting rhizobacteria (PGPR) *Physiological Molecular Plant Pathology* **56** 13-23

51. Conn KL, Nowak J, Lazarovits G 1997 A gnotobiotic bioassay for studying interactions between potato and plant growth-promoting rhizobacteria, *Canadian Journal of Microbiology* **43** 801–808

52. Cook RJ 2002 Advances in plant health management in the twentieth century, *Annual Review of Phytopathology* **38** 95-116

53. Costacurta A and Vanderleyden J 1995 Synthesis of phytohormones by plant-associated bacteria, *Critical Review of Microbiology* **21** 1-18

54. Cronin D, Moënne-Loccoz Y, Fenton A, Dunne C, Dowling DN, O'gara F 1997 Role of 2,4-Diacetylphloroglucinol in the interactions of the biocontrol Pseudomonad strain F113 with the potato cyst nematode *Globodera rostochiensis, Applied Environmental Microbiology* **63** 1357-1361

55. Crowley DE, Gries D 1994 Modeling of iron availability in the plant rhizosphere In *Biochemistry of metal micronutrients in the rhizosphere* (Eds JA Manthey, DE Crowley, DG Luster) Lewis Publishers, Boca Raton, Florida 199-224 pages

56. Crowley DE 2006 Microbial siderophores in the plant rhizosphere In *Iron Nutrition in Plants and Rhizospheric Microorganisms* (Eds LL Barton, J Abadía) Springer, Netherlands 169-198 pages

57. Czarnes S, Zemrany HE, Hallett PD, Alamercery S, Bally R, Monrozier LJ 2007 Early changes in root characteristics of maize (*Zea mays*) following seed inoculation with the PGPR *Azospirillum*

*lipoferum* CRT1 *Plant Soil* **291** 109–118

58. Davies PJ 1995. The plant hormones their nature, occurrence, and functions In Davies PJ (Ed) Plant hormones physiology, biochemistry, and molecular biology, 2nd edn Kluwer, Dordrecht 1–12 pages
59. de Freitas JR, Banerjee MR, Germida JJ 1997 Phosphate-solubilizing rhizobacteria enhance the growth and yield but not phosphorus uptake of canola (*Brassica napus* L) *Biology and Fertility of Soils* **24** 358-364
60. De Meyer G, Ho¨fte M 1997 Salicylic acid produced by the rhizobacterium *Pseudomonas aeruginosa* 7NSK2 induces resistance to leaf infection by *Botrytis cinerea* on bean *Phytopathology* **87** 588-593
61. de Salamone IEG, Hynes RK, Nelson LM 2001 Cytokinin production by plant growth promoting rhizobacteria and selected mutants *Canadian Journal of Microbiology* **47** 404–411
62. Dekkers LC, Mulders IH, Phoelish CC, Chin-A-Woeng TF, Wijifies AH, Lugtenburg BJJ 2000 The sss colonization gene of tomato-*Fusarium oxysporum* fsp *radicisly copersici* biocontrol strain *Pseudomonas fluorescens* WCS365 can improve root colonization of the wild type *Pseudomonas* spp bacteria *Molecular Plant-Microbe Interaction* **131** 177-1183
63. Devi KK, Seth N, Kothamasi S, Kothamasi D 2007 Hydrogen cyanide producing rhizobacteria kill subterranean termite *Odontotermes obesus* (Rambur) by cyanide poisoning under *in vitro* conditions *Currunt Microbiology* **54** 74-78
64. Dey R, Pal KK, Bhatt DM, Chauhan SM 2004Growth promotion and yield enhancement of peanut (*Arachis hypogaea* L) by application of plant growth promoting rhizobacteria *Microbiologial Research* **159** 371–394
65. Dobbelaere S, Croonenborghs A, Thys A, Vande Broek A, Vanderleyden J 1999 Phytostimulatory effect of *Azospirillum brasilense* wild type and mutant strains altered in IAA production on wheat *Plant Soil* **212** 155-164
66. Dobbelaere S, Vanderleyden J, Okon Y 2003 Plant growth-promoting effects of diazotrophs in the rhizosphere *Critical Review Plant Science* **22** 107–149
67. Duijff BJ, Meijer JW, Bakker PAHM, Schippers B 1993 Siderophore mediated competition for iron and induced resistance in the suppression of fusarium wilt of carnation by fluorescent *Pseudomonas*

*spp Netherlands, Journal of Plant Pathology* **99** 277-289

68. Dunbar J, Ticknor LO, Kuske CR 2000 Assessment of microbial diversity in four southwestern united states soil by 16S rRNA gene terminal restriction fragment analysis *Applied Environm-ental Microbiology* **66** 2943-2950

69. Elcoka E, Kantar F, Sahin F 2008 Influence of nitrogen fixing and phosphorus solubilizing bacteria on the nodulation, plant growth, and yield of chickpea *Journal of Plant Nutrition* **31** 157–171

70. Erdogan U, Donmez MF, Cakmakci R 2007 The effect of plant growth promoting rhizobacteria on barley seedling growth, nutrient uptake, some soil properties, and bacterial counts, *Turkish Journal Agricultral Forest* **31**(3)189-199

71. Erturk Y, Ercisli S, Sekban R, Haznedar A, Donmez MF 2008 The effect of plant growth promoting rhizobacteria (PGPR) on rooting and root growth of tea (*Camellia sinensis* var *sinensis*) cuttings *Roum Biotechnology Letters* **13** 3747-3756

72. Fatima Z, Saleemi M, Zia M, Sultan T, Aslam M, Rehman RU, Chaudhary MF 2009 Antifungal activity of plant growth-promoting rhizobacte ria isolates against *Rhizoctonia solani* in wheat *African Journal Biotechnology* **8** 219–225

73. Fernando WGD, Nakkeeran S, Zhang Y 2006 Biosynthesis of antibiotics by PGPR and its relation in biocontrol of plant diseases In *PGPR Biocontrol and Biofertilization* (Ed ZA Siddiqui) Springer, Netherlands 67-109 pages

74. Foster RC, Rovira AD, Cock TW 1983 Ultrastructure of the Root– Soil Interface *The American Phytopathological Society*, St Paul, MN 157 page

75. Fravel DR, Rhodes DJ, Larkin RP 1999 Production and commercialization of biocontrol products In *Integrated Pest and Disease Management in Greenhouse Crops* Albajes R, Gullino ML, van Lenteren JC, Elad Y Eds Kluwer Academic Publishers, Dordrecht 365-376 pages

76. Gaind S, Gaur AC 1999 Microbial phosphate solubilization as influenced by sodium chloride *Indian Journal of Experimental Biology* **37** 209-210

77. Ganeshamoorthia P, Ananda T, Prakasama V, Bharanib M, Ragupathia N, Samiyappana R 2008 Plant growth promoting rhizobacterial (PGPR) bioconsortia mediates induction of defense-related proteins against infection of root rot pathogen in mulberry plants, *Journal Plant Interaction* **3** 233-244

78. Gaudin V, Vrain D, Jouanin L 1994 Bacterial genes modifying hormonal balance in plant *Plant Physiology Biochemistry* **32** 11-29

79. Ghosh S, Penterman JN, Little RD,Chavez R, Glick BR 2003 Three newly isolated plant growth-promoting bacilli facilitatethe seedling growth of canola, *Brassica campestris Plant Physiology Biochemistry* **41** 277-281

80. Ghosh S, Sengupta C, Maiti TK, Basu PS 2008 Production of 3-indolylacetic acid in root nodules and culture by a *Rhizobium* species isolated from root nodules of the leguminous pulse, *Phaseolus mungo Folia Microbiology* **53** 351-355

81. Ghosh U, Subhashini P, Dilipan E, Raja S, Thangaradjou T, Kannan L 2012 Isolation and Characterization of Phosphate-solubilizing bacteria from sea grass rhizosphere soil, *Joiurnal Ocean University of China* **11(1)** 86-92

82. Glick BR, Bashan Y 1997 Genetic manipulation of plant growth-promoting bacteria to enhance biocontrol of phytopathogens, *Biotechnology Advance* **15**(2) 353-378

83. Glick BR, Penrose DM, Li J, 1998 A model for the lowering of plant ethylene concentrations by plant growth promoting bacteria, *Journal Theoritical Biology* **190** 63 -68

84. Glick BR, Todorovic B, Czarny J, Cheng Z, Duan J, McConkey B 2007 Promotion of plant growth by bacterial ACC deaminase, *Critical Review Plant Science* **26** 227-242

85. Glick BR1995The enhancement of plant growth by free-living bacteria, *Canadian Journal Microbiology* **41** 109-117

86. Goldstein AH, Rogers RD 1999 Biomediated continuous release phosphate fertilizer (Lockheed Idaho Technologies Co, USA) USUS 5912398 A 15 June 19 page

87. Gopalakrishnan S, Humayun P, Kiran BK, Kannan IGK, Vidya MS, Deepthi K, Rupela O 2011 Evaluation of bacteria isolated from rice rhizosphere for biological control of charcoal rot of sorghum caused by *Macrophomina phaseolina* (Tassi) Goid *World Journal Microbiology and Biotechnology* 27 1313–1321

88. Gray EJ, Smith DL 2005 Intracellular and extracellular PGPR commonalities and distinctions in the plant–bacterium signaling processes, *Soil Biology Biochemistry* **37** 395–412

89. Greer CW, Whyte LG, Lawrence JR, Masson L , Brousseau R 2001 Genomics technologies for environ-

mental science, *Environmental Scienceand Technology* **35** 364-370

90. Grichko VP, Glick BR 2001 Amelioration of flooding stress by ACC deaminase-containing plant growth-promoting bacteria, *Plant Physiology and Biochtechnology* **39** 11-17

91. Gupta RP, Vyas MK, Pandher MS 1998 Role of phosphorus solubilising microorganisms in P-economy and crop yield In *Soil–Plant–Microbe Interaction in Relation to Nutrient Management* (Ed B D Kaushik) Venus Printers & Publishers, New Delhi 95-101 pages

92. Gutierrez-Manero FJ, Ramos-Solano B, Probanza A, Mehouachi J, Tadeo FR, Talon M 2001 The plant-growth promoting rhizobacteria *Bacillus pumilus* and *Bacillus licheniformis* produce high amounts of physiologically active gibberellins, *Plant Physiology* **111** 206–211

93. Haas D, Defago G 2005 Biological control of soil-borne pathogens by fluorescent pseudomonads, *Natural Review of Microbiology* **3** 307-319

94. Haas D, Keel C, Reimmann C 2002 Signal transduction in plant beneficial rhizobacteria with biocontrol properties, *Anton Leeuwen* **81** 385-395

95. Haas D, Keel C 2003 Regulation of antibiotic production in root-colonizing *Pseudomonas spp* and relevance for biological control of plant disease, *Annual Review of Phytopathology* **41** 117–153

96. Halda-Alija L 2003 Identification of indole-3- acetic acid producing fresh-water wetland rhizosphere bacteria associated with *Juncus effusus* L, *Canadian Journal of Microbiology* **49** 781-787

97. Hallmann J, Quadt-Hallmann A, Mahaffee WF, Kloepper W 1997 Bacterial endophytes in agricultural crops *Canadian Journal of Microbiology* **43** 895–914

98. Hariprasad P, Niranjana SR 2009 Isolation and characterization of phosphate solubilizing rhizobacteria to improve plant health of tomato *Plant Soil* **31** 613-624

99. Hase S, van Pelt JA, van Loon LC, Pieterse CMJ 2003 Colonization of *Arabidopsis* roots by *Pseudomonas fluorescens* primes the plant to produce higher levels of ethylene upon pathogen infection, *Physiology and Molecular Plant Pathology* **62** 219-226

100. Hassen AI, Labuschagne N 2010 Root colonization and growth enhancement in wheat and tomato by rhizobacteria isolated from the rhizoplane of grasses *World Journal Microbiology Biotechnology* **26** (10) 1837-1846

101. Hoffland E, Pieterse C, Bik L, van Pelt JA 1995 Induced systemic resistance in radish is not associated with accumulation of pathogenesis related proteins *Physiology and Molecular Plant Pathology* **46** 309-320

102. Höfte M 1993Classes of microbial siderophores, In L L Barton and B C Hemming (Eds) *Iron Chelation in Plants and Soil Microorganisms,* Academic Press, San Diego, CA, USA 1-27 pages

103. Honma M, Shimomura T 1978 Metabolism of 1-aminocyclopropane- 1-carboxylic acid *Agricultual Biology Chemisty* **42** 1825-1831

104. Hontzeas N, Zoidakis J, Glick BR, Abu-Omar MM 2004 Expression and characterization of 1-aminocyclopropane-1-carboxylate deaminase from the rhizobacterium *P putida* UW4 a key enzyme in bacterial plant growth promotion, *Biochimistry and Biophysics Acta* **1703** 11-19

105. Hussain A, Hasnain S 2009 Cytokinin production by some bacteria Its impact on cell division in cucumber cotyledons *African Journal of Microbiological Research* **3** (11) 704-712

106. Hussin ASM, Farouk AE, Greiner R, Salleh HM, Ismail AF 2007 Phytatedegrading enzyme production by bacteria isolated from Malaysian soil, *World Journal of Microbial Biotechnology* **23** 1653-1660

107. Idris A, Labuschagne N, Korsten L 2009 Efficacy of rhizobacteria for plant growth promotion in sorghum under greenhouse conditions and selected modes of action studies *Journal of Agricultural Science* **147** 17–30

108. Idris EE, Bochow H, Ross H, Borriss R 2004 Use of *Bacillus subtilis* as biocontrol agent VI Phytohormonelike action of culture filtrates prepared from plant growth promoting *Bacillus amyloliquefaciens* FZB24, FZB42, FZB45 and *Bacillus subtilis* FZB37, *Journal of Plant Disease Protection* **111** 583–597

109. Jalili F, Khavazi K, Pazira E, Nejati A, Asadi RH, Rasuli SH, Miransari M 2009 Isolation and characterization of ACC deaminase producing fluorescent pseudomonads, to alleviate salinity stress on canola (*Brassica napus* L) growth, *Journal of Plant Physiology* **166** 667-674

110. Janzen RA, Rood SB, Dormaar JF, McGill WB 1992 *Azospirillum brasilense* produces gibberellins in pure culture on chemically defined medium and in co-culture on straw, *Soil Biology and Biochemistry* **24** 1061-1064

111. Jha CK, Patel B, Saraf M 2012 Stimulation of the growth of *Jatropha curcas* by the plant growth promoting bacterium *Enterobacter cancerogenus* MSA2 *World Journal Microbiology and Biotechnology* **28** 891–899

112. Joo GJ, KimYM, Lee IJ, Song KS, Rhee IK 2004 Growth promotion of red pepper plug seedlings and the production of gibberellins by *Bacillus cereus, Bacillus macroides* and *Bacillus pumilus, Biotechnological Lettters* **26** 487-491

113. Jorquera MA, Hernández MT, RengelZ, Marschner P, Mora ML 2008a Isolation of culturable phosphobacteria with both phytatemineralization and phosphate-solubilization activity from the rhizosphere of plants grown in a volcanic soil, *Biology and Fertility of Soils* **44** 1025-1034

114. Jorquera MA, Martínez O, Maruyama F, Marschner P, Mora ML 2008b Current and future biotechnological applications of bacterial phytases and phytase-producing bacteria, *Microbes Environment* **23** 182-191

115. Kerry BR 2000 Rhizosphere interactions and the exploitation of microbial agents for the biological control of plant-parasitic nematodes, *Annual Review of Phytopathology* **38** 423-441

116. Kilic-Ekici O, Yuen GY 2004 Comparison of strains of *Lysobacter enzymogenes* and PGPR for induction of resistance against *Bipolaris sorokiniana* in tall fescue *Biological Control* **30** 446-455

117. Kim SG, Khan Z, Jeon YH, Kim YH 2009 Inhibitory effect of *Paenibacillus polymyxa* GBR-462 on *Phytophthora capsici* causing *Phytophthora* Blight in Chili Pepper, *Journal of Phytopathology* **157** 329–337

118. Kloepper JW, Ryu CM, Hu CH, Locy RD 2005 Study of mechanisms for plant growth promotion elicited by rhizobacterial in *Arabidopsis thaliana Plant Soil* **268** 285–292

119. Kobayashi, M, Suzuki T, Fujita T, Masuda M, Shimizu S, 1995 Occurrence of enzymes involved in biosynthesis of indole-3-acetic acid from indole-3-acetonitrile in plant-associated bacteria, *Agrobacterium* and *Rhizobium, Proceedings of the National Academy of Sciences* **92** 714–718

120. Kole SC, Hajra JN 1997 Isolation and evaluation of tricalcium and rock phosphate solubilizing microorganisms from acidic terai and lateritic soils of West Bengal, *Journal of Interacademicia* **1** 126-128

121. Kuffner M, Puschenreiter M, Wieshammer G, Gorfer M, Sessitsch A 2008 Rhizosphere

bacteria affect growth and metal uptake of heavy metal accumulating willows, *Plant Soil* **30** 435-44

122. Kumatani T, Yoshimi Y, Nakayashiki H, Aino M 2009 Phylogenetic analyses of plant-growth-promoting rhizobacteria isolated from tomato, lettuce, and Japanese pepper plants in Hyogo Prefecture, Japan, *Journal of General Plant Pathology* **75** 316–321
123. Lambert B, Joos J 1989 Fundamental aspects of rhizobacterial plant growth promotion research, *Trends Biotechnology* **7** 215-219
124. Landa BB, Mavrodi DM, Thomashow LS, Weller DM 2003 Interactions between strains of 2,4-diacetylphloroglucinolproducing *Pseudomonas fluorescens* in the rhizosphere of wheat, *Phytopathology* **93** 982– 994
125. Lavania M, Chauhan PS, Chauhan SV, Singh HB, Nautiyal CS 2006 Induction of plant defense enzymes and phenolics by treatment with plant growth-promoting rhizobacteria *Serratia marcescens* NBRI 1213, *Currunt Microbiology* **52** 363-368
126. Leeman M, Den Ouden FM, Van Pelt JA, Dirkx FPM, Steijl H, Bakker PAHM, Schippers B 1996b Iron availability affects induction of systemic resistance to fusarium wilt of radish by *Pseudomonas fluorescens, Phytopathology* **86** 149-155
127. Leeman M, Denouden FM JA, van Pelt JA, Dirkx F, Steijl H, Bakker P, Schippers B 1996 Iron availability affects induction of systemic resistance to fusarium wilt of radish by *Pseudomonas fluorescens, Phytopathology* **861** 49-155
128. Leeman M, van Pelt JA, Denouden FM, Heinsbroek M, Bakker P, Schippers B 1995 Induction of systemic resistance against Fusarium wilt of radish by lipopolysaccharides of *Pseudomonas fluorescens, Phytopathology* **85** 1021-1027
129. Li Q, Saleh-Lakha S, Glick BR 2005 The effect of native and ACC deaminase-containing *Azospirillum brasilense* Cd1843 on the rooting of carnation cuttings, *Canadian Journal of Microbiology* **51** 511-514
130. Lim BL, Yeung P, Cheng C, Hill JE 2007 Distribution and diversity of phytatemineralizing bacteria, *ISME Journal* **1** 321-330
131. Liu X, Zhao H, Chen S 2006 Colonization of Maize and Rice Plants by Strain *Bacillus megaterium* C4, *Currunt Microbiology* 52 186–190
132. Loper JE, Gross H 2007 Genomic analysis of antifungal metabolite production by *Pseudomonas fluorescens*

Pf-5, *Eur Journal of Plant Pathology* **119** 265-278

133. Lynch JM, Whipps JM 1991 Substrate flow in the rhizosphere In *The rhizosphere and plant growth* (Eds D L Keister and B Cregan) Beltsville Sympos in Agric Res 14 Kluwer, Dordrecht, The Netherlands 15-24 pages

134. Lynch JM 1990 The Rhizosphere John Wiley and Sons, New York

135. M'Piga P, Belanger RR, Paulitz TC, Benhamou N 1997 Increased resistance to *Fusarium oxysporum* f sp*lycopersici* in tomato plants treated with the endophytic bacterium *Pseudomonas fluorescens* strain 63-28, *Physiology and Molecular Plant Pathology* **50** 301-320

136. Ma W, Sebestianova SB, Sebestian J, Burd GI, Guinel FC, Glick BR2003 Prevalence of 1-aminocyclopropane-1-carboxylate deaminase in *Rhizobium* spp, *Anton Leeuwen* **83** 285-291

137. Madhaiyan M, Poonguzhali S, Ryu J, Sa T 2006 Regulation of ethylene levels in canola (*Brassica campestris*) by 1-aminocyclopropane- 1-carboxylate deaminase-containing *Methylobacterium fujisawaense, Planta* **2242** 68-78

138. Mahaffee WF, Kloepper JW 1997 Temporal changes in the bacterial communities of soil, rhizosphere and endorhiza associated with field grown cucumber (*Cucumis sativus* L), *Microbiology Ecology* **34** 210-223

139. Malhotra M, Srivastava S 2008 An *ipdC* gene knock-out of *Azospirillum brasilense* strain SM and its implications on indole-3-acetic acid biosynthesis and plant growth promotion, *Anton Leeuw* **93** 425-433

140. Mamatha G, Shivayogi M S, Bagyaraj DJ, Suresh CK 2003 Effects of *Bacillus coagulans* and *Pseudomonas fluorescens* (PGPR's) on growth and nutrition of *Santalum album* L, *Geobios-(Jodhpur)* **30** 221-224

141. Martinez-Romero E, Wang ET 2000 *Sesbania herbacea–Rhizobium huautlense* nodulation in flooded soils and comparative characterization of *S herbaces* nodulating rhizobia in different environments, *Microbialogical Ecology* **41** 25–32

142. Martínez-Viveros O, Jorquera MA, Crowley DE , Gajardo G, Mora ML 2010 Mechanisms and practical considerations involved in plant growth promotion by rhizobacteria, *Journal of Soil Science and Plant Nutrition* **10**(3) 293-319

143. Marulanda-Aguirre A, Azcon R, Ruiz-Lozano JM, Aroca R 2008 Differential effects of a *Bacillus megaterium* strain on *Lactuca sativa* plant

growth depending on the origin of the arbuscular mycorrhizal fungus coinoculated physiologic and biochemical traits, *Journal of Plant Growth Regulation* **27** 10-18

144. Maurhofer M, Hase C, Meuwly P, Metraux JP, Defago G1994 Induction of systemic resistance of tobacco to tobacco necrosis virus by the root-colonizing *Pseudomonas fluorescens* strain CHAO influence ofthe gacA gene and of pyoverdine production, *Phytopathology* **84** 139-146
145. Mayak S, Tirosh T, Glick BR 1999 Effect of wild-type and mutant plant growth- promoting rhizobacteria on the rooting of mung bean cuttings, *Journal of Plant Growth Regulation* **18** 49–53
146. Mayak S, Tirosh T, Glick BR 2004 Plant growth-promoting that confer resistance to water stress in tomatoes and peppers, *Plant Science* **166** 525-530
147. Miethling R, Wieland G, Backhaus H, Tebbe CC 2000 Variation of microbial rhizosphere communities in response to crop species, soil origin and inoculation with *Sinorhizobium meliloti* L33, *Microbiological Ecology* **41** 43-56
148. Miransari M, Saleh-Rastin N, Asadi-Rahmani H, Khavazi K, Soltani A, Shoary-Nejati AR, Abbas-Zadeh P 2010 Plant growth-promoting activities of fluorescent pseudomonads, isolated from the Iranian soils, *Acta Physiologiae Plantarum* **32** 281–288
149. Mirza MS, Ahmad W, Latif F, Haurat J, Bally R, Normajd P, Malik KA 2001 Isolation, partial characterization, and the effect of plant growth-promoting bacteria (PGPB) on micropropagated sugarcane *in vitro, Plant Soil* **23** 747–54
150. Mishra RK, Prakash O, Alam M, Dikshit A 2010 Influence of plant growth promoting rhizobacteria (PGPR) on the production of *Pelargonium graveolens* L Herit, *Recent Research in Science and Technology* **2**(5) 53-57
151. Mostajeran A, Amooaghaie R, Emtiazi G 2007 The participation of the cell wall hydrolytic enzymes in the initial colonization of *Azospirillum brasilense* on wheat roots, *Plant Soil* **291** 239-248
152. Muthukumarasamy R, Revathi G, Vadivelu M 2000 *Acetobacter diazoptrophicus* prospects and potentialities – An overview In *Recent Advances in Biofertilizer Technology* (Eds A K Yadav, M R Motsara and S Ray Chauduri) Society for Promotion & Utilization of Resources and Technology, New Delhi 126-153 pages
153. Nakayama T, Homma Y, Hashidoko JM, Tahara S

1999 Possible role of xanthobaccins produced by *Stenotrophomonas sp strain SB-K88* in suppression of sugar beet damping-off disease, *Applied Environmental Microbiology* **65** 4334-4339

154. Naz I, Bano A, Hassan TU 2009 Isolation of phytohormones producing plant growth rhizobacteria from weeds growing in Khewra salt range, Pakistan and their implication in providing salt tolerance to *Glycine max* L, *African Journal of Biotechnology* **8** (21) 5762-5766

155. Noel TC, Sheng C, Yost CK, Pharis RP, Hynes MF 1996 *Rhizobium leguminosarum* as aplant growth-promoting rhizobacterium direct growth promotion of canola and lettuce, *Canadian Journal of Microbiology* **42** 279-283

156. Nowak J 1998 Benefits of *in vitro* 'biotization' of plant tissue cultures with microbial inoculants *In Vitro Cell & Developmental Biology* **34** 122–130

157. Ortiz Castro R, Valencia Cantero E, Lopez Bucio J 2008 Plant growth promotion by *Bacillus megaterium* involves cytokinin signaling, *Plant Signal Behaviour* **3** 263-265

158. Pal KK, Tilak KVBR, Saxena AK, Dey R, Singh CS 2000 Antifungal characteristics of a fluorescent *Pseudomonas* strain involved in the biological control of *Rhizoctonia solani, Microbiological Research* **155** 233-242

159. Pal KK, Tilak KVBR, Saxena AK, Dey R, Singh CS 2000 Enhancement of phosphate solubilization and siderophore production by Tn5 mutagenesis of a biocontrol rhizobacterium *Pseudomonas* spp Em85, *Journal of Microbiological World* **2** 9-15

160. Pandey P, Kang SC, Maheshwari DK 2005 Isolation of endophytic plant growth promoting *Burkholderia* sp MSSP from root nodules of *Mimosa pudica, Currunt Science* **89** 170-180

161. Paula MA, Reis VM, Dobereiner J 1991 Interaction of *Glomus clarum* with *Acetobacter diazotrophicus* in infection of sweet potato (*Ipomoea batatas*), sugarcane (*Saccharum* spp) and sweet sorghum (*Sorghum vulgare*), *Biology and Fertility of Soils* **11** 111-115

162. Penrose DM, Glick BR 2001 Levels of 1-aminocyclopropane- 1-carboxylic acid (ACC) in exudates and extracts of canola seeds treated with plant growth-promoting bacteria, *Canadian Journal Microbiology* **47** 368-372

163. Pettersson M, Bååth E 2004 Effects of the properties of the bacterial community on pH adaptation during recolonization of a humus soil, *Soil*

*Biology and Biochemistry* **36** 1383-1388

164. Pierson LS, Wood DW, Pierson EA 1998 Homoserine lactone-mediated gene regulation in plant-associated bacteria, *Annual Review of Phytopathology* **36** 207–225

165. Pieterse CMJ, van Loon LC 1999 Salicylic acid-independent plant defense pathways, *Trends in Plant Science* **4** 52-58

166. Pieterse CMJ, van Wees SCM, van Pelt JA, Knoester M, Laan R, Gerrits H, Weisbeek PJ, van Loon LC 1998 A novel signaling pathway controlling induced systemic resistance in *Arabidopsis, Plant Cell* **10** 1571-1580

167. Ping L, Boland W 2004 Signals from the underground bacterial volatiles promote growth in *Arabidopsis Trends in Plant Science* **9** 263-269

168. Pinton R, Varanini Z, Nannipieri P 2001 The rhizosphere as a site of biochemical interactions among soil components, plants and microorganisms In Pinton R, Varanini Z, Nannipieri P (Eds), The Rhizosphere Marcel Dekker, Inc, New York, USA 1-18 pages

169. Prithiviraj B, Zhou X, Souleimanov A, Kahn W, Smith DL 2003 A host-specific bacteria-to-plant signal molecule (Nod factor) enhances germination and early growth of diverse crop plants, *Planta* **216** 437–445

170. Raaijmakers JM, Bonsall RF, Weller DM 1999 Effect of population density of *Pseudomonas fluorescens* on production of 2,4- diacetyl-phloroglucinol in the rhizosphere of wheat, *Phytopathology* **89** 470-475

171. Raaijmakers JM, Vlami M, de Souza JT 2002 Antibiotic production by bacterial biocontrol agents, *Anton Leeuw* **81** 537-547

172. Rajarathinam K, Balamurugan T, Kulasekarapandian R, Veerasami S, Jayabalan M 1995 Isolation and screening of phosphate solubilizers from soil of Kamarajar district (Tamil Nadu), *Journal of Ecotoxicological Environment Monitor* **5** 155-157

173. Ramamoorthy V, Viswanathan R, Raguchander T, Prakasam V, Samiyappan R 2001 Induction of systemic resistance by plant growth-promoting rhizobacteria in crop plants against pests and diseases, *Crop Protection* 201–11

174. Ramarathnam R, Fernando WGD, Kievit T 2011 The role of antibiosis and induced systemic resistance, mediated by strains of *Pseudomonas chlororaphis*, *Bacillus cereus* and *B amyloliquefaciens*, in controlling blackleg disease of canola, *Biocontrol* **56** 225-235

175. Reed MLE, Glick BR 2005 Growth of canola (*Brassica napus*) in the presence of plant growth-promoting bacteria and either copper or polycyclic aromatic hydrocarbons *Canadian Journal of Microbiology* **51** 1061-1969

176. Rengel Z 2008 Bioavailability of phosphorus and micronutrients in the soil-plant-microbe continuum, *Journal of Soil Science and Plant Nutrition* **8** 84-91

177. Rezzonico F, Zala M, Keel C, Duffy B, Moënne-Loccoz Y, Défago G 2007 Is the ability of biocontrol fluorescent pseudomonads to produce the antifungal metabolite 2,4-diacetylphloroglucinol really synonymous with higher plant protection? *New Phytology* **173** 861-872

178. Richardson AE, Hadobas PA, Hayes JE, O'Hara CP, Simpson RJ 2001 Utilization of phosphorus by pasture plants supplied with myoinositol hexaphosphate is enhanced by the presence of soil microorganisms, *Plant Soil* **229** 47-56

179. Richardson AE, Hadobas PA 1997 Soil isolates of *Pseudomonas* spp that utilize inositol phosphates *Canadian Jounal of Microbiology* **43** 509-516

180. Rodriguez H, Vessely S, Shah S, Glick BR 2008 Effect of a Nickel-tolerant ACC deaminase producing Pseudomonas strain on growth of nontransformed and transgenic Canola plants *Currunt Microbiology* **57** 170–174

181. Roesti D, Gaur R, Johri BN, Imfeld G, Sharma S, Kawaljeet K, Aragno M 2006 Plant growth stage, fertiliser management and bio-inoculation of arbuscular mycorrhizal fungi and plant growth promoting rhizobacteria affect the rhizobacterial community structure in rainfed wheat fields, *Soil Biology and Biochemistry* **38** 1111–1120

182. Rudrappa T, Splaine RE, Biedrzycki ML, Bais HP 2008 Cyanogenic pseudomonads influence multitrophic interactions in the rhizosphere *PLOS ONE* 3 e2073doi 101371/journalpone0002073

183. Ryan PR, Delhaize E, Jones DL 2001 Function and mechanism of organic anion exudation from plant roots, *Annual Review of Plant Physiology and Plant Molecular Biology* **52** 527-560

184. Ryu CM, Farag MA, Hu CH, Reddy MS, Kloepper JW, Pare PW 2004 b Bacterial volatiles induce systemic resistance in *Arabidopsis, Plant Physiology* **134** 1017-1026

185. Ryu CM, Kim J, Choi O, Kim SH, Park CS 2006 Improvement of biological control capacity of *Paenibacillus polymyxa* E681 by seed pelleting

on sesame *Biological Control* **39** 282–289

186. Ryu CM, Murphy JF, Mysore KS, Kloepper JW 2004a Plant growth-promoting rhizobacterial systemically protect *Arabidopsis thaliana* against Cucumber mosaic virus by a salicylic acid and NPR1-independent and jasmonic acid-dependent signaling pathway *The Plant Journal* **39** 381-392

187. Safronova VI, Stepanok VV, Engqvist GL, Alekseyev YV, Belimov AA 2006 Root-associated bacteria containing 1-aminocyclopropane- 1-carboxylate deaminase improve growth and nutrient uptake by pea genotypes cultivated in cadmium supplemented soil *Biology and Fertility of Soils* **42** 267-272

188. Saikia N, Brezbaruah B 1995 Iron-dependent plant pathogen inhibition through *Azotobacter* RRLJ 203 isolated from iron-rich acid soils, *Indian Journal of Experimental Biology* **33** 571–575

189. Sattar MA, Gaur AC1985 Characterization of phosphate dissolving microorganisms isolated from some Bangladesh soil samples, *Bangle Journal of Microbiology* **2** 22-28

190. Saxena AK, Tilak KVBR 1998 Free-living nitrogen fixers Its role in crop production In *Microbes for Health, Wealth and Sustainable Environment* (Ed A K Verma) Malhotra Publ Co, New Delhi 25-64 pages

191. Schippers B, Bakker AW, Bakker PAHM, Van Peer R 1990 Beneficial and deleterious effects of HCN-producing pseudomonads on rhizosphere interactions, *Plant Soil* **129** 75-83

192. Schmalenberger A, Tebbe CC 2002 Bacterial community composition in the rhizosphere of a transgenic herbicide-resistant maize (*Zea mays*) and comparison to its non-transgenic cultivar, *FEMS Microbiological Ecology* **40** 29-37

193. Schwieger F, Tebbe CC 2000 Effect of field inoculation with *Sinorhizobium meliloti* L33 on the composition of bacterial communities in rhizosphere of a target plant (*Medicago sativa*) and a non target plant (*Chenopodium album*)-linking of 16S rRNA gene based single strand conformation polymorphisms community profiles to the diversity of cultivated bacteria, *Appiedl Environmental Microbiology* **66** 3556-3565

194. Sessitsch A, Coenye T, Sturz AV, Vandamme P, Barka E, Wang- Pruski G, Faure D, Reiter B, Glick BR, Nowak J 2005 *Burkholderia phytofirmins* sp Nov, a novel plant-associated bacteriumwith plant beneficial properties

*Internation Journal of Systematic Evolution and Microbiology* **55** 1187-1192

195. Shaharoona B, Arshad M, Zahir ZA 2006 Performance of *Pseudomonas* spp containing ACC-deaminase for improving growth and yield of maize (*Zea mays* L) in the presence of nitrogenous fertilizer, *Soil Biology and Biochemistry* **38** 2971-2975
196. Sharma A, Johri BN 2003 Growth promoting influence of siderophore-producing *Pseudomonas* strains GRP3A and PRS9 in maize (*Zea mays* L) under iron limiting conditions, *Microbiological Research* **158** 243-248
197. Sharma VK, Nowak J1998 Enhancement of verticillium wilt resistance in tomato transplants by in vitro coculture of seedlings with a plant growth-promoting rhizobacterium (*Pseudomonas* sp strain PsJN), *Canadian Journal of Microbiology* **44** 528–536
198. Shedova E, Lipasova V, Velikodvorskaya G, Ovadis M, Chernin L, Khmel I 2008 Phytase activity and its regulation in a rhizospheric strain of *Serratia plymuthica, Folia Microbiology* **53** 110-114
199. Shoebitz M, Ribaudo CM, Pardo MA, Cantore L, Ciampi L, Curá JA 2009 Plant growth promoting properties of a strain of *Enterobacter ludwigii* isolated from *Lolium perenne* rhizosphere, *Soil Biol Biochemistry* **411** 768-1774
200. Siddiqui ZA, Shaukat SS, Hussain Sheikh I, Khan A 2006 Role of cyanide production by *Pseudomonas fluorescens* CHA0 in the suppression of root-knot nematode, *Meloidogyne javanica* in tomato, *World Journal of Microbiology and Biotechnology* **22** 641-650
201. Siddiqui ZA 2006 PGPR Prospective biocontrol agents of plant pathogens In *PGPR Biocontrol and Biofertilization* (Ed ZA Siddiqui) Springer, Netherlands 67-109 pages
202. Son SH, Khan Z, Vim SG, Kim YH 2008 Effects of seed treatment with rhizobacterium, *Paenibacillus* species on management of root-knot nematode-*Fusarium* wilt fungus disease complex in tomato plants, *Russian Journal of Nematology* **16** 97–105
203. Spaepen S, Vanderleyden J, Remans R 2007 Indole-3-acetic acid in microbial and microorganism-plant signaling, *FEMS Microbiological Review* **31** 425-448
204. Stiens M, Schneiker S, Keller M, Kuhn S, Pühler A, Schlüter A 2006 Sequence analysis of the 144-kilobase accessory plasmid psmesmlla, isolated from a dominant *Sinorhizobium meliloti* strain

identified during a long-term Weld release experiment, *Applied Environmental Microbiology* **72** 3662-3672

205. Strigul NS, Kravchenko LV 2006 Mathematical modeling of PGPR inoculation into the rhizosphere, *Environmental Modeller Software* **21** 1158–1171

206. Svercel M, Duffy B, Défago G 2007 PCR amplification of hydrogen cyanide biosynthetic locus *hcnAB* in *Pseudomonas spp, Journal of Microbiological Method* **70** 209-213

207. Swain MR, Naskar SK, Ray RC 2007 Indole-3-acetic acid production and effect on sprouting of Yam (*Dioscorea rotundata* L) minisetts by *Bacillus subtilis* isolated from culturable cowdung microflora, *Polish Journal of Microbiology* **56** 103-110

208. Tilak KVBR, Ranganayaki N, Pal KK, De R, Saxena K, Nautiyal CS, Mittal S , Tripathi A K, Johri BN 2005 Diversity of plant growth and soil health supporting bacteria, *Currunt Science* **89** 136-150

209. Tilak KVBR, Saxena AK 1999 Potentials and prospects of *Rhizobium* biofertilizer In *Current Trends in Life Sciences, Agromicrobes* (Ed MN Jha) Today & Tomorrow Printers & Publishers, New Delhi 51-78 pages

210. Tilak KVBR, Saxena AK 2001 *Azospirillum* – Its impact on crop production In *Recent Advances in Biofertilizer Technology* (Eds A K Yadav, M R Motsara and S Ray Chauduri) Society for Promotion & Utilization of Resources and Technology, New Delhi 176-189 pages

211. Timmusk S, Wagner GH 1999 The plant-growth-promoting rhizobacterium *Paenibacillus polymyxa* induces changes in *Arabidopsis thaliana* gene expression a possible connection between biotic and abiotic stress responses, *Molecular Plant Microbe Interaction* **12** 951–959

212. Trivedi P, Pandey A 2008 Plant growth promotion abilities and formulation of *Bacillus megaterium* strain B 388 (MTCC6521) isolated from a temperate Himalayan location, *Indian Journal of Microbiology* 48 342–347

213. Tsavkelova EA, Klimova SY, Cherdyntseva TA, Netrusov AI 2006 Microbial producers of plant growth stimulators and their practical use a review, *Applied Biochemistry and Microbiology* **42** 117-126

214. Turner BL, Papházy MJ, Haygarth PM, McKelvie ID 2002 Inositol phosphates in the environment, *Philosophical Transactions of the Royal Society* B **357** 449-469

215. Uchiumi T, Oowada T, Itakura M, Mitsui H, Nukui N, Dawadi P, Kaneko T, Tabata S, Yokoyama T, Tejima T, Saeki K, Oomori H, Hayashi M, Maekawa T, Sriprang R, Murooka Y, Tajima S, Simomura K, Nomura M, Suzuki A, Shimoda S, Sioya K, Abe M, Minamisawa K 2004 Expression islands clustered on symbiosis island of *Mesorhizobium loti* genome, *Journal of Bacteriology* **186** 2439-2448

216. UnnoY, Okubo K, Wasaki J, Shinano T, Osaki M 2005 Plant growth promotion abilities and microscale bacterial dynamics in the rhizosphere of lupin analysed by phytate utilization ability, *Environmental Microbiology* **73** 96-404

217. Van Elsas JD, Boersma FGH 2011 A review of molecular methods to study the microbiota of soil and the mycosphere, *European ournal of JSoil Biology* **47** 77-87

218. Van Loon LC, Bakker PAHM, Pieterse CMJ 1998 Systemic resistance induced by rhizosphere bacteria, *Annual Review of Phytopathology* **36** 453-483

219. van Loon LC 2007 Plant responses to plant growth-promoting rhizobacteria, *Europian Jourtnal of Plant Pathology* **119** 243-254

220. Van Peer R, Niemann GJ, Schippers B 1991 Induced resistance and phytoalexin accumulation in biological control of *Fusarium* wilt of carnation by *Pseudomonas* sp strain WCS417r *Phytopathology* **81** 728-734

221. van Wees S, Pieterse C, Trijssenaar A, Van't Westende Y, Hartog F, van Loon LC 1997 Differential induction of systemic resistance in *Arabidopsis* by biocontrol bacteria, *Molecular Plant-Microbe Interaction* **10** 716-724

222. Vassilev N, Vassileva M, Nikolaeva I 2006 Simultaneous P-solubilizing and biocontrol activity of microorganisms potentials and future trends, *Applied Microbiology Biotechnology* **71** 137-144

223. Velusamy P, Immanuel JE, Gnanamanickam SS, Thomashow L 2006 Biological control of rice bacterial blight by plant-associated bacteria producing 2,4- diacetylphloroglucinol, *Canadian Journal of Microbiology* **52** 56–65

224. Venkateshwarlu B, Rao AV, Raina P 1984 Evaluation of phosphorus solubilization by microorganisms isolated from arid soil, *Journal of the Indian Society of Soil Science* **32** 273-277

225. Vessey JK 2003 Plant growth promoting rhizobacteria as biofertilizers, *Plant Soil* 255 571- 586

226. Viswanathan R, Samiyappan R 1999 Induction of systemic resistance by plant growth-

promoting rhizobacteria against red rot disease caused by *Colletotrichum falcatum* went in sugarcane *In Proceedings of the Sugar Technology Association of India*, vol 61 Sugar Technology Association, New Delhi, India 24–39 pages

227. Viveros OM, Jorquera MA, Crowley DE, Gajardo G, Mora ML 2010 Mechanisms and practical considerations involved in plant growth promotion by rhizobacteria, *Journal of Soil Science and Plant Nutrtrion* **10** (3) 293 – 319

228. Voisard C, Keel C, Haas D, Défago G 1989 Cyanide production by *Pseudomonas fluorescens* helps suppress black root rot of tobacco under gnotobiotic conditions, *EMBO Journal* **8** 351-358

229. Wachowska U, Majchrzak B, Borawska M, Karpinska Z 2004 Biological control of winter wheat pathogens by bacteria Acta fytotech zootech, Vol 7, Special Number, Proceedings of the XVI Slovak and Czech Plant Protection Conference organized at Slovak Agricultural University in Nitra, Slovakia

230. Wei L, Kloepper JW, Tuzun S 1991 Induction of systemic resistance of cucumber to *Colletotrichum orbiculare* by select strains of plant growth-promoting rhizobacteria, *Phytopathology* **81** 1508-1512

231. Whipps JM 2001 Microbial interactions and biocontrol in the rhizosphere, *J Exp Bot* 52 487–511

232. Woyessa D, Assefa F 2011 Effects of Plant Growth Promoting Rhizobacteria on Growth and Yield of Tef (*Eragrostis tef* Zucc Trotter) under Greenhouse Condition, *Res Journal Microbiology* 6(4) 343-355

233. Yamada Y, Hoshino K, Ishikawa T 1998 Taxonomic studies of acetic acid bacteria and allied organisms XII, The phylogeny of acetic acid bacteria based on the partial sequences of 16S ribosomal RNA, *International Journal of Systematic Bacteriology* **48** 3270-3280

234. Yang J, Kloepper J, Ryu C 2009 Rhizosphere bacteria help plants. tolerate abiotic stress, *Trends in Plant Science* 14 1-4

235. Zabihi HR, Savaghebi GR, Khavazi K, Ganjali A, Miransari M 2010 *Pseudom-onas* bacteria and phosphorous fertilization, affecting wheat (*Triticum aestivum* L) yield and P uptake under greenhouse and field conditions, *Acta Physiologiae Plantarum* in press

236. Zahir AA, Arshad M, Frankenberger WT 2004 Plant growth promoting rhizobacteria applications and persp-

ectives in agriculture, *Advanced Agronomy* **8** 197-168

237. Zahir ZA, Abbas SA, Khalid M, Arshad M 2000 Substrate dependent microbially derived plant hormones for improving growth of maize seedlings, *Pakisthan Journal of Biological Science* **32** 89-291

238. Zahir ZA, Ghani U, Naveed M, Nadeem SM, Asghar HN 2009 Comparative effectiveness of Pseudomonas and Serratia sp containing ACC-deaminase for improving growth and yield of wheat (*Triticum aestivum* L) under salt-stressed conditions, *Archive of Microbiology* **191** 415–424

239. Zehnder GW, Murphy JF, Sikora EJ, Kloepper JW 2001 Application of rhizobacteria for induced resistance, *Europian Journal of Plant Pathology* **107** 39-50

240. Zumstein E, Moletta R, Godon JJ 2000 Examination of two years of community dynamics in an anaerobic bioreactor using fluorescens polymerase chain reaction (PCR) single strand conformation polymorphisms analysis, *Environmental Microbiology* **2** 69-78.

*Rev. Plant Pathol. Vol. 6, 2014*
*Indian Society of Mycology and Plant Pathology*
*Scientific Publishers (India), Jodhpur pp. 269-297*

# ACTINOMYCETES: DIVERSITY, DISTRIBUTION, ECONOMIC IMPORTANCE AND THEIR ROLE IN PLANT PROTECTION

***M. Krishnaraj*[1], *V.R. Prabavathy*[2] *and N. Mathivanan*[1*]**

[1]Biocontrol and Microbial Metabolites Laboratory, Centre for Advanced Studies in Botany, University of Madras, Guindy Campus, Chennai – 600 025, TN, India
[2]Microbiology Lab, M.S. Swaminathan Research Foundation, 3[rd]Cross Street, Taramani Institutional Area, Chennai – 600113, TN
* E-mail: prabhamathi@yahoo.com

**KEY WORDS :** Actinomycetes, Biocontrol agents, Plant pathogens

## ABSTRACT

Actinomycetes are being considered as sustained mines for various enzymes, antibiotics and other bioactive metabolites till date. The global drug discovery programs are still searching novel bioactive leads from the microbial products especially from actinomycetes. Apart from the medical field, handful exploitation of actinomycetes towards the successful crop protection in agriculture sector is made every part of the world. Many research and developments are coming out regularly, which empower the farmer as well as agriculture industries towards the integrated crop management. Many actinomycetes have been discovered and used; lot more remains in adjacent environment to be recovered. In this review article, we enlighten the master beast "actinomycetes" credential as enzyme producer, antibiotic and other active metabolites producer, biocontrol agents, plant pathogens and bioremediation agents. This review also emphasizes the biogeography, isolation and culture practices, economic importance and future directions of the actinomycetes research.

## INTRODUCTION

The Class *Actinobacteria* is gram positive filamentous bacteria with a high guanine + cytosine (G+C) content and is delineated into four Subclasses (104). From the Subclass *Actinobacteridae*the Order Actinomycetales are commonly referred as actinomycetes. Actinomycetes are the most studied of the actinobacteria thus include their diverse nature of habitants such as plant symbionts and human pathogens (11).Recently, there is rising interest in the discovery of eco-friendly bio-herbicidal compounds and biocontrol agent (BCA) for pest and weed control in sustainable agriculture (80) as well as in plant growth promotion(32). Being an antibiotic producers or hyperparasites of fungi, actinomycetes have played a pivotal role in controlling soil-borne plant pathogens (37). The moderndiscovery targeted to multiple activity of actinomycete which includes, showing pathogenic activity against the nematode may indicate the potential for it to be used as a biocontrol agent of parasitic nematodes, in addition to its ability to suppress fungal pathogens (86) may be encouraged. Bio-prospecting program directed towards discovery of soil microorganisms that produce herbicidal and insecticidal compounds especially from actinomycetes, several herbicidal (cycloheximide, geldanamycin, nigericin, (+)-epiepoformin, 3-hydroxybenzyl alcohol, 2-methylhydroquinone) or insecticidal (avermectin, valinomycin) compounds produced by these microorganisms have already been isolated, identified, and evaluated for potential as pesticides. Actinomycetes constitute a prevalent microbial component in most of the soil biota because of their metabolic diversity and growth characteristics, mycelial forms and relatively rapid colonization of selective substrates, indicate them are well suited for bioremediation of heavy metals and other organic compounds (90).The global crop protection market relay on the natural products which exhibit low toxic or nil have progressed significantly and actinomycetes, especially from the genus *Streptomyces*, prospects to be a suitable candidate to find alternative methods to control plant diseases (9).

## DIVERSITY AND DISTRIBUTION

Natural products with industrial applications can be produced from primary and secondary metabolism of living organisms especially from microbes. Actinomycetes are targeted for their metabolites owing superior applicability and abundant nature in the entire environmental niche. Many researchers reported regularly, actinomycetes have been isolated from rhizosphere soils (57) for the management of various plant diseases caused by fungal, bacterial and insect phytopathogens.A total of five thousand actinomycetes were isolated from soil samples collected from rainforests in Singapore and identified based on the polyphasic taxonomical approaches. Actinomycetes belonging to a total of 36 genera were identified. The most abundant isolates are members of *Streptomyces*, *Micromonospora*, *Actinoplanes*, *Actinomadura*, *Nonomuria*, *Nocardia* and *Streptosporangium* (124). Thirty-five rare actinomycete strains belonged to one of the following genus, *Micromonospora*, *Microbispora*, *Actinoplanes* and *Streptosporangium* were isolated from soil samples were collected from the humus layer of forest soil on Mt. Baekun, Chonnam, Korea (99).Recently, many researchers have isolated actinomycetes from different locations of the Bay of Bengal starting from Pulicat lake to Kanyakumari (63,89,95). A halotolerant alkaliphilic actinomycete, Kut-8, was isolated from saline desert of Kutch, Western India and it has also been identified as *Streptomycesaburaviensis* based on the chemotaxonomic characteristics, including cell wall constituents (113). Seventeen different media have been used for the recovery of this microbial group from marine sediments samples collected in the Gulf of California and the Gulf of Mexico and isolated nearly 300 actinomycetes from the selective isolation plates (73).A total of thirty-two isolates of actinomycetes were isolatedand identified, most of the isolates belonged to the genus *Streptomyces*. Some less frequent isolates were classified in the genera *Nonomuraea*, *Nocardiopsis* and *Actinomadura* frombee hives in Thailand (93).Thirty-eight actinomycetes were isolated from sediment collected from the Mariana Trench (10,898 m) using marine agar and media selective for actinomycetes, notably raffinose-histidine agar. The isolates were assigned to the class *Actinobacteria* using primers specific for members of this taxon. The phylogenetic analysis based on 16S rRNA gene sequencing showed that the isolates belonged to the genera *Dermacoccus*, *Kocuria*, *Micromonospora*, *Streptomyces*, *Tsukamurella* and *Williamsia* (87). Interestingly, 52 actinobacteria were isolated from the marine sponge *Iotrochota* sp. collected in shallow water of the South China Sea using selective media for actinobacteria (60).A thermophilic actinomycetes *Thermasporomycescomposti* gen. nov., sp. nov. was isolated from compost (127). On the other hand, *Nocardioidesalpines* sp. nov., a psychrophilic actinomycete was isolated from alpine glacier cryoconite (132). Obviously, bio-prospecting of these actinomycetes still warrant newer isolation technology and accurate assay methodologies for gains more from these microbes.

## ISOLATION, CULTURING AND MAINTENANCE

The development and application of new methodologies with which to isolate less-explored actinomycete taxa is an important for improving our knowledge about their taxonomy, ecology and industrial applications. You and Park (130) developed a new combination of coal-vitamin medium was developed to isolate actinomycetes from soil (130) which showed superior to existing media and increased the number of actinomycetes and inhibited the growth of other soil bacteria. In addition, the pretreatment of soil suspension with peptone (6%) and lauryl sulfate (0.05%) at 50°C for 10 min, greatly increased the number of actinomycetes isolates. Isolation of endophytic actinomycetes from plant samples has been described (6) which includes, plant samples after washing in running water in order to remove soil particles followed by sequential immersion in 70% ethanol for 5 min and a solution of sodium hypochlorite (0.9% available chlorine) for 20 min, once again samples to be washed in sterile water thrice to remove surface sterilization agents before being soaked in 10% $NaHCO_3$ solution to disrupt the plant tissues and inhibit the growth of fungi. Each sample was divided into small fragments and plated on S medium with 15 µg nalidixic acid $ml^1$ to suppress the growth of bacteria, and incubated at 26°C.A moderately halophilic actinomycete, designated HA-9 was isolated from a saltern in Kunsan, Republic of Korea by a relatively simple amendment of seawater in the Bennet medium yielded a newer actinomycetes *Nocardiopsiskunsanensis* sp. nov. (22). Various pretreatment procedures and selective media have been used to assess the optimal conditions for the isolation of rare actinomycetes from soil (99) and it was found among the pretreatment, wet-heating for 15 min at 70°C and phenol treatment of soil suspension were the most effective methods for the isolation and among the media hair hydrolysate vitamin agar (HHVA) was the best medium for the isolation of rare actinomycetes. Sucrose gradient differential centrifugation has been developed for successful and efficient isolation of *Nocardia* populations that are suppressed by conventional dilution plating approaches from rhizosphere soil (128). Another preferential desert actinomycetes isolation medium containing glucose, yeast extract and mineral salts has been described (51) which showed that the highest actinomycete diversity count with more selective for actinomycetes and significantly decreased the number of fungal colonies and to a lower extent the number of bacterial colonies. A selective medium containing mupirocin and metronidazole has been evaluated (71) for the isolation actinomycetes from clinical material and this medium significantly isolated more actinomycetes than the nonselective media. Another technique to reduce the numbers of thermophilic bacteria on isolation plates has been developed, which in turn facilitated the detection and isolation of thermophilic actinomycetes (64) by using a novel approach to exposing the test material to bacteriophage suspensions prior to inoculation on isolation plates which yielded more actinomycetes such as *Thermomonospora*, *Saccharopolysporarectivirgula*, and thermophilic *Streptomyces* spp. A novel technique called rehydration and centrifugation (RC) method to isolate and relatively simple enrichment technique which enables rapid and selective isolation of diverse zoosporic actinomycete genera directly from soil and plant litter has been described (48), which consists of immersing the air-dried source material in 10 mM phosphate buffer containing 10% soil extract, letting the preparation stand at 30°C for 90 min, followed by centrifugation of the fluid at 1500 × g for 20 min. Portions of the supernatant containing actinomycete zoospores are plated on the humic acid-vitamin agar which is supplemented with nalidixic acid and trimethoprim as the selective inhibitors for Gram-negative bacteria and bacilli. The treatment of soil with 3minutes electric pulses at a field intensity of 12 kV/cm$^2$ resulted in a significant increase rare actinomycetes isolation (17). From this treatment efficiently isolated representatives of the genera *Actinoplanes*, *Micromonospora*, *Actinomadura*, *Microtetraspora*, *Streptosporangium*, *Saccharotrix* and *Nocardioides*. Selective isolation of Kitasatospora strains using novobiocin or Actinoplanes strains using chemotactic (KCl) and the use of gellan gum as a solidifying agent 9 novel compounds have been documented (111) and two new genera, five new species and one new subspecies have been proposed.

## ECONOMIC IMPORTANCE

Actinomycetes are pivotal source of a very huge number of antibiotics and other bioactive metabolites. They used as biological agents to control the devastating plant pathogen which includes fungi, bacteria and insects. They also helpful in converting agricultural wastes into useful industrial products like alcohol, proteins, etc. Actinomycetes especially from the genus *Streptomyces* offered more than 100 antibiotics to combat the serious outbreaks across the world caused by various pathogen mainly multidrug resistant pathogens which are currently in the market (12). Actinomycetes have also been imparted as potential veterinary pathogens affecting many different animal species. *Dermatophiluscongolensis* is the causative agent of dermatophilosis (also called streptothricosis, lumpy wool disease, rain rot, senkob skin disease) of cattle. Economic losses to farmers due to this disease is serious, notably through loss of animal's physical vitality, reduction of milk and meat production and reduced reproductive performance and loss of wool quality in sheep(107). They may cause significant morbidity and mortality, in particular in highly susceptible severely immunocompromised patients, including transplant recipients and patients infected with human immunodeficiency virus and cancer patients (4,77). Some of the genus of the actinomycetes causes diseases in crop plants, which are economically important. Predominantly, the genus *Streptomyces*, *Rhodococcus*, *Clavibacter* and *Liefsonia* are most devastating plant pathogens and made huge economic losses to the farming community (10,70,120). Actinobacteria are playing a crucial role in the recycling of refractory biomaterials by decomposing complex mixtures of polymers in dead plant, animal and fungal materials. They are important in soil biodegradation and humus formation by the recycling of nutrients associated with recalcitrant polymers such as keratin, lignocelluloses and chitin (46,75,103). Actinomycetes are being used in integrated crop management strategy to control the plant diseases across the world which prevent the both crop plant and economic losses to the farmer community as well (28,36,54,86,91).

## INDUSTRIAL ENZYMES

Industrialization revolutionized entire world past five decades, thus produces a lots of revenues along with environmental pollutions. Nowadays, the microbial enzymes are replacing the chemical catalysts in manufacturing chemicals, leather goods, food, paper, pharmaceuticals and textiles (96). Seventeen strains of actinomycetes antagonistic to yeast and moulds have been found to produce amylase, lipase, gelatinase and caseinase using solid media containing starch, Tween-20, gelatin and skimmed milk, respectively(59). Actinomycetes are capable to produce various industrial enzymes includes, alkaline proteases, L-glutaminase, ligninases, amylases, cellulases, lipases,xylanases, pectinases, and chitinases have been reported (1,7,20,44,65,76,82,96,105).

## ANTIBIOTICS

Versatility of antibiotics has now been isolated from various cultures of actinomycetes. Some of them namely, actinomycetin, micromonosporin, mycetin, and actinomyces lysozyme, have been only partly purified, whereas others, including actinomycin, proactinomycin, streptothricin, and streptomycin, have been isolated and crystallized. These substances are differing greatly in their chemical structure, antimicrobial properties, toxicity to animals, and *in vivo* activity (122). The fame of streptomycetes as versatile producers of secondary metabolites started with the discovery of actinomycin in 1940, followed by streptomycin in 1943. Two-thirds of the marketed microbial drugs are produced by streptomycetes (74). More than 10,000 antibiotic compounds were registered under MICrobial Metabolite Database from actinomycetes alone (12).The novel antibiotics are being discovered from actinomycetes frequently, which are biologically active chemotypes, such as thienamycin, daptomycin and echinocandins (81). It has been estimated that more than 120 bioactive molecules from actinomycetes especially from genus *Streptomyces* reached clinical use as natural products per se or as semisynthetic derivatives.

## OTHER BIOACTIVE COMPOUNDS

A potent immunosuppressive along with antifungal 32-membered macrolide compound, brasilinolide A was successfully isolated from the cultured broth of the actinomycete *Nocardiabrasiliensis* IFM 0406 (101).The production of tetrodotoxin (TTX), known otherwise as puffer fish toxin, was first time isolated marine actinomycetes (55). Actinomycetes are the bioactive factories for a large number of unusual natural products with different biological activities, including antitumor properties. These antitumor compounds diversified to several structural classes (i.e. anthracyclines, enediynes, indolocarbazoles, isoprenoides, macrolides, non-ribosomal peptides and others) and they exert antitumor activity by inducing apoptosis through DNA cleavage mediated by topoisomerase I or II inhibition, mitochondria permeabilization, inhibition of key enzymes involved in signal transduction like proteases, or cellular metabolism and in some cases by inhibiting tumor-induced angiogenesis. An isolate *Streptomycesendus* N40 produces an anti-algal compound against *Anabaena* sp. *Anabaenaflos-aqueae*, *Nostoc* sp. and *Anacystisnidulans* and identified as Niromycin A based on the spectral analyses (35). A novel anti-HIV protein, referred to as actinohivin, that inhibits syncytium formation in HIV has been successfully cloned and sequenced the gene encoding actinohivin from the actinomycete strain K97-0003, and expressed in *Escherichia coli* (56). Pure compound of salinosporamide A, produced by the *Salinisporatropica*, showed a strong inhibitory activity against the erythrocytic stages of the malarial parasite cycle and biochemical experiments support the likely inhibition of the parasite 20S proteasome (94).

## BIOREMEDIATION

Environmental clean up by microorganisms are being carried out through out world due to the urge posed by those pollutant effects on climate change and habitat destruction. Researchers trend now been changed to think of Gram positive bacterial populations to use in the refreshing field of bioremediation. Recently, an actinomycete recovered from chronically oil-polluted coastal region which degraded aliphatic and aromatic fractions of three different crude oils (100). An *Streptomyces* growing in chromium amended medium without any difficulties and reduce almost 100% of 0.96 mM Cr(VI), both in complex media (29,67). More than 10 actinomycete strains isolated from contaminated soils which experimentally challenged with Cr(VI) and has been successfully removed (90). Among actinomycete species, *Microbacterium*, *Arthrobacter* and *Streptomyces* showed Cr(VI) reduction ability (50,67,88). The removal of radionuclides such as uranium and thorium from aqueous solutions, especially from contaminated sources, is an important topic in environmental control. High uranium adsorption ability was exhibited by actinomycetes. Among actinomycetes, *Streptomyceslevoris* cells could adsorb about 380 μmol of uranium per gram dry cells from the solution (115). Pyrethroid insecticide cypermethrin was degraded by a new strain of *Streptomyces parvulus* which was isolated from waste water sludge (72), while pesticides removal by actinomycetes isolates were also observed with chlordane (40).

## ASSOCIATION WITH PLANT

*Frankia* strains, symbionts of actinorhizal plants, can induce $N_2$-fixing root nodules on certain non-leguminous plants and were identified as actinomycetes. Some endophytic *Streptomyces* synthesize cytokinin-like metabolites that inhibit rice seed germination and rooting (133). Endophytic actinomycete *Streptomyces griseorubiginosus* showed antagonistic activity against *Fusarium oxysporum* f. sp. *cubense*in banana leaf extract agar (18) and banana plant tissues and produce antifungal metabolites *in situ*. However, these endophytic actinomycetes can penetrate the sterile plant tissue easily. A root-colonizing *Streptomyces* species, *Streptomyceslydicus* WYEC108 has association with legume *Pisumsativum* and this interaction is potentially of great importance to the health and growth in nature of this nodulating legume. This *S. lydicus* WYEC108 influences pea root nodulation by increasing root nodulation frequency, possibly at the level of infection by *Rhizobium* spp. and also colonizes and then sporulates within the surface cell layers of the nodules (114).

## PLANT PATHOGENS

Only few of the actinomycetes genera include *Clavibacter*, *Streptomyces*, *Leifsonia* and *Rhodococcus*are plant pathogenic forms. Among these genera *Streptomyces* and *Rhodococcus* have wide host range whereas, *Clavibacter* and *Liefsonia* are host-specific at the species or subspecies level (49). *Streptomyces scabies* can infect vide range of crops, but is most commonly encountered causing disease on tuber and tap root crops. It causes common scab on potato (*Solanum tuberosum*), beet (*Beta vulgaris*), carrot (*Daucus carota*), parsnip (*Pastinaca sativa*), radish (*Raphanus sativus*), rutabaga (*Brassicanapo brassica*) and turnip (*Brassicarapa*) (70). A new biosynthetic gene clusters from *Streptomyces scabies* has been reported (13) that produces coronafacic acid, part of the plant toxin coronatine which mimics the plant hormone jasmonate, contributing to virulence. The gram-positive coryneform bacteria cause economic losses in several crops worldwide, the general genetic intractability of these bacteria have long been deterrents to successful identification of the specific molecular mechanisms by which they cause diseases in plants (16). A pathogenic adaptation niche region in their genome of *Clavibacter michiganensis* subsp. *sepedonicus* has been discovered (10), which is a plant-pathogenic bacterium and the causative agent of bacterial ring rot, a devastating agricultural disease under strict quarantine control and zero tolerance in the seed potato industry. The genus *Leifsonia* includes xylem-limited, fastidious bacterial pathogens. The best known of these pathogens is *L. xyli* subsp. *xyli*, (Lxx) the causative agent of a systemic disease called ratoon stunting of sugarcane. *Rhodococcus fascians* infects both monocot and dicot plants, many of which are economically important crops (120).

## BIOCONTROL AGENTS

Actinomycetes have been evaluated as biocontrol agents against various plant pathogens (3,28,33,45,85,134). The antagonistic activity of actinomycetes against *Alternaria mali*, *Colletotrichum gleosporiodes*, *Fusarium oxysporum* f. sp.

*cucumerinium* (3), *Magnoporthe grisea*, *Phytophthora capsici* and *Rhizoctonia solani* (24,68) has already been reported. Antagonistic effects of *Streptomyces violaceusniger* on *Fusarium oxysporum* proved that the inhibition of pathogen by actinomycete was due to the production of antibiotic (42), the strain was used for biological control of Fusarium wilt disease of banana. Endophytic actinomycetes strains belonged to *Streptomyces*, *Streptoverticillium*, *Streptosporangium* were isolated from surface-sterilized banana roots (19), among them 18.3% of the isolates inhibited the growth of pathogenic *Fusarium oxysporum* f. sp. *cubense*.

*Streptomyces lydicus* as a potential biocontrol agent against fungal root rot and seed rot diseases have been reported (131). The streptomycetes have been shown to inhibit the soil fungi (117) and this ability may be related to the secretion of extracellular chitinase during host antagonist interaction (30). Many actinomycetes were active against different species of *Fusarium* and *Rhizoctonia* (28) of which, *Streptomyces spiroverticillatus* was effective in controlling wilt disease of French bean and banana. *Streptomyces* spp. have been shown to protect several plants from different soil-borne pathogens in glasshouse experiments (14,53,78,97,98,108,116,117,131) and also in field experiments (108,109). *Streptomyces hygroscopicus* was found to be an effective antagonist to *Fusarium graminearum*, the causal organism of fusariosis in wheat (43).

*Streptomyces* strain5406 has been successfully used for several decades in the treatment of cottonseeds in China (129). *Streptomyces* species were tested for their ability to control fungal diseases in Finland (109) and a strain of *Streptomyces griseoviridis* has been developed as biofungicide and tested on a wide range of glass house and field crops (66,110). Later a commercial product in the name of Mycostop has been developed using *S. griseoviridis* K61 isolated from Sphagnum peat and commercially exploited as biocontrol agents against various fungal pathogens (27). Different formulations of *Streptomyces* sp have been developed (98) and among them, the talc formulation was highly effective in controlling Rhizoctonia damping-off in tomato compared to the formulations made with alginate beads and starch granule. Interestingly, several *Streptomyces* spp. are known to form intimate relationship with plants, at times they are believed to prevent the infection from plant pathogens. The presence of non-pathogenic *Streptomyces* in the root tissue of healthy wheat plants has been reported .

Soil actinomycetes particularly *Streptomyces* spp. enhances the soil fertility and exhibited antagonistic activity against wide range of soil-borne plant pathogens (2). They isolated 110 different actinomycetes from Iranian soils, of which, only 14 of them exhibited antifungal activity. Further, they reported that the *Streptomyces pilicatus* strain 101, *Frankia* sp. strain 103 and *Streptomyces* sp. strain 44 had the widest antifungal activity against *Alternaria alternata, Alternaria solani, Fusarium solani, Phytophthora megasperma* and *Verticillium dahliae*. Endophytic streptomycetes strains isolated from tomato plants inhibited the growth of *R. solani* (18). Application of *Streptomyces melanosporofaciens* EF-76 spores incorporated beads of chitosan and polyphosphate using the entrapment technique (complex coacervation) to seed potato tubers protected common potato scab (61). *Streptomyces padanus* SS-07 and *Streptomyces xantholiticus* SS-09 exhibited strong antagonistic effect against *R. solani* (23). Treating the soil with *S. padanus* or *S. xantholiticus* alone or in combination with 1% (w/w) PBGG significantly reduced the colonization of *R. solani* in cabbage seeds compared to untreated controls. They also suggested that soil application with *S. padanus* or *S. xantholiticus* either alone or in combination with 1% bio-fungicide formulation significantly reduced the damping-off of Chinese cabbage caused by *R. solani*. Commericial formulation of *S. griseoviridis*, Mycostop has been commercially used for the control soil-borne pathogens in greenhouse grown tomato (78).

Actinomycetes are quantitatively and qualitatively abundant in the rhizosphere, where they may influence on plant growth promotion and protect plant roots against invasion by root pathogenic fungi (28). A *Streptomyces* from rhododendron produced a novel antifungal antibiotic, fistupyrone, which protected Chinese cabbage from infection by *Alternaria brassicicola* (54). Among 53 *Streptomyces* strains isolated from soils of Minnesota, Nebraska, and Washington (125), eight isolates having the greatest pathogen-inhibitory capabilities against *Phytophthora medicaginis* and *Phytophthora sojae in vitro*. were subsequently tested for their ability to control *Phytophthora* root rots on alfalfa and soybean in sterilized vermiculite and naturally infested field soil and significantly reduced root rot severity in alfalfa and soybean caused by *P. medicaginis* and *P. sojae*, respectively (P<0.05). An actinomycete, *Actinoplanes missouriensis*, derived from lupin roots, inhibited the growth of *Plectosporium tabacimum*, the causal pathogen of lupin root rot (36). An isolate from Western Australia, *Streptoverticillium albireticuli* showed strong nematicidal activity against *Caenorhabditis elegans* which pathogenic activity was also evident with the colonized nematode by the isolate on fishmeal extract agar. It also showed remarkable antifungal activity against certain fungal pathogens such as *Rhizoctonia solani*, *Phytophthora cinnamomi* and *Fusarium oxysporum* (86). A potent *Streptomyces* sp. was isolated from chilli rhizosphere with broad spectrum antifungal property against four phytopathogens such as *Alternaria brassiceae* OCA3, *Collectotrichum gleosporioides* OGC1, *Rhizoctonia solani* MTCC 4633 and *Phytophthoracapsici*. This isolate was capable to produces chitinaseand β-1, 3 glucanase along with cellulase, lipase and protease. It also produced metabolites like siderophore, IAA which showed 87.5% germination index when coated onto chilli seeds (102). Two *Streptomyces* spp. strains effectively suppressed the development of *Aspergillus* spp., *Curvularia lunata*, and *Drechslera maydis, Fusarium subglutinans* and *Cephalosporium acremonium*in stored maize grains (15). A collection of 105 strains of actinomycetes isolated from the soils of Riyadh, Saudi Arabia were examined against an array of plant pathogenic fungi *Fusarium oxysporum*, *Alternaria alternata*, *Aspergillus niger*, and *Aspergillus flavus* in vitro. About 61.90% isolates were able to effectively inhibited the growth of *A. alternata*, followed by 61.8, 52.24 and 50.9% inhibiting *A. flavus*, *F. oxysporum*and *A. niger*, respectively (5).

## ACTINOMYCETES METABOLITES FOR PLANT DISEASE CONTROL

Although chemical pesticides have been used widely as plant protection agents, their indiscriminate and constant application causes serious environmental problems that eventually lead to severe health disorders in human and animals. Therefore, in biological control programmes, the natural products particularly the microbial metabolites have included as supplement or an alternative to chemical control (27,39). Several antimicrobial metabolites were identified from actinomycetes and used as agrochemicals for the control of various plant diseases and some of them are listed in Table 1. Streptomycin produced by *S. griseus* was the first antibiotic used for the control of plant diseases caused by *E. amylovora* (8,119) and many other bacterial diseases of crop plants (25,26). At present, several economically important plant diseases are being managed by the use of *Streptomyces* metabolites. Blasticidin S (79,118), kasugamycin (118), polyoxins (58), validamycin A and mildiomycin (47) are some of the noteworthy antibiotics obtained from *Streptomyces* spp. used to control various diseases in agricultural crops. Irumamycin, a 20 membered macrolide from *Streptomyces flavus* sub sp. *irumaensis* was discovered (83,84) with potent *in vivo* activity against *Pyricularia* and *Botrytis*. Polyoxins represent the class of fungicides, which target on the cell wall of the fungi and have been extensively used in agriculture fields. Avermectins, the notably insecticidal metabolites produced by *Streptomyces avermitilis* also inhibited the pathogenic *Streptomyces scabies* and fungal pathogens in green house and field trials (125,126). A thermostable low molecular weight substance of *Streptomyces halstedii* suppressed the growth of the phytophthora blight pathogen of red pepper, *Phytophthora capsici* (62). Two antimicrobial metabolites were isolated from *Streptomyces* sp. PM5. Among them, a compound coded as SPM5C-1 significantly controlled the blast and sheath blight of rice (91).

**Table 1.** The important *Streptomyces* metabolites used for the control of plant diseases

| Compound | Species | Target disease | Reference |
|---|---|---|---|
| Blasticidin S | *S. griseochromogenes* | Blast of rice | 79 |
| Tautomycetin | *S. griseochromogenes* | | |
| Polyoxin | *S. cacaoi* sp. *asoensis* | Blast of rice | 58 |
| Kasugamycin | *S. kasugaensis* | Blast of rice | 118 |
| Validamycin | *S. hygroscopicus* sp. *limoneus* | Sheath blight of rice | |
| Fumaramid-mycin | *S. kurssanovii* | Downy mildew of vine | |
| Mildiomycin | *S. rimofaciens* | Powdery mildews | 47 |
| Streptomycin | *S. griseus* | Bacterial diseases | 8 |

| | | | |
|---|---|---|---|
| Irumamycin | *S. flavus* sub sp. *irumaensis* | Blast of rice & *Botrytis* diseases | 84 |
| Abamectin | *S. avermitilis* | Scabies | |
| SPM5-1 | *Streptomyces* sp. PM5 | Blast and sheath blight of rice | 91 |

An antibiotic-producing *Streptomyces* sp. J2 effectively controlled damping-off of sugar beet seeds caused by *Sclerotium rolfsii* showed evident that of control plants. In addition to that of disease control it also increase ($P \leq 0.05$) in seedling development compared to the control (38). Two antifungal compounds, phenylethyl alcohol and (+)-epi-bicycle sesquiphellandrene produced by *Streptomyces platensis* F-1 effectively controlled the leaf blight/ seedling blight of rice caused by *Rhizoctonia solani*, leaf blight of oilseed rape caused by *Sclerotinia sclerotiorum* and fruit rot of strawberry caused by *Botrytis cinerea* (123).

## INSECT CONTROL

Awareness of environmental contamination by hazardous chemical increases worldwide, alternative approaches for controlling pest populations have become research priorities. The seriousness of these chemical approaches end up with either biological or ecological control methods for limiting the destructive impacts of pest populations, especially in agriculture sectors. The secondary metabolite, Quinomycin A from a forest strain of *Streptomyces* sp KN-0647 gave displayed growth inhibition on the test pathogenetic insects, such as *Spodopteraexigua*, *Dendrolimus punctatus*, *Plutellaxylo stella*, *Aphisglycines* and *Culexpipiens* (52). Seven Streptomycetes species includes, *Streptomycesfungicidicus*, *Streptomycesgriseus*, *Streptomyces albus*, *Streptomycesrochei*, *Streptomyces violaceus*, *Streptomyces alboflavus* and *Streptomyces griseofuscus* secondary metabolites showed 100% total mortality of larvae of mosquitoes *Culexpipiens* (34). Two *Streptomyces* species were recovered from Western Ghats of Agumbe region of Karnataka, of which, isolate 2 showed 100% mortality against second instar larvae of *Aedes aegypti* (92). Thirty eight actinomycetes species were isolated from United Arab Emirates; based on the chitinase enzyme production only three isolates were selected and identified as *Actinoplanes philippinensis*, *Actinoplanes missouriensis* and *Streptomycescla vuligerus*. Chitinolytic activities of the three isolates were tested against *Drosophila melanogaster* development as percentage of successful pupal formation. The combination of *A. philippinensis*, *A. missouriensis* gave an indication of antagonism of 50.05% shown successful pupal formation (41).

## WEED CONTROL

Sustainable weed management have been broadly made in recent days due to the compatibility shown when microbial biological agents with small amount of herbicides have proven that combinations of these agents can certainly improve the biocontrol efficacy (21). A comprehensive procedure has been developed for the screening of herbicides activity of actinomycetes isolated from rhizosphere soil (31). Bio-herbicidal compounds are being used directly as herbicides or herbicidal adjuvant to develop biological agents that are very selective, easily degradable and environment friendly (69). Phytotoxins produced by *Streptomyces* spp. was commercially successfully developed as a herbicide for many crop plants and have been considered as potential herbicides (106) and the use of phytotoxin produced by BCA can enhance the efficacy of bio-herbicide (121). *Streptomyces* sp. 8E-12 produced a selective herbicidal metabolite which was identified as methoxyhygromycin, which showed stronger *in vivo* activity against monocotyledonous plants than dicotyledonous plants, and caused a bleaching (albino symptom) on some weeds including *Digitaria sanguinalis* and *Echinochloa crus-galli* (69).

## CONCLUSION

Among the bioactive natural products, microbial origin stand alone the best source, among them actinomycetes hold predominant position for the source of bioactives. Because of their complex metabolic system and its adaptability to any environmental niches they produce various metabolites to withstand the adverse condition posed by the environment. The proper selective isolation and screening methodologies are required to culminate the never isolated actinomycetes, which are hided in the environment. All the actinomycetes are antagonist to their vicinity microbiota, thus characteristic features should be acclimatize for the development of biocontrol agents without compromise of their consistent performance. Molecular variations of these microbes are meant of successful introduction to any environment to offer valuable outputs with chemical outlooks. Further, there is a sea of research arena available in the actinomycetes taxa, to take up the nowadays preliminary research to widen the in-depth research to bag the fruitful outcome with industrial and agricultural values.

## FUTURE PROSPECTS

Actinomycetes still remains an excellent source for the novel bioactive molecules especially antibiotics and other agrochemicals. Newer selective isolation and high throughput screening technology would offer more insights of actinobacterial taxa. The whole genome analysis of this group of organisms shall greatly helpful in searching of biosynthetic gene cluster and its application in the field of medical and agriculture. Taxonomical relationship between the pathogenic species and non pathogenic groups are to be identified and well documented. The biocontrol potential actinomycetes traits are developed for commercial use by enhancing the consistency of their performance in the field. Genetically engineered actinomycetes strains are developed by putting more biocontrol genes in one strain to control the multiple pathogenicity of the crop plant. Actinomycetes are hyperactive microorganisms available in the earth and thus primary and secondary metabolites contributed more to the mankind. Till date only 1.2% of the actinomycetes populations are isolated and much of them are hided in nearby environment. To enjoy the apart of the actinomycetes research priority and grants are generously given to actinomycetes search.

## REFERENCES

1. Abrami, M, Le, I, Koricab, T, Vitalea, L, Saengerc, W and Pigac J 1999 Purification and properties of extracellular lipase from *Streptomyces rimosus Enzyme Microbial Technology* **25** 522–529
2. Aghighi, S, Shahidi, BGH and Saadoun, I 2004 First report of antifungal properties of a new strain of *Streptomyces plecatus* (strain 101) against four Iranian phytopathogenic isolates of *Verticillium dahliae*, a new horizon in

biocontrol agents, *Biotechnology* **3** 90-97

3. Ahn, SJ and Hwang, BK 1992 Isolation of antibiotic producing actinomycetes antagonistic to Phytopthora capsici from pepper growing soil, *Korean Journal of Mycology* **20** 259-268
4. Al Akhrass, F, Al Wohoush, I, Chaftari, AM, Reitzel, R, Jiang, Y, Ghannoum, M, Tarrand, J, Hachem, R and Raad, I 2012 Rhodococcus bacteremia in cancer patients is mostly catheter related and associated with biofilm formation *PLoS One* 7 e32945
5. Ara, I, Bukhari, NA, Perveen, K and Bakir, MK 2012 Antifungal activity of some actinomycetes isolated from Riyadh soil, Saudi Arabia An evaluation for their ability to control *Alternaria* caused tomato blight in green house pot trial, *African Journal of Agricultural Research* **7** 2042–2050
6. Baker, D 1990 Methods for the isolation, culture and characterization of the Frankiaceae soil actinomycetes and symbionts of actinorrhizal plants In Labeda, DP (editor) Isolation of Biotechnological Organisms from Nature New York McGraw-Hill, 213–236 pages
7. Balagurunathan, R, Radhakrishnan, M and Somasundaram, ST 2010L-glutaminase producing actinomycetes from marine sediments – selective isolation, semi quantitative assay and characterization of potential strain, *Australian Journal of Basic Applied Science* 4 698–705
8. Beer, SV, Rundle, JR and Norielli, JL 1984 Recent progress in the development of biological control for fire blight *Acta Horticulturae* 151 195
9. Behal, V 2000 Bioactive products from *Streptomyces, Advances in Applied Microbiology* 47 113–157
10. Bentley, SD, Corton, C, Brown, SE, Barron, A, Clark, L, Doggett, J, Harris, B, Ormond, D, Quail, MA, May, G, Francis, D, Knudson, D, Parkhill, J, and Ishimaru, CA 2008 Genome of the actinomycete plant pathogen *Clavibacter michiganensis* subsp *sepedonicus* suggests recent niche adaptation *Journal of Bacteriology* 190 2150–2160
11. Bérdy, J 2005 Bioactive microbial metabolites A personal view *Journal of Antibiotics* 581–26
12. Bérdy, J 2012 Thoughts and facts about antibiotics Where we are now and where we are heading *Journal of Antibiotics* **65** 385–395
13. Bignell, DR, Seipke, RF, Huguet-Tapia, JC, Chambers,

AH, Parry, RJ and Loria, R 2010 *Streptomyces scabies* 87-22 contains a coronafacic acid-like biosynthetic cluster that contributes to plant-microbe interactions *Molecular Plant Microbe Interaction* **23** 161–175

14. Bolton, AT 1980 Control of Pythium aphanidermatum in poinsettia in a soiless culture by *Trichoderma viride* and a *Streptomyces* sp *Canadian Journal of Plant Pathology* **2** 93-95
15. Bressan, W 2003 Biological control of maize seed pathogenic fungi by use of actinomycetes *Biocontrol* **48** 233–240
16. Brumbley, SM, Petrasovits, LA, Hermann, SR, Young, AJ and Croft BJ 2006 Recent advances in the molecular biology of *Leifsonia xyli* subsp *xyli,* causal organism of ratoon stunting disease *Australian Plant Pathology* **35** 681–689
17. Bulina, TI, Terekhova, LP and Tiurin, MV 1998 Use of electrical impulses for selective isolation of actinomycetes from soil *Mikrobiologia* **67** 556–560
18. Cao, L, Qiu, Z, Dai, X, Tan, H, Lin, Y and Zhou, S 2004 Isolation of endophytic actinomycetes from roots and leaves of banana (*Musaacuminata*) plants and their activities against *Fusarium oxysporum* f sp *cubense World Journal of Microbiology and Biotechnology* **20** 501–504
19. Cao, L, Qiu, Z, You, J, Tan, H and Zhou, S 2005 Isolation and characterization of endophytic streptomycete antagonists of fusarium wilt pathogen from surface-sterilized banana roots *FEMS Microbiological Letters* **247** 147-152
20. Chang, CC, Ng, CC, Wang, CY and Shyu, YT 2009 Activity of cellulase from thermoactinomycetes and *Bacillus* spp isolated from *Brassica* waste compost *Science in Agricaltual* (Piracicaba, Braz) **66** 304–308
21. Christy, AL, Herbst, KA, Kostka, SJ, Mullen, JP and Carlson, PS 1993 Synergizing weed biocontrol agents with chemical herbicides In Pest Control with Enhanced Environmental Safety eds Duke, SO, Menn, JJ and Plimmer, JR Washington DC American Chemical Society, 87–100 pages
22. Chun, J, Bae, KS, Moon, EY, Jung, SO, Lee, HK and Kim, SJ 2000 *Nocardiopsiskunsanensis* sp nov, a moderately halophilic actinomycete isolated from a saltern, *International Journal of Systematic Evolution Microbiology* **50** 1909–1913
23. Chung, WC, Huang, JW and Huang, HC 2005 Formulation of a soil biofungicide for

control of damping-off of Chinese cabbage (*Brassica chinensis*) caused by *Rhizoctonia solani, Biolcontrol* **32** 278-294

24. Ciampi, L and Tewari, JP 1990 Evaluation of soil microorganisms with inhibitory activity against *Rhizoctonia solani* causal agent of the damping-off of canola *Archives of Biological Medical Experiment* (Santiago) **23**(2) 101-12

25. Colins, JE and Chafik, Z 1986 Comparison of biological and chemical treatments for control of bacterial speck of tomato under field conditions in Morocco *Plant Disease* **70** 1048-1050

26. Colins, KC and Mc Carter, SM 1983 Effectiveness of selected chemicals in inhibiting *Pseudomonas syringe*pv tomato in vitro and in controlling bacterial specks *Plant Disease* **67** 639-644

27. Copping, LG 1996 Crop Protection Agents from Nature Natural products and Analogues The Royal Society of Chemistry, Cambridge, UK501 p

28. Crawford, DL, Lynch, JM, Whipps, JM and Ousley, MA 1993 Isolation and characterization of actinomycete antagonists of a fungal root pathogen *Applied Environment Microbiology* **59** 3899–3905

29. Das, S and Chandra, AL 1990 Chromate reduction in *Streptomyces Experientia* **46** 731–733

30. De Boer, W, Gunnewick, PJAK, Lafeber, P, Janse, JD, Sujit, DE and Woldendrop, JW 1998 Antifungal properties of chitinolytic dune soil bacteria *Soil Biology Biochemistry* **30** 193-203

31. DeFrank, J and Putnam, AR 1985 Screening procedures to identify soil-borne actinomycetes that can produce herbicidal compounds *Weed Science* **33** 271–274

32. Doumbou, CL, Hamby-Salove, MK, Crawford, DL and Beaulieu, C 2001 Actinomycetes, promising tools to control plant diseases and to promote plant growth, *Phytoprotection* **82** 85–102

33. El-Abyad, MS, El-Sayed, MA, El-Shanshoury, AR and El-Bagh, SM 1996 Antimicrobiol activities of *Streptomyces pulcher, Streptomyces canescens* and *Streptomyces citreofluorescens* against fungal and bacterial pathogen of tomato *in vitro*, *Folia Microbiologica* **41** 321-328

34. El-Khawagh, M A, Hamadah, Kh Sh and El-Sheikh, TM 2011 The insecticidal activity of actinomycete metabolites, against the mosquitoe *Culexpipiens, Egyptian Academy of Journal of Biological Science* **4** 103–113

35. El-Shirbiny, SA, Ghaly, MF, El-Ayoty, YM and Fleafil, NS 2007 Niromycin A An antialgal substance produced by *Streptomyces endus* N40, *Research of Journal Microbiology* **2** 606–618
36. El-Tarabily, KA 2003 An endophytic chitinase-producing isolate of *Actinoplanes missouriensis*, with potential for biological control of root rot of lupin caused by *Plectosporium tabacinum*, *Australian Journal of Botany* **51** 257–266
37. El-Tarabily, KA, Soliman, MH, Nassar, AH, Al-Hassani, HA, Sivasithamparam, K, McKenna, F and Hardy, GESt J 2000 Biological control of *Sclerotinia minor* using a chitinolytic bacterium and actinomycetes, *Plant Pathology* **49** 573–583
38. Errakhi, R, Bouteau, F, Lebrihi, A, Barakate, M 2007 Evidences of biological control capacities of *Streptomyces* spp against *Sclerotium rolfsii* responsible for damping-off disease in sugar beet (*Beta vulgaris* L) *World Journal of Microbioogy and Biotechnology* **23** 1503–1509
39. Fravel, DR 1988 Role of antibiotics in the biocontrol of plant diseases *Annual Review of Phytopathology* **26** 75-92
40. Fuentesa, MS, Benimelia, CS, Cuozzoa, SA and Amorosoa, MJ 2010 Isolation of pesticide-degrading actinomycetes from a contaminated site Bacterial growth, removal and dechlorination of organochlorine pesticides *International Biodeterioration & Biodegradation* **64** 434–441
41. Gadelhak, GG, El-Tarabily, AM, and Al-Kaabi, FK 2005 Insect control using chitinolytic soil actinomycetes as biocontrol agents *International Journal of Agricultural Biology* **7** 627–633
42. Geetha, K and Vikineswary, S 2002 Antagonistic effects of *Streptomyces violaceusniger* strain G10 on *F oxysporum* f sp *cubense* race 4 Indirect evidence for the role of antibiosis in the antagonistic process, *Journal Indan Microbiol Biotechnology* **28** 303-310
43. Gesheva, R, Panajotov, P, Darakchieva, M, Naumova, MT, Sankov, H, Mladenov, M and Karajova, J 1977 Actinomycetes antagonistic of *Fusarium graminearum*Schw, responsible for fusariosis of wheat, *Prilojna Mikrobiology* **7** 76-81
44. Gomes, RC, Semêdo, LTAS, Soares, RMA, Alviano, CS, Linhares, LF and Coelho, RRR 2000 Chitinolytic activity of actinomycetes from a cerrado soil and their potential in biocontrol, *Letters in Applied Microbiology* **30** 146–150

45. Gonzalez, I, Miebra, A, Lemus, M, Gonalez, L, Ofero, I, Iznaga, Y, Perez, ME and Vallin, C 1999 Ecological approach of macrolide, lincosamide streptogramin producing actonomyceytes from Cuban soil *Letters in Applied Microbiology* **29** 147-150
46. Goodfellow, M and Williams, ST 1983 Ecology of actinomycetes *Annual Review of Microbiology* **37** 189– 216
47. Harada, S and Kishi, T 1978 Isolation and characterization of mildiomycin, a new nucleoside antibiotic *Journal of Antibiotics* **31** 519
48. Hayakawa, M, Otoguro, M, Takeuchi, T, Yamazaki, T and Iimura, Y 2000 Application of a method incorporating differential centrifugation for selective isolation of motile actinomycetes in soil and plant litter *Antonie Van Leeuwenhoek* **78** 171–85
49. Hogenhout, SA and Loria, R 2008 Virulence mechanisms of Gram-positive plant pathogenic bacteria *Current Opinion in Plant Biology* **11** 449–456
50. Horton, RN, Apel, WA, Thompson, VS and Sheridan, PP 2006 Low temperature reduction of hexavalent chromium by a microbial enrichment consortium and a novel strain of *Arthrobacter aurescens BMC Microbiology* **6** 5
51. Hozzein, WN, Ali, MIA and Rabie, W 2008A new preferential medium for enumeration and isolation of desert actinomycetes *World Journal of Microbiology and Biotechnology* **24** 1547–1552
52. Huamei, L, Sheng, Q, Yongxia, W, Wenjun, L and Jie, Z 2008 Insecticidal action of Quinomycin A from *Streptomyces* sp KN-0647, isolated from a forest soil *World Journal of Microbiology and Biotechnology* **24** 2243-2248
53. Hussain, S, Gghaffar, A and Asiam, M 1990 Biological control of *Macrophomina phaseolina* charcoal rot of sunflower and mung bean *Journal of Phytopathology* **130** 157-160
54. Igarashi, Y, Ogawa, M, Sato, Y, Saito, N, Yoshida, R, Kunoh, H, Onakam H and Furumai, T 2000 Fistupyrone, a novel inhibitor of the infection of Chinese cabbage by *Alternaria brassicicola*, from *Streptomyces* sp TP-A0569 *Journal of Antibiotic* (Tokyo) 53 1117–1122
55. Imada, C 2005 Enzyme inhibitors and other bioactive compounds from marine actinomycetes *Antonie van Leeuwenhoek* **87** 59–63
56. Inokoshi, J, Chiba, H, Asanuma, S, Takahashi, A, Omura, S and Tanaka, H 2001 Molecular cloning of actinohivin, a novel anti-HIV

protein from an actinomycete, and its expression in *Escherichia coli, Biochemical and Biophysics Research Communication* **28** 11261–1265

57. Intra, B, Mungsuntisuk, I, Nihira, T, Igarashi, Y and Panbangred, W 2011 Identification of actinomycetes from plant rhizospheric soils with inhibitory activity against *Colletotrichum* spp, the causative agent of anthracnose disease *BMC Research Notes* **4** 98
58. Isono, K, Nagatsu, J, Kawashima, Y and Suzuki, S 1965 Studies on polyoxins, antifungal antibiotics Part I Isolation and characterization of polyoxins A and B *Agricultural Biological Chemistry* **29** 848
59. Jain, PK, Jain, R and Jain, PC 2004 Production of industrially important enzymes by some actinomycetes producing antifungal compounds Hindustan, *Antibiotic Bulletin* **45** 4629–33
60. Jiang, S, Li, X, Zhang, L, Sun, W, Dai, S, Xie, L, Liu, Y and Lee, KJ 2008 Culturable actinobacteria isolated from marine sponge *Iotrochota* sp, *Marine Biology* **15** 3945–952
61. Jobin, G, Couture, G, Goyer, C, Brzezinski, R and Beaulieu, C 2005 *Streptomycete* spores entrapped in chitosan beads as a novel biocontrol tool against common scab potato *Applied Microbiology and Biotechnology* **68** 104-110
62. Joo, JG, 2005 Purification and characterization of an extracellular chitinase from the antifungal biocontrol agent *Streptomyces halstedii Biotechnology Letters* **27** 1483-1486
63. Krishnaraj, M and Mathivanan, N 2011 Diversity of marine actinomycetes in the Bay of Bengal and their antibacterial activity against human pathogens *Journal of Marine Biology Association of India* **53** 135–138
64. Kurtböke, DI, Murphy, N E and Sivasithamparam, K 1993 Use of bacteriophage for the selective isolation of thermophilic actinomycetes from composted eucalyptus bark *Canadian Journal Microbiology* **39** 46–51
65. Ladjama, A, Taibi, Z and Meddour, A 2007 Production of pectinolytic enzymes using *Streptomyces* strains isolated from palm grove soil in Biskra area (Algeria) *African Crop Science Conference Proceedings* **8** 1155–1158
66. Lahdenpera, ML 1987 The control of Fusarium wilt on carnation with a *Streptomyces* preparation *Acta Horticultural* **216** 85-92
67. Laxman, SR and More, S 2002 Reduction of hexavalent chromium by *Streptomyces*

*griseus Minerals Engineering* **15** 831–837

68. Lee, JY and Hwang, BK 2002 Diversity of antifungal actinomycetes in various vegetative soils of Korea *Canadian Journal Microbiology* **48** 407-416
69. Lee, HB Kim, C-J, Kim, J-S, Hong, K-Sand Cho, KY 2003 A bleaching herbicidal activity of methoxyhygromycin (MHM) produced by an actinomycete strain *Streptomyces* sp 8E-12 *Letters in Applied Microbiology* **36** 387–391
70. Lerat, S, Simao-Beaunoir, A and Beaulieu, C 2009 Genetic and physiological determinants of *Streptomycesscabies* pathogenicity *Molecular Plant Pathology* **10** 579–585
71. Lewis, R, McKenzie, D, Bagg, J and Dickie, A 1995 Experience with a novel selective medium for isolation of *Actinomyces* spp from medical and dental specimens, *Journal of Clinical Microbiology* **33** 1613–1616
72. Lin, QS, Chen, SH, Hu, MY, Ul Haq, MR, Yang, L and Li, H 2011 Biodegradation of cypermethrin by a newly isolated actinomycetes HU-S-01 from wastewater sludge, *International Journal of Environmental Science and Technology* **8** 45–56
73. Maldonado, LA, Fragoso-Yanez, D, Pe´rez-Garcı´a, A,Rosello´n-Druker, J and Quintana, ET 2008 Actinobacterial diversity from marine sediments collected in Mexico *Antonie van Leeuwenhoek* **95** 111–120
74. Marinelli, F 2009 Antibiotics and *Streptomyces* The future of antibiotic discovery *Microbiology Today* **36** 21–22
75. McCarthy, AJ and Williams, ST 1992 Actinomycetes as agents of biodegradation in the environment-a review *Gene* **115** 189–192
76. McCarthy, AJ, Peace, E and Broda, P 1985 Studies on the extracellular xylanase activity of some thermophilic actinomycetes *Applied Microbiology and Biotechnology* **21** 238–244
77. McNeil, MM and Brown, JM 1994 The medically important aerobic actinomycetes epidemiology and microbiology *Clinical Microbiology Review* **7** 357–417
78. Minuto, A, Spadaro, D, Garibaldi, A and Gullino, ML 2006 Control of soilborne pathogens of tomato using a commercial formulations of *Streptomyces griseoviridis* and solarization *Crop Protection* **25** 468-475
79. Misato, T, Okimoto, Y, Ishii, I, Asakawa M and Fukunaga, K 1961 Antibiotics as protectant fungicides against rice blast 1V Effect of blasticidin S on the metabolism of *Pyricularia oryzae, Annual Phyto-*

*pathological Society of Japan* **26** 25

80. Muller-Scharer, H, Scheepens, PC and Greaves, MP 2000 Biological control of weeds in European crops recent achievement and future work *Weed Research* **40** 83–98
81. Newman, DJ and Cragg, GM 2007 Natural products as sources of new drugs over the last 25 years *Journal of Natural Products* **70** 461–477
82. Niladevi, KN and Prema, P 2005 Mangrove actinomycetes as the source of ligninolytic enzymes *Actinomycetologica*, **19** 40–47
83. Omura, S, Nakagawa, A and Tanaka, Y 1982Structure of a new antifungal antibiotic, irumamycin *Journal Organic Chemistry* **47** 5413-5415
84. Omura, S, Tanaka, Y, Takahashi, Y, Chias, I and Inoue, M 1984 Irumamycin, an antifungal 20 membered macrolide produced by *Streptomyces* Taxonomy fermentation and biological properties *Journal of Antibiotics* **37** 157-158
85. Ouhdouch, Y, Barakate, M and Finanse, C 2001 Actinomycetes of Moroccan habitats Isolation and screening for antifungal activities *Europian Journal Soil Biology* **37** 69-74
86. Park, JO, El-Tarabily, KL, Ghisalberti, EL and Sivasithamparam, K 2002 Pathogenesis of *Streptoverticillium albireticuli* on *Caenorhabditis elegans* and its antagonism to soil-borne fungal pathogens, *Letters in Applied Microbiology* **35** 361–365
87. Pathom-aree, W, Stach, JEM, Ward, AC, Horikoshi, K, Bull, AT and Goodfellow, M 2006 Diversity of actinomycetes isolated from Challenger Deep sediment (10,898 m) from the Mariana Trench *Extremophiles* **10** 181–189
88. Pattanapipitpaisal, P, Brown, NL and Macaskie, LE 2001 Chromate reduction by *Microbacterium liquefaciens* immobilised in polyvinyl alcohol, *Biotechnology Letters* **23** 61-65
89. Peela, S Bapiraju Kurada, VVSN and Terli, R 2005 Studies on antagonistic marine actinomycetes from the Bay of Bengal *World Journal of Microbiology and Biotechnology* **21** 583–585
90. Polti, MA, Amoroso, MJ and Abate, CM 2007 Chromium (VI) resistance and removal by actinomycete strains isolated from sediments *Chemosphere* **67** 660–667
91. Prabavathy VR, Mathivanan N, Murugesan K (2006) Control of blast and sheath blight diseases of rice using antifungal metabolites produced by *Streptomyces* sp PM5 *Bio Control* **39** 313–319

92. Prashith Kekudal, TR, Shobha, KS and Onkarappa, R 2010 Potent insecticidal activity of two *Streptomyces* species isolated from the soils of the western ghats of Agumbe, Karnataka *Journal of Natural Pharmaceuticals* 130–320

93. Promnuan, Y, Kudo, T and Chantawannakul, P 2009 Actinomycetes isolated from beehives in Thailand *World Jounal Microbiology and Biotechnol* **25** 1685–1689

94. Prudhomme, J, McDaniel, E, Ponts, N, Bertani, S, Fenical, W, Jensen, P and Le Roch, K 2008 Marine Actinomycetes A new source of compounds against the human malaria parasite *PLoS ONE* 3 e2335

95. Ramesh, S and Mathivanan, N 2009 Screening of marine actinomycetes isolated from the Bay of Bengal, India for antimicrobial activity and industrial enzymes *World Journal of Microbiology and Biotechnology* **25** 2103–2111

96. Ramesh, S, Rajesh, M and Mathivanan, N 2009 Characterization of a thermostable alkaline protease produced by marine *Streptomyces fungicidicus* MML1614 *Bioprocess Biosyst Eng* **32** 791–800

97. Reddi, GS and Rao AS 1971 Antogonism of the soil actinomycetes to some soil–borne plant pathogenic fungi, *Indian Phytopathology* **24** 649-657

98. Sabaratna, S and Traquair, JA 2002 Formulation of a *Streptomyces* biocontrol agent for the suppression of *Rhizoctonia* damping off in tomato transplants *Biological Control* **23** 245-253

99. Seong, CN, Choi, JH and Baik, KS 2001An improved selective isolation of rare actinomycetes from forest soil, *The Journal of Microbiology* **39** 17–23

100. Sharma, SL and Pant, A 2001 Crude oil degradation by a marine actinomycete *Rhodococcus* sp *Indian Journal of Marine Science* **30** 146–150

101. Shigemori, H, Tanaka, Y, Yazawa, K, Mikami, Y and Kobayashi, J 1996 Brasilinolide A, new immunosuppressive macrolide from actinomycete *Nocardiabrasiliensis Tetrahedron* **52** 9031–9034

102. Srividya, S, Thapa, A, Bhat, DV, Golmei, K and Dey, N 2012 *Streptomyces* sp 9p as effective biocontrol against chilli soilborne fungal phytopathogens *Europian Journal of Experrimental Biology* **2** 163–173

103. Stach, JE and Bull, AT 2005 Estimating and comparing the diversity of marine actinobacteria *Antonie van Leeuwenhoek* **87** 3–9

104. Stackebrandt, E, Rainey, FA and Ward-Rainey, NL 1997 Proposal for a new hierarchic classification system, Actinobacteria classis nov *International Journal of Systematic Bacteriology* **47** 479–491

105. Stamford, TLM, Stamford, NP, Coelho, LCBB and Araujo, JM 2002 Production and characterization of a thermostable glucoamylase from *Streptosporangium* sp endophyte of maize leaves *Bioresource Technoogyl* **83** 105–109

106. Strobel, GA, Kenfield, D, Bunkers, G, Sugawara, F and Clardy, J 1991 Phytotoxins as potential herbicides *Experientia* **47** 819–826

107. Sullia, SB and Shantharam, S 2005 General Microbiology Oxford and IBH Publishing Co Pvt Ltd New Delhi, India p 475

108. Sutherland, ED and Papavizas, GC 1991 Evaluation of oospore hyperparasites for the control of Phytophthora crown rot of pepper *Journal of Phytopathology* **131** 33-39

109. Tahavonen, R 1982 Preliminary experiments into the use of *Streptomyces* spp isolated from peat in the biological control of soil and seed borne disease in peat culture *Journal of Agricultural Society of Finland* **54** 357-369

110. Tahovonen, R and Avikainen, H 1987 The biological control of seed-borne *Alternaria brassicicola* of cruciferous plants with a powdery preparation of *Streptomyces* sp *Journal of Agricultural Society of Finland* 59 199-208

111. Takahashi, Y and Ômura, S 2003 Isolation of new actinomycetes strains for the screening of new bioactive compounds *Journal of General Applied Microbiology* **49** 141–154

112. Tanaka, N, Sakagami, Y, Yamaki, H and Umezawa, H 1961 Activity of cytomycin and blasticidin S against transplatable animal tumors *Journal of Antibiot Resistance* **14** 1223

113. Thumar, JT, Dhulia, K and Singh, SP 2010 Isolation and partial purification of an antimicrobial agent from halotolerant alkaliphilic *Streptomyces aburaviensis* strain Kut-8 *World Journal of Microbiology and Biotechnology* **26** 2081–2087

114. Tokala, RK, Strap, JL, Jung, CM, Crawford, DL, Salove, MH, Deobald, LA, Bailey, JF and Morra, MJ 2002 Novel plant–microbe rhizosphere interaction involving *Streptomyces lydicus* WYEC108 and the pea plant (*Pisum sativum*) *Applied Environmental Microbiology* **68** 2161–2171

115. Tsuruta, T 2004 Adsorption of uranium from acidic solution by microbes and effect of thorium on uranium adsorption by *Streptomyces levoris* *Journal of Bioscience Bioengineering* **97** 275–277

116. Tu, JC 1986 Hyperparasitism of *Streptomyces albus* on a destructive mycoparasite *Nectria inventa, Journal of Phytopathology* **117** 71-76

117. Turhan, G and Turhan, K 1989 Suppression of damping-off on pepper caused by *Pythium ultimum* Trow and *Rhizoctonia solani* Khun by some new antagonists in comparison with *Trichoderma harzianum* Rifai *Journal of Phytopathology* **126** 175-182

118. Umezawa, H, Okami, Y, Hashimoto, T, Suhara, Y, Hamada, M and Takeuchi, T 1965 A new antibiotic, kasugamycin *Journal of Antibiotic Ser A* 18 101

119. Vanneste, JL, Yu, J and Beer, SV 1992 Role of antibiotic production by *Erwinia herbicola* Eh 252 in biological control of *Erwinia amylovora, Journal of Bacteriology* 174 2785

120. Vereecke, D, Burssens, S, Simon-Mateo, C, Inze, D, Van Montagu, M, Goethals, K and Jaziri, M 2000 The *Rhodococcusfascians*-plant interaction morphological traits and biotechnological applications *Planta* **210** 241–251

121. Vurro, M, Zonno, MC, Evidente, A, Andolfi, A and Montemurro, P 2001 Enhancement of efficacy of *Ascochyta caulina* to control *Chenopodium album* by use of phytotoxins and reduced rates of herbicides *Biological Control* **21** 182–190

122. Waksman, SA, Schatz, A and Reynolds, DM 2010 Production of antibiotic substances by actinomycetes Ann NY *Academy of Science* **1213** 112–124

123. Wan, M, Li, G, Zhang, J, Jiang, D, and Huang, HC 2008 Effect of volatile substances of *Streptomyces platensis* F-1 on control of plant fungal diseases *Biological Control* **46** 552–559

124. Wang, Y, Zhang, Z S, Ruan, JS Wang, YM and Ali, SM 1999 Investigation of actinomycete diversity in the tropical rainforests of Singapore, *Journal of Indusrtrial Microbiology and Biotechnology* **23** 178–187

125. Xiao, K, Kinkel, LL and Samac, DA 2002 Biological control of *Phytophthora* root rots on alfalfa and soybean with *Streptomyces* *Biological Control* **23** 285–295

126. Xu, LH, Li, QR and Jiang, CL 1996 Diversity of soil actinomycetes in Yunnan, China *Applied Environmental Microbiology* **62** 244-248

127. Yabe, S, Aiba, Y, Sakai, Y, Hazaka, M and Yokota, A 2011 *Thermasporomyces composti* gen nov, sp nov, a thermophilic actinomycete isolated from compost *International Journal of Systematic Evolutional Microbiology* **61** 86–90

128. Yamamura, H, Hayakawa, M and Iimura, Y 2003 Applica--tion of sucrose-gradient centrifugation for selective isolation of *Nocardia* spp from soil, *Journal of Applied Microbiology* **95** 677–685

129. Yin, SY, Chang, JK and Xun, PC 1965 Studies in the mechanisms of antagonistic fertilizer '5406' IV The distribution of the antagonist in soil and its influence on the rhizosphere *Acta Microbiologica Sinica* 11 259

130. You, KM and Park, YK 1996 A new method for the selective isolation of actinomycetes from soil *Biotechnological Techniques* 10 541–546

131. Yuan, W and Crawford, DL 1995 Characterization of *Streptomyces lydicus* WYEC 108 as a potential biocontrol agent against fungal root and seed rots *Applied Environmental Microbiology* **61** 3119-3128

132. Zhang, DC, Schumann, P, Redzic, M, Zhou, YG, Liu, HC, Schinner, F and Margesin, R 2012 *Nocardioidesalpinus* sp nov, a psychrophilic actino mycete isolated from alpine glacier cryoconite *International Journal of Systematic Evolutionary Microbiology* **62** 445-450

133. Zhang, YB 2003 Physiologically active metabolite produced by actinomycetes living in plant *World Pesticide* **25** 9–12

134. Zheng, Z, Zeng, W, Huang, Y, Yang, Z, Li, J, Cai, H and Su, W 2000 Detection of antitumour and antimicrobial activities in marine organisms associated actinomycetes isolated from the Taiwan strait, China *FEMS Microbiology Letters* **188** 87-91.

*Rev. Plant Pathol. Vol. 6, 2014*
*Indian Society of Mycology and Plant Pathology*
*Scientific Publishers (India), Jodhpur pp. 299-334*

# ENDOPHYTIC FUNGI AS POTENTIAL SOURCES OF BIOACTIVE NATURAL PRODUCTS: PROSPECTS AND CHALLENGES

***K. Tayung*[1] *and D.K Jha*[2]**

[1] Department of Botany, North Orissa University,
Baripada, Orissa, 757 003
[2] Microbial Ecology Laboratory, Department of Botany,
Gauhati University, Guwahati 781 014, Assam

**KEY WORDS:** Fungal endophytes, Medicinal plants, Bioactive substances, Taxol, Campothecin, Crytocandin, Phomol and Helvolic acid.

## ABSTRACT

Fungal endophytes are microorganisms that grow inside living plant tissues without causing disease symptoms. They may establish symbiotic or mutualistic relationship with their respective hosts. Endophytes have been found in every plants studied so far and are to be found in virtually every plant on earth. Therefore, they are important constituent of microbial diversity. Endophytes are rich sources of functional metabolites because of their ability to produce good number of new and interesting bioactive molecules, which are of pharmaceutical, industrial and agricultural importance. In the recent years, fungal endophytes have been studied and exploited for bioactivity. Some important bioactive substances discovered from fungal endophytes associated are taxol, campothecin, crytocandin, phomol and helvolic acid. One of the major problems facing the future of endophyte biology and natural product discovery is the rapid disappearance of plant diversity, which holds the greatest possible resources of acquiring novel microorganism and their novel product. Over exploitation of medicinal plants in particular and plants in general have threatened the very existence many important plant species. The study of fungal endophytes is the need of the hour because once such plants get lost or become extinct so is the entire suite of associated endophytes and their novel products.

## INTRODUCTION

Microorganisms are found in diverse habitat and can survive in all sorts of inhospitable environment. Microbial diversity constitutes an infinite source of novel compound, making up a valuable source for innovative biotechnology. The most recent estimates suggest that till now, only approximately 5% of the total species of fungi and may be as little as 0.1% of the bacteria are currently known (54). And among the ones already described, only a small fraction have been studied for their metabolites profile. Therefore, in spite of lot of microbes being studied there are still lots of them in the environment because many of them are overlooked or couldn't have been cultured in the laboratory . Over the years, a great deal of scientific attention has been given to economically and important medicinal plants due to the role-played by them as source of human food and drugs in the world of pharmacopoeias. However, it is now known that these plants serve as a reservoir for an untold number of microbes known as endophytes (7,37). The term endophyte include all organisms that grow inside plant tissues without causing disease symptoms (74, 25). Bacon and White gave an inclusive and widely accepted definition of endophytes—"microbes that colonize living, internal tissues of plants without causing any immediate, overt negative effects". This particular definition implies a symbiotic or mutualistic relationship between the host plants and the endophytic microbes. However, there have been differences in opinion about the exact relationship that exist with their respective hosts. Stone *et al.* in their review, suggest that relationships other than mutualistic interactions in endophytes are recognized and should be anticipated (98). Similarly, Strobel and others are of the opinion that the relationship can range from mutualistic to bordering on pathogenic (86,99). Both fungi and bacteria are the most common microbes existing as endophytes. It seems that other microbial forms, e.g., mycoplasmas and archaebacteria, most certainly exist in plants as endophytes, but no evidence for them has yet been presented. The most frequently encountered endophytes are fungi, and it has been estimated that there may be over a million different fungal species on this Earth, of which only a small fraction has been identified (67,86). Endophytes have been found in every plants studied so far and are to be found in virtually every plant on the earth. Almost all vascular plants examined to date were found to harbor either endophytic bacteria and/or fungi (5,12). Moreover, colonization of endophytes in marine algae, mosses and ferns has also been recorded (75,94). Several to hundreds of endophytes species can be isolated from a single plant, among them, at least one species showing host specificity. As a matter of fact endophytes are important component of microbial diversity. Endophytes as a rich source of functional metabolites have been well studied and documented (108). In recent years special attention have been paid to endophytic microbes because of its ability to produce good number of new and interesting bioactive metabolites, which are of pharmaceutical, industrial and agricultural importance (33,97, 102,108,120). Considering the present health problem of the world owing to cancer, multi drugs resistance in clinically important bacteria, parasitic protozoan, fungi and newly immerging diseases, new and effective agents to deal with these problems is urgently required. The ability of endophytes to produce new and functional metabolites proves as a potent candidature in solving these growing human problems. This review discusses the endophytes as an important source of bioactive substances and their Bioprospecting with special references to the fungal endophytes associated with medicinal plants and its natural products.

## THE ENDOPHYTIC ROLE

The types of biological associations that endophytic microbes may have developed with higher plants range from borderline pathogenic, to commensal, and to symbiotic (95,99). Therefore, it is believed that plants might contribute nutrition to the endophyte or compound critical for the completion of its life cycle or essential for growth or self-defense. Although the role of endophytes on host plants is less studied, different function have been described. Like mycorrhizae, endophytic fungi are thought to interact mutualistically with their host plant mainly by increasing host resistance to herbivores (20,21,27) and have been termed "acquired plant defenses"(26). Endophytes are also believed to initiate the biological degradation of the dead or dying host plant, thus recycling the nutrients 7. The endophytic fungi of grasses appear to have a pronounced role in biological protection of the host by production of antibiotic substances24. Indeed, some agronomic grass species infected with systemic endophytes show striking toxic and noxious effects on vertebrate and invertebrate herbivores (16,28,29,92) and pathogens (35) by virtue of alkaloids such as pyrrolizidine alkaloids, ergot alkaloids and peramine produced by the fungi (58,76,91). Endophytes, at least systemic ones in agronomic grasses, may also increase host grass competitive abilities, by increasing germination success, resistance to drought and water stress and resistance to seed predators (52,118). Other studies of agronomic grasses show that endophytes provide other fitness-enhancing properties for their host, such as increasing plant competitive abilities, mainly by increasing efficiency of water use (6). A detailed analysis of this endophyte on plant diversity in successional fields have been carried out (30) and observed that in a four-year period, plant species richness declined whereas tall fescue dominance increased in infected plots, even though there was no distinct difference in total plant productivity. Similar effects have been suggested for many fungal species that inhabit trees (19,41,49,82). Endophytes of woody plants provide a defensive role for the host plant because they produce a wide array of mycotoxins and enzymes that can inhibit growth of microbes and invertebrate herbivores (20). As with their fungal counterparts, many root-associated bacteria act also to protect the host from pathogens (107). The rhizobacteria are the best studied in this respect, and they apply different strategies to control the pathogens; by producing antibiotics and siderophores (51,53) or by nutrient competition(63). In addition to the protective function, grass endophytes affect the growth and survival of the plant, although the means of these effects have not yet been determined(38,44). One fungal endophyte, *Colletotrichum* sp. has been reported to produce the plant growth hormone indole-acetic acid48. Unlike the tall fescue fungal endophyte that moves through plants via vertical transmission, several plant species including woody trees harbor such endophytes through horizontal transmission and can help plants sustain thermal resistance and attack by phytopathogens (4,81). In an interesting study of mutualistic endophyte and associated plant adaptation, 200 plants of *Dichanthelium lanuginosum* were collected from geothermal sites in Lassen Volcanic National Park (LVNP) and Yellowstone National Park (YNP), where average annual temperature varied between 20 and 50°C. A fungal endophyte, characterized by rDNA sequencing as *Curvularia* sp., was consistently recovered from the roots, crowns, leaves and seed coats of all plants examined. Since soils in the study region were devoid of a *Curvularia* sp., this indicated horizontal entry of the fungus in the host plant. The fungus did not grow axenically and thus appeared exclusively mutualistic with *D. lanuginosum*. Artificially inoculated plants with spores of this *Curvularia* sp. could resist constant soil temperatures of 50°C for 3 days and intermittent temperatures as high as 65°C for 10 days. The endophyte could be recovered from surface-sterilized roots and leaves, suggesting protection from thermal stress to both plants and the fungus(49). Endophytes may also increase host fitness by altering attack by natural enemies of herbivores or by increasing fire resistance or resistance to competing microorganisms (17). In fact, endophyte-carrying plants grow more vigorously and are toxic to herbivores. Furthermore, such plants are more drought-tolerant than non-infected plants (50).

## FUNGAL ENDOPHYTES AND MEDICINAL PLANTS

The most frequently isolated endophytes are the fungi. Endophytic fungi are an important, yet relatively unstudied group of microbial plant symbionts. Endophytic fungi live asymptomatically and sometime systemically, within plant tissue. They are defined as fungi colonizing healthy plant tissue without causing overt symptoms in or apparent injury to the host. When compared to the bacterial endophytes, the research of endophytic fungi has a long history and their diversity among plants has been found to be considerably large. Endophytic fungi were reported for the first time in grasses (34) and in trees in *Picea canadiensis* (59). Since then, fungal endophytes have been found in the leaves, bark, and xylem of almost all plant species examined (73,80,89). Although the meaning of endophytes have undergone various transformation during the last decades, with considerable disagreement as what constitutes an endophytes, they are referred broadly to fungi that live for all, or at least significant part, of their life cycle internally and asymptomically in plant parts. Studies of fungal endophytes in trees, shrubs, and ferns show that individual species and even individual plants typically harbor scores of fungal species (22,40,42). Despite the great diversity and abundances of endophytes in woody plants, these endophytes and interaction with their host plants have received less attention than those of grass endophytes. The endophytic fungi of grasses are well studied, especially species belonging to *Acremonium*, found within the grass tall fescue (*Festuca arundinacea*). Other well-known endophytic fungi are members of the genus *Epichloë* and their asexual relatives of *Neotyphodium*, infecting the *Bromus* grass species(116). Besides in grasses, fungal endophytes have been discovered to a great extent in trees, particularly within the leaves. The interactions between fungi and trees are less known in detail, but the endophytic fungal diversity has been found to be very large in trees. The species richness of fungi has been extensively studied especially in conifers (8,35,46,70). Fungal endophytes have been found within the wood and bark tissues of several tree species (11,31). However, when compared to the bark tissue, endophytic fungi are rarely found in the wood (8,13,117). An exception is the endophyte *Ceratocystis erinaceus*, which can colonize the heartwood of oak (*Quercus robur* L.) (14).

In recent years fungal endophytes have been investigated and reported from wide varieties of tree species, shrubs and medicinal plants. The most common isolated endophytic fungi of rain forest are *Pestalotiopsis* sp (79). In fact, it is rare not to isolate one of these species from any tropical plant. Thus, that this fungus could be considered as the ' *Escherichia coli* of the rain forest'. Studies on fungal endophytes gain tremendous importance after the discovery of taxol producing endophytic fungus *Taxomyces andreanae* from *Taxus breviflolia* (96). At present, fungal endophytes has been thoroughly investigated from medicinal plants because of the facts that these plants harbors some distinct and rare microbes that mimic the chemistry of their respective host plants and make the same bioactive natural product or derivatives that are more bioactive than those of their respective host. It is also believed that bioactivity of a medicinal plants may be due to associated endophytes that live on the plant tissues. There has been much speculation regarding the functional metabolites of endophytes and the host plant and it is believed that transfer of such metabolites might be due to genetic exchange between the hosts and the endophytes living in close association with it, as exemplified by the discovery of an endophytic fungus isolated from bark of *Taxus brevifolia* that have been reported to produces taxol. The assumption was further justified after genetic level study of *Pestalotiopsis microspora* an another taxol producing fungus, explaining how genes for taxol might have acquired by this fungus (64). Values of plant medicines have been established since antiquity. A large variety of drugs have already been provided by plant kingdom to mankind to alleviate suffering from diseases. About 34% of all pharmaceutical preparation are reported to be obtained from higher plant alone (71). Inspite of considerable progress made in synthetic drugs, plant constituent are still considered to be major sources of valuable medicaments and many of the plant products still occupy an honorable and formidable place in world of pharmacopoeias. The flora of India comprises about 3000 medicinal plants. Almost 75% of the plant drugs mentioned in the various pharmacopoeias grow in this sub-continent(48). Many of the reputed plant drugs of potential pharmaceutical value and trade grow in wild condition of this region with many more other plant species being used as crude drug in indigenous system of medicine. In the recent years, several fungal endophytes have been reported and studied from medicinal plants and most of these endophytes have produces many important bioactive metabolites of agricultural, industrial and pharmaceutical importance (Table 1).

There may be certain advantages in studying the associated fungal endophytes of medicinal plants:

(i) The associated fungal endophytes may produce the same bioactive molecules as the host plant. Moreover, extraction of bioactive substances from plants origin is rather difficult and complicated procedure as compared to that of the microbial origin.

(ii) If microbial origin of bioactive substance would be available it would eliminate the need to harvest, extract and amount of the substance from the plants as the compound would conceivably be produce by fermentation and price of the drug would be highly reduced.

(iii) Exploitation of rare and endanger medicinal plants would be reduced and gene pool of those plants could be conserved, as microbial sources of the drug obtained from those plants would be available.

(iv) Biotechnological approach could be made on the microbes. The microbes could be manipulated and molecular studies can be done. Attempts could be made to enhance the production and activity of the metabolites by genetic engineering techniques.

**Table 1.** Some important plant species, the endophytes and their bioactive metabolites

| **Plant source** | **compound** | **endophytic fungus** | **Biological activity** | **Reference** |
|---|---|---|---|---|
| *Taxus brevifolia* | Taxol | *Taxomyces andreanae* | anticancer | 96 |
| *Tripete-rigeum wilfordii* | Crytocandin | *Crytosporiopsis quercinia* | antifungal | 103 |
| *Torreya taxifolia* | Pestaloside | *Pestalotiopsis microspora* | antifungal | 56 |
| *Artemisia mongolica* | Colletotric acid | *Colletotrichum gloeosporioides* | antifungal, antibacterial | 120 |
| *Atremisia annua* | Active metabolites | *Colletotrichum sp.* | antifungal, antimalarial | 65 |
| *Cinnamomum zeylanicum* | Volatile antibiotics | *Muscodor albus* | antibacterial, antifugal | 39 |
| *Paullina paullinioides* | Naphthalene | *Muscodor vitigenus* | insect repellent | 33 |
| *Taxus wallichiana* | Taxol | *Pestalotiopsis microspora* | anticancer | 105 |
| *Taxus mairei* | Taxol | *Tubercularia sp.* | anticancer | 113 |
| *Terminalia morobensis* | Pestacin, Isopestacin | *Pestalotiopsis microspora* | antioxidant | 100 |
| *Bontia daphnoides* | Nodulis-poric acid | *Nodulisporium sp.* | insecticidal activities | 97 |

| | | | | |
|---|---|---|---|---|
| *Tripeterigeum wilfordii* | Subglutinol A and B | *Fusarium subglutinans* | immuno-suppressive | 57 |
| *Wollemia nobilis* | Taxol | *Pestalotiopsis guepini* | anticancer | 104 |
| *Cassia spectabilis* | Phomopsi-lactone | *Phomopsis sp.* | antibacterial, antitumor | 93 |
| *Taxus wallichinia* | Taxol | *Trichothecium sp.* | anticancer | 90 |
| *Spondias mombin* | Active metabolites | *Guignardia sp.* | antifungal, antibacterial | 82 |
| *Torreya grandis* | Brefeldin A | *Aspergillus clavatus* | Cytotoxin | 114 |
| *Cynodon dactylon* | Helovic acid | *Aspergillus sp.* | antifungal, antibacterial | 62 |
| *Melia azedarach* | Janthinone | *Penicillium janthinellum* | antibacteria | 84 |
| *Taxus baccata* | Leucino-statin A | *Acremonium sp.* | antifungal, anticancer | 106 |
| *Erythrina crista-galli* | Phomol | *Phomopsis* sp. | Antibacterial, Antifungal Anti-inflammatory | 115 |

It is a fact that, some plants generating bioactive natural products have associated endophytes that produce the same natural products. Moreover it is believed that the healing process, as discovered by indigenous people might be facilitate by compounds produced by one or more specific plant-associated endophytes as well as the plant products themselves. The recent exploitation of medicinal plants for various therapeutic values has cause extinction and endangered many important plants species. One of the major problems facing the future of endophyte biology and natural product discovery is the rapid diminishment of plant diversity, which holds the greatest possible resources of acquiring novel microorganism and their product. Considering the vast potential of medicinal plants in general and in Indian sub-continent at particular a through and comprehensive study of fungal endophytes of medicinal plants is urgently required. Because it is believed that these plant species may harbor quite distinct and potential fungal endophytes and disappearance of any of these plant species will also disappear the entire suite of associated endophytes.

## ISOLATION OF FUNGAL ENDOPHYTES FROM PLANTS

Most procedures for isolating fungal endophytes are relatively simple and routine for anyone skilled in basic microbiological technique. One of the critical needs for isolating endophytic fungi is obtaining fresh plant material. Prolonged transport in sealed plastic bags should be avoided if possible. Sturdy paper bags, wax paper bags, or zip-locked, perforated bags designed for vegetable storage work well for transport and temporary storage of most types of plant tissues. If plants are to be stored for long periods of time, especially in frost-free refrigerators, tissue desiccation will occur. However, it is possible to isolate a surprising number of fungal species even from desiccated woody tissues after freezer storage of more than a year.

To acquire fungal endophytes of medicinal plants, selection of specific plants of medicinal importance having special properties can be made. Usually, small stem pieces are cut from the plant and placed in plastic bags after any excess moisture is removed. A standard method employed for isolation of endophytic fungi requires surface sterilization of plant materials. The commonly used surface disinfectants are 70% ethanol for few minutes, then treatment in 0.5-3.5% sodium hyphochoride solutions for 2-3 min and rinsed in sterile distilled water. However different surface treatment have been adopted and followed (Table 2). In case of herbaceous and epiphytic plants, roots and leaves were washed under running water to remove dirt, surface-sterilized in a sequence of 75% ethanol for 1 min, 65% Clorox® (= 3·4% NaOCl) for 10 min and 75% ethanol for 30 S, and rinsed in sterile distilled water (83). In some cases, to minimize the risks of isolating epiphytic microorganisms during internal tissue sampling, plant organs were surface sterilized by dipping them in sodium hypochlorite solution for 5 minutes (7% Cl active) or by exposing them to propylene oxide vapours for 1 hour (88). Sometime the stem material is rinsed in running water and then successive surface sterilization in 75% ethanol and 40% formalin for 3 min each and rinsed three times in sterile distilled water 48. In some hard and woody plants ethanol immersion followed by flaming also yields good results with larger diameter twigs, limbs and wood fragments. This method is particularly effective with canker-forming fungi. A chip of wood and bark from the margin of the canker is dipped in ethanol and briefly flamed before plating out on agar. For very delicate tissues or tissues not amenable to surface sterilization, serial washing in sterile, deionized water, leaving the tissues in a running stream of water for several hours, or other modifications is often substituted for sterilizing agents. It is usually necessary to use a medium amended with antibiotics if no sterilant is used. One way to cut down on the amount of washing time, or to remove surface debris, is to sonicate the tissue. The homogenized tissues were then plated in appropriate media amended with antibiotics. After the surface sterilization processes are completed the sterilized segments (barks, stem, twigs and leaves) are air dried under laminar airflow chamber. Then with a sterile blade or scalpel the outer part of surface treated material (bark, twig and stem) is removed and inner tissue are carefully dissected out and plated on different mycological media amended with antibiotics. For leaves, 3-4mm diameter disc were cut out from veins or distal, central and proximate parts of the blade. Each leaf disc is then plated on isolation medium (10). The commonly used media for isolation of fungal endophytes are water agar (WA), Potato dextrose agar (PDA) and Malt extract agar (MEA). These media may or may not be supplemented with antibiotics. The commonly used antibiotics are streptomycin and tetracycline. However different combination of media and antibiotics are used (Table 2). But in our recent studies (109) on fungal endophytes of Himalayan yew we found that more endophytes were isolated on media amended with bark extract of the host plant (Table 3). It indicates that compound from the bark extract might have favored the growth of the endophytes. Moreover it has been reported that certain fungal endophytes sporulate only when autoclaved host tissue are added on the cultured media This suggests that compounds found in certain host plants appear to be necessary for the fungus to complete its life cycle. It seems difficult to study the fungal endophytes without isolating them in cultural medium because structural tools such as X-ray and electron diffraction are not very useful in this context, and it remains to be seen whether more specialized analytical methods can be developed that will enable scientists to successfully probe under the outer surface of a plant (72). Traditionally, staining techniques (*e.g.* thionin/phenol) have been used so that fungal hyphae could be distinguished from the cell tissues (15). Once the fungal endophytes are grown in cultural medium after several days of incubation, the hypal tips of the fungi are removed and transferred to mycological slant preferably in potato dextrose agar slants. The endophytes are encouraged to sporulate on different media under different condition and on specific plant materials and are eventually identified via standard morphological and molecular biological techniques and methods. Most endophytes will sporulate after a few weeks at 15 to 21°C, either in darkness or in daylight on 1% malt extract agar (MEA). For nonsporulating isolates, various methods to induce sporulation may be used. Exposure at 8°C under fluorescent or near-ultraviolet light with a 12 h dark-light cycle or incubation at 4°C in the darkness may induce sporulation in some fungi. Transferring nonsporulating isolates to natural media such as corn meal agar, V8 agar, potato carrot agar, or on chemically defined media such as Czapek's agar, peptone dextrose agar, etc. may induce sporulation (73). Sometimes it is necessary to grow the fungus on host substrate to induce sporulation. This can be accomplished several different ways. The easiest method is to sterilize pieces of host tissue for 20 min at 120°C and place these in the medium. For some host tissues it is desirable to sterilize using irradiation or propylene oxide. A decoction of host leaves, brewed up in a manner similar to making tea, may be added to the medium. Endophytic fungi can be identified most reliably using polymerase chain reaction (PCR) and DNA sequencing. PCR can be performed on cultured endophytes using primers that amplify DNA encoding ribosomal RNA. Since many of the fungi isolated from plants are likely to be novel, it is helpful to have primers covering as many taxonomic groups as possible. Sequencing the amplified DNA and comparing it with the database can help determine whether or not the specimen is novel; if the sequence is known, the species can be identified. Eventually, when an endophyte is acquired in pure culture, it is tested for its ability to be grown in shake or still culture by the use of various media and growth conditions. It is also immediately placed in storage under various conditions, including 15% glycerol at –70°C. Ultimately, once appropriate growth conditions are found, the microbe is fermented and extracted and the bioactive compound(s) is isolated and characterized.

**Table 2.** Some important protocols for the isolation of fungal endophytes from medicinal and woody plants

| Plant species | Parts/ tissue | Surface sterilization | Media | Reference |
|---|---|---|---|---|
| *Taxus baccata* | Stem, Bark | 70% ethanol for few seconds | Water agar (WA) | 90,109 |
| *Artemisia annua* | Stem, twigs | 70% ethanol and 40% formalin for 3 sec each | Potato dextrose Agar (PDA),with100mg/l[1] Ampicillin and 150mg/l$^{-1}$ Streptomycin sulphate | 65 |
| *Theobroma cacao* | Stem, twigs | 70% ethanol and 3% NaOCl for 3 mins each | Potato dextrose agar (PDA) supplement with 50µg/ml tetracycline | 85 |
| *Lepanthes* sp. | Roots, leaves | 70% ethanol for 1 mins, 3.4% NaOCl and 75% ethanol for 30 secs | Malt extract agar (MEA), supplement with 35 µg/ml rose Bengal and Rhizoctonia isolation with 35 µg/ml rose Bengal and *Rhizoctonia* isolation medium with 50µg/ml Streptomycin | 10 |
| *Azadirachta indica* | Bark | 70% ethanol for 1 min and 3.5% NaOCl for 2min | Water agar (WA), supplement 100mg/ml Streptomycin | 66 |
| *Kennedia nigriscans* | Stem, twig | 70% ethanol | Potato dextrose agar (PDA) | 24 |
| *Erythrina crista-galli* | Stem, twig | 70% ethanol and NaOCl | Potato dextrose agar (PDA), Water agar (WA) | 115 |
| *Taxus* spp. | Stem, bark, leaves, twigs, leaves | NaOCl (7% Cl active) for 5min, Propylene oxide Vapour for 1 hour | Malt agar (Difco) and malt agar + 50ppm rose Bengal and Tetracycline | 23 |
| *Alnus* | Bark, xylem | 35% paracetic acid | Malt extract agar (MEA) with 10 mg/l cyclosporin | 12 |

| | | | | |
|---|---|---|---|---|
| Thai medicinal Plants | Leaves and branches | 75% ethanol for 2 mins.3.5% NaOCl for 5 mins (for branches) | Potato dextrose agar, supplement with 50 mg/l Chloramphenicol | 78 |
| *Melia azedarach* | Root bark | Wash with water then ethanol and sterilized with 11% NaOCl for1min. | Potato dextrose agar (PDA) | 84 |

**Table 3.** Recovery of endophytes from Himalayan yew bark on various media combination.

| Sample | Media* | Incubation Time (in weeks) | No. of colonies recovered |
|---|---|---|---|
| | PDA | 3 | 4 |
| | MEA | 2 | 6 |
| | WA | 2-3 | 3 |
| Bark | | | |
| | PDA+BE | 2 | 6 |
| | MEA+BE | 2 | 8 |
| | WA+BE | 3 | 6 |

*Media: PDA-Potato dextrose agar; MEA-Malt extract agar; WA-Water agar; BE-Bark extract

## EXTRACTION AND SEPARATION OF ACTIVE COMPONENT

Isolation and separation of active metabolites from fungal endophytes requires different cultural, solvent extraction and analytical techniques (Fig. 1). For extraction and testing of bioactivity in fungal endophytes, obtaining a pure culture is necessary. However, sometime pure fungal culture is not crucial at first because in some instances greater activity is generated when endophytes stimulate and interact with each other. The endophytes are screened for their ability to produce any biological activity. Here it must be stressed that having appropriate bioassays available in order to check the activity of an endophyte is crucial for the isolation and purification of the active components. The active microbial fungi are eventually cultivated on appropriate liquid medium. The most commonly used liquid medium is potato dextrose broth (PDB). However, different fermentation medium have been used for cultivation of fungal endophytes (Table 4). The endophytes are then grown in Erlenmeyer flasks at room temperature or incubated at 25°C-28°C for weeks or more at 150rpm-250rpm in a rotary shaker. Sometime periodic shaking is also necessary. After the fermentation process is completed, the mycelia mats are separated and culture filtrate are obtained or sometime centrifuged to collect the supernatant. The culture filtrate or supernatant so obtained is extracted using various organic solvents. The commonly used organic solvents are ethyl acetate and methanol (62,84,115). The extracted solvent is then collected and evaporated under reduced pressure or using a rotary evaporator. The crude extract can be concentrated using $Na_2SO_4$ or $MgSO_4$. Isolation and purification of the bioactive compounds is attained through chromatographic techniques such as thin layer chromatography (TLC) and high-pressure liquid chromatography (HPLC). After obtaining a sample of pure bioactive material, the final goal is to determine its structure and composition. The two main structure-determining techniques used for this work are mass spectrometry (MS) and nuclear magnetic resonance spectroscopy (NMR). Both of these methods have seen vast improvements in sensitivity and technological advances over the past ten years, and when taken in tandem they represent very powerful means of solving unknown molecular structures. In addition, X-ray diffraction is also available as a powerful method if the sample can be crystallized. Utilizing high-resolution mass spectrometry can, in most instances, provide the molecular mass to such a precision that the molecular formula can also be determined. The exact number of hydrogen and carbon atoms can usually be verified from 1-D NMR data, and various 1-D and 2-D experiments can give information about connectivity between magnetic nuclei in the sample, leading to establishing the full structure. It should be noted however that not all experiments lead to the discovery of a totally new chemical compound; occasionally the structural probing reveals a known bioactive chemical, although its source might be novel.

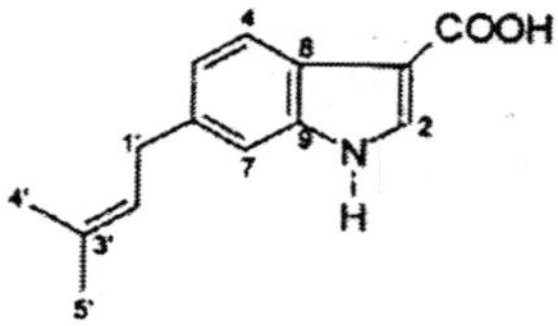

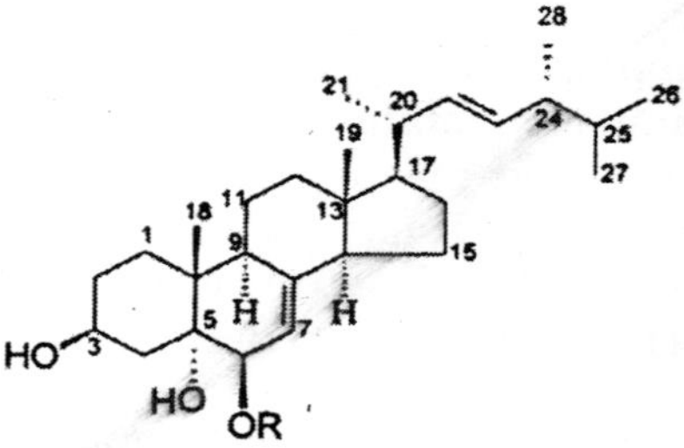

2 R = $COCH_3$
3 R = $COCH_2C_6H_5$

**Fig. 1**: Isolation and separation of active metabolites from *Collectotricum* sp, an endophytic fungus isolated from *Atremisia annua*

**Table 4**: Some important fermentation medium for cultivation of fungal endophytes

| Endophytic | Fermentati on medium Fungi | composition | | Growth factor | References |
|---|---|---|---|---|---|
| *Phomopsis* sp. | KSA | Dried mashed potato | 0.4% | Incubated at room temp. at 120rpm for 39 days | 32 |
| | | Glucose | 2% | | |
| | | pH | 5.5 | | |
| *Penicillium* sp. | Sterile rice | 90g rice (uncles Ben's) | 75ml | Incubated at 25°C for 20 days | 87 |
| | | Water | | | |
| *Alternaria* sp. | | Corn steep liquor | 10g/L | | |
| *Phomopsis* sp. | Liquid seed | $(NH_4)_2SO_4$ | 1 g/L | Incubated at 28°C | 23 |
| *Gliocladium* sp. | medium (VS) | $CaCO_3$ | 5 g/L | 250rpm for 7-9days | |
| *Fusarium* sp. etc | | $K_2HPO_4$ | 0.1g/L | | |
| *Penicillium* | White corn | White corn | 90 g | Incubated at 25°C for 20 days | 84 |
| *Janthinellum* | (Yoki) | Water | 75 ml | | |

| | | | | | |
|---|---|---|---|---|---|
| *Colletotrichum* sp. | Potato dextrose broth (PDB) | Peel potatoes | 200 g | Incubated at 28°C for 10 days at 150 rpm | 65 |
| | | Glucose | 20 g | | |
| | | Distilled water | 500 ml | | |

## BIOASSAYS

The isolated fungal endophytes are cultured and screened for their ability to produce any biological activity. There must be an appropriate and accurate bioassays procedure available in order to check the activity of an endophyte for the isolation and purification of the active components. The types of bioassay depend largely on the interest of the investigator. However, most of the bioassay is conducted to test the antibacterial and antifungal activity targeting multi-drugs resistant strains and looking for new and effective Antimicrobial metabolites. Fungal endophytes are also bioassay to test the cytotoxic properties because of their ability to produce number of anticancer compounds. Antiviral properties could also be investigated in the fungal endophytes but one of the major limitations in this regard is the absence of appropriate antiviral screening systems in most compound discovery programs.

## ANTI-BACTERIAL ASSAY

Anti-bacterial activity can be determined in the culture filtrate of the fungal endophytes. The cultural broth is centrifuge and supernatants are collected for bioassay or the broth is extracted by organic solvent, vaporized and concentrated to get the crude extract. The crude extract can be dissolved in Dimethly Sulphoxide (DMSO) for Antimicrobial bioassay. Normally two methods can be employed for anti-bacterial activity; the first is paper disc method and second is agar cup diffusion method (43). In the former, a lawn of the test bacterial strain is prepared in Mueller Hinton agar petri-plates with a sterile cotton swab. Paper disc impregnated with the crude extract or the supernatant are then placed over the plates with a sterile forceps. The plates are incubated at 35°C for 16 to 18 hours. The activity is determined by a visual zone of inhibition surrounding the discs (Fig. 2). In agar cup diffusion method, overnight culture of standard test organism is inoculated in Mueller Hinton agar petri-plates and evenly spread over the entire surface with a sterile glass rod, then agar cups are made with a cork borer and each cup is loaded with 100 to 200µl of the crude extract and incubated. Activity is determined by measuring the zone of inhibition surrounding the cups (Fig. 2). Another method employed is the liquid anti-bacterial method involves adding the bacterial strain in a broth to a 96-well plate and adding endophytes extracts at varying concentrations to the wells. Inhibition of growth in the wells is determined by optical transmittance through the wells (72).

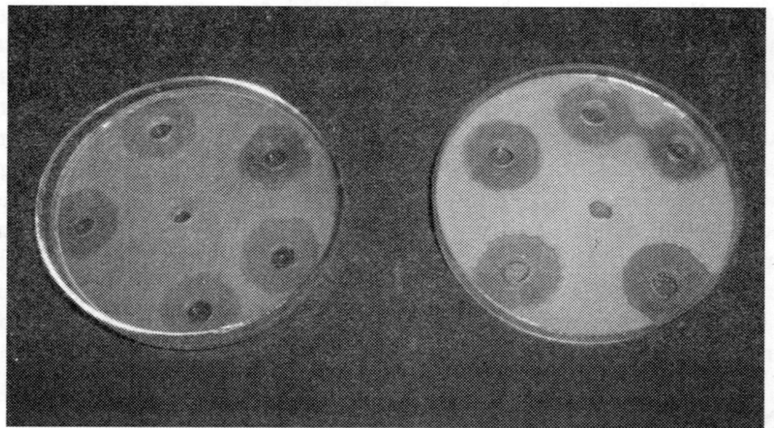

**Fig. 2**. Anti-bacterial activity of crude metabolites of five fungal endophytes against *E. coli* and *Bacillus subtilis*

## ANTI-FUNGAL ASSAY

Fungal infections are often divided into superficial mycoses and systemic mycoses. Treatment of these diseases is very different and mostly done by commercially available synthetic and semi-synthetic antibiotic. Few natural derived antibiotics such as Nystatin and Griseofulvin are available as antifungal agent. Therefore anti-fungal assay is determined by targeting some important human pathogens such as *Candida* sp. and *Trichophyton* spp. Anti-fungal bioassay for these pathogens are similar to anti-bacterial bioassay, employing the paper disc and agar cup diffusion method. The test fungal organism is grown in Sabouraud's agar plate and endophytic fungal extract are placed or loaded over the plate. Activity is determined by observing the zone of inhibition (Fig. 3). Again there are two possible approaches to an anti-fungal assay. In a competitive assay, two fast-growing fungi (*e.g. Pythium and Geotricium*) are placed on opposite sides of a petri dish, and a plug containing a drop of the endophyte liquid is placed at the center of the plate. If the endophyte has any anti-fungal activity, the center spot will have a strong zone of inhibition surrounding it. In a non-competitive assay, the liquid extracted from the endophyte is placed in a 6-well plate containing M1D solution, and *Geotricium* or *Pythium* is added to the wells. The wells may be inspected at 24 hr intervals and the growth of the added fungus compared with that in a control medium. Certain plant pathogenic fungi such as *Pythium, Phytophthora* and *Aphanomyces* are also used as screening tools for detection of anticancer compound like taxol, because these fungi are extremely sensitive to taxol. Therefore, fungal endophytes extract producing anticancer compounds shows strong anti-fungal activity against these fungi.

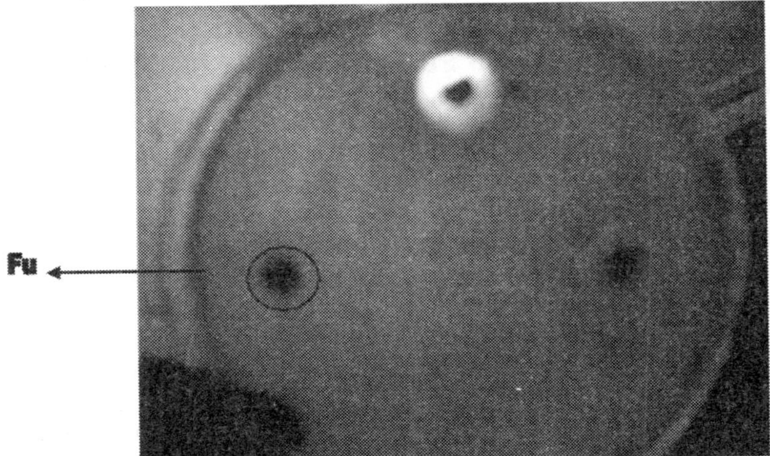


**Fig. 3**. Anti-fungal activity of endophytic fungal extract against *Candida krusei*, a pathogen resistant to fluconazole (Fu) antibiotic (10 mcg/disc)

## CYTOTOXIC ASSAY

Many fungal endophytes have been reported to produce anticancer agents. Most of the anticancer metabolites have been investigated from fungal endophytes of yew plants, which produces an anticancer compound paclitaxel and its derivatives.

The presence of these metabolites can be detected in the liquid fluid of the endophytes by electrospray mass spectroscopy and comparing with the authentic paclitaxel peaks. However many other fungal endophytes may have the ability to produce unknown anticancer metabolites. For this, cytotoxic assay should be conducted in several cancer cell lines. This assay measures the ability of the endophytes extract to inhibit the growth of cancer cells in vitro. Immortalized HeLa cervical cancer cells (or any other available cancer cells) can be used for this purpose. The cells are incubated for 18 hr in 96-well plates before the endophytes extract is added. After 24 hr with the extract, the assay is stopped. Sulfurhodamine is added as a dye, and the color identifies living cells as opposed to those that have been killed by the extract. The microbial extracts can also be assayed by a competitive inhibition enzyme immunoassay (CIEIA) using a commercial kit (Hawaii Biotechnology Group, Inc., Aiea, Hawaii) employing a polyclonal antibody against the taxane molecules (23).

## NATURAL PRODUCTS FROM FUNGAL ENDOPHYTES OF MEDICINAL PLANTS

Medicinal plants have been used for treatment of various human diseases since time immemorial. But in the recent years, these plants have been found to harbor quite distinct and potential fungal endophytes that have produces some important bioactive metabolites. Moreover these endophytes were also found to generate the same bioactive natural produce as that of the host plants. Some of the important discoveries of natural products from fungal endophytes associated with medicinal plants have been discussed below:

## FUNGAL ENDOPHYTES OF MEDICINAL PLANTS AS ANTICANCER AGENTS

Yew plant (*Taxus* spp.) worldwide is an important medicinal plant because its leaves and bark are prime source of taxol, a potent anticancer drug (7). Taxol compounds are unusual in their anticancer effect because that they bind to microtubulin in the cells and thus inhibit mitosis (2,68). Therefore, it is used to treat variety of cancer including breast, lung, ovary carcinomas and other human tissue proliferation diseases as well. Later it was discovered that fungal endophytes associated with yew plants around the world have the ability to produce taxol *in vitro*. The first report on taxol production from such fungi was from Taxomyces andreanae, an endophytic fungus isolated from bark of Pacific yew (*Taxus brevifolia*). Thereafter several endophytic fungi isolated from different Taxus spp. have been reported to produce taxol. An endophytic fungus, *Tubercularia* sp. isolated from southern Chinese yew (*Taxus mairei*) in the Fujian province of southeastern China was also reported to produce paclitaxelz (113). Taxol production have been reported from three different fungal endophytes of Himalayan yew (*Taxus wallichinia*) namely *Sporormia minima*, *Trichothecium* sp., and an unidentified dimorphic fungus (90). Evidences of Taxane production have been reported in several endophytic fungi isolated from *Taxus baccata* and *Taxus brevifolia*. *Pestalotiopsis* spp. are the most commonly isolated endophytic fungus of rain forest plants. One of the most commonly isolated endophytic species is *P. microspora*. This fungus was isolated as endophytes from *Taxus wallichinia*, which was reported to produce paclitaxel (105). Furthermore, several other *P. microspora* isolates were obtained from bald cypress in South Carolina and also shown to produce paclitaxel (60). This was the first indication that endophytes residing in plants other than *Taxus* spp. were producing paclitaxel. Thereafter several other plants species other than yew plant have also been found to harbor specific fungal endophytes that have produces paclitaxel (103,104). Furthermore, fungal paclitaxel production has also been noted in a *Periconia* sp.(61) and in *Seimatoantlerium nepalense*, another novel endophytic fungal species (9). Simply, it appears that the distribution of those fungi making paclitaxel is worldwide and not confined to endophytes of yews.

One of the interesting discoveries of anticancer compound other than paclitaxel from an endophytic fungus is torreyanic acid, a selectively cytotoxic quinone dimer (anticancer agent), isolated from a *P. microspora* strain. This strain was originally obtained as an endophyte associated with the endangered tree *Torreya taxifolia* (55). Torreyanic acid was tested in several cancer cell lines, and it demonstrated 5 to 10 times more potency in those lines that are sensitive to protein kinase C agonists and causes cell death by apoptosis. Another anticancer compounds are the alkaloids, commonly found in endophytic fungi. Such fungal genera as *Xylaria*, *Phoma*, *Hypoxylon*, and *Chalara* are representative producers of a relatively large group of substances known as the cytochalasins, of which over 20 are now known (112). Many of these compounds possess antitumor and antibiotic activities, but because of their cellular toxicity they have not been developed into pharmaceuticals.

Another example of anticancer drugs derived from fungal endophytes, is the discovery of Campothecin from endophytic extract of *Entrophospora infrequens* isolated from the inner bark of *Nothapodytes foetida* (3). Campothecin and its anologues are naturally occurring group of quinoline alkaloids occupying an important position among the plant based anticancer drugs. Various plant species such as *Camptotheca acuminata*, *Ophiorrhiza mungo*, *Ervatomia hyneana* and *Nothapodytes foetida* are known sources of this phytochemical(1). Overexploitation of these plants rendered them as endangered species all over the globe, especially in China. The recent discovery of this compound in fungal endophytes has not only conserved the gene pool of those plants but has made microbial sources available.

These discoveries exemplified that plants used for anticancer agent may harbor specific fungal endophytes that might also produce anticancer metabolites. There have been several reports on medicinal plants used for anticancer therapy and also different indigenous community uses many other plant species as crude anticancer drugs. Therefore study of fungal endophytes associated with those plants may leads to the discovery of important anticancer metabolites. However there has been much speculation about the transfer such metabolites from plants to the associated endophytes. Further it is not known, who might have synthesized the compound first, by the plant or the fungus? Moreover it is being question that whether microbes are capable of communicating with each other and communicating genetic information with other organisms. Although complete answer is not at hand, it is believed that production and transfer of such metabolites might have arisen due to genetic exchange and lateral genes transfer between the endophytes and plants living in close association. Such evidences have been established in the genetic level studies on *Pestalotiopsis microspora* explaining how genes for taxol might have acquired by this fungus.

## FUNGAL ENDOPHYTES OF MEDICINAL PLANTS AS ANTIBIOTIC PRODUCERS

Antibiotics are defined as low-molecular-weight organic natural products made by microorganisms that are active at low concentration against other microorganisms (36). Fungal endophytes from medicinal plants have been reported to produce some new and important antibiotics. Natural products from these endophytes have been observed to inhibit or kill a wide variety of harmful disease-causing agents including, but not limited to, phytopathogens, as well as bacteria, fungi, viruses, and protozoan that affect humans and animals. A good example of such antibiotic is the discovery a unique peptide antimycotic, termed Crytocandin which contains a number of peculiar hydroxylated amino acids and a novel amino acid: 3-hydroxy-4-hydroxy methyl proline. This compound was isolated from a fungus, *Cryptosporiopsis quercina,* as endophytes on a medicinal plant, *Tripterigeum wilfordii* native to Eurasia (103). On petri plates, *C. quercina* demonstrated excellent antifungal activity against some important human fungal pathogens—*Candida albicans* and *Trichophyton* spp. Cryptocandin is also active against a number of plant-pathogenic fungi, including *Sclerotinia sclerotiorum* and *Botrytis cinerea.* Cryptocandin and its related compounds are currently being considered for use against a number of fungi causing diseases of skin and nails.

A previously unknown compound showing significant bioactivity has been extracted and isolated from an endophytic fungus found in the leaves of a plant from the highlands of Papua New Guinea. The plant (*Desmodium uncinatum*) is used by the indigenous people for healing wounds and body sores. The compound shows anti-fungal and anti-bacterial effects and has been shown to destroy HeLa cervical cancer cells with an IC50 value of 0.9 $\mu g/\mu L$.

A new bio-active compound from a fungal endophyte of *Desmodium uncinatum,* a medicinal plant growing in Papua New Guinea. Another important discovery of bioactive substances from fungal endophytes associated with medicinal plants is the isolation of colletotric acid, a metabolite of *Colletotrichum gloeosporioides*, an endophytic fungus in *Artemisia mongolica.* This metabolite displayed antimicrobial activity against bacteria as well as against the fungus *Helminthsporium sativum* (120). Another colletotrichum sp. isolated from isolated from *Artemisia annua*, produces bioactive metabolites that showed varied antimicrobial activity against gram positive and gram-negative bacteria as well as against plant pathogenic fungi. One component of the active metabolites was also been shown to produce plant growth hormone, indole acetic acid (IAA). These plants are used as traditional medicines in China. *Artemisia annua*, a traditional Chinese medicinal herb is well recognized for its synthesis of artemisinin, which is an important antimalarial drug.

*Erythrina crista-galli* (Fabaceae) is widely distributed in tropical and subtropical regions of the American continent and is a popular ornamental plant in subtropical areas. In Argentina the wood of this medicinal plant is used in infusions or decoctions as astringent, narcotic, and sedative (110). Antibacterial and anti-inflammatory activities have been reported for this plant (69). An endophytic fungus, Phomopsis sp. was isolated from this plant, which has produced new anti-inflammatory metabolites, Phomol. This compound exhibited antibacterial and antifungal in wide variety of microorganisms in very low concentration. Another important metabolite Phomopsichalasin has also been reported from this endophytes (47). This metabolite mainly exhibits antibacterial activity in disk diffusion assays (at a concentration of 4 µg/disk) against *Bacillus subtilis* (12-mm zone of inhibition), *Salmonella enterica* serovar *Gallinarum* (11-mm zone of inhibition), and *Staphylococcus aureus* (8-mm zone of inhibition). It also displays a moderate activity against the yeast *Candida tropicalis* (8-mm zone of inhibition)

Leucinostatin A an antifungal and anticancer metabolites is derived from Acremonium sp. an endophytic fungus isolated from *Taxus baccata* (European yew) (106). In India this plant is widely distributed in the temperate zone of Himalaya between 1800 and 3300m above the sea level and has been used as beverages and medicine by people inhabiting Himalayan region (77). Leucinostatin A is active against certain oomycetous plant pathogenic fungus – *Pythium utimum* with an effective 1 day 50% inhibitory concentration of <1 $\mu$mol. Leucinostatin A also possesses activity against certain human cancer cell lines, for instance, its IC50 value is 2.3 nM for breast cancer cell line BT-20 contrasted with 640 nM for a normal mammary cell line. Since antifungal agents such as taxol and leucinostatin A are produced by some endophytic fungi, oomycetes such as *P. ultimum* may serve as initial screening tools for anticancer agents.

Another interesting examples of Antimicrobial agent produce by fungal endophytes, is the isolation of anti-helicobacter pylori substances from endophytic fungus, *Aspergillus* sp. CY725 associated with a Chinese medicinal herb *Cynodon dactylon* (Poaceae). The human pathogenic bacterium *Helicobacter pylori* have been ascertained to be an aetiological agent for chronic active gastritis and a significant determinant in peptic and duodenal ulcer diseases (111). The Ethyl acetate extracts of the 16 fungal endophytes out of 32 isolates from *Cynodon dactylon* have shown to produce anti-helicobacter activities. The most potent strain identified as Aspergillus sp. CY725 was further characterized for its active metabolites. Based on combination of spectral and chemical methods (IR, MS, 1H- and 13C-NMR) four active compounds were identified to be helvolic acid, monomethyl-sulochrin, ergosterol and 3$\beta$-hydroxy-5$a$,8$a$-epidioxy- ergosta-6,22-diene with corresponding MICs of 8.0, 10.0, 20.0 and 30.0 $\mu$g/ml, respectively. The MIC of ampicillin co-assayed as a reference drug against *H. pylori* was 2.0 $\mu$g/ml. Furthermore, preliminary examination of the antimicrobial spectrum of helvolic acid, the most active anti - *H. pylori* metabolite characterized from the endophyte culture, showed that it was inhibitory to the growth of *Sarcina lutea*, *Staphylococcus aureus*and *Candida albicans* with MICs of 15.0, 20.0 and 30.0 $\mu$g/ml, respectively.

Moreover, there have been several reports on Antimicrobial evaluation of fungal endophytes of medicinal plants active against wide variety of gram positive and gram-negative bacteria as well as pathogenic fungi (32,78,82). Although details structures of the active metabolites were not worked in most of the cases, preliminary screenings have suggested the great potential of these endophytes as Antimicrobial agents. Such work have been done by Radu and Kqueen where they conducted preliminary screening of endophytic fungi from medicinal plants in Malaysia for Antimicrobial and antitumor activity (78). Similar Antimicrobial evaluations of fungal extract produce by endophytic fungus Phomopsis sp. isolated from the leaves of *Aspidosperma tomentosum* and twigs of *Spondias mombin* have been reported (32). Tayung and Jha have evaluated the Antimicrobial activity of some fungal endophytes of

Himalayan yew (*Taxus baccata*) bark. Fungal endophytes isolated from medicinal plants (Orchidacea) have been also been reported to produce Plant growth hormone (119).

## CONCLUSION

Recent evidences as exemplified by various discoveries, fungal endophytes associated with medicinal plants are important sources of unusual secondary metabolites of pharmaceutical, agricultural and industrial importance. Considering the vast potential of medicinal plants in our country like India, a through study of plant-microbes relation, as endophytes is urgently required because it is believed that such studies may leads to the discovery of important endophytic strains that might produce new and interesting bioactive metabolites useful for the mankind. One of the major problems facing the future of endophyte biology and natural product discovery is the rapid diminishment of plant diversity, which holds the greatest possible resources of acquiring novel microorganism and their product. Over exploitation of medicinal plants have threatened many important plants species. Therefore, study of fungal endophytes associated with these medicinal plants is the need of the hour because once such plant get lost or extinct so is the entire suite of associated endophytes. Besides, it has also been hypothesized by some workers that the endophytes might have direct or indirect role in the production of metabolites extracted from medicinal plants. This belief is further made strong by the discovery of some endophytes that produce metabolites like the host does. Lot of works are to be carried out before we are in a position to generalize this observation. This property of endophyte would help us to stop over exploitation of plants for extraction of drugs and consequently help in our conservation strategies.

## REFERENCES

1. Aiyama, R, Nagai, Kokata, K, Shinonara, C and Swada, S, 1988 Camptothecin derivatives from *Nothapodytes foetida* *Phytochemistry,* **27** 3663-3664

2. Altmann, KH 2001 Microtubule-stabilizing agents a growing class of important anticancer drugs *Current Opinion in Chemical Biology* **5** 424–431

3. Amna, T, Khajuria,RK, Puri, SC, Verma, V and Qazi, GN 2006 Determination and quantification of camptothecin in an endophytic fungus by liquid chromatography-positive mode electrospray ionization tandem mass spectrometry *Current Science* **91** 208-212

4. Arnold, A E, Meffa, L C, Kyollo, D, Rojas, E I, Maynard, Z, Robbins, N and Herre, E A, *Proceeding of National Academy of Science USA*, 2003, **100**, 15649–15654

5. Arnold, A E, Z Maynard, G S Gilbert, P D Coley, and T A Kursar 2000 Are tropical endophytes hyperdiverse? *Ecology Letters* **3** 267–274

6. Bacon, C W 1993 Abiotic stress tolerances (moisture, nutrients) and photosynthesis in endophyte-infected tall fescue, *Agricultural Ecosystem and Environment* **44** 123–141

7. Bacon, CW, and White, JF (2000) Microbial endophytes Marcel Decker, Inc, New York

8. Barklund P & Kowalski T (1996) Endophytic fungi in branches of Norway spruce with particular reference to *Tryblidiopsis pinastri*, *Canadian Journal of Botany* **74** 73-678

9. Bashyal, B, J Y Li, G A Strobel, and W M Hess 1999 *Seimatoantlerium nepalense*, an endophytic taxol producing coelomycete from Himalayan yew (*Taxus wallichiana*), *Mycotaxon* **72** 33-42

10. Bayman, P, Lebron, LL, Trembley, RL and Lodge and DJ 1997 Variation in endophytic fungi from roots and leaves of *Lepanthes* (Orchidaceae), *New Phytology* **135** 143-149

11. Bettucci L, Alonso R and Tiscornia S 1999 Endophytic mycobiota of healthy twigs and the assemblage of species associated with twig lesions of *Eucalyptus globulus* and *E grandis* in Uruguay *Mycological Research* **103** 468-472

12. Bills G F 1996 Isolation and analysis of endophytic fungal communities from woody plants pp 31-65 In Systematics, Ecology and Evolution of Endophytic Fungi in Grasses and Woody Plants, S Redlin and LM Carris, eds APS Press, St Paul, MN

13. Bissegger M and Sieber TN 1994 Assemblages of endophytic fungi in coppice shoots of *Castanea sativa, Mycologia* **86** 648-655

14. Bohár G 1996 *Ceratocystis erinaceus* a new endophyte in the heartwood of oak, *Acta Phytopathology Entomology Hung* **31** 213-218

15. Boursnell J G 1950 The Symbiotic seed-borne fungus in the *Cistaceae, Annals ofBotany* **14** 217-243

16. Breen JP 1994 Acremonium endophytes interactions with enhanced plant resistance to insects *Annual Review of Entomology* **39** 401-23

17. Bultman, T L, K L Borowicz, R M Schneble, T A Coudron, and L P Bush 1997 Effect of a fungal endophyte on the growth and survival of two *Euplactrus parasitoids, Oikos* **78**170–176

18. Bush, LP Wilkinson, HH, and Schardl, CL, 1997 Bioprotective alkaloids of grass-fungal endophyte symbioses, *Plant Physiology* **114** 1-7

19. Calhoun LA, Findlay JA, Miller JD and Whitney NJ 1992 Metabolites toxic to spruce budworm from balsam fir needle endophytes, *Mycological Research* **96** 281-286

20. Carroll G 1988 Fungal endophytes in stems and leaves from latent pathogen to mutualistic symbiont, *Ecology* **69** 2-9

21. Carroll, G C 1991 Beyond pest deterrence Alternative strategies and hidden costs of endophytic mutualisms in vascular plants In J H Andrews and S S Monano (eds), *Microbial ecology of leaves*, Springer-Verlag, New York 358–378 pages

22. Carroll GC, Carroll FE 1978 Studies on the incidence of coniferous needle endophytes in the Pacific Northwest, *Canadian Journal of Botany* **56** 3034-43

23. Caruso, M, Colombo, AL, Fedeli, L, Pavesi, A, Quaroni, SM, Saracchi, M and Ventrella G 2000 Isolation of endophytic fungi and actinomycetes taxane producers, *Annals of Microbiology* **50** 3-13

24. Castillo, U F, G A Strobel, E J Ford, W M Hess, H Porter, J B Jensen, H Albert, R Robison, M A Condron, D B Teplow, D Stevens, and D Yaver 2002 Munumbicins, wide-spectrum antibiotics produced by *Streptomyces* NRRL 30562, endophytic on *Kennedia nigriscans, Microbiology* **148** 2675-2685

25. Chanway, CP 1996 Endophytes they're not just fungi! *Canadian Journal of Botany* **73** 321-322

26. Cheplick GP, Clay K 1988 Acquired chemical defenses of grasses the role of fungal endophytes, *Oikos* **52** 309-18

27. Clay K 1988 Fungal endophytes of grasses a defensive mutualism between plants and fungi *Ecology* **69** 10-16

28. Clay K 1990 Fungal endophytes of grasses, *Annual Review of Ecological System* **21** 275-79

29. Clay K, 1992 Fungal endophytes of plants biological and

chemical diversity, *Nat Toxins* **1** 147-49

30. Clay, K Holah, J, 1999 *Science* **285** 1742–1744
31. Collado J, Platas G, Gonzáles I and Pélaez F 1999 Geographical and seasonal influences on the distribution of fungal endophytes in *Quercus ilex, New Phytology* **144** 525-532
32. Corrado M Rodrigues KF 2004 Antimicrobial evaluation of fungal extracts produced by endophytic strain of *Phomopsis* sp, *Journal of Basic Microbiology* **44** (2) 157-60
33. Daisy, BH, Strobel, GA, Castillo, U and Ezra D 2002 Naphthalene, an insect repellent, is produced by *Muscodor vitigenus*, a novel endophytic fungus, *Microbiology* **148** 3737-3741
34. De Bary A 1866 Morphologie and Physiologie der Pilze, Flenchten and Myxomyceten Leipzig, Germany Engelmann
35. Deckert RJ and Peterson RL 2000 Distribution of foliar fungal endophytes of *Pinus strobus* between and within host trees, *Canadian Journal For Research* **30** 1436-1442
36. Demain, A L 1981 Industrial microbiology, *Science* **214** 987-994
37. Ellis MB, Ellis JP 1985 Microfungi on land plants Macmillan Publishing Co, New York, 1985
38. Elmi AA & West CP 1995 Endophyte infection effects on stomatal conductance, osmotic adjustment and drought recovery of tall fescue, *New Phytology* **131** 61-67
39. Ezra, D, Hess WM and Strobel GA 2004 New endophytic isolates of *Muscodor albus*, a volatile-antibiotic-producing fungus, *Microbiology* **150** 4023-4031
40. Faeth, S H and T L Bultman 2002 Endophytic fungi and interactions among host plants, herbivores, and natural enemies *In* T Tscharnkte and B A Hawkins (eds), *Multitrophic level interactions* Cambridge Univ Press, Cambridge
41. Findlay JA Li G and Johnson JA 1997 Bioactive compounds from an endophytic fungus from eastern larch (*Larix laricina*) needles, *Canadian Journal of Chemistry* **75** 716-719
42. Fisher PJ and Petrini O 1990 A comparative study of fungal endophytes in xylem and bark of *Alnus* species in England (UK) and Switzerland *Mycological Research* **94** 313-319
43. Grammer, A 1976 Antibiotic sensitivity and Assay test In *Microbiological methods*, Collins, C H and Lyne, PM

(Eds), Butterworths, London, 235 page

44. Groppe K, Steigner T, Sanders I, Schmid B, Wiemken A and Boller T 1999 Interaction between the endophytic fungus *Epichloë bromicola* and the grass *Bromus erectus* effects of endophyte infection, fungal concentration and environment on grass growth and flowering, *Molecular Ecology* **8** 1827-1835
45. Gwinn K D, Gavin A M 1992 Relationship between endophytic infection level of tall fescue seed lots and *Rhizoctonia zeae* seedling disease *Plant Disease* **76** 911–914
46. Hata K, Futai K 1996 Variation in fungal endophyte populations in needles of the genus *Pinus, Canadian Journal of Botany* **74** 103-114
47. Horn, W S, M S J Simmonds, R E Schwartz, and W M Blaney 1995 Phomopsichalasin, a novel antimicrobial agent from an endophytic *Phomopsis* sp, *Tetrahedron* **14** 3969-3978
48. Iyengar, MA 1988 Our green wealth vis a vis health, *Bulletin of Science* Jan/Feb 21-23
49. Johnson JA, Whitney NJ 1989 A study of fungal endophytes of needles of balsam fir (*Abies balsamea*) and red spruce (*Picea rubens*) in New Brunswick, Canada, using culture and electron mircoscope techniques, *Canadian Journal of Botany* **67** 3513-3516
50. Johri BN 2006 Endophytes to the rescue of plants *Current Science*, **90** 1315-1316
51. Kloepper JW, Leong J, Tientze M and Schroth MN 1980 Enhanced plant growth by siderophores produced by plant growth promoting rhizobacteria, *Nature* **286** 885-886
52. Knoch T R, Faeth SH and Arnott DL 1993 Endophytic fungi alter foraging and dispersal by desert seed-harvesting ants, *Oceologia* **95** 470–475
53. Lambert B, Leyns F, Van Rooyen L, Gosselé F, Papon Y and Swings J 1987 Rhizobacteria of maize and their fungal activities, *Applied Environmental Microbiology* **53** 1866-1871
54. Lange, L, 1996 Microbial metabolites- an infinite source of novel chemistry *Pure and Applied Chem*, **68** 745-748
55. Lee JC, Strobel GA, Lobkovsky E and Clardy J C 1996 Torreyanic acid a selectively cytotoxic quinone dimer from the endophytic fungus *Pestalotiopsis microspora, Journal of Organic Chemistry* **61** 3232-3233
56. Lee, J C, X Yang, M Schwartz G Strobel, and J Clardy 1995

The relationship between an endangered North American tree and an endophytic fungus, *Chemical Biology* **2** 721-727

57. Lee, J, E Lobkovsky, N B Pliam, G A Strobel, and J Clardy 1995 Subglutinols A and B immunosuppressive compounds from the endophytic fungus *Fusarium subglutinans, Journal of Organic Chemistry* **60** 7076-7077

58. Leuchtmann, A, D Schmidt, and L P Bush 2000 Different levels of protective alkaloids in grasses with stroma-forming and seed transmitted *Epichloe Neotyphodium* endophytes, *Journal of Chemical Ecology* **26** 1025–1036

59. Lewis FJ 1924 An endotrophic fungus in the Coniferae *Nature* **114** 860

60. Li, J Y, G A Strobel, R Sidhu, W M Hess, and E Ford 1996 Endophytic taxol producing fungi from Bald Cypress, *Taxodium distichum, Microbiology* **142** 2223-2226

61. Li, J Y, R S Sidhu, E Ford, W M Hess and G A Strobel 1998 The induction of taxol production in the endophytic fungus *Periconia sp* from *Torreya grandifolia, Journal of Indian Microbiology* **20** 259-264

62. Liu, JY, Song, YC, Zhang, Z, Wang L, Guo, ZJ, Zou, WX and Tan, RX, 2004 *Aspergillus fumigatus* CY018, an endophytic fungus in *Cynodon dactylon* as a versatile producer of new and bioactive metabolites, *Journal of Biotechnology* **114** 279-287

63. Lockwood JL 1990 Relation of energy stress to behaviour of soil-borne plant pathogens and to disease development In Hornby D (ed) Biological control of soil-borne plant pathogens CAB International, Wallingford, Oxon, UK 197-214 pages

64. Long, DM, Smidmansky, ED, Archer, AJ, and Strobel, GA 1998 *In vivo* addition of telomeric repeats to foreign DNA generates chromosomal DNAs in the taxol-producing fungus *Pestalotiopsis microspora, Fungi Genetic Biology* **24** 335–344

65. Lu H, Zou WX, Meng JC, Hu J and Tan RX 2000 New bioactive metabolites produced by *Colletotrichum* sp, and endophytic fungus in *Artemisia annua, Plant Science* **151** 67-73

66. Mahesh, V, Tejesvi, MV, Nalini, MS, Prakash, HS, Kini, KR, Subbhiah, V and Shetty, HS, Endophytic mycoflora of inner bark of *Azadirachta indica* A Juss, *Current Science* **83** 218-219

67. Manoharachary, C, Sridhar, K, Singh, R, Adholeya, A, Suryanarayanan, TS, Rawat, S and Johri, BN 2005 Fungal diversity Distribution, conservation and prospecting

of fungi from India, *Current Science* **89** 58-71

68. Miglietta, A, Bocca, C and Gabriel, L 2002 Comparative studies on biological activity of certain microtubule interacting taxanes, *Chemico–Biology Interactions* **139** 283–299

69. Mitscher, LA, Gollapudi, RSR, Gerlach, S, Drake, D, Veliz, EA and Ward, J 1988 Erycristin, a new antimicrobial petrocarpan from *Erythrina crista-galli, Phytochemistry* **27** 381-385

70. Müller MM & Hallaksela AM (1998) Diversity of Norway spruce needle endophytes in various mixed and pure Norway spruce stands, *Mycological Research* **102** 1183-1189

71. Nath SC 2000 Diversity of the higher plants of medicinal value growing in Northeast India in relation to conservation and sustainable use, *Journal of Assam Science Society* **41** 267-288

72. Owen NL and Hundley N 2004 Endophytes-the chemical synthesizers in plants, *Science Progress* **87** 79-99

73. Petrini O 1986 Taxonomy of endophytic fungi of aerial plant tissues In Microbiology of the Phyllosphere N J Fokkema and J van den Heuvel, eds Cambridge University Press, Cambridge 175-187

74. Petrini, O 1991 Fungal endophytes of tree leaves In Andrews J, Hirano, SS, eds Microbial ecology of leaves New York Springer Verlag, 179-197 pages

75. Petrini, O, T H Sieber, L Toti, and O Viret 1992 Ecology, metabolite production, and substrate utilization in endophytic fungi, *Natural Toxins* **1** 185–196

76. Powell R G, R J Petroski 1992 Alkaloid toxins in endophyte infected grasses, *Natural Toxins* **1**(3)163–170

77. Purohit, A, Maikhuri, RK, Rao, KS and Nautiyal, S 2001 Impact of bark removal on survival of *Taxus baccata* L (Himalayan yew) in Nanda Devi Biosphere Reserve, Garwhal Himalaya, India, *Current Science* **81** 586-590

78. Radu, S Kqueen, CY 2002 Preliminary screening of endophytic fungi from medicinal plants in Malaysia for antimicrobial and antitumour activity, *Malayasian Journal of Medical Science* **9**(2) 23-33

79. Raj, N 1993 Coelomycetous anamorphs with appendage bearing conidia Edwards Brothers Publishing Co, Ann Arbor, Mich

80. Redlin SC, Carris LM 1996 Endophytic fungi in grasses and woody plants systematics, ecology and evolution APS Press, St Paul, MN, USA

81. Redman, R S, Sheehan, K B, Stout, R G, Rodriguez, R J and Henson, J M, 2002 *Science*, **298**, 1581

82. Rodrigues KF, Hesse M and Werner C 2000 Antimicrobial activities of secondary metabolites produced by endophytic fungi from *Spondias mombin, Journal of Basic Microbiology* **40** 261-267

83. Rodrigues KF 1994 The foliar fungal endophytes of the Amazonian palm, *Euterpe oleracea Mycologia* **86** 376-385

84. Rosario M, Edson AM, Rodrigues-Filho, E, Moitinho, MLR and Santos LS 2005 Biologically Active Polyketides Produced by *Penicillium janthinellum* Isolated as an Endophytic Fungus from Fruits of *Melia azadarachta, Journal of Brazilian Chemical Society* **16** 28—283

85. Rubini MR 2005 Diversity of endophytic fungal community of cacao (*Theobroma cacao* L) and biological control of *Crinipellis perniciosa,* causal agent of Witches' Broom Disease, *International Journal of Biological Science* **1** 24–33

86. Saikkonen, K, S H Faeth, M Helander, and T J Sullivan 1998 Fungal endophytes A continuum of interactions with host plants, *Annual Review of Ecological System* **29** 319–343

87. Santos, RMG and Rodrigues-Filho, E, 2003Structures of Meroterpenes Produced by *Penicillium* sp, an Endophytic Fungus Found Associated with *Melia azedarach Journal of Brazilian Chemical Society* **14** 722-727

88. Sardi P, Saracchi M, Quaroni S, Petrolini B, Borgonovi GE and Merli S 1992 Isolation of endophytic *Streptomyces* strains from surface-sterilized roots, *Applied Environment Microbiology* **58** 2691-2693

89. Schulz B, Wanke U, Draeger S and Aust HJ 1993 Endophytes from herbaceous plants and shrubs effectiveness of surface sterilization methods, *Mycological Research* **97** 1447-1450

90. Shrestha, K, Strobel, GA, Shrivastava, SP, and Gewali, MB 2001 Evidence for paclitaxel from three new endophytic fungi of Himalayan yew of Nepal, *Planta Medica* **67** 374-376

91. Siegel, M R and L P Bush 1996 Defensive chemicals in grass fungal endophyte associations *In* J T Romeo (ed), Phytochemical diversity and redundancy in ecological interactions *Recent Advances in Phytochemistry* 3081–119 pages

92. Siegel, MC, Latch, GCM, and Johnson MC 1987 Fungal endophytes of grasses,

*Annual Review of Phytopathology* **25** 293-315

93. Silva, GH 2005 New bioactive metabolites produced by *Phomopsis cassiae*, an endophytic fungus in *Cassia spectabilis, Journal of Brazilian Chemical Society* 16

94. Smith, CS, Chand, T, Harris, RF and Andrews, JH 1989 *Applied Environmental Microbiology* **55** 2326

95. Stanley HF, William FF 2002 Fungal Endophytes Common Host Plant Symbionts but Uncommon Mutualists, *Integrated And Computa-tional Bioogy l***42** 360-368

96. Stierle, A, Strobel, GA, and Stierle, D 1993 Taxol and taxane production by *Taxomyces andreanae, Science* (Wash, DC) **260** 214–216

97. Stobel GA, Daisy B 2003 Bioprospecting for Microbial Endophytes and Their Natural Products, *Microbiology and Molecular Biology Review* **67** 491

98. Stone, J K, Bacon, C W White JF 2000 An overview of endophytic microbes endophytism defined In Bacon, C W and White, J F (eds), *Microbial Endophytes*, chap 1, Marcel Dekker New York 3-29 pages

99. Strobel GA 2002 Rainforest Endophytes and Bioactive Products *Critical Review of Biotechnoogyl* **22** 315-333

100. Strobel, G A, E Ford, J Worapong, J K Harper, A M Arif, D M Grant, P C W Fung, and K Chan 2002 Ispoestacin, an isobenzofuranone from *Pestalotiopsis microspora*, possessing antifungal and antioxidant activities *Phytochemistry* **60** 179-183

101. Strobel, G A, J Y Li, F Sugawara, H Koshino, J Harper, and W M Hess 1999 Oocydin A, a chlorinated macrocyclic lactone with potent anti-oomycete activity from *Serratia marcescens, Microbiology* **145** 3557-3564

102. Strobel GA 2002 Microbial gifts of rain forests, *Canadian Journal of Plant Pathology* **24** 14-20

103. Strobel, G A, R V Miller, C Miller, M Condron, D B Teplow, and W M Hess 1999 Cryptocandin, a potent antimycotic from the endophytic fungus *Cryptosporiopsis cf quercina, Microbiology* **145** 1919-1926

104. Strobel, G A, W M Hess, J Y Li, E Ford, J Sears, R S Sidhu, and B Summerell 1997 *Pestalotiopsis guepinii*, a taxol producing endophyte of the Wollemi Pine, *Wollemia nobilis, Australian Journal of Botany* **45** 1073-1082

105. Strobel, G, X Yang, J Sears, R Kramer, R S Sidhu, and W M Hess 1996 Taxol from *Pestalotiopsis microspora*, an endophytic fungus of *Taxus*

*wallichiana, Microbiology* **142** 435-440

106. Strobel, GA, Torczynski, R and Bollon, A 1997 *Acremonium* sp-a leucinostatin A producing endophyte of European yew (*Taxus baccata*), *Plant Science* **128** 97-108

107. Sturz AV, Christie BR and Nowak J 2000 Bacterial endophytes potential role in developing sustainable systems of crop production, *Critical Review of Plant Science* **19** 1-30

108. Tan, R X and Zou WX 2001 Endophytes a rich source of functional metabolites, *Natural Product Report* **18** 448-459

109. Tayung, K, Jha, DK 2006 Antimicrobial evaluation of some fungal endophytes isolated from bark of Himalayan yew, *World Journal of Agricultural Sciences*, **2** (4) 489-494

110. Toursarkissian, M, 1980 Plantas Medicinales de la Argentina, Ed Hemisferio Sur, Buenos Aires, 70 page

111. Tsuji T, Y Kohli, S Fukumitsu, K Morita, H Kaneko, T Ohkawara, *J Gastroenterol* **1999**, *34*, 455-460

112. Wagenaar, M, J Corwin, G A Strobel, and J Clardy 2000 Three new chytochalasins produced by an endophytic fungus in the genus *Rhinocladiella, Journal of Natural Product* **63** 1692-1695

113. Wang, J, G Li, H Lu, Z Zheng, Y Huang, and W Su 2000 Taxol from *Tubercularia* sp strain TF5, an endophytic fungus of *Taxus mairei, FEMS Microbiology Lettets* **193** 249-253

114. Wang, J, Huang, Y, Fang, M, Zhang, Y, Zheng, Z, Zhao, Y and Su, W 2002 Brefeldin A, a cytotoxin produced by *Paecilomyces* sp and *Aspergillus clavatus* isolated from *Taxus mairei* and *Torreya grandis*, FEMS *Immunology and Medical Microbiology* **34** 51-57

115. Weber, D, Sterner, O, Anke, T, Gorzalczancy, S, Martino, V and Acevedo, C, 2004 Phomol, a New Anti-inflammatory Metabolite from an Endophyte of the Medicinal Plant *Erythrina crista-galli The Journal of Antibiotics* **59** 559-563

116. White JF Jr 1987 Widespread distribution of endophytes in the *Poaceae, Plant Disease* **71** 340-342

117. Wilson D, Carroll G 1994 Infection studies of *Discula quercina*, and endophyte of *Quercus garryana, Mycologia* **86** 635-647

118. Wolock-Madej C, K Clay 1991 Avian seed preference and weight loss experiment The role of fungal-infected fescue seeds, *Oecologia* **88** 296–302

119. Zhang, J, Wang, C, Gua, S, Chen, J and Xioa, P 1999

Studies on the plant hormones produced by 5 species of endophytic fungi isolated from medicinal plants (Orchidacea*), Zhongguo Yi Xue Ke Xue Yuan Xue Bao*, **21** 460-465

120. Zou, MX, Meng, JC, Lu, H, Chen, GX, Shi, Gx, Zhang, TY, and Tan, RX 2000 Metabolites of *Colletotrichum gloeosporioides*, an endophytic fungus in *Artemisia mongolica, Journal of Natural Product* **63**(11)1529-1530.

*Rev. Plant Pathol. Vol. 6, 2014*
*Indian Society of Mycology and Plant Pathology*
*Scientific Publishers (India), Jodhpur pp. 335-365*

# MOLECULAR INTERACTIONS AMONG PARTNERS OF ARBUSCULAR MYCORRHIZAL SYMBIOSIS

***E. Sharma[a], S. Pal[a], B. Giri[b] and R. Kapoor*[a]***

[a] Department of Botany, University of Delhi, Delhi-110007, India
[b] Swami Shraddhanand College, University of Delhi- 110036, India

**KEY WORDS:** Arbuscular mycorrhizal fungi, symbiosis, nutrient exchange, signaling pathway mycorrhizal associated bacteria

## ABSTRACT

Arbuscular mycorrhiza (AM) is a root endosymbiosis between fungi from a monophyletic group, the phylum Glomeromycota and terrestrial plants. It is the most widespread terrestrial plant symbiosis, improving plant's uptake of water and mineral nutrients. Advances in molecular and genetic tools, coupled to high throughput sequencing and advanced microscopy, have opened new avenues to address early communications between the two partners to enable development of functional symbiosis. This review focuses briefly on the advances in our understanding on development of the symbiosis emerging from these studies and briefly considers the mechanism underlying nutrient transport. Recent studies have revealed increased complexity of the system by the presence of third component of mycorrhizal association with helper bacteria. The interaction of these "helper bacteria" with AM fungi and their likely role in mycorrhizal function is discussed.

## INTRODUCTION

Arbuscular mycorrhiza (AM) represent a symbiosis between terrestrial plant roots and fungi of phylum Glomeromycota (95). The AM symbiosis derive its name from the Latin word *arbusculum* (=little tree) specifying typical tuft-like structures formed by fine dichotomously branched fungal hyphae, and the Greek word for fungus roots. The symbiosis entails a carbon cost for the plant host, which receives in return improved access to nutrients, protection from pathogens and a range of other benefits (102). To date, it is the most widespread mutualistic symbiosis which is formed by the vast majority of phototrophic terrestrial plants including both gametophytic mosses - liverworts and hornworts as well as sporophytic club mosses, horsetails, gymnosperms and angiosperms (102). The physiological responses of plants to AM inoculation are numerous and include an increase in size (30), higher plant tissue phosphorous concentrates, and higher photosynthetic rate (102).

Relatively little is known about AM fungi, because these species are obligate symbionts with no confirmed sexual stage, although the hyphae of genetically distinct strains can anastomose and exchange genetic material (23). Spores are multinucleate, containing thousands of nuclei and heterokaryotic. In the syncytial mycelia of these fungi, nuclei are constantly translocated by rapid cytoplasmic streaming (4). The lack of a uninucleate cell stage in the life cycle of AM fungi along with above peculiar traits have contributed to the difficulties in standard molecular approaches such as genetic transformation, mutant generation and characterization, as well as large scale transcriptomic and genomic analysis (19, 73).

The introduction of advanced technologies *in vitro* root-organ culture system and methods based on polymerase chain reaction has offered opportunity to conduct a molecular investigation of the fungal partners too, thus overcoming the problem of the limited availability of biological material due to the obligate biotrophic status of the mycosymbiont. Genetic tools and genomic information have helped in investigating several aspects of AM interactions such as root colonization and communication between symbionts (79). Gene expression studies during the establishment of AM symbiosis have targeted specific genes, especially those known to be involved in nutrient uptake and in plant-fungus interactions. In addition, plant mutants have been used as key tools for the genetic dissection of AM development and have been instrumental in the identification of plant genes controlling specific steps (62, 68, 77). The existence of plant mutants having $Nod^-$ and $Myc^-$ phenotypes has suggested a common genetic program for root nodulation symbiosis and AM symbiosis. The molecular probes developed for nodules have been used to investigate gene expression in AM hosts. Consequently, over past one and half decade there has been an explosion of molecular, genetic and biochemical analysis of the AM fungi and AM symbiosis. This review focuses briefly on the advances in our understanding on development of the symbiosis emerging from these studies (19, 47, 79) and briefly considers the mechanism underlying nutrient transport. Recent studies have revealed increased complexity of the system by the presence of a third component of mycorrhizal association – bacteria (18, 35). The interaction of these "helper bacteria" with AM fungi and their likely role in mycorrhizal function is discussed.

## EVOLUTION AND SYSTEMATICS

Arbuscular mycorrhizal fungi have been engaged in symbiotic relationship with species from many terrestrial plant families and completely reliant on their hosts to obtain photosynthetic product (102). In exchange, the fungi may enhance host fitness through a variety of mechanisms, including enhanced nutrient acquisition and increased biotic and abiotic stress tolerance. They have been reported to exist from early Devonian era (83). Their widespread geographical and biological distribution imparts the fact that this kind of mutualism was pivotal in the origin of terrestrial flora (104). The morphological similarities of fossil spores to chlamydospores of a living genus, *Glomus* and sequence divergence data, based on a molecular study, provide additional evidence for this fact (99).

Till the early nineties, AM fungi were considered the members of Zygomycota. Recent studies based on nuclear-encoded rRNA gene markers supported that these fungal symbionts form a monophyletic group of true fungi, the phylum Glomeromycota (89, 95), which comprises about 200 described morphospecies (84). Glomeromycotan fungi produce spores (about 40-800 $\mu$m) with layered walls (important to distinguished species) containing numerous nuclei (8). The mode of spore formation has been important to circumscribe genera and families. Their spores may produce singly or in clusters (sporocarps) (39). Molecular studies have described no genetic recombination or either very low level (66); therefore it was assumed that spores are formed asexually. The members of this group also lack formation of septa.

The phylum Glomeromycota is currently divided into four orders namely, Diversisporales, Glomerales, Archaesporales and Paraglomerales (95). Three families; Acaulosporaceae and Diversisporaceae and Gigasporaceae have been described within the order Diversisporales, of these, the Acaulosporaceae and Gigasporaceae contain the greatest number of described species. The Glomerales consists of monotypic family, the Glomeraceae with genus *Glomus*, although species have been recognized in two distinct clades (84, 97).

Until Gerdemann and Trappe (39) placed AM fungi in four different genera, *Glomus, Sclerocystis, Gigaspora* and *Acaulospora* in the order Endogonales, the most arbuscular mycorrhizal fungi (AM) were classified in the genus *Endogone*. A new order, Glomales was established by Morton and Benny (76) in the phylum Zygomycota, which comprises six genera (Table 1). Since then the classification of AM fungi was repeatedly refined by several taxonomists (84, 89, 95, 97). Emphasis was given on the structure of spores/sporocarps and their wall for the taxonomy of Glomales (105). Later more evidence accumulated in support of the view that arbuscular mycorrhizal fungi are distinct from other zygomycotaen, as they do not seem to develop characteristic zygospores. Molecular studies based on r DNA phylogenies revealed that AM fungi are the sister group of Asco- and Basidiomycota and not monophyletic with any part of the Zygomycota. Imparting to the evidences, Schüßler *et al.* (95) raised the order Glomales to the rank of a phylum Glomeromycota. They also corrected the grammatically incorrect order Glomales to 'Glomerales' and several new orders were established (Table 1). Molecular studies conducted by researchers have shown much higher glomeromycotan diversity at phylum and genus level than expected through microscopic examinations and observations of spore morphology (87, 97). It is speculated that some of the morphological characters that were used previously could have evolved multiple times independently.

**Table 1.** Evolutionary representation of AM fungi classification

| **Gerdemann and Trappe (39), Benjamin (9), Warcup (109)** | **Morton and Benny (76)** | **Present Classification Schubler *et al.* (95), Redecker *et al.* (87), Schwarzott *et al.* (97), Walker and Schubler *et al.* (106)** |
|---|---|---|
| Phylum: Zygomycota | Phylum: Zygomycota | Phylum: Glomeromycota |
| Order: Endogonales | Order: Endogonales | Order: Diversisporales |
| Family: Endogonaceae | Family: Endogonaceae | Family: Acaulosporaceae |
| Genera: *Endogone* | Genera: *Endogone* | Genus: *Acaulospora* |
| *Sclerogone* | *Sclerogone* | Family: Diversisporaceae |
| *Glomus* | | Genus: Diversispora |
| *Sclerocystis* | Order: Glomales | Family: Gigasporaceae |
| *Acaulospora* | Sub-order: Glomineae | Genus: *Gigaspora* |
| *Entrophosphora* | Family: Glomaceae | |
| *Gigaspora* | Genera: *Glomus* | Order: Glomerales |
| | *Sclerocystis* | Family: Glomaceae |
| | Family: Acaulosporaceae | Genus: *Glomus* |
| | Genera: *Acaulospora* | (with two distinct clades) |
| | *Entrophosphora* | |
| | | Order: Archaesporales |
| | | Family: Archaesporaceae |
| | | Genus: *Archaespora* |
| | | Family: Geosiphonaceae |
| | | Genus: *Geosiphon* |
| | | |
| | | Order: Paraglomerales |
| | | Family: Paraglomeraceae |
| | | Genus: *Paraglomus* |

## Phylum Glomeromycota

The Glomeromycota includes ten distinct genera (Table 1)

i. *Glomus:* The genus comprises more than 70 morpho-species. Spores are formed by budding from a hyphal tip and are found typically with layered wall structure. Subtending hypha often remains attached to the mature spores. The spore germination occurs through hyphal attachment and mode of spore formation is symplesio-morphic. Some species form sporocarps.

ii. *Gigaspora*: It is closely related to *Scutellospora* and both are placed in the family Gigasporaceae. Spores are formed on a bulbous sporogeneous cell and germinate through a newly formed opening in the spore wall after the formation of a papillate layer on the inside of the spore wall around the point of penetration. This warty layer has also referred as germinal wall. The genus lacks vesicles formation in roots. The extraradical mycelium bears so-called auxiliary cells.

iii. *Scutellospora*: The genus is closely related to *Gigaspora.* Spores are formed on a bulbous sporogeneous cell. They possess flexible inner spore walls, which are present permanently in mature spores, and referred to as germinal walls and a germination shield, which are involved in germination process. The flexible walls are not naturally pigmented. The definite sub-layers often can be stained with Melzer's reagent.

*iv.* *Acaulospora*: It is characterized by the formation of a sporiferous saccule which eventually disappears during spore maturation. The germinal walls (flexible inner walls) are conspicuous and a "germination orb" is formed on the inner walls at the time of spore germination. Germination orb helps in penetrating the outer spore wall. Sporiferous saccule is produced laterally.

v. *Entrophospora*: It is also characterized by the formation of a sporiferous saccule formed within the subtending hypha. Sporiferous saccule collapses eventually and disappears during spore maturation. A "germination orb" is formed on the inner walls during spore germination helping in penetrating the outer spore wall. Flexible inner walls are also formed.

vi. *Pacispora*: The spores are typically like *Glomus*, the only difference is they do have flexible inner walls and a germination orb.

vii. *Diversispora*: On the bases of ribosomal DNA phylogenies, the genus is placed in the family Diversisporaceae (*Glomus* group C), which is closely related to the family Acaulosporaceae than to *Glomus* groups A and B. one of its species, *D. spurca* is more closely related to *Glomus versiforme*.

viii. *Archaeospora:* The genus was established for those AM fungi which deeply divergent within phylum in ribosomal phylogenies. The species viz., *A. leptoticha* and *A. gerdemannii* are dimorphic. They produce both glomoid and acaulosporoid types of spores. The acaulosporoid spores of *Archaeospora* exihibit several typeset distinct from *Acaulospora* such as the sporiferous saccule is separated from the hyphae by a short of stalk, referred as pedicel. The germinal walls are much thicker (*A. leptoticha* and *A. gerdemannii*), which do not stain with Melzer's regent and do not appear to be involved in the germination process. The basal position of *Archaeospora* is supported by presence of unique fatty acids not found in other glomeromycotan fungi.

ix. *Geosiphon:* The genus is the only member of the Glomeromycota that is known to engage in a different type of symbioses (fungus-cyanobacterium symbiosis). It comprises a coenocytic soil fungus, *Geosiphon pyriforme* that lives in endosymbiotic association with a *Nostoc punctiforme*. The fungus lives together with the cyanobacterium on the surface and in the upper layer of wet soils poor in inorganic nutrients, particularly phosphate (95). When a fungal hypha comes into contact with free-living *Nostoc* cells, the latter are incorporated by the fungus at the hyphal tip, which thereafter swells and forms a unicellular "bladder", upto 2 mm. It forms glomoid spores and was identified as a basal relative of AM fungi by rDNA phylogeny (36). The latest sequence of *Geosiphon* in the databases supports this fungus to be closer to *Archaeospora leptoticha* and *A. gerdemannii* (88, 97).

x. *Paraglomus:* The genus is basal to the phylum. The basal position of this genus has been supported by presence of unique fatty acids not found in other glomeromycotan fungi. Ribosomal DNA phylogenies clearly distinct *Paraglomus* from other *Glomus* species. The intraradical structures of *Paraglomus* and *Archaeospora* stain weekly or even not stained if acidic stains like trypan blue are used which are commonly adapted to view mycorrhizal colonization. Species form small hyaline spores.

## MYCORRHIZAL DEVELOPMENT

Roots are colonized by AM fungal hyphae originating from germinating spores, other mycorrhizal roots, or dead roots serving as inoculum. After the formation of a hyphopodium (=AM fungal appressorium) on the root surface, hyphae cross the epidermal cell layer and penetrate into the cortex where they grow intracellularly and intercellularly and form arbuscules, coiled hyphae, and vesicles. New spores are formed by the extraradical hyphae. An overview of this mycorrhizal life cycle is shown in Fig.1. Establishment of a functional AM symbiosis thus requires completion of consecutive developmental stages implicating both symbionts and embracing presymbiotic interactions before physical contact between the two partners to the point of intracellular accommodation of the mycobiont, and triggering of AM-specific plant gene expression in the symbiosome enabling nutrient exchange (21).

## MOLECULAR INTERACTIONS IN DEVELOPMENT OF SYMBIOSIS

The formation of the symbiosis requires the co-ordinate development of both the fungus and the plant and is assumed to require an interchange of signals. AM fungal spores are capable of germination (only) and limited hyphal growth in the absence of the plant, the plant signals are essential for the critical stages of the symbiosis. The release of soluble signals in the rhizosphere is thus an easy solution to allow both partners to be timely informed of their presence, even before physical contact (19).

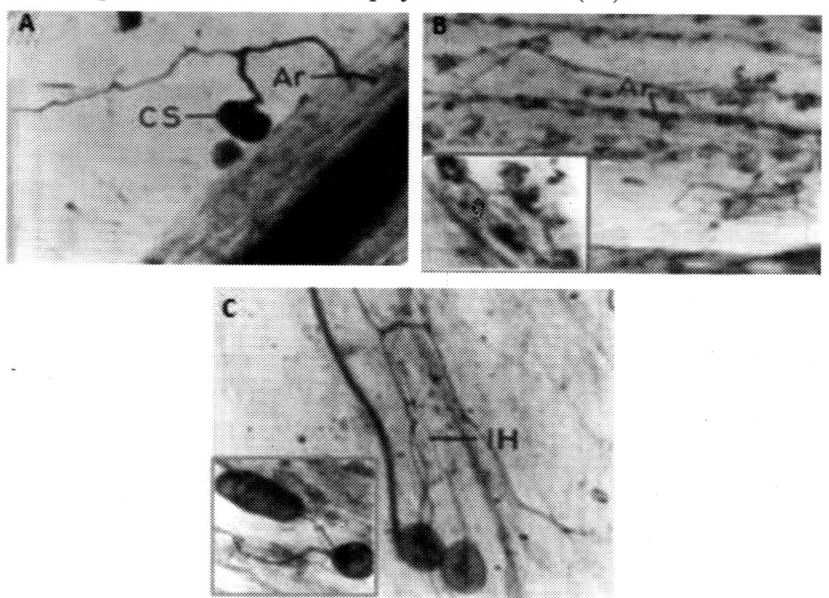


**Fig. 1.** Various stages of AM colonization in plant roots
**A.** Germinating chlamydospore (CS) forming single entry pointy; **B.** Heavily colonized root segment with internal hyphae (IH) and arbuscules. Inset showing an arbuscule with dichotomus branching; **C.** Internal hyphae with 'H' and 'Y' connections. inset showing vesicles with oil globules

## Strigolactones

The stimulatory effect of plant root exudates on AM fungal hyphae has been recognized for a long time, but the molecular identity of the 'branching factors' has only recently been identified. In two landmark papers, strigolactones were found to be responsible for the induction of branching (1) and alterations in fungal physiology and mitochondrial activity (10). Strigolactones can also stimulate spore germination in some AM fungi. Strigolactone - mediated signaling is necessary for a normal level of root colonization as demonstrated by using strigolactone – deficient mutant (41). These root exuded compounds also play an important role *in planta*, acting as hormone that contributes to the regulation of shoot branching (41). Strigolactones are short-lived in the rhizosphere owing to a labile ether bond that spontaneously hydrolyses in water. This ephemeral compound forms a steep concentration gradient, and therefore, its perception has been suggested to be a reliable indicator of the proximity of a host root (78). Strigolactone perception by the fungus induces the so-called presymbiotic stage, which is characterized by continued hyphal growth, increased physiological activity and profuse branching of hyphae. Phosphorous availability influence strigolactone release into the soil. Several studies have demonstrated a strong negative effect of high P supply on strigolactone production and exudation in various species (71).

## Myc factor

In analogy with rhizobial Nod factors, the AMF diffusible molecules that assist the plant in recognizing them are named as Myc factors that trigger the expression of symbiosis related genes such as *ENOD-11* (63). The Myc factor of *Glomus intraradices* has recently been shown to consist of lipochitoligosaccharides (LCOs) that are closely related to Nod factors. Considering the relatively recent evolution of rhizobium nodulation symbiosis, this suggests that (LCOs) evolved first as Myc factors in AM and became secondarily copied as Node factors in rhizobia. This also suggests similar perception mechanism in the plant host.

## Common symbiosis signaling pathway

As in root nodulation symbiosis, the common signaling pathway is required for successful intracellular colonization of epidermal cells by AM fungi (Fig.2). Mutations in common *SYM* genes generally leads to an early arrest of hyphal infection (79); however, AM fungi can in many cases overcome this barrier and later form normal arbuscules in the cortex (26, 43). These studies suggest that *SYM* gene signaling is crucial for epidermal infection.

***Calcium.*** Calcium is the most widespread intracellular messenger. It couples a wide array of extracellular stimuli to specific physiological responses (51). In analogy with nodules, $Ca^{2+}$ oscillations were hypothesized to be involved in AMF signal transduction. However, the first experimental evidence of $Ca^{2+}$ as a transducer of mycorrhizal signals came from measurements of $Ca^{2+}$ oscillations in *Medicago truncatula* root hairs in the presence of, but not in contact with AM hyphae (64). The use of a nuclear cameleon probe has provided two additional crucial contributions: firstly, the nucleus is the main target of the signaling pathway; secondly, the nuclear location of $Ca^{2+}$ oscillations is an indispensable step for the activation of the nuclear kinase DMI3 (22). AMF exudates, as well as hyphopodium contact, are able to induce $Ca^{2+}$ oscillations in the epidermal cells of legume and non-legume plants. Furthermore, the root regions of that responsive to AM signals were located between1and 2 cm from the tip of young lateral roots (22). Like the rhizobium – legume symbiosis, $Ca^{2+}$ oscillations were detected in *dmi3* but not in *dmi1* and *dmi2*. Medicago mutants in response to AMFs that support the hierarchial distribution of the SYM pathway with the nuclear kinase DMI3 located downstream as the decoder of the calcium spiking (22, 64). It seems evolutionary convenient that the same subset of genes, which are common in plant taxa (108) and have been shown to function in mycorrhizal signaling in non- legumes (44), could control two diverse biological systems in legumes.The legumes distinguish between AMFs and rhizobia and organize appropriate accommodation of the microbe inside the root tissues on the basis of different calcium signatures in response to different stimuli. AMF related nuclear calcium spikes are less frequent and more irregular than those induced by Nod factors (22, 45, 64)

***CASTOR and POLLUX.*** CASTOR and POLLUX are predicted to be membrane-integrated channel proteins (57). They are very similar to each other and show structural homology with Ca-gated potassium channels such as in the *Methanobacterium* MthK. Despite their high similarity, CASTOR and POLLUX do not complement each other. In addition, the temporal and spatial expression patterns of CASTOR and POLLUX are not identical during nodulation, suggesting that they possess functional differentiation and play distinct roles in symbiosis signaling.

***Nucleoporins.*** NUP133 (60) and NUP85 (93) are predicted as nucleoporins, which are the components of the nucleopore complex. Ca spiking is induced around the nucleus, and some of the proteins involved in the early symbiotic signaling pathway are localized on the nucleus. Hence, it is speculated that NUP85 and NUP133 are involved in trafficking and/or localization of factors essential for induction of Ca signals.

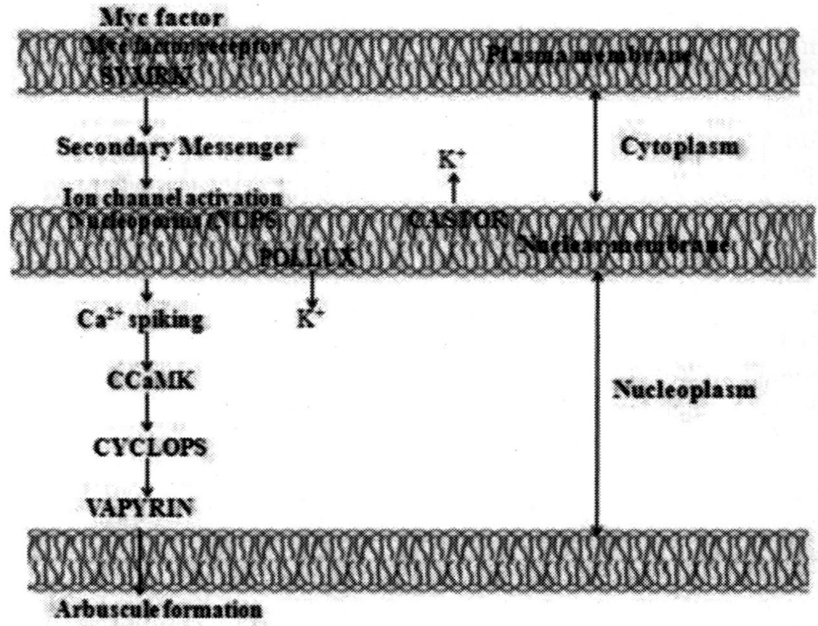


**Fig. 2.** The common symbiosis pathway in AM:
The Myc factor is perceived by the host plant by a still-to-be- identified Myc factor receptor (MFR). A second membrane protein, SYMRK codes for a receptor like kinase that has the potential to perceive AM fungal signals and transduces the signal to cytoplasm by phosphorylating a unknown substrate through its kinase domains. The signal is readily transduced to the nucleus resulting in calcium spiking in and around the nucleus. Perinuclear calcium spiking involves the release of calcium from a storage compartment. A role for the nuclear envelope is strongly suggested by involvement of two nucleoporins (NUPs). The potassium permeable channels CASTOR and POLLUX might compensate for the charge imbalance. The calcium-calmodulin-dependent protein kinase (CCaMK) forms a complex with CYCLOPS, a phosphorylation substrate, within the nucleus. Together with calmodulin, this complex decode the symbiotic calcium signatures

***CCaMK.*** It is generally assumed that a calcium-calmodulin-dependent protein kinase (CCaMK) is the essential decoder of the calcium signal (68, 75). The calmodulin - binding domain and calcium - binding EF hand motifs of CCaMK allow the protein to serve calcium, which make it a prime candidate for the response to calcium signatures that are induced by AM fungi (64).

***CYCLOPS.*** CYCLOPS is indispensable for epidermal infection as well as for cortical infection; and arbuscule formation (29). CYCLOPS encodes a protein with no overall sequence similarity to proteins with known function but contains a functional nuclear localization signal and a carboxy-terminal coiled-coil domain (79). CYCLOPs interacts with CCaMK in Yeast and in plant and can be phosphorylated by CCaMK *in vitro*.

***VAPYRIN.*** Recently, VAPYRIN has been discovered as a new component required for intracellular accommodation of AM fungi and rhizobia in epidermal, as well as cortical cells (32, 85, 111). VAPYRIN consists of a major sperm protein domain and an ankyrin domain, both known to mediate protein–protein interactions. The large ankyrin domain with 11 repeats is unique in plants (33), and closely resembles animal ankyrins which serve to connect integral membrane proteins to elements of the spectrin cytoskeleton (74). VAPYRIN and the identification of its potential interaction partners will provide an entry point to the elusive molecular mechanisms involved in intracellular accommodation of endosymbionts.

## PREPENETRATION APPARATUS

Plants have evolved a cellular infection structure, the prepenetration apparatus (PPA), which guides AM fungi through epidermal cells (38). The PPA is a transcellular tunnel of cytoplasm containing microtubules and microfilaments, which together with endoplasmic reticulum cisternae form a hollow tube across the vacuole of the host cell, that connects the leading nucleus with the site of appresorial contact. Arbuscule formation in cortical cells also involves PPA formation.

## ARBUSCULE FORMATION

As the fungus grows through the apoplast, hyphal branches are formed that penetrate inner cortical cells to initiate arbuscule formation. The fungal hyphae branch repeatedly to produce the tree-shaped arbuscule structure. The exact structure that is formed can vary depending on the fungal and host genotype (102). The branches of the fungi are excluded from the host cytoplasm by a plant-derived periarbuscular membrane (PAM). Nutrients, and perhaps signals, are exchanged across the symbiotic interface between the fungus and the plant which constitutes the PAM, the fungal plasma membrane and the periarbuscular space that exists between these two membranes. The PAM is continuous with the plant plasma membrane, but has a distinct protein composition, as revealed by the observation that the *Medicago truncatula* phosphate transporter PT4 is present in the PAM but absent from the plasma membrane (47). Ultrastructural analyses have detected molecules in the periarbuscular space that are typically found in the plant primary cell wall, including β-1,4-glucans, non-esterified homogalacturonans, xyloglucans, proteins that are rich in hydroxyproline and arabinogalactan proteins (17).

A special role in the development of arbuscules has been ascribed to lysophosphatidylcholine (LPc). LPc is a normal product of phospholipid metabolism and was recently described as a signalling molecule that activates the expression of phosphate-transporter genes, including the potato gene *PT3* (27). This type of phosphate transporter was found to be required for arbuscule maintenance (58). A phosphate-containing molecule, such as LPc, might be a cell autonomous molecular measure for how much phosphate is made locally available to the plant.

Arbuscule development is accompanied by plastid proliferation and the formation of a plastidial network in close physical contact with the arbuscule (103). The plastid is involved in numerous biosynthetic activities, including the production of apocarotenoids that specifically accumulate in AM roots (107). Given the involvement of hormones in almost all plant developmental processes, it is thought that hormones have key roles during the development of AM. This is still a developing area of research, but abscisic and jasmonic acids have emerged as potential regulators of AM (49, 52).

## NUTRIENT UPTAKE ACROSS THE INTERFACES

The transfer of nutrients across the interfaces in AM symbiosis is diagrammatically summarized in Fig. 3.

### Carbon uptake

The transfer of carbohydrates is thought to be the main benefit for the fungal symbiotic partner. One major factor explaining the obligate biotrophy of AM fungi was the finding that extraradical hyphae of these fungi are unable to take up carbohydrates. This has been shown by the application of various $^{13}C$-labelled compounds to split petri dishes harbouring Ri T-DNA transformed *Daucus carota* roots colonized by *G. intraradices* in one compartment and extraradical mycelium of the fungus alone in the other compartment (81). The plant can control the flux of sucrose directed to the root, including the fungus. Jasmonic acid has been proposed to be involved in the regulation of sink strength of AM roots (48). Sucrose that is delivered to the AM root is cleaved either by symbiosis-induced sucrose synthases or invertases (56). Fungal $H^+$-ATPases, possibly involved in the fungal uptake of carbohydrates from the apoplast, are found to be predominantly located on arbuscular trunks and intercellular hyphae, supporting an uptake of carbohydrate by these structures (40). After longer incubation periods, glucose is either used directly for lipid biosynthesis or enters the pentose phosphate pathway, thus providing the reduction equivalents necessary for lipid biosynthesis (81). Lipids and glycogen are then transferred to the extraradical mycelium (7), where the bidirectional movement of lipid bodies can be observed *in vivo* (6). A member of the novel clade of hexose transporters has been identified from the symbiotic organ of the glomeromycotan *Geosiphon pyriformis* may also be expressed in the fungal interface membrane of the arbuscule (96). In a typical AM, no carbon transport from the fungus to the plant has been detected (82).

## Phosphorous uptake

Phosphate is a mineral nutrient limiting plant growth at many natural stands due to its poor solubility. AM fungi transport phosphate from distant reservoirs to the plant, extending the reach of plant root systems. Using $^{33}P$ provided in soil compartments only admissible to fungal hyphae and not to plant roots, it could be shown that even under non-limiting phosphate supply and without apparent growth effects of mycorrhizal colonization, colonized plant roots reduce the activity of their own phosphate uptake system and rely mainly on their fungal symbionts for phosphate provision (101). The characterization in *Glomus versiforme* of a high affinity Pi transporter (PT), expressed in extraradical hyphae (46) and thus involved in the uptake of P from the soil, was a milestone in AM research. The gene showed a regulation pattern typical of fungal and plant high-affinity PT, with induction at the transcriptional level by low amounts of Pi, comparable to the micromolar concentration found in soil. Another PT homolog (GmosPT) has recently been described in *G. mosseae*. GmosPT showed an expression pattern similar to the *G. intraradices* PT gene.

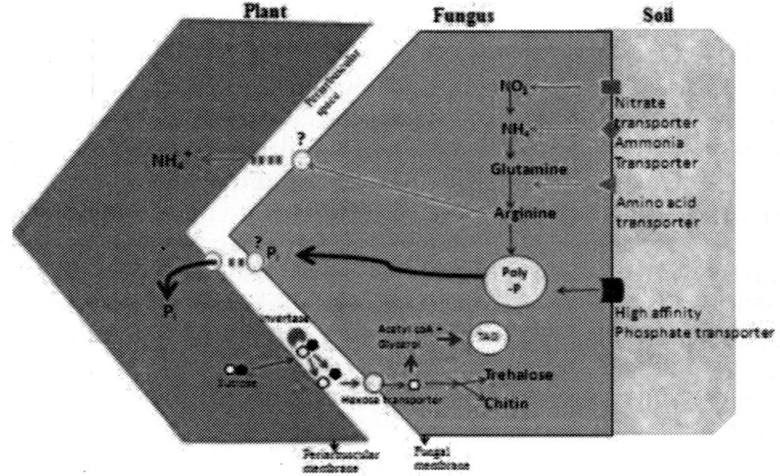


**Fig. 3.** Nutrient exchange in AM roots: **Carbohydrates:** Sucrose is cleaved by apoplastic invertases and taken up by fungal hexose transporters. The fungus transforms hexoses into trehalose. Hexoses are also converted into glycogen and triacylglycerol (TAG) lipid droplets which serve as suitable units for long-distance transport through the hyphal network. **Phosphorus**: Phosphate is taken by extra radical mycelium via high affinity phosphate transporters on the fungal membrane. Phosphate is transported towards the root and internal hyphae in the form of polyphosphate granules, which reside in membrane-enclosed vesicles. Polyphosphate carry arginine due to its negative charge. Phosphate is released from polyphosphate granules within internal hyphae and transferred to periarbuscular space by a mechanism yet unknown. Plants take phosphates by plant transporters on periarbuscular membrane. **Nitrogen**: Nitrogen is taken up by ammonium, nitrate or amino-acid transporters in extra-radical hyphae. In AM fungal hyphae, nitrogen is mainly transported as Arginine. Within the internal hyphae, nitrogen is released from arginine and transported to the plant directly or after cleavage to ammonium

Interestingly, a relatively high expression level, independent of external Pi concentrations, was observed in intraradical fungal structures (11). This presence of PT transcripts suggests that inside the root the fungus may exert control over the amount of phosphate delivered to the plant. On the plant side, phosphate transporters operating at the root–soil interface are downregulated, and the plant relies largely on phosphate delivered by the fungal symbionts (101). Investigations performed on host plants such as potato, rice, and *M. truncatula* highlighted the presence of Pi transporters exclusively expressed during symbiosis (47, 61, 80, 86, 91). In particular, MtPT4 from *M. truncatula* was shown to be located in the periarbuscular membrane, at the interface between arbuscules and the invaginated plant cell membrane, generally assumed as the site of nutrient exchanges (5, 47). Here, it could be involved in the uptake of phosphate released from the arbuscule. It is also worth mentioning that a gene (Mtha1) for a plasma membrane $H^+$ATPase is upregulated in *M. truncatula* during AM symbiosis (72), and Mtha1 transcripts accumulate in arbuscule-containing cells (65). In addition to phosphate transporters, genes encoding alkaline phosphatase in *G. intraradices* and *G. margarita* and a gene encoding a mycorrhiza-responsive acid phosphatase in *Tagetes patula* have been described (2, 31).

## Nitrogen uptake

Advances have also been made in understanding the movement of nitrogen in the AM symbiosis, another nutrient whose availability frequently limits plant growth. AM fungi directly take up and transfer N to their host plants (3, 50, 59) and increase the utilization of different forms of N by plants (55). A fungal amino-acid transporter (98) and an ammonium transporter that might be involved in nitrogen uptake by extraradical hyphae have been cloned (70). Long-distance transport to the plant probably proceeds mainly through arginine (24, 42). N is then transferred from the fungus to the plant as ammonium without any loss of C skeleton, thanks to the catabolic arm of the urea cycle that converts arginine into ammonium. In fact, quantitative real-time PCR assays have shown that a gene of primary nitrogen assimilation (glutamine synthase) was preferentially expressed in the extraradical hyphae, whereas genes involved in arginine breakdown (urease accessory protein and ornithine amino transferase) and $NH4^+$ transfer (ammonium transporter) were more highly expressed in the intraradical mycelium (42). It is worth mentioning that different N sources modulated the activity but not the expression of the fungal glutamine synthase (20). Another fungal gene (*Gi-1*) possibly involved in the synthesis of nitrogen compounds was shown to be expressed only during the symbiotic phase and upregulated under N fertilization (92).

## BACTERIA: THE THIRD COMPONENT IN MYCORRHIZAL ASSOCIATIONS

Mycorrhizal roots offer excellent ecological niches for other microbes, some rhizosphere bacteria adhere strongly to fungal hyphae; whereas others are directly associated with the root surfaces (14). In addition, mycorrhizal fungi may host bacteria that complete their life cycle within fungal cells. Bacteria seem to represent the third component of mycorrhizal associations, as they are closely associated with mycorrhizal fungi and most likely to play a role in mycorrhizal function (18). Duponnios and Garbaye (28) were first to observe a significant stimulation of ectomycorrhiza formation by *Pseudomonas flourescens*. This was experimental evidence for the "helper effect". Bacteria involved in mycorrhizal establishment and/or its functioning were therefore defined as mycorrhizal helper bacteria (MHB) by Garbaye (37) and have since then become the most investigated group among bacteria interacting with mycorrhizas (35). Mycorrhiza-associated bacterial communities have been investigated upon culturing and direct molecular approaches that avoid a cultivation step – PCR denaturing gradient gel electrophoresis analysis of the 16S rRNA gene. These investigations have revealed a repertoire of microbes, including several bacterial taxa with a predominance of species from the genera *Pseudomonas*, *Burkholderia* and *Bacillus* (25).

### Arbuscular mycorrhizal spore associated bacteria (AMB)

AM fungi are unique in hosting the bacteria in their cytoplasm (94). The composition of communities of AMB is more dependent on AM fungal identity (12, 90). One major factor controlling the abundance and occurrence of AMB could be traits of the spore themselves. The surface structures as well as attachment mechanisms of AMB to surface depending on surface molecules would affect the population sizes and composition of

AMB on spores. On the basis of the 16S rDNA sequences, the bacterial endosymbionts living in *Gigaspora margarita* (BEG 34) were first identified as belonging to the genus *Burkholderia* (13). As a further step, on the basis of the 16S rDNA amplified from *Scutellospora persica*, *S. castanea* and *G. margarita*, a strongly supported clade was obtained, which contained all endosymbiotic bacteria so far sequenced in Gigasporaceae. It was located close to the genus *Burkholderia* as well as to the genera *Ralstonia* and *Pandorea*. A new bacterial taxon was therefore proposed: Candidatus Glomeribacter gigasporarum (15). The results demonstrate that endobacteria are widespread, and represent a stable cytoplasmic component in Gigasporaceae. The bacteria move along with the fungi from one generation to the next, following a vertical transmission mechanism (16).

## Nutritional interaction between AM fungi and their AMB

The nutritional interactions between mycorrhizal fungi and their associated bacteria have long been ignored and only recently it is becoming clear that their nutritional dynamics may be ecologically important (18). Two classes represent their interactions: bacteria feeding on fungi (25, 69); and fungi feeding on bacteria (53, 54).

Leveau and Preston (69) recently identified three types of bacterial mycophagy (consumption of fungus-derived substrates) namely (i) extracellular necrotrophy; (ii) extracellular biotrophy; and (iii) endocellular biotrophy. There is evidence for necrotrophic and extracellular biotrophic activities of bacteria towards mycorrhizal fungi. *Burkholderia* spp. penetrated the spores of AM fungi *Gigaspora decipiens* colonizing senescing spores and attaching to fungal hyphae – as detected in a spore lysis assay using GFP-tagged bacteria (67). Electron microscopy has revealed bacteria that probably feed on the outer hyaline spore layer of AM fungi – as they represent several taxonomic groups of biodegraders (90). The third type of mycophagy i.e. endocellular biotrophy, includes bacterial endosymbionts and free-living bacteria that penetrate and live inside mycorrhizal fungi.

## Role AMB in mycorrhizal function

Frey-Klett *et al.* (35) proposed some mechanisms that may explain AMB success. These mechanisms involve the production of growth factors that might stimulate fungal spore germination, mycelia growth, increased root branching and greater root colonization, and reduction of soil-mediated stress through detoxification of antagonistic substances and inhibition of competitors and antagonists. The AM-spore-associated bacteria (AMB) has been found to stimulate the germination of spores and the colonization of roots (110) and stimulate the growth of AM fungi up to the formation of fertile spores in the absence of a host (54). This influence has been suggested to be due to the production of stimulating volatile compounds such as $CO_2$, erosion of the spore walls and phosphorous acquisition by bacteria (12).

There are some reports on multifunctional characteristics of AMB indicating that some of these can be proteolytic, chitinolytic and/or cellulolytic (34, 90) and could therefore help in AM fungal colonization by degrading cell walls. The presence of cellulolytic bacteria on the spore surface indicates that microorganisms attached to the spores may also degrade plant material around them. The presence of active biopolymer – degrading bacterial populations on the spore surface could also support spore germination by releasing nutrients or degrading toxic compounds that inhibit germination (90).

Large proportion of the AMB isolates have the ability to inhibit the growth of pathogenic fungi including *Rhizoctonia solani*. The inhibitory effect may be due to the production of antibiotics or siderophores or both (12). Bacterial siderophores bind to ferric iron faster than siderophores of fungal pathogens.

The constant finding of bacteria and the new roles attributed to them represent a paradigm shift in our way of viewing mycorrhizal symbiosis. Although, the exact nature of the interactions remains unclear, much evidence supports the current trend to view mycorrhizas as tripartite associations.

## REFERENCES

1. Akiyama K, Matsuzaki K, and Hayashi H 2005 Plant sesquiterpenes induce hyphal branching in arbuscular mycorrhizal fungi, *Nature* **435** 824–827

2. Aono T, Maldonado-Mendoza IE, Dewbre GR, Harrison MJ, and Saito M 2004 Expression of alkaline phosphatase genes in arbuscular mycorrhizas, *New Phytologist* **162** 525–553

3. Bago B, Chamberland H, Goulet A, Vierheilig H, Lafontaine JG, and Piché Y 1996 Effect of nikkomycin Z, a chitin-synthase inhibitor, on hyphal growth and cell wall structure of two arbuscular–mycorrhizal fungi, *Protoplasma* **192** 80–92.
4. Bago B, Pfeffer PE, Douds DD, Brouillette J, Bécard G, and Shachar-Hill Y 1999 Carbon metabolism in spores of the arbuscular mycorrhizal fungus *Glomus intraradices* as revealed by nuclear magnetic resonance spectroscopy, *Plant Physiology* **121** 263–271.
5. Bago B, Shachar-Hill Y, and Pfeffer PE 2000 Dissecting carbon pathways in arbuscular mycorrhizas with NMR spectroscopy, in *Current advances in mycorrhizae research* (ed) GK Podila, DD Douds, APS Press, St. Paul, USA. Vol. 124 111–126 pages
6. Bago B, Warren Zipfe, Rebecca M. Williams, Jeongwon Jun, Raoul Arreola, Peter J. Lammers, Philip E. Pfeffer, and Yair Shachar-Hill 2002 Translocation and utilization of fungal storage lipid in the arbuscular mycorrhizal symbiosis, *Plant Physiology* **128** 108–124
7. Bago B, Philip E. Pfeffer, Jehad Abubaker, Jeongwon Jun, James W. Allen, Janine Brouillette, David D Douds, Peter J. Lammers, and Yair Shachar-Hill 2003 Carbon export from arbuscular mycorrhizal roots involves the translocation of carbohydrate as well as lipid. *Plant Physiology* **1311** 496–1507
8. Becard G and Pfeffer PE 1993 Status of nuclear division in arbuscular mycorrhizal fungi during in vitro development, *Protoplasma* **174** 62–68
9. Benjamin RK 1979 Zygomycetes and their spores, in *The Whole Fungus: the sexual-asexual synthesis,* (ed) B Kendrick, National Museums of Canada: Ottawa, Canada. 2. 573–622 pages
10. Besserer A, Besserer A, Puech-Pagès V, Kiefer P, Gomez-Roldan V, Jauneau A, and Roy S 2006 Strigolactones stimulate arbuscular mycorrhizal fungi by activating mitochondria, *Plosbiology* **4** 1239–1247
11. Benedetto A, Magurno F, Bonfante P, and Lanfranco L 2005 Expression profiles of a phosphate transporter gene (GmosPT) from the endomycorrhizal fungus *Glomus mosseae*, *Mycorrhiza* **15** 620–627
12. Bharadwaj DP, Lundquist P, Person P, and Alstrom S 2008. Evidence of specificity of cultivable bacteria associated with arbuscular mycorrhizal fungal spores,

*FEMS Microbial Ecology* **65** 310–322

13. Bianciotto V, Minerdi D, Perotto S, and Bonfante P 1996. Cellular interactions between arbuscular mycorrhizal fungi and rhizosphere bacteria, *Protoplasma* **193** 123–131
14. Bianciotto V and Bonfante P 2002 Arbuscular mycorrhizal fungi: a specialized niche for rhizospheric and endocellular bacteria, *Antonie Leeuwenhoek* **81** 365–375
15. Bianciotto V, Lumini E, Bonfante P, and Vandamme P 2003. 'Candidatus Glomeribacter gigasporarum' an endosymbiont of arbuscular mycorrhizal fungi, *International Journal of Systematic and Evolutionary Microbiology* **53** 121–124
16. Bianciotto V, Genre A, Jargeat P, Lumini E, Becard G, and Bonfante P 2004. Vertical transmission of endobacteria in the arbuscular mycorrhizal fungus *Gigaspora margarita* through generation of vegetative Spores, *Applied Environmental Microbiology* **70** 3600–3608
17. Bonfante P 2001. At the interface between mycorrhizal fungi and plants: the structural organization of cell wall, plasma membrane and cytoskeleton, in *The Mycota, IX Fungal Associations*. (ed. Esser K, B Hock) Springer Verlag, Berlin, 45–91 pages
18. Bonfante P and Anca IA 2009. Plants, mycorrhizal fungi, and bacteria: A network of interactions, *Annual Review of Microbiology* **63** 363–383
19. Bonfante P and Genre A 2010 Mechanisms underlying beneficial plant–fungus interactions in mycorrhizal symbiosis, *Nature Communication* 1 doi: 10.1038/ncomms1046
20. Breuninger M, Trujillo CG, Serrano E, Fischer R, and Requena N 2004 Different nitrogen sources modulate activity but not expression of glutamine synthetase in arbuscular mycorrhizal fungi, *Fungal Genetics and Biology* **41** 542–552
21. Bucher M 2007 Functional biology of plant phosphate uptake at root and mycorrhiza interfaces, *New Phytologist* **173** 11–26
22. Chabaud M, Genre A, Sieberer BJ, Faccio A, Fournier J, Novero M, Barker DG, and Bonfante P 2011 Arbuscular mycorrhizal hyphopodia and germinated spore exudates trigger $Ca^{2+}$ spiking in the legume and non legume root epidermis, *New Phytologist* **189** 347–355
23. Croll D 2009 Nonself vegetative fusion and genetic exchange in the arbuscular

mycorrhizal fungus *Glomus intraradices, New Phytologist* **181** 924 – 937

24. Cruz C, Egsgaard H, Trujillo C, Ambus P, Requena N, Martins-Loucao MA, and Jakobsen 2007 Enzymatic evidence for the key role of arginine in nitrogen translocation by arbuscular mycorrhizal fungi, *Plant Physiology* **144** 782–792
25. De Boer W, Folman LB, Summerbell RC, and Boddy L 2005 Living in a fungal world: impact of fungi on soil bacterial niche development, *FEMS Microbiology Reviews* **29** 795–811
26. Demchenko K, Winzer T, Stougaard J, Parniske M, and Pawlowski K 2004 Distinct roles of *Lotus japonicus* SYMRK and SYM15 in root colonization and arbuscule formation. *New Phytologist* **163** 381–392
27. Drissner D, Kunze G, Callewaert N, Gehrig P, Tamasloukht M, Boller T, Felix G, Amrhein N, and Bucher M 2007 Lyso-phosphatidylcholine is a signal in the arbuscular mycorrhizal symbiosis, *Science* **318** 265–268
28. Duponnois R and Garbaye J 1991 Mycorrhization helper bacteria associated with the Douglas fir–*Laccaria laccata* symbiosis: effects in aseptic and in glasshouse conditions, *Annals of Forest Science* **48** 239–251
29. Ercolin F and Reinhardt D 2011 Successful joint ventures of plants: arbuscular mycorrhiza and beyond, *Trends in Plant Science* doi: 10.1016/ j.tplants.
30. Evelin H, Kapoor R, and Bhoopander G 2009 Arbuscular mycorrhizal fungi in alleviation of salt stress: a review, *Annals of Botany* **104** 1263–1280
31. Ezawa T, Hayatsu M, and Saito M 2005 A new hypothesis on the strategy for acquisition of phosphorus in arbuscular mycorrhiza: up-regulation of secreted acid phosphatase gene in the host plant, *Molecular Plant-Microbe Interactions* **18** 1046–1053
32. Feddermann N, Reddy R, Muni D, Zeier T, Stuurman J, Ercolin F, S Martine, and Reinhardt D 2010 The PAM1 gene of petunia, required for intracellular accommodation and morphogenesis of arbuscular mycorrhizal fungi, encodes a homologue of VAPYRIN, *Plant Journal* **64** 470–481
33. Feddermann N and Reinhardt D 2011 Conserved residues in the ankyrin domain of VAPYRIN indicate potential protein-protein interaction surfaces, *Plant*

*Signaling and Behavior* **6(5)** 680–684

34. Filippi C, Bagnoli G, Citernesi AS, and Giovannetti M 1998 Ultrastructural spatial distribution of bacteria associated with sporocarps of *Glomus mosseae*, *Symbiosis* **24** 1–12
35. Frey-Klett P, Garbaye J, and Tarkka M 2007 The mycorrhiza helper bacteria revisited, *New Phytologist* **176** 22–36
36. Gehrig H, Schüßler A, and Kluge M 1996 Geosiphon pyriforme fungus forming endocytobiosis with Nostoc (Cyanobacteria) is an ancestral member of the glomales: evidence by SSU rRNA analysis, *Journal of Molecular Evolution* **43** 71–81
37. Garbaye J 1994 Helper bacteria: a new dimension to the mycorrhizal symbiosis, *New Phytologist* **128** 197–210
38. Genre A, Chabaud M, Timmers T, Bonfante P, and Barker DG 2005 Arbuscular mycorrhizal fungi elicit a novel intracellular apparatus in *Medicago truncatula* root epidermal cells before infection, *Plant Cell* **17** 3489–3499
39. Gerdemann JW and Trappe JM 1974 Endogonaceae in the Pacific North West, *Mycol Mem* **5** 1–76.
40. Gianinazzi-Pearson V, Smith SE, Gianinazzi S, and Smith FA 1991 Enzymatic studies on the metabolism of vesicular-arbuscular mycorrhizas V. Is $H^+$-ATPase a component of ATP-hydrolysing enzyme activities in plant–fungus interfaces? *New Phytologist* **117** 61–76
41. Gomez-Roldan V, Fermas S, Brewer P B, Puech-Pagès V, Dun E A, Pillot JP, Letisse F, Matusova R, Danoun S, Portais JC, Bouwmeester H, Bécard G, Beveridge CA, Rameau C, and Rochange SF 2008 Strigolactone inhibition of shoot branching, *Nature* **455**189–194
42. Govindarajulu M, Pfeffer PE, Jin H, Abubaker J, Douds DD, Allen JW, Bücking H, Lammers PJ, and Shachar-Hill Y 2005 Nitrogen transfer in the arbuscular mycorrhizal symbiosis, *Nature* **435** 819–823
43. Groth M, Takeda N, Perry J, Uchida H, Dräxl S, Sato S, Tabata S, Kawaguchi M, Wang TL, and Parniske M 2010 NENA, a *Lotus japonicus* homolog of Sec13, is required for rhizodermal infection by arbuscular mycorrhiza fungi and rhizobia but dispensable for cortical endosymbiotic development, *Plant Cell* **22** 2509–2526
44. Gutjahr C, Banba M, Croset V, An K, Miyao A, An G,

Hirochika H, Imaizumi-Anraku H, and Paszkowski U 2008 Arbuscular mycorrhiza-specific signaling in rice transcends the common symbiosis signaling pathway, *Plant Cell* **20** 2989-3005

45. Harper JF and Harmon A 2005 Plants, symbiosis and parasites: a calcium signaling connection, *Nature Reviews Molecular Cell Biology* **6** 555–566

46. Harrison MJ and van Buuren ML 1995 A phosphate transporter from the mycorrhizal fungus *Glomus versiforme*, *Nature* **378** 626–629

47. Harrison MJ, Dewbre GR, and Liu J 2002 A phosphate transporter from *Medicago truncatula* involved in the acquisition of phosphate released by arbuscular mycorrhizal fungi, *Plant Cell* **14** 2413–2429

48. Hause B, Walter M, Otto M, Robert K, and Dieter S 2002 Induction of jasmonate biosynthesis in arbuscular mycorrhizal barley roots, *Plant Physiology* **130** 1213–1220

49. Hause B, Mrosk C, Isayenkov S, and Strack D 2007 Jasmonates in arbuscular mycorrhizal interactions, *Phytochemistry* **68** 101–110

50. He XL, Mouratov S, and Steinberger Y 2002 Temporal and spatial dynamics of vesicular–arbuscular mycorrhizal fungi under the canopy of *Zygophyllum dumosum Boiss.* in the Negev Desert, *Journal of Arid Environments* **52** 379–387

51. Hetherington AM and Brownlee C 2004 The generation of $Ca^{2+}$ signals in plants, *Annual Review of Plant Biology* **55** 401–427

52. Herrera-Medina MJ, María José H, Siegrid S, Horst V, Juan A, Ocampo B, José M, and García G 2007 Abscisic acid determines arbuscule development and functionality in the tomato arbuscular mycorrhiza, *New Phytologist* **175** 554–564

53. Hildebrandt U, Janetta K, and Bothe H 2002 Towards growth of arbuscular mycorrhizal fungi independent of a plant host, *Applied Environmental Microbiology* **68** 1919–1924

54. Hildebrandt U, Ouziad F, Marner FJ, and Bothe H 2006 The bacterium *Paenibacillus validus* stimulates growth of the arbuscular mycorrhizal fungus *Glomus intraradices* up to the formation of fertile spores, *FEMS Microbiology Letters* **254** 258–267

55. Hodge A, Campbell CD, and Fitter AH 2001 An arbuscular mycorrhizal fungus accelerates decomposition and acquires nitrogen directly

from organic material, *Nature* **413** 297–299

56. Hohnjec N, Perlick AM, Puhler A, and Kuster H 2003 The *Medicago truncatula* sucrose synthase gene *MtSucS1* is activated both in the infected region of root nodules and in the cortex of roots colonized by arbuscular mycorrhizal fungi, *Molecular Plant Microbe Interaction* **16** 903–915
57. Imaizumi-Anraku H, Takeda N, and Cherpentier M 2005 Plastid proteins crucial for symbiotic fungal and bacterial entry into plant roots, *Nature* **433** 527–531
58. Javot H, Varma Penmetsa R, Terzaghi N, Cook DR, and Harrison MJ 2007 *Medicago truncatula* phosphate transporter indispensable for the arbuscular mycorrhizal symbiosis, *Proceedings of National Academy of Science* USA, **104** 1720–1725
59. Johansen A, Jakobsen I, and Jensen ES 1993 Hyphal transport by a vesicular–arbuscular mycorrhizal fungus of N applied to the soil as ammonium or nitrate, *Biology & Fertility of Soils,* **16** 66–70
60. Kanamori, N, Madsen L H, and Stougaard J 2006 A nucleoporin is required for induction of $Ca^{2+}$ spiking in legume nodule development and essential for rhizobial and fungal symbiosis, *Proceedings of National Academy of Science* USA, **103** 359–364
61. Karandashov V and Bucher M 2005 Symbiotic phosphate transport in arbuscular mycorrhizas, *Trends in Plant Science* **10** 22–29
62. Kistner C, Winzer T, Pitzschke A, Mulder L, Sato S, Kaneko T, Tabata, S Sandal N, Stougaard J, Webb KJ, Szczyglowski K, and Parniske M 2005 Seven *Lotus japonicus* genes required for transcriptional reprogramming of the root during fungal and bacterial symbiosis, *Plant Cell* **17** 2217–2229
63. Kosuta S, Chabaud M, and Bécard G 2003 A diffusible factor from arbuscular mycorrhizal fungi induces symbiosis-specific MtENOD11 expression in roots of *Medicago truncatula, Plant Physiology* **131** 952–962
64. Kosuta S, Hazledine S, Sun J, Miwa H, Morris RJ, Downie JA, and Oldroyd GE 2008 Differential and chaotic calcium signatures in the symbiosis signaling pathway of legumes, *Proceedings of National Academy of Science* USA, **105** 9823–9828
65. Krajinski F, Hause B, Gianinazzi-Pearson V, and Franken P 2002 Mtha1, a plasma membrane $H^+$ ATPase gene from *Medicago truncatula*, shows arbuscule specific

induced expression in mycorrhizal tissue, *Plant Biology* **4** 754–761

66. Kuhn G, Hijri M, and Sanders IR 2001 Evidence for the evolution of multiple genomes in arbuscular mycorrhizal fungi, *Nature* **414** 745–748

67. Levy A, Chang BJ, Abbott LK, Kuo J, Harnett G, and Inglis T JJ 2003 Invasion of spores of the arbuscular mycorrhizal fungus *Gigaspora decipiens* by *Burkholderia* spp, *Applied Environmental Microbiology* **69** 6250–6256

68. Levy J, Bres C, Geurts R, Chalhoub B, Kulikova O, Duc G, Journet EP, Ane JM, Lauber E, and Bisseling T 2004 A putative $Ca^{2+}$ and calmodulin-dependent protein kinase required for bacterial and fungal symbioses, *Science* **303** 1361–1364

69. Leveau JH and Preston GM 2008 Bacterial mycophagy: definition and diagnosis of a unique bacterial-fungal interaction, *New Phytologist* **177** 859–876

70. Lopez-Pedrosa A, Gonzalez-Guerrero M, Valderas A, Azcon-Aguilar C, and Ferrol N 2006 *GintAMT1* encodes a functional high-affinity ammonium transporter that is expressed in the extraradical mycelium of *Glomus intraradices*, *Fungal Genetics and Biology* **43** 102–110

71. Lopez-Raez Gonzalez-Guerrero M, Valderas A, Azcon-Aguilar C, and Ferrol N 2010 Does abscisic acid affect strigolactone biosynthesis? *New Phytologist* **187** 343–354

72. Manthey K, Manthey K, Krajinski F, Hohnjec N, Firnhaber C, Pühler A, Perlick AM, and Küster H 2004 Transcriptome profiling in root nodules and arbuscular mycorrhiza identifies a collection of novel genes induced during *Medicago truncatula* root endosymbiosis, *Molecular Plant-Microbe Interaction* **17** 1063–1077

73. Martin F, Aerts A, Ahr D, Brun A, Danchin E G J, Duchaussoy F, Gibon J, Kohler A, Lindquist E, Pereda V, Salamov A, Shapiro H J, Wuyts J, Blaudez D, Bu M, Brokstein P, Canb B, Cohen D, Courty PE, Coutinho P M, Delaruelle C, Detter J C, Deveau A, DiFazio S, Duplessis S, Fraissinet-Tachet L, Lucic E, Frey-Klett P, Fourrey C, Feussner I, Gay G, Grimwood J, Hoegger P J, Jain P, Kilaru S, Labb J, Lin Y C, Legu V, Tacon F Le, Marmeisse R, Melayah D, Montanini B, Muratet M, Nehls U, Niculita-Hirzel H, Oudot-Le Secq M P, Peter M, Quesneville H, Rajashekar B, Reich M, Rouhier N, Schmutz

J, Yin T, Chalot M, Henrissat B, Kes U, Lucas S, Van de Peer Y, Podila G K, Polle A, Pukkila P J, Richardson PM, Rouz P, Sanders IR, Stajich JE, Tunlid A, Tuskan G, and Grigoriev IV 2008 The genome of *Laccaria bicolor* provides insights into mycorrhizal symbiosis, *Nature* **452** 88–92

74. Michaely P, Tomchick DR, and Anderson RGW 2002 Crystal structure of a 12 ANK repeat stack from human ankyrinR, *EMBO Journal* **21** 6387–6396

75. Mitra RM, Gleason CA, Edwards A, Hadfield J, Downie JA, Oldroyd GED, and Long SR 2004 A $Ca^{2+}$ calmodulin-dependent protein kinase required for symbiotic nodule development: Gene identification by transcript-based cloning, *Proceedings of National Academy of Science* USA, **101** 4701–4705

76. Morton JB and Benny GL 1990 Revised classification of arbuscular mycorrhizal fungi (Zygomycetes): a new order, glomales, two new suborders Glomineae and Gigasporineae and two new families Acaulosporaceae and Gigasporaceae with an emendation of Glomaceae, *Mycotaxon* **37** 471-491

77. Parniske M 2004 Molecular genetics of the arbuscular mycorrhizal symbiosis, *Current Opinion in Plant Biology* **7** 414–421

78. Parniske M 2005 Cue for the branching connection, *Nature* **435** 750–751

79. Parniske M 2008 Arbuscular mycorrhiza: the mother of plant root endosymbiosis, *Nature Reviews Microbiology* **6** 763–775

80. Paszkowski U, Kroken S, Roux C, and Briggs SP 2002 Rice phosphate transporters include an evolutionarily divergent gene specifically activated in arbuscular mycorrhizal symbiosis, *Proc-eedings of National Academy of Science* USA, **99** 13324–13329

81. Pfeffer PE, Douds DD, Becard G, and Shachar-Hill Y 1999 Carbon uptake and the metabolism and transport of lipids in an arbuscular mycorrhiza, *Plant Physiology* **120** 587–598

82. Pfeffer PE, Douds DD, Bucking H, Schwartz DP, and Shachar-Hill Y 2004 The fungus does not transfer carbon to or between roots in an arbuscular mycorrhizal symbiosis, *New Phytologist* **163** 617–627

83. Pirozynski KA and Dalpe Y 1989 Geological history of the Glomaceae with particular reference to mycorrhizal symbiosis, *Symbiosis* **7** 1–36

84. Powell JR, Parrent JL, Hart MM, Klironomos JN, Rillig

MC, and Maherali H 2009 Phylogenetic trait conservatism and the evolution of functional tradeoffs in arbuscular mycorrhizal fungi, *Proceedings of the Royal Society* **276** 4237–4245

85. Pumplin N, Mondo SJ, Topp , Starker CG, Gantt JS, and Harrison MJ 2010 *Medicago truncatula* Vapyrin is a novel protein required for arbuscular mycorrhizal symbiosis, *Plant Journal* **61** 482–494

86. Rausch C, Daram P, Brunner S, Jansa J, Laloi M, Leggewie G, Amrhein N, and Bucher M 2001 A phosphate transporter expressed in arbuscule-containing cells in potato, *Nature* **414** 462–466

87. Redecker D, Morton JB, and Bruns TD 2000 Ancestral lineages of arbuscular mycorrhizal fungi (Glomales), *Molecular Phylogenetics and Evolution* **14** 276–284

88. Redecker D 2002 Molecular identification and phylogeny of arbuscular mycorrhizal fungi, *Plant Soil* **244** 67–73

89. Redecker D and Raab P 2006 Phylogeny of the Glomeromycota (arbuscular mycorrhizal fungi): recent developments and new gene markers, *Mycologia* **98** 885–895

90. Roesti D, Ineichen K, Braissant O, Redecker D, Wiemken A, and Aragno M 2005 Bacteria associated with spores of the arbuscular mycorrhizal fungi *Glomus geosporum* and *Glomus constrictum*, *Applied Environmental Microbiology* **71** 6673–6679

91. Rosewarne G, Barker SJ, Smith SE, Smith FA, and Schachtman D P 1999 A *Lycopersicon esculetum* phosphate transporter (LePT1) involved in phosphorus uptake from a vesicular–arbuscular mycorrhizal fungus, *New Phytologist* **144** 507–516

92. Ruiz-Lozano JM, Collados C, Porcel R, and Azcón Barea JM 2002 Identification of a cDNA from the arbuscular mycorrhizal fungus *Glomus intraradices* that is expressed during mycorrhizal symbiosis and up-regulated by N fertilization, *Molecular Plant-Microbe Interaction* **15** 360–367

93. Saito K, Yoshikawa M, and Kawaguchi M 2007 NUCLEOPORIN 85 is required for calcium spiking, fungal and bacterial symbiosis and seed production in *Lotus japonicas*, *Plant Cell* **19** 610–624

94. Scannerini S and Bonfante P 1991 Bacteria and bacteria like objects in endomycorrhizal fungi (Glomaceae), in: *Symbiosis as Source of Evolutionary Innovation: Speciation and Morphogenesis.* (ed. Margulis L, Fester R) MA: MIT Press, Cambridge, 273–87 pages

95. Schüßler A, Mollenhauer D, Schnepf E, and Kluge M 2001 A new fungal phylum, the Glomeromycota: phylogeny and evolution, *Mycological Research* **105** 1413–1421

96. Schüßler A, Martin H, Cohen D, Fitz M, and Wipf D 2006 Characterization of a carbohydrate transporter from symbiotic glomeromycotan fungi, *Nature* **444** 933–936

97. Schwarzott D, Walker C, and Schüßler A 2001 *Glomus*, the largest genus of the arbuscular mycorrhizal fungi (Glomales), is non monophyletic, *Molecular Phylogenetics and Evolution* **21** 190–197

98. Siciliano V, Genre A, and Bonfante P 2007 Transcriptome analysis of arbuscular mycorrhizal roots during development of the prepenetration apparatus, *Plant Physiology* **144** 1455–1466

99. Simon L, Bousquet J, Levesque RC, and Lalonde M 1993 Origin and diversification of endomycorrhizal fungi and coincidence with vascular land plants, *Nature* **363** 67–69

100. Smith SE and Read DJ 1997 *Mycorrhizal Symbiosis,* ed 2. Academic Press, London.

101. Smith SE, Smith AF, and Jakobsen I 2003 Mycorrhizal fungi can dominate phosphate supply to plants irrespective of growth responses, *Plant Physiology* **133** 16–20

102. Smith SE and Read DJ 2008 *Mycorrhizal Symbiosis,* ed 3. Academic Press, London.

103. Strack D and Fester T 2006 Isoprenoid metabolism and plastid reorganization in arbuscular mycorrhizal roots, *New Phytologist* **172** 22–34

104. Trappe JM 1987 Phylogenetic and ecological aspects of mycotrophy in the angiosperms from an evolutionary standpoint, in: *Ecophysiology of VA Mycorrhizal Plants.* (ed. G. Safir) CRC Press, Boca Raton, Florida, 5–25 pages

105. Walker C 1992 Systematics and taxonomy of the arbuscular endomycorrhizal fungi (Glomales) - a possible way forward, *Agronomie* **12** 887–897

106. Walker C, Blaszkkowski JDS, and Schüßler A 2004 Gerdemannia gen. nov., a genus separated from *Glomus* and Gerdemanniaceae fam. nov., a new family in the Glomeromycota, *Mycological Research* **108** 707–718

107. Walter MH, Floß DS, Hans J, Fester T, and Strack D 2007 Apocarotenoid biosynthesis in arbuscular mycorrhizal roots: contributions from methylerythritol phosphate pathway isogenes and tools for its manipulation, *Phytochemistry* **68** 130–138

108. Wang B, Yeun LH, Xue JY, Liu Y, Ané JM, and Qiu YL 2010 Presence of three mycorrhizal genes in the common ancestor of land plants suggests a key role of mycorrhizas in the colonization of land by plants, *New Phytologist* **186** 514–525

109. Warcup JH 1990 Taxonomy, culture and mycorrhizal associations of some zygosporic Endogonaceae, *Mycological Research* **94** 173–178

110. Xavier LJC and Germida JJ 2003 Bacteria associated with *Glomus clarum* spores influence mycorrhizal activity, *Soil Biology and Biochemistry* **35** 471–478

111. Zhang Q, Blaylock L, and Harrison M 2010 Two *Medicago truncatula* half-ABC transporters are essential for arbuscule development in arbuscular mycorrhizal symbiosis, *Plant Cell* **22** 1483–1497

*Rev. Plant Pathol. Vol. 6, 2014*
*Indian Society of Mycology and Plant Pathology*
*Scientific Publishers (India), Jodhpur pp. 367-386*

# *TRICHODERMA VIRENS*- AN OLD WARHORSE IN THE BATTLE AGAINST PLANT PATHOGENS

***Mala Mukherjee*[1]*, Amar N. Mukhopadhyay*[2*] *and Prasun K. Mukherjee*[1]**

[1]Central Institute for Cotton Research, Shankar Nagar, Nagpur 440010
[2]Sangini, 151 Akansha, Udhay-II, Raibareilly Road, Lucknow, 226025
*amar.mukhopadhyay@gmail.com

**KEYWORDS**: *Trichoderma virens*, biological control, mycoparasitism, antibiosis, induced resistance, genomics

## ABSTRACT

*Trichoderma* spp. are the most widely used biocontrol fungi the world over. *Trichoderma virens* (= *Hypocrea virens*) is one of the most extensively researched biocontrol species. Back in 1930s, it was demonstrated to control plant pathogens in pot experiments, and, apart from biocontrol, it was during this period that the mechanisms of biocontrol *viz.* mycoparasitism and antibiosis were described rather meticulously. Since then, a wealth of literature has accumulated on the biological control potential of *Trichoderma virens* resulting in some commercial formulations that are still in use. More interestingly, this species enjoyed great attention of researchers for the basic studies on the mechanisms of biocontrol. This species has another distinction that the geographical population has been classified into two distinct groups ("P" and "Q") based on the secondary metabolites profile. *T. virens* is a mycoparasite on many plant pathogenic fungi, both on actively growing hyphae and the resting structures, which are otherwise resistant to microbial degradation. Role of signal transduction in mycoparasitic interactions has been established using molecular tools. Many secondary metabolism related genes have been identified and the functions of some of these gene clusters have been studied. *T. virens* also has the ability to induce resistance in plants, and two molecules, a small elicitor-like protein Sm1 and a peptaibol have been identified to be involved in induced resistance. Two genes from *T. virens* i.e., an endochitinase (*ech42*) and a glutathione transferase, when transferred to plants imparted tolerance to biotic and abiotic stresses, respectively.

## INTRODUCTION

"After 25 years of working with bacteria, actinomycetes, and fungi as biocontrol agents of cotton seedling and root diseases, I believe Trichoderma virens to be one of the most versatile and effective biocontrol agents we have studied"- CR Howell (2006).

The above quote by one of the most famous *Trichoderma* researchers of modern time underlines the importance of *Trichoderma virens* (earlier known as *Gliocladium virens)* as a plant disease biocontrol agent. As a matter of fact, *Trichoderma virens* is the first fungus demonstrated to have biocontrol potential in pot culture. Back in 1930s, R. Weindling and colleagues carried out a series of experiments to provide scientific evidence that *Trichoderma lignorum* (synonym *T. virens*) suppresses *Rhizoctonia solani* in citrus seedlings through mycoparasitism and the production of antimicrobial antibiotics (99-103). This species of *Trichoderma* is thus historically important (and hence we chose this title for this article). Ever since, a wealth of literature has been added covering all aspects of *T. virens*- as a commercial biofungicide, as a source of secondary metabolites, as source of transgenes, as a model for understanding the molecular mechanisms of biocontrol, and finally culminating into the whole genome sequencing. In this brief review we will outline the developments that took place between 1932 and 2012 with particular focus on the taxonomy, biocontrol and molecular genetics.

## TAXONOMY

*Trichoderma* taxonomy has been extremely confusing until the recent time. The taxonomy "stabilized" with the advent and deployment of molecular tools to study the phylogenetic status of these economically and ecologically important fungi. *Trichoderma virens* was initially described by Weindling as *Trichoderma lignorum* (99,100,103). Later, he realized that the species was wrongly identified as *T. lignorum* and described it as *Gliocladium fimbriatum* (101, 102). Bisby (5) subsequently merged *Hypocrea gelatinosa* (thought to be the perfect stage of *T. virens*) with *Hypocrea rufa* (perfect stage of *Trichoderma viride*). Miller et al. (54) described it as *Gliocladium virens* and Arx (3) placed this fungus back in *Trichoderma virens*, which was confirmed by Rehner and Samuels (77) based on rDNA sequences. Subsequently, Chaverri et al. (12) described the sexual stage as *Hypocrea virens*. A detailed description of this species can be found in Chaverri et al. (12) and will not be repeated here. *Trichoderma virens* is easily identifiable by the characteristic "spore balls", the production of plenty of chlamydospores and biosynthesis of viridin and viridiol (Fig. 1). Unlike most other *Trichoderma* spp, *T. virens* isolates can further be subdivided into two well-defined "strains"- the gliotoxin-producing "Q" strains and the gliovirin-producing "P" strains (36).

## BIOLOGICAL CONTROL

*Trichoderma virens* has been widely used as biological control agents of many plant pathogens (Table 1).

**Table 1.** Select examples of successful biological control by *Trichoderma virens*

| Crop(s) | Pathogen(s) | References |
|---|---|---|
| Cotton | *Pythium ultimum, Rhizoctonia solani* | 31 |
| Cabbage | *P. ultimum, R. solani* | 49 |
| Sesame | *Fusarium oxysporum* | 39 |
| Apple | *Phytophthora cactorum* | 87 |
| Cotton | *P. ultimum* | 32 |
| Cotton | *R. solani* | 45 |
| Tomato | *Sclerotium rolfsii* | 80 |
| Various pulses and oilseed crops | *S. rolfsii, R. solani, F. oxysporum* | 56 |

| | | |
|---|---|---|
| Field pea | *R. solani* | 38 |
| Tomato | *P. aphanidermatum* | 15 |
| Poinsettias | *Pythium sp., R. solani* | 68 |
| Tomato, Carrot | *S. rolfsii* | 79 |
| Lettuce | *Sclerotinia sclerotiorum* | 8 |
| Cotton | *F. oxysporum* | 107 |
| Eggplant | *R. solani* | 46 |
| Bell pepper | *S. rolfsii* | 81 |
| Sunflower | *Sclerotinia minor* | 9 |
| Cotton | *Verticillium dahliae* | 26 |
| Gladiolus | *Fusarium oxysporum* | 55 |
| Lentil | *S. rolfsii, R. solani, F. oxysporum* | 86 |
| Catharanthus | *P. ultimum* | 10 |
| Soybean | Seed and seedling rot pathogens | 69 |
| Peanut | *Macrophomina phaseolina* | 52 |
| Chickpea | *S. rolfsii, R. solani, F. oxysporum* | 88,89 |
| Sunnhemp, Mungbean | *Pythium aphanidermatum, S. rolfsii* | 74 |
| Soybean | *R. solani* | 25 |
| Rice | *R. solani* | 48 |
| Tomato | *F. oxysporum* | 13 |

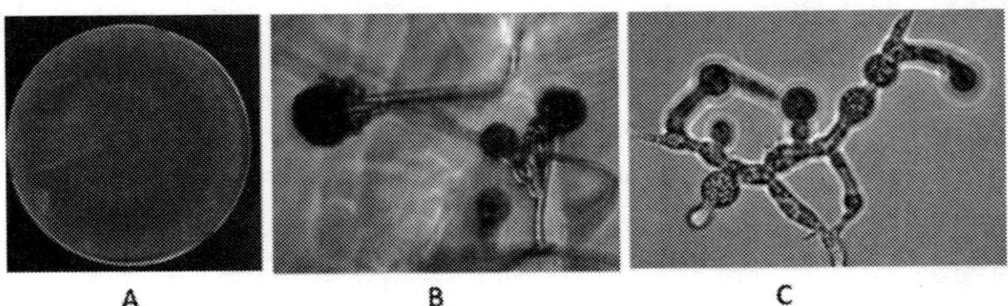


**Fig. 1**. Colony (A), conidiophores (B) and chlamydospores of *Trichoderma virens*

Even though *T. virens* have been used mostly against seed and seedling diseases, there are successful examples to establish the fact that this versatile fungus is effective against wood rot, foliar and post-harvest pathogens too. Highley (30) used *T. virens* for control of wood rot. Philip et al. (72) demonstrated the efficacy against *Colletotrichum gloeosporioides* and *Pestalotiopsis disseminate* in mulberry. Similarly, Anitha and Murugesan (2) showed that this antagonist is effective against *Alternaria helianthi* in sunflower. Recently, Etebarian and Mohammadifar (24) and Rahman et al. (76) demonstrated bioefficacy of *T. virens* against *Bipolaris spicifera* in wheat and *Ceratocystis paradoxa* in sugarcane, respectively. There is at least one example each for the biocontrol of a post-harvest pathogen *Penicillium expansum* in apple (6) and a nematode (*Meloidogyne incognita* (53). Control of pine needle blast with T. virens was reported by Tianmin (90).

One of the earliest *Trichoderma* spp. to be sold commercially is *T. virens* (GL-21) under the trade name GlioGard™ an alginate prill preparation, sold by WR Grace and Co., USA (50). This product is presently available under a new trade name SoilGard™, as a granular formulation for soil application (51). Recently, two strains from USA have been reported to be commercially produced ("AU Derma" based on *T. virens* strain Gv29-8 and "Sanjeevni" based on *T. virens* strain G-6) and used against a foliar pathogen *Botrytis cinerea* in India (40).

## MECHANISMS OF BIOCONTROL

Biological control is the culmination of a complex interactions between the antagonist, the pathogen, the plants and the biotic and abiotic environments, to the benefit of plants. It is difficult to pin-point the exact mechanisms by which the biological control works in vivo and hence, several theories have been proposed that include mycoparasitism, antibiosis, competition, induced resistance and metabolism of germination stimulants (33).

### Mycoparasitism

Mycoparasitism is the ancestral trait in *Trichoderma* (22, 41) and has been the major driving force behind its success as biofungicides. Mycoparasitism of *R. solani* by *T. virens* was first reported by Weindling et al. (99). Subsequently several studies have appeared in literature on the details of mycoparasitic interactions between *T. virens* and plant pathogens. In general, the mycoparasitism involves steps like chemotrophic attraction of the mycoparasite to the host fungus, attachment, coiling, and lysis brought about by hydrolytic enzymes (Fig. 2). The mycoparasitism could be on actively growing hyphae as well as resting structures. Aluko et al. (1) reported mycoparasitism on *T. virens* on *R. solani*, however, they concluded that antibiosis, rather than mycoparasitism is the principal mechanism of biocontrol. Tu (91) and Lee and Wu (44) studied the mycoparasitism of *T. virens* on the hyphae and sclerotia of *S. sclerotiorum*. Using light and scanning electron microscopy, Tu and Vaartaja (92) studied in detail the parasitism of *T. virens* on *R. solani*. The mycoparasitism of *T. virens* on the sclerotia of several plant pathogens (*S. sclerotiorum, S. minor, Botrytis cinerea, S. rolfsii* and *Macrophomina phaseolina*) was studied by Phillips (73). Papavizas and Collins (70) observed that *T. virens* reduced the germinability and decayed the sclerotia of *S. rolfsii*. Mukherjee et al. (60) noticed that *T. virens* is able to colonize *S. rolfsii* in natural soil. Heller and Theiler-Hedtrich (29) demonstrated that *T. virens* is capable of parasitizing four *Phytophthora* spp. Detailed studies on the parasitism of sclerotia of *S. rolfsii* and *R. solani* was done by Mukherjee et al. (61) and the mycoparasite was found to extensively grow and produce conidia inside the sclerotia. The antagonism of *T. virens* on *R. solani* and *P. ultimum* was observed in non-sterile potting medium (28). Using GFP, the detailed of colonization of sclerotia of phytopathogenic fungi was studied by Sarrocco et al. (82). Using gene deletion, the role of the 42-KDa endochitinase in biocontrol of *T. virens* was established by Baek et al. (4). Subsequently, the same group overexpressed a serine-protease encoding gene *tvsp1* in *T. virens* that resulted in superior biocontrol strains (75). Similarly, constitutive overexpression of two beta glucanases improved the biocontrol potential of *T. virens* against *P. ultimum*, *Rhizopus oryzae* and *Rhizoctonia solani* (19). The role of hydrolytic enzymes in parasitism of sclerotia has not been studied except of a laccase gene *lcc1* (11). Since *T. virens* is a mycoparasite on several fungi, it would be interesting to know if these are able to parasitize the mycorrhizal fungi too. Paulitz and Linderman (71) demonstrated the lack of antagonism between *T. virens* and *Glomus etunicatum*. Similarly, the interactions between an ectomycorrhiza (*Laccaria laccata*) with *Trichoderma* was studied and it was observed that *L. laccata* acts as a mycoparasite and destroys the protoplast of *T. harzianum* and *T. virens* (104, 108).

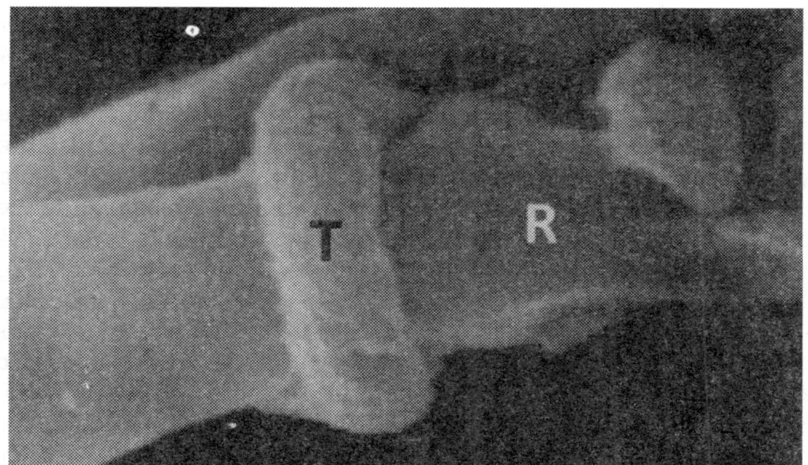


**Fig. 2.** Mycoparasitism of *Trichoderma virens* on *Rhizoctonia solani*

Like in many other host-parasite interactions, the role of signal transduction in mycoparasitism has been studied in *T. virens*- pathogen interactions. In one of the early studies, Mukherjee et al. (62) established the role of a MAP kinase in parasitism of sclerotia of *S. rolfsii* and *R. solani*. Interestingly, the MAPK mutants retained its ability to parasitize hyphae of *R. solani*. Subsequently, a small cysteine rich secreted protein Mrsp1 was identified to be transcriptionally regulated by this MAP kinase TmkA (64). This MAPK mutant also had reduced ability to induce systemic resistance in cucumber (97). The deletion of two G-proteins TgaA and TgaB did not affect the mycoparasitism on *R. solani*, but TgaA deletion reduced mycoparasitism on sclerotia of *S. rolfsii*. The deletion of *T. virens* adenylate cyclase gene *tac1* drastically reduced the growth and mycoparasitic ability (67). The deletion of another MAP kinase TmkB had similar effects as TmkA deletion, and the mutants had impaired mycoparasitism (42).

## Antibiosis

*Trichoderma* spp. are rich sources of secondary metabolites (78). Weindling (100) was the first to demonstrate that biocontrol efficacy of *T. virens* could be due to the production of antibiotic substances (designated as a "lethal principle"). Subsequently, he purified and named the antibiotic as gliotoxin, one of the most extensively studied secondary metabolites from *Trichoderma* spp. (102). Other known secondary metabolites from *T. virens* with antimicrobial properties are gliovirin, viridin and the peptaibols. Even though some of these secondary metabolites exhibit strong in vitro antifungal properties, the exact role that these compounds play in biocontrol in vivo is not very clear and the literature have been controversial. For example, the gliovirin-deficient mutants lost ability to suppress *Pythium* damping-off in cotton, but the over-producers were not more effective than the wild type (35). Howell et al. (36), based on the ability to produce either gliotoxin ("Q" group) or gliovirin ("P" group) classified *T. virens* strains into two groups and proposed that the "P" groups are effective against *Pythium* damping off in cotton while the "Q" groups are effective in suppressing *R. solani*. Wilhite et al. (106), using gliotoxin deficient mutants proved that gliotoxin is involved in suppression of *P. ultimum*. Subsequently, Howell and Stipanovic (37), using gliotoxin-deficient mutants demonstrated that gliotoxin is not important in biological suppression of *R. solani* in cotton. One of the major drawbacks of such mutational studies is the associated non-target mutations. Using gene knockout, Mukherjee and Kenerley (57) provided evidence that the *velvet* protein Vel1 regulates gliotoxin biosynthesis and expression of many other secondary metabolism related genes. Viridin is a fungistatic compound but its role in biocontrol has not been studied yet (66). Similarly, even though several new peptaibols have recently been discovered in *T. virens*, their role in biocontrol is yet to be assessed (65). The genome of *T. virens* is rich in secondary metabolism related gene clusters and time is not far when we will see evidences for secondary metabolism in biology of this important fungus (59).

## Induced resistance

One of the advantages of using *Trichoderma* as seed treatment or soil application is that these fungi are generally rhizosphere competent and can penetrate roots, thus inducing resistance in plants, a phenomenon known as induced systemic resistance or ISR (27, 85). It has recently been demonstrated that *T. virens* produces "auxins" that promote root growth, and also take-up sucrose from roots using a sucrose transporter and an invertase (14, 94, 95). In an earlier study, Howell et al. (34)

demonstrated that treating seeds with *T. virens* induces peroxidase activity and terpenoid biosynthesis in cotton roots. *T. virens* may secrete elicitors (small cysteine-rich secreted proteins) that induce resistance in plants including production of reactive oxygen species and induction of defense related genes. One of the most well-known elicitor proteins from *T. virens* is Sm1, a small hydrophobin-like protein regulated by sucrose (18, 93). Using gene deletion, it was proved to be involved in ISR in maize (20). Subsequently it was demonstrated that only the monomeric form possesses elicitation properties (93). This protein has been expressed in yeast and the recombinant protein retained elicitation properties (8). A peptaibol, designated as Tex1 and an unknown metabolite produced by a PKS/NRPS gene cluster have also been proved to be having elicitation properties (58, 94, 105).

## GENES FROM *T. VIRENS* CAN IMPART BIOTIC AND ABIOTIC STRESS TOLERANCE IN PLANTS

The 42-KDa endochitinase is the most studied *Trichoderma* gene for developing transgenic plants resistant to pathogens. Emami et al. (23) transferred *T. virens* chitinase gene ech42 to cotton and the plants exhibited tolerance to the root pathogen *R. solani* and foliar pathogen *Alternaria alternata*. Subsequently, they provided evidence that the transgenics showed rapid/greater induction of ROS, expression of several defense-related genes, and activation of some PR enzymes and the terpenoid pathway (43). The same construct was also transferred to rice to enhance tolerance to sheath blight caused by *R. solani* (83). This gene was cloned from an Indian isolate of *T. virens* and expressed in tobacco and tomato. The transgenic plants exhibited enhanced tolerance to three pathogenic fungi namely *Alternaria alternata*, *Botrytis cinerea* and *Sclerotinia sclerotiorum* (84). Genes from *T. virens* not only improved biotic stress tolerance, but also the abiotic stress tolerance. For example, TvGST, encoding a glutathione transferase made tobacco plants tolerant to the heavy metal cadmium (Fig. 3). The same plants also were shown to tolerate the toxic PAH anthracene. The transgenic plants degraded anthracene to the non-toxic naphthalenes (17). With the genome sequence published, it is likely that more such genes will be mined form the genome to improve crop plants for tolerance to biotic and abiotic stresses.

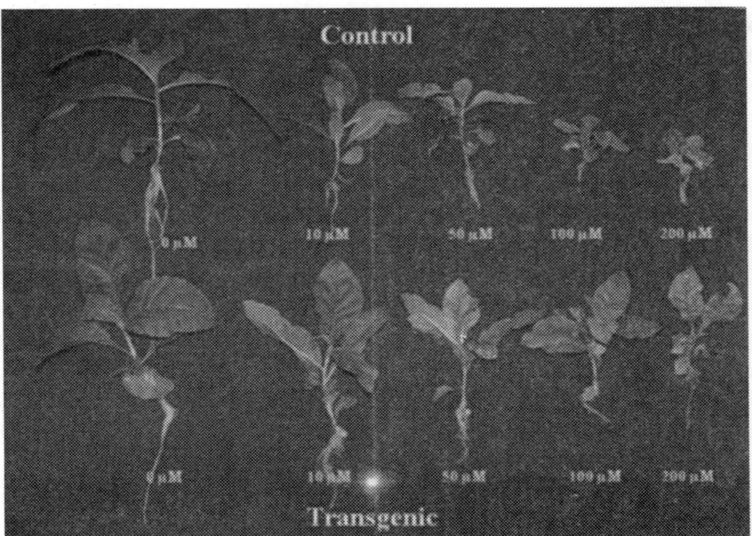


**Fig. 3.** Tolerance of transgenic tobacco plants harboring *Trichoderma virens* GST gene to cadmium (Source: ref. 16)

## CONCLUSION AND OUTLOOK

As discussed above, *T. virens* is one of the most ancient biocontrol fungi to be studied in detail. Researches over the past eight decades have proved beyond doubt the high level of efficacy of this fungus in biological control of many different pathogens in various crops. The mechanisms of biocontrol have been well-studied by using both classical and molecular techniques, and the genome has been sequenced and analyzed. However, ironically, only one product is sold commercially (apart from two products being promoted by Aligarh Muslim University, Aligarh), as compared to many formulations based on other species like *T. harzianum*, *T. asperellum* and *T. viride* (96). The reason could be that just a handful of research groups have been working on *T. virens*, compared to other species that are very widely studied. The good thing about this species is that lots of basic research has been done with this fungus, and researchers should now take advantage of the information generated to commercially exploit *T. virens* for better crop management. This will surely be facilitated by the whole genome sequence data that have been generated and being analyzed.

## REFERENCES


1. Aluko MO and Hering TF 1970 The mechanism associated with the antagonistic relationship between *Corticium solani* and *Gliocladium virens*, *Transactions of the British Mycological Society* **55** 173-179
2. Anitha R and Murugesan K 2001 Mechanism of action of *Gliocladium virens* on *Alternaria helianthi*, *Indian Phytopathology* **54** 449-452
3. Arx JA 1987 Plant pathogenic fungi, *Nova Hedwigia Beihefte* **87** 1-288
4. Baek JM, Howell CR, and Kenerley CM 1999 The role of an extracellular chitinase from *Trichoderma virens* Gv29-8 in the biocontrol of *Rhizoctonia solani*, *Current Genetics* **35** 41-50
5. Bisby GR 1939 *Trichoderma viride* Pers ex Fries, and notes on *Hypocrea*, Transactions of the British Mycological Society **23** 149-168
6. Bordbar, FT, Etebarian HR, Sahebani N, and Rohani H 2010 Control of postharvest decay of apple fruit with *Trichoderma virens* isolates and induction of defense responses, *Journal of Plant Protection Research* **50** 146-152
7. Buensanteai N, Mukherjee PK, Horwitz BA, Cheng C, Dangott LJ, and Kenerley CM 2010 Expression and purification of biologically active *Trichoderma virens* proteinaceous elicitor Sm1 in *Pichia pastoris*, *Protein Expression and Purification* **72** 131-138
8. Budge SP, Mcquilken MP, Fenlon JS, and Whipps JM 1995 Use of *Coniothyrium minitans* and *Gliocladium virens* for biological control of *Sclerotinia sclerotiorum* in glasshouse lettuce, *Biological Control* **5** 513–522
9. Burgess DR and Hepworth G 1996 Biocontrol of sclerotinia stem rot (*Sclerotinia minor*) in sunflower by seed treatment with *Gliocladium virens*, *Plant Pathology* **45** 583–592
10. Burns JR and Benson DM 2000 Biocontrol of damping-off of *Catharanthus roseus* caused by *Pythium ultimum* with *Trichoderma virens* and binucleate *Rhizoctonia* fungi, *Plant Disease* **84** 644-648
11. Catalano V, Vergara M, Hauzenberger JR, Seiboth B, Sarrocco S, Vannacci G, Kubicek CP, and Seidl-Seiboth V 2011 Use of a non-homologous end-joining-deficient strain (delta-ku70) of the biocontrol fungus *Trichoderma virens* to investigate the function of the laccase gene *lcc1* in sclerotia degradation, *Current Genetics* **57** 13-23

12. Chaverri P, Samuels GJ, and Stewart EL 2001 *Hypocrea virens* sp nov, the teleomorph of *Trichoderma virens*, *Mycologia* **93** 1113-1124

13. Christopher DJ, Raj TS, Usha Rani S, and Udhayakumar R 2010 Role of defense enzymes activity in tomato as induced by *Trichoderma virens* against *Fusarium* wilt caused by *Fusarium oxysporum* f sp *Lycopersici*, *Journal of Biopesticides* **3** 158 – 162

14. Contreras-Cornejo HA, Macías-Rodríguez L, Cortés-Penagos C, and López-Bucio J 2009 *Trichoderma virens*, a plant beneficial fungus, enhances biomass production and promotes lateral root growth through an auxin-dependent mechanism in *Arabidopsis*, *Plant Physiology* **149** 1579-1592

15. De RK and Mukhopadhyay AN 1994 Biological control of tomato damping-off by *Gliocladium virens*, *Journal of Biological Control* **8** 34-40

16. Dixit P, Mukherjee PK, Ramachandran V, and Eapen S 2011 Glutathione transferase from *Trichoderma virens* enhances cadmium tolerance without enhancing its accumulation in transgenic *Nicotiana tabacum PLoS One* 6 e16360

17. Dixit P, Mukherjee PK, Sherkhane PD, Kale, SP, and Eapen S 2011 Enhanced tolerance and remediation of anthracene by transgenic tobacco plants expressing a fungal glutathione transferase gene, *Journal of Hazardous Materials* **192** 270– 276

18. Djonović S, Pozo MJ, Dangott LJ, Howell CR, and Kenerley CM 2006 Sm1, a proteinaceous elicitor secreted by the biocontrol fungus *Trichoderma virens* induces plant defense responses and systemic resistance, *Molecular Plant Microbe Interactions* **19** 838-853

19. Djonović S, Pozo MJ, and Kenerley CM 2006 Tvbgn3, a beta-1,6-glucanase from the biocontrol fungus *Trichoderma virens*, is involved in mycoparasitism and control of *Pythium ultimum*, *Applied Environmental Microbiology* **72** 7661-7670

20. Djonovic S, Vargas WA, Kolomiets MV, Horndeski M, Wiest A, and Kenerley CM 2007 A proteinaceous elicitor Sm1 from the beneficial fungus *Trichoderma virens* is required for induced systemic resistance in maize, *Plant Physiology* **145** 875-889

21. Djonović S, Vittone G, Mendoza-Herrera A, and Kenerley CM 2007 Enhanced biocontrol activity of *Trichoderma virens* transformants constitutively coexpressing beta-1,3- and beta-1,6-

glucanase genes, *Molecular Plant Pathology* **8** 469-480

22. Druzhinina IS, Seidl-Seiboth V, Herrera-Estrella A, Horwitz BA, Kenerley CM, Monte E, Mukherjee PK, Zeilinger S, Grigoriev I, and Kubicek CP 2011 *Trichoderma* the genomics of opportunistic success, *Nat Reveiws Microbiology* **9** 749-759
23. Emani C, Garcia JM, Lopata-Finch E, Pozo MJ, Uribe P, Kim DJ, Sunilkumar G, Cook DR, Kenerley CM, and Rathore KS 2003 Enhanced fungal resistance in transgenic cotton expressing an endochitinase gene from *Trichoderma virens*, *Plant Biotechnology Journal* **1** 321-336
24. Etebarian HR and Mohammadifar M 2007 Evaluation of some *Trichoderma* isolates for biological control of crown and root rot of wheat caused by *Bipolaris spicifera* (Bainier) Subram, *Seed Plant* **23** 343-356
25. Haikal NZ 2008 Control of *Rhizoctonia solani* in soybean (*Glycine max* L) by seed-coating with *Tricho-derma viride* and *Glioclad-ium virens* spores, *Journal of Applied Bioscience* **2** 34 – 39
26. Hanson L 2000 Reduction of verticillium wilt symptoms in cotton following seed treatment with *Trichoderma virens*, *Journal of Cotton Science* **4** 224-231
27. Harman GE, Howell CR, Viterbo A, Chet I, and Lorito M 2004 *Trichoderma* species-opportunistic, avirulent plant symbionts, *Nature Reviews in Microbiology* **2** 43-56
28. Harris AR and Lumsden RD 1997 Interactions of *Gliocladium virens* with *Rhizoctonia solani* and *Pythium ultimum* in non-sterile potting medium, *Biocontrol Science Technology* **7** 37-48
29. Heller WE and Theiler-Hedtrich R 1994 Antagonism of *Chaetomium globosum, Gliocladium virens* and *Trichoderma viride* to four soil-borne *Phytophthora* species, *Journal of Phytopathoogy* **141** 390–394
30. Highley TL 1997 Control of wood decay by *Trichoderma* (*Gliocladium*) *virens* I Antagonistic properties, *Mat Organismen* 79-89
31. Howell CR 1982 Effect of *Gliocladium virens* on *Pythium ultimum, Rhizoctonia solani*, and damping-off of cotton seedlings, *Phytopathology* **72** 496-498
32. Howell CR 1991 Biological control of *Pythium* damping-off of cotton with seed coating preparations of *Gliocladium virens*, *Phytopathology* **81** 738-741
33. Howell CR 2006 Understanding the mechanisms employed by *Trichoderma virens* to effect biological control of

cotton diseases, *Phytopathology* **96** 178-180

34. Howell CR, Hanson LE, Stipanovic RD, and Puckhaber LS 2000 Induction of terpenoid synthesis in cotton roots and control of *Rhizoctonia solani* by seed treatment with *Trichoderma virens*, *Phytopathology* **90** 248-252
35. Howell CR and Stipanovic RD 1983 Gliovirin, a new antibiotic from *Gliocladium virens*, and its role in the biological control of *Pythium ultimum*, *Canadian Journal of Microbiology* **29** 321-324
36. Howell CR, Stipanovic RD, and Lumsden RD 1993 Antibiotic production by strains of *Gliocladium virens* and its relation to the biocontrol of cotton seedling diseases, *Biocontrol Science Technology* **3** 435-441
37. Howell CR and Stipanovic RD 1995 Mechanisms in the biocontrol of *Rhizoctonia solani*-induced cotton seedling disease by *Gliocladium virens* Antibiosis, *Phytopathology* **85** 469-472
38. Hwang SF and Chakravarty P 1993 Integrated chemical and biological control of *Rhizoctonia* root rot of field pea by *Gliocladium virens* and a fungicide, *Journal of Plant Disease and Protection* **100** 308-316
39. Kang SW and Kim HK 1989 *Gliocladium virens*, a potential biocontrol agent against damping-off and fusarium wilt of sesame (*Sesamum indicum* L) in problem field in Korea, *The Research Reports of The Rural Development Administration* **31** 19-26
40. Khan MR, Anwer MA, and Shahid S 2011 Management of gray mold of chickpea, *Botrytis cinerea* with bacterial and fungal biopesticides using different modes of inoculation and application, *Biological Control* **57** 13-23
41. Kubicek CP, Herrera-Estrella A, Seidl-Seiboth V, Martinez DA, Druzhinina IS, Thon M, Zeilinger S, Casas-Flores S, Horwitz BA, Mukherjee PK, and Mukherjee M 2011 Comparative genome sequence analysis underscores mycoparasitism as the ancestral life style of *Trichoderma*, *Genome Biology* **12** R40
42. Kumar A, Scher K, Mukherjee M, Pardovitz-Kedmi E, Sible GV, Singh US, Kale SP, Mukherjee PK, and Horwitz BA 2010 Overlapping and distinct functions of two *Trichoderma virens* MAP kinases in cell-wall integrity, antagonistic properties and repression of conidiation, *Biochemistry & Biophysics Research Communications* **398** 765-770
43. Kumar V, Parkhi V, Kenerley CM, and Rathore KS 2009 Defense-related gene expression and enzyme activities in transgenic

cotton plants expressing an endochitinase gene from *Trichoderma virens* in response to interaction with *Rhizoctonia solani*, *Planta* **230** 277-291

44. Lee YA and Wu WS 1984 The antagonisms of *Trichoderma* spp and *Gliocladium virens* against *Sclerotinia sclerotiorum*, *Plant Protection Bulletin (Taiwan)* **26** 293-304

45. Lewis JA and Papavizas GC 1991 Biocontrol of cotton damping-off caused by *Rhizoctonia solani* in the field with formulations of *Trichoderma* spp and *Gliocladium virens*, *Crop Prot*ecton **10** 396–402

46. Lewis JA and Larkin RP 1997 Extruded granular formulation with biomass of biocontrol *Gliocladium virens* and *Trichoderma* spp to reduce damping-off of eggplant caused by *Rhizoctonia solani* and saprophytic growth of the pathogen in soil-less mix, *Biocontrol Science and Technology* **7** 49-60

47. Lewis JA, Lumsden RD, and Locke JC 1996 Biocontrol of damping-off diseases caused by *Rhizoctonia solani* and *Pythium ultimum* with alginate prills of *Gliocladium virens*, *Trichoderma hamatum* and various food bases, *Biocontrol Science and Technology* **6** 163-174

48. Liu LN, Tu YL, and Zhang JZ 2010 Biocontrol potential of *Trichoderma virens* Strain TY009 against rice sheath blight and other main fungal diseases, *Scientific Agricultural Sinica* 2010-10

49. Lumsden RD and Locke JC 1989 Biological control of damping-off caused by *Pythium ultimum* and *Rhizoctonia solani* with *Gliocladium virens* in soilless mix, *Phytopathology,* **79** 361-366

50. Lumsden RD, Locke JC, and Walter JF 1991 Approval of *Gliocladium virens* by the US Environmental Protection Agency for biological control of *Pythium* and *Rhizoctonia* damping-off, *Petria* **1** 138

51. Lumsden RD and Walter JF 1996 Development of *Gliocladium virens* for damping-off disease control, *Canadian Journal of Plant Patho*logy **18** 463-468

52. Maheshwari DK, Dubey RC, and Sharma VK 2001 Biocontrol effects of *Trichoderma virens* on *Macrophomina phaseolina* causing charcoal rot of peanut, *Indian Journal of Microbio*logy **41** 251-256

53. Meyer SLF, Massoud SI, Chitwood DJ, and Roberts DP 2000 Evaluation of *Trichoderma virens* and *Burkholderia cepacia* for antagonistic activity against root-knot nematode, *Meloidogyne*

*incognita*, *Nematology* **2** 871-879

54. Miller JH, Giddens JE, and Foster AA 1957 A survey of the fungi of forest and cultivated soils of Georgia, *Mycologia* **49** 779-808

55. Mishra PK, Mukhopadhyay AN, and Fox RTV 2000 Integrated and biological control of gladiolus corm rot and wilt caused by *Fusarium oxysporum* fsp *gladioli*, *Annual Applied Biology* **137** 361-364

56. Mukhopadhyay AN, Shrestha SM, and Mukherjee PK 1992 Biological seed treatment for the control of soil-borne plant pathogens, *FAO Plant Protection Bulletin* **40** 21-30

57. Mukherjee PK and Kenerley CM 2010 Regulation of morphogenesis and biocontrol properties in *Trichoderma virens* by a VELVET protein Vel1 *Applied Environmental Microbiology* **76** 2345- 2352

58. Mukherjee PK, Buensanteai N, Moran-Diez M, Druzhinina IS, and Kenerley CM 2012 Functional analysis of NRPSs in *Trichoderma virens* reveals a PKS/NRPS hybrid enzyme involved in induced systemic resistance response in maize, *Microbiology* **158** 155-165

59. Mukherjee PK, Horwitz BA, and Kenerley CM 2012 Secondary metabolism in *Trichoderma*- a genomic perspective, *Microbiology* **158** 35-45

60. Mukherjee PK, Shrestha SM, and Mukhopadhyay AN 1993 Baiting with *Sclerotium rolfsii* for selective isolation of *Gliocladium virens* from natural soil, *Biocontrol Science and Technology* **3** 101-104

61. Mukherjee PK, Mukhopadhyay AN, Sarmah DK, and Shrestha SM 1995 Comparative antagonistic properties of *Gliocladium virens* and *Trichoderma harzianum* on *Sclerotium rolfsii* and *Rhizoctonia solani* - its relevance to understanding the mechanisms of biocontrol, *Journal of Phytopathology* **143** 275-279

62. Mukherjee PK, Latha J, Hadar R, and Horwitz BA 2003 TmkA, a mitogen activated protein kinase of *Trichoderma virens*, is involved in biocontrol properties and repression of conidiation in the dark, *Eukaryotic Cell* **2** 446-455

63. Mukherjee PK, Latha J, Hadar R, and Horwitz BA 2004 Role of two G-protein alpha subunits, TgaA and TgaB, in the antagonism of *Trichoderma virens* against plant pathogens, *Applied Environmental Microbiology* **70** 542-549

64. Mukherjee PK, Hadar R, Pardovitz-Kedmi E, Trushina N, and Horwitz BA 2006 MRSP1, encoding a

novel *Trichoderma* secreted protein is negatively regulated by MAPK, *Biochemistry and Biophysics Research Communications* **350** 716-722

65. Mukherjee PK, Wiest A, Ruiz N, Keightley A, Moran-Diez ME, McCluskey K, Pouchus YF, and Kenerley CM 2011 Two classes of new peptaibols are synthesized by a single non-ribosomal peptide synthetase of *Trichoderma virens*, *Journal of Biological Chemistry* **286** 4544-4554
66. Mukherjee M, Horwitz BA, Sherkhane PD, Hadar R, and Mukherjee PK 2006 A secondary metabolite biosynthesis cluster in *Trichoderma virens* evidence from analysis of genes under-expressed in a mutant defective in morphogenesis and antibiotic production, *Current Genetics* **50** 193-202
67. Mukherjee M, Mukherjee PK, and Kale SP 2007 cAMP signaling is involved in growth, germination, mycoparasitism and secondary metabolism in *Trichoderma virens*, *Microbiology* **153** 1734-1742
68. Newman SE, Quarrels JR, Cáceres J, and Batson WE 1994 Biological control of media-borne pathogens of poinsettias with *Gliocladium virens*, *Horticultural Science* **29** 737
69. Pant and Mukhopadhyay AN 2001 Integrated management of seed and seedling rot complex of soybean, *Indian Phytopathology* **54** 346-350
70. Papavizas GC and Collins DJ 1990 Influence of *Gliocladium virens* on germination and infectivity of *Sclerotium rolfsii*, *Phytopathology* **80** 627-630
71. Paulitz TC and Linderman RG 1991 Lack of antagonism between the biocontrol agent *Gliocladium virens* and vesicular arbuscular mycorrhizal fungi, *New Phytologist* **117** 303–308
72. Philip T, Sharma DD, and Shivapratap HR 2000 Antagonistic effect of *Trichoderma* and *Gliocladium virens* on two foliar pathogens of mulberry, *Séricologia* **40** 109-121
73. Phillips AJL 1986 Factors affecting the parasitic activity of *Gliocladium virens* on sclerotia of *Sclerotinia sclerotiorum* and a note on its host range, *Journal of Phytopathology* **116** 212–220
74. Piriyaprin S, Manoch L, Sunantapongsuk V, and Somrang A 2007 Biological control of *Pythium aphanidermatum* and *Sclerotium rolfsii* root rot of Sunn-hemp and mungbeen by *Gliocladium virens*, *Proceedings of the 45th Kasetsart University Annual Conference*, Bangkok, Thailand, 694-701

75. Pozo MJ, Baek JM, García JM, and Kenerley CM 2004 Functional analysis of *tvsp1*, a serine protease-encoding gene in the biocontrol agent *Trichoderma virens, Fungal Genetics Biology* **41** 336-348

76. Rahman MA, Begum MF, and Alam MF 2009 Screening of *Trichoderma* isolates as a biological control agent against *Ceratocystis paradoxa* causing pineapple disease of sugarcane, *Mycobiology* **37** 277-285

77. Rehner SA and Samuels GJ 1994 Taxonomy and phylogeny of *Gliocladium* analyzed by large subunit ribosomal DNA sequences, *Mycological Research* **98** 625-634

78. Reino JL, Guerrero, RF, Hernandez-Galan R, and Collado IG 2008 Secondary metabolites from species of the biocontrol agent *Trichoderma, Phytochemistry Reviews* 7 89-123

79. Ristaino JB, Perry KB, and Lumsden RD 1994 Influence of isolates of *Gliocladium virens* and delivery systems on biological control of Southern blight on carrot and tomato in the field *Plant Disease* **78** 153-156

80. Ristaino JB, Perry KB, and Lumsden RD 1991 Effect of solarization and *Gliocladium virens* on sclerotia of *Sclerotium rolfsii*, soil microbiota, and the incidence of southern blight of tomato *Phytopathology* **81** 1117-1124

81. Ristaino JB, Perry KB, and Lumsden RD 1996 Soil solarization and *Gliocladium virens* reduce the incidence of Southern blight *Sclerotium rolfsii* in bell pepper in the field, *Biocontrol Science and Technology* **6** 583-594

82. Sabrina S, Mikkelsenb L, Vergarac M, Jensenb DF, Lübeckb M, and Vannaccia G 2006 Histopathological studies of sclerotia of phytopathogenic fungi parasitized by a GFP transformed *Trichoderma virens* antagonistic strain, *Mycological Research* **110** 179–187

83. Shah JM, Raghupathy V, and Veluthambi K 2009 Enhanced sheath blight resistance in transgenic rice expressing an endochitinase gene from *Trichoderma virens, Biotechnology Letters* **31** 239-244

84. Shah M, Mukherjee PK, and Eapen S 2010 Expression of a fungal endochitinase gene in transgenic tomato and tobacco results in enhanced tolerance to fungal pathogen, *Physiology and Molecular Biology of Plants* **16** 39-51

85. Shoresh M, Harman GE, and Mastouri F 2010 Induced systemic resistance and plant responses to fungal biocontrol agents, *Annual Review of Phytopathology* **48** 21-43

86. Singh and Mukhopadhyay AN 2000 Biocontrol potential of mutants of *Gliocladium*

*virens* for wilt complex of lentil, *Legume Research* **23** 133-135

87. Smith VL, Wilcox WF, and Harman GE 1990 Potential for biological control of *Phytophthora* root and crown rots of apple by *Trichoderma* and *Gliocladium* spp, *Phytopathology* **80** 880-885

88. Tewari AK, Mukhopadhyay AN, and Aggarwal R 2002 Colonization of *Gliocladium virens* on biologically treated chickpea seeds for biocontrol of chickpea wilt complex, *Acta Phytopatholica Entomological Hungarica* **38** 79-85

89. Tewari AK and Mukhopadhyay AN 2003 Management of chickpea root rot and collar rot by integration of biological and chemical seed treatment, *Indian Phytopathology* **56** 39-42

90. Tianmin Q 2006 Study on the biological control of pine needle blast with *Gliocladium virens* and *Bacillus firmus, Forestry Science Technology* 2006-01

91. Tu JC 1980 *Gliocladium virens*, a destructive mycoparasite of *Sclerotinia sclerotiorum, Phytopathology* **70** 670-674

92. Tu JC and Vaartaja O 1981 The effect of the hyperparasite (*Gliocladium virens*) on *Rhizoctonia solani* and on *Rhizoctonia* root rot of white beans, *Canadian Journal of Botany* **59** 22-27

93. Vargas WA, Djonović S, Sukno SA, and Kenerley CM 2008 Dimerization controls the activity of fungal elicitors that trigger systemic resistance in plants, *Journal of Biological Chemistry* **283** 19804-19815

94. Vargas WA, Mandawe JC, and Kenerley CM 2009 Plant-derived sucrose is a key element in the symbiotic association between *Trichoderma virens* and maize plants, *Plant Physiology* **151** 792-808

95. Vargas WA, Crutcher FK, and Kenerley CM 2011 Functional characterization of a plant-like sucrose transporter from the beneficial fungus *Trichoderma virens* Regulation of the symbiotic association with plants by sucrose metabolism inside the fungal cells, *New Phytologist* **189** 777-789

96. Verma M, Brar SK, and Tyagi RD 2007 Antagonistic fungi, *Trichoderma* spp panoply of biological control, *Biochemical Engginеering Journal* **37** 1-20

97. Viterbo A, Harel M, Horwitz BA, Chet I, and Mukherjee PK 2005 *Trichoderma* mitogen-activated protein kinase signaling is involved in induction of plant systemic resistance, *Applied Environmental Microbiology* **716** 241-6246

98. Viterbo A, Wiest A, Brotman Y, Chet I, and Kenerley CM

2007 The 18mer peptaibols from *Trichoderma virens* elicit plant defence responses, *Molecular Plant Patholgy* **8** 737-746

99. Weindling R 1932 *Trichoderma lignorum* as a parasite of other soil fungi, *Phytopathology* **22** 837-845

100. Weindling R 1934 Studies on a lethal principle effective in the parasitic action of *Trichoderma lignorum* on *Rhizoctonia solani* and other soil fungi, *Phytopathology* **241** 153-1179

101. Weindling R 1937 isolation of toxic substances from the culture filtrates of *Trichoderma* and *Gliocladium*, *Phytopathology* **27** 1175-177

102. Weindling R 1941 Experimental consideration of the mold toxin of *Gliocladium* and *Trichoderma*, *Phytopathology* **3** 1991-1003

103. Weindling R and Emerson OH 1936 The isolation of a toxic substance from the culture filtrate of *Trichoderma*, *Phytopathology* **26** 1068-1070

104. Werner A, Zadworny M, and Idzikowska K 2002 Interaction between *Laccaria laccata* and *Trichoderma virens* in co-culture and in the rhizosphere of *Pinus sylvestris* grown *in vitro*, *Mycorrhiza* **12** 139-145

105. Wiest A, Grzegorski D, Xu BW, Goulard C, Rebuffat S, Ebbole DJ, Bodo B, and Kenerley C 2002 Identification of peptaibols from *Trichoderma virens* and cloning of a peptaibol synthetase, *Journal of Biological Chemistry* **277** 20862-20868

106. Wilhite SE, Lumsden RD, and Straney DC 1994 Mutational analysis of gliotoxin production by the biocontrol fungus *Gliocladium virens* in relation to suppression of *Pythium* damping-off, *Phytopathology* **84** 816

107. Zhang J, Howell CR, and Starr JL 1996 Suppression of *Fusarium* colonization of cotton roots and Fusarium wilt by seed treatments with *Gliocladium virens* and *Bacillus subtilis Biocontrol Science and Technology* **6** 175-187

108. Zadworny M, Smoliński DJ, Idzikowska K, and Werner A 2007 Ultrastructural and cytochemical aspects of the interaction between the ectomycorrhizal fungus *Laccaria laccata* and the saprotrophic fungi, *Trichoderma harzianum* and *T virens*, *Biocontrol Science and Technolology* **17** 921-932.

*Rev. Plant Pathol. Vol. 6, 2014*
*Indian Society of Mycology and Plant Pathology*
*Scientific Publishers (India), Jodhpur pp. 387-412*

# THE PLANT DESTROYER GENUS *PHYTOPHTHORA* IN THE 21ST CENTURY

***S. Guha Roy[1] and N.J. Grünwald[2]***

[1]Department of Botany, West Bengal State University, Barasat, Kolkata - 700126, WB, India
[2]USDA Agricultural Research Service, Corvallis, OR, USA

**KEYWORDS:** Aquatic Phytophthoras, Crop/Plantation Phytophthoras, Forest Phytophthoras integrated morphological & phylogenetic key, multi-host species, putative new species, genome sequencing

## ABSTRACT

The 21st century has already seen a major paradigm shift in our understanding of the biology, evolution, and genetics of the genus *Phytophthora* as well as the tools and approaches used to develop novel approaches for disease management. The combination of novel tools and approaches provided by the convergence of genome sequencing, computing power and novel genomic/biotechnological tools paints a promising picture of the future of Phytophthora disease management. At the same time *Phytophthora* pathogens continue to emerge at an accelerated rate due to increased global travel and trade. This article provides a brief review of the impact, biology and the current genome-enabled era that will characterize progress in the 21st century.

## INTRODUCTION

The genus *Phytophthora* has a long history in modern science. The scientific discipline of plant pathology was born in the early 1860s when Anton de Bary recognized *Phytophthora infestans* as the pathogen causing potato late blight responsible for the Irish potato famine in the 1840s (4,5). It also brought about the first formulated use of a fungicide. In addition to this substantial social and historical impact, even considering merely a handful of *Phytophthora* spp. (e.g., *P. sojae, P. infestans*) documented to cause significant economic impacts, costs amount to anywhere between 2-7 billions of dollars per crop per year worldwide in combined crop losses and management costs not even considering the less quantifiable but equally large impacts to natural ecosystems severely affected by some species (e.g., *P. cinnamomi, P. ramorum*).

## TAXONOMY AND BIOLOGY

The genus *Phytophthora* is presently placed in the kingdom Straminipila (7), under the phylum Heterokonta, sub-phylum Pernosporomycotina, class Pernosporomycetes (Oomycetes), subclass Pernosporomycetidae, order Pythiales and family Pythiaceae. The taxonomy of *Phytophthora* has undergone an evolution in the way it has been studied; from the era of six morphospecies groups (119, 95, 109) to the era of phylogenetic clades (23) and ITS fingerprinting based keys (37) and now to a new era of an integrated morphological and phylogenetic key (102).

Validation of new *Phytophthora* taxa in recent years has been supported with phylogenies generated by sequence analysis of the internal transcribed spacer region (ITS) of the ribosomal DNA gene repeat, the translation elongation factor 1-α, β-tubulin and mitochondrial encoded cytochrome oxidase (cox) I, II, and spacer regions (23, 74, 86, 45). The multilocus phylogeny constructed for 82 *Phytophthora* species with seven molecular markers (8) confirmed the presence of the 10 well-supported clades for the genus previously observed (23). Until 2010, 108 recognized *Phytophthora* taxa have been recognized (94), but many more have been described since then, perhaps an indication of the furious pace of discovery of *Phytophthora* species in recent times (Table 1).

**Table 1:** Newly discovered/re-described species of *Phytophthora* in the last decade

| Year | Name of species | Remarks | Reference |
|---|---|---|---|
| 2001 | *P. tropicalis* Aragaki & Uchida | formerly referred to as *P. palmivora* MF4 [=*P. capsici*] | 5 |
| | *P. oryzo-bladis* Wang and Lu (ex H H Ho) | | 57 |
| | *P. pistaceae* Mirab. | formerly referred to as *P. megasperma* s. l. [sensu lato] | 90 |
| | *P. ramorum* Werres, de Cock & Man In't veld | | 120 |
| 2002 | *P. europaea* Hansen & Jung | | 71 |
| | *P. psychrophila* Jung & Hansen | | 71 |
| | *P. uliginosa* Jung & Hansen | | 71 |
| | *P. brassicae* de Cock & Man in 't Veld | formerly referred to as *P. porri* | 81 |
| | *P. ipomoeae* Flier & Grünwald | | 35 |
| 2003 | *P. inundata* Brasier, Sanchez-Hernadez & Kirk | formerly referred to as *P.* sp. O-group | 15 |
| | *P. pseudosyringae* Jung & Delatour | | 72 |
| | *P. nemorosa* Hansen & Reeser | | 54 |
| 2004 | *P. alni* with subspp. *alni, multiformis,* and *uniformis* Brasier & Kirk | formerly referred to as alder *Phytophthora,* a species hybrid [Brasier et al., 1999 | 13 |
| | *P. hedraiandra* de Cock & Man in't Veld | | 26 |

| | | | |
|---|---|---|---|
| 2005 | *P. kernoviae* Brasier, Beales & Kirk | formerly referred to as *P.* taxon C | 14 |
| 2006 | *P. captiosa* Dick & Dobbie | | 27 |
| | *P. fallax* Dick & Dobbie | | 27 |
| | *P. foliorum* Donahoo & Lamour | | 29 |
| | *P. polonica* Belbahri, Melgarejo & Leford | | 7 |
| 2007 | *P. alticola* Maseko, Coutinho & M.J. Wingf. | | 88 |
| | *P. austrocedrae* Gresl. & Hansen | | 42 |
| | *P. frigida* Maseko, Coutinho & MJ Wingf | | 88 |
| | *P. siskiyouensis* Reeser & EM Hansen | | 101 |
| | *P. rubi* (Wilcox and Duncan) Man in 't Veld, comb. nov. | formerly referred to as *P. fragariae* var. *rubi* | 81 |
| 2008 | *P. bisheria* Abad ZG, Abad JA & Louws | | 2 |
| | *P. asparagi* | formerly referred to as *P. megasperma* s. l., *P.* taxon asparagi | 105 |
| | *P. gallica* T Jung & J Nechwatal | formerly referred to as *P.* taxon G | 68 |
| | *P. irrigata* C Hong & M Gallegly | formerly referred to as *P. drechsleri* s. l. [Dre I subgroup] | 58 |
| | *P. parsiana* Mostowfizadeh, Cooke & Banihashemi | | 91 |

| | | | |
|---|---|---|---|
| | *P. pinifolia*. AJ. Durán, Gryzenh. & MJ Wingf | formerly referred to as *P. cryptogea* or *P. drechsleri* | 31 |
| | *P. quercetorum* Balci & Balci | | 6 |
| 2009 | *P. mengei* G.T. Browne, M.E. Gallegly & C.X. Hong | formerly referred to as *P. citricola* s. l. [avocado subgroup] | 59 |
| | *P. multivora* PM Scott & T Jung | formerly referred to as *P. citricola* s. l. / '*P. citricola* complex' | 108 |
| | *P. plurivora* T. Jung and T.I. Burgess | formerly referred to as *P. citricola* s. l. / '*P. citricola* complex'or *P. inflata* | 67 |
| | *P. rosacearum* EM Hansen & Wilcox | formerly referred to as *P. megasperma* s. l. [AC subgroup] | 55 |
| | *P. sansomeana* EM Hansen & Reeser | formerly referred to as *P. megasperma* s. l. [DF subgroup] | 55 |
| | *Phytophthora* X *pelgrandis* W.F. Gerlach, Nirenberg & Gra¨fenhan | | 96 |
| 2010 | *P. andina* Adler & Flier | formerly referred to as *P. infestans* s. l., | 97 |
| | *P. chrysanthemi* Naher M, Watanabe, H., Chikuo, Y., & Kageyama | | 93 |
| | *P. hydropathica* Hong,C. and Gallegly, M. | Previously termed taxa Dre II. Closest relatives, *P. parsiana* and *P. irrigata* (Dre I). High temperature tolerant and isolated from irrigation water | 61 |
| | *P. morindae* Abad ZG & S Nelson | In ITS clade 10 and most closely related to *P. kernoviae* but differ in having umbellate sporangiophores. Host: Indian mulberry, *Morinda citrifolia* L. var. *citrifolia* | 94 |

| | | | |
|---|---|---|---|
| 2011 | *P. glovera* Z.G. Abad & H.D. Shew | In ITS clade 2 and most closely related to *P. capsici* but differs in being homothallic and produces both paragynous and amphigynous antheridia.<br>Host :Tobacco | 3 |
| | *P. gibbosa* T. Jung, M.J.C. Stukely & T.I. Burgess,<br>sp. nov. | In ITS clade 6 and associated with dying vegetation in swampy or water logged areas | 69 |
| | *P. gregata* T. Jung, M.J.C. Stukely & T.I. Burgess,<br>sp. nov | | 69 |
| | *P. litoralis* T. Jung, M.J.C. Stukely & T.I. Burgess,<br>*sp. nov.* | In clade 6 and present in water ways | 69 |
| | *P. thermophila* T. Jung, M.J.C. Stukely & T.I. Burgess, *sp. nov* | In clade 6 and present in water ways; previously been misidentified as *P. drechsleri* and is referred to as *P.* sp. 3 (Burgess et al. 2009) | 69 |
| | *P. himalsilva* Vettraino, Brasier & Vannini, *sp. nov* | In ITS Clade 2 and forms part of a species cluster defined as Clade 2c by Blair et al. (2008) | 116 |
| | *P. pini* Leonian, 1925 Emend. Gallegly, Hong, Richardson & Kong = *Phytophthora citricola* I Gallegly & Hong 2008 | Formerly known as *Phytophthora citricola* Cil I type (Kong et al. 2003, Gallegly and Hong 2008) | 60 |
| | *P. obscura* Grünwald & Werres, sp. Nov. | Defines a novel subclade 8d with close relatives *P. Syringae* and *P. Austrocedrae* | 43 |

Although there have been considerable advances in the taxonomy of these organisms, the identification to species level using molecular characters remains a major challenge for plant pathologists. Several species "complexes" can be observed in phylogenetic trees showing the presence of potential cryptic species. Presently, powerful molecular techniques combined with morphological characterisation and a renewed interest in probing of the environments have led to discovery of new species, novel variants within species, hybrids and provided a better resolution of species "complexes" with differentiation of the species therein (46).

There are numerous putative new species, which are in progress for official description in different laboratories around the world. These include *Phytophthora (P.)* taxa, such as *P.* taxon Pgchlamydo, *P.* taxon walnut, *P.* taxon oaksoil, *P.* taxon forestsoil, *P.* taxon riversoil, *P.* taxon salixsoil, *P.* taxon raspberry, *P.* taxon Banksia, *P.* taxon *meadii*-like, *P.* taxon chicory, *P.* taxon Acer, *P.* taxon Agathis, *P.* taxon orphan. Most of these originate from natural ecosystems and are shown to belong to the ITS clade 6 (7, 10, 23). Others are P. taxon paludosa (69), P. sp. personii (8) *P.* taxon rosacearum-like, *P.* taxon kwongan, *P.* taxon hungarica and *P.* taxon sulawesiensis. Many of these belong to clades 7 or 11 and are so far found only in Australian natural ecosystems where they are associated with plant mortalities (19). The discoveries of natural hybrids have added another dimension (9, 13, 82, 84, 38). And this is not the end; the following provisionally named species are awaiting their validation: *P. cuyabensis, P. hungarica, P. lagoariana, P. niederhauserii, P. parvasperma, P. sulawesiensis* and *P. sylvatica* (=*P.* taxon forestsoil?). Some, like *P. niederhauserii, P. asparagi, Phytophthora gemini* although not yet described formally, have been proposed as new species (4, 105, 82).

With the advent of molecular techniques, the most rapid resolution and differentiation of species has been in the ITS clades 2, 3, 6 & 7 (23). This, in some cases, has necessitated division into sub-clades; *viz.* clade 6, which previously consisted of only three species: *P. gonapodyides*, *P. humicola* and *P. megasperma* (33) has now radiated to the extent that it can be subdivided into 3 sub clades with 11 described species, many *P.* taxons and undescribed species (69). With description of more species in ITS clade 2, it has now been suggested that Clade 2c (8) has evolved within Asia (116).

Considering that there are 200- 600 extant *Phytophthora* species (10), a large number of species therefore remain to be discovered. Once more species are discovered and described, clearer and well resolved species boundaries will emerge, resulting in revised phylogeny reconstructions. For example, it has become clear that *P. citricola* formerly recognized as a single taxon in phylogenetic reconstructions (23, 8) is now split into a group of species including *P. multivora* and *P. plurivora* (67). As more sequence data become available and new species are described, a clearer picture of species boundaries will emerge.

Host ranges of *Phytophthora* species can vary from one extreme of being very diverse (*P. ramorum*, *P. cinnamomi*, *P. nicotianae*) to that of a single host (*P. sojae*). Host ranges of new invasive forest or other *Phytophthora* species are being determined but the range continues to expand with passage of time (48, 107). Molecular validation and renaming of species are now often increasingly changing host boundaries for each species (e.g. *P. palmivora* MF4 [=*P. capsici*] - *P. tropicalis*). A comparative proteomics study between *P. ramorum* (broad host range) and *P. sojae* (single host range) identified some candidate proteins that might be factors involved in determination of host range (105), but more such comparisons are needed with other broad host range species whose genomes have been sequenced like *P. capsici* and *P. nicotianae* for validation and refinement. As host–pathogen (*A. thaliana-P. nicotianae*) interaction models for broad host range are also available (118), the focus perhaps needs to shift to multi-host species and finding the determinants for host specificity which would provide insights for control of these species which cause the most economic impact worldwide but have been difficult to control. Also, as proposed (89), species diverge by geographic isolation (allopatric) or, in the case of pathogens, by a switch in host (sympatric e.g. *P. infestans* - *Phytophthora mirabilis* - *Phytophthora ipomoeae*) and hence it is highly unlikely that two closely related species are able to diverge if they share a common host in the same geographical area. The study of host ranges will therefore become increasingly important in the future to answer questions of potential hosts of invasive pathogens and their speciation and emergence.

## IMPACT

*Phytophthora* pathogens have a large impact on native ecosystems, forests and agricultural crops. The past decade has seen the discovery of a large number of Phytophthoras especially from aquatic and forest ecosystems and a wealth of information is now available on some of the Phytophthoras attacking crops and plantations. These advances and their possibilities will only be discussed here as exhaustive reviews are already available on the subject (75, 51). Here, we selected a few major examples from three distinct habitats: Crop/plantations, forests, and aquatic environments.

*Crop / Plantation Phytophthoras*

For the most studied of all Phytophthoras, namely *Phytophthora infestans*, the population structure has changed dramatically in the 21[st] century due to migrations and establishment of sexual populations in select areas. In fact, at very few locations is the population structure similar to what it was in the 20[th] century. The populations in Mexico (which is thought to be the source of recent migrations) are sexual and this region represents the centre of origin of this pathogen (49, 35, 45). Populations in Asia, North America, South America and north of Mexico have changed and remain mostly clonal while population in parts of Europe have become sexual (36). Recent analysis of simple sequence repeats has revealed diversity within US clonal lineages (for e.g. US 22) (Fry, personal communication) and some new types have also been identified (92). Eucablight and now Euroblight are two websites of the concerted EU networks that rely on extensive multiplexed SSR analyses (78) for surveillance. These efforts have established that while 'blue 13' is a new virulent 'SSR genotype' found in most places in Europe, in France there has been a spatiotemporal change in population structures and a change in allele frequencies in the UK, suggesting a suspected gradual change in populations. A new set of multiplexed SSR primers (79) have revealed that some of the 'SSR genotypes' obtained by using the Eucablight primers could further be subdivided into more genotypes; the converse was also true (20). Targeting of different loci may bring about finer resolution of some of the SSR genotypes; however, if the resultant "groupings/clusters/SSR genotypes" is to be of practical field value towards control of the disease, then each of these SSR genotypes have to be correlated with phenotypic characters like mating type, fungicide sensitivity, growth temperature etc. Implementation of effective control strategies requires knowledge about the genetic structure of population of plant pathogens (122), as control strategies must target a population instead of an individual if they are to be effective. Defining the genetic structure of a population is a logical first step in studies of fungal population genetics because the genetic structure of a population reflects its evolutionary history and its potential to evolve: aspects important for formulating disease management strategies. Genetic diversity analyses of populations at these locations have therefore become a priority.

*P. capsici* in many parts of the world are the most important factors limiting vegetable production (52, 75, 123). An excellent review on the population structures and host ranges in both the Americas and Korea on *P. capsici* is available (75). Interestingly the populations of *P. capsici* are very diverse in the US and clonal in neighbouring countries in contrast to the clonal nature of *P. infestans* found in the USA. *P. capsici* additionally affects plantation crops like bell pepper, betelvine and cocoa in countries like India and Vietnam. Populations in India have been found to have intraspecies variation depending on host and techniques used (21) while population diversity is lower in Vietnam (112). Cocoa isolates were however of a single clonal lineage in both countries.

*P. nicotianae* affects over 255 plant genera in 90 families (22) and is relatively neglected (by scientists and policy makers) considering that this species is one of the most common pathogens on ornamental plants, the cultivation and trade of which has been recognized as a principal pathway for the introduction and spread of invasive plant pathogens (cited in 80). Also it is a major pathogen of different vegetables in some tropical economies. The neglect is potentially influenced by having an extremely broad host range that makes estimation of economic impact difficult. The intraspecific variation of *P. nicotiane* was studied variously by RAPD [tobacco isolates, (124); vegetable and fruit isolates, (52)]; AFLP [floricultural hosts, (77) tobacco, (110)] and variable mitochondrial intergenic regions [various hosts, (80)]. The results were similar showing that intraspecific variability exists and in some cases were related to hosts. The future course of action for both *P. capsici* and *P. nicotianae* would be to find out the extent of the variability with special reference to host specificity.

*P. sojae* with a single host is considered to be a low risk pathogen because of limited outcrossing and low mutation and has been extensively reviewed (118). However, it is evident that pathotype structures, in clonal populations, are changing (30).

*Forest Phytophthoras*

There are many prominent forest *Phytophthora* species. Perhaps most notable is the sudden oak death epidemic caused by the pathogen *P. ramorum* (48, 104). This pathogen emerged simultaneously on rhododendrons in gardens, woodlands and nurseries in Germany and forests and nurseries in the USA. Sudden oak death is only found in the USA where coast live oak and tanoak forests are severely impacted. However, a recent significant epidemic on Japanese larch has been documented in the UK that is of global concern (12). In both Europe and North America the pathogen exhibits a very broad host range that makes control exceedingly difficult. *P. ramorum* is obviously an exotic pathogen that has been introduced into the US and Europe from elsewhere. Currently, three clonal lineages are being recognized that have been named after the continent (NA = North America; EU = Europe) on which they were first found (47). EU1 only affects Europe, while all three lineages are found in the US (40,39). It has also become evident that the EU1 clonal lineage was moved from Europe to the Pacific Northwest, either British Columbia, Canada or Washington, US, most likely via movement of ornamental plants such as *Rhododendron* spp. (39). The origin of this pathogen is not well established, but recent coalescent analysis determined that the three lineages are either reproductively isolated or geographically isolated even though both mating types exist and have diverged before the rise of modern agriculture (40). Several studies have monitored and described establishment, migration and clonal divergence within each of the continents on which *P. ramorum* is observed (40, 39, 115, 87, 99, 98). Despite the occasional presence of both mating types in nursery environments of the Western US, sexual reproduction has to date not been found (38).

Another prominent forest *Phytophthora* is *P. lateralis* affecting Port Orford Cedar in the Pacific North Western US (56). Like *P. ramorum* this pathogen is introduced to the US and has become a new pathogen on a new host. Unlike *P. ramorum, P.*

*lateralis* has only one known host. Recent work indicated that this pathogen originated in Asia (11). It is thought that *P. ramorum* might also have originated from Asia, but this remains to be established.

There exist numerous other forest *Phytophthora* species of which notable examples include *P. kernoviae* (14) and *P. alni* (13).

*Aquatic Phytophthoras*

Though *Phytophthora* species are commonly referred to as water moulds, little is known about their aquatic ecology and the species present in these habitats are poorly characterized relative to those present on crops. More than 25 species of *Phytophthora*, (including *P. ramorum*, the sudden oak death pathogen), have been isolated from irrigation reservoirs and natural waterways in riparian ecosystems and as well as from swampy or water-logged areas (17, 111, 121, 63,101, 58, 61, 69). A number of previously unknown taxa also have been documented in aquatic environments (15, 58, 68, 69). Systematic sampling efforts have been undertaken for early detection of these pathogens and because of the renewed interest in the assemblage of *Phytophthora* species already resident in streams in natural ecosystems: in western North America (101), south Eastern America (64, 65), South China (123) and Australia (69). However, such efforts are lacking in the Indian subcontinent.

Most of the species found in aquatic environments belong to clade 6 and show a strong association with both forest and riparian ecosystems and, with the exceptions of *P.* taxon asparagi, *P. gonapodyides*, *P. megasperma* and *P. rosacearum*, which have limited association with agriculture and horticulture. However all ITS Clade 6 members are not aquatic, such as *P. pinifolia*, *P. inundata*, *P.* taxon PgChlamydo and *P. gonapodyides* which can be opportunistic and sometimes aggressive tree pathogens (18, 31, 68, 70). The function of most of these taxa within the ecosystems is unclear. It has been hypothesized (15, 17) that a saprotrophic lifestyle and sexual degeneration for taxa in this clade may be connected to their aquatic habitat. Furthermore, due to the almost complete absence of heterothallism within Clade 6 and a lack of records of Clade 6 taxa from the tropics, it has been hypothesised (69) that a tropical origin seems unlikely. Whatever their origin might be, the fact that recycling of water for irrigation is becoming increasingly important, baiting or filtering irrigation water for pathogenic *Phytophthora* species will increasingly become a useful monitoring tool in agricultural systems, including nurseries (63).

## GENOME-ENABLED BIOLOGY

Three *Phytophthora* genomes have been sequenced to date and these include *P. sojae, P. ramorum* and *P. infestans* (114, 53). Several other genomes are considered, being or have been sequenced including *P. parasitica*, *P. cinnamomi* and *P. capsici* among others. Comparative genomics of these genomes has revealed a remarkable diversity of effectors that include avirulence genes now referred to as RxLR genes for a conserved amino acid motif characterizing this unique class of small secreted proteins.

Availability of whole genome sequences provides novel tools and approaches for research and translational applications of research. For example, discovery of mating type gene, metalaxyl resistance genes, and other functional and or adaptive traits is within grasp. With a novel class of functional genetic markers plant pathologists will be able to characterize *Phytophthora* populations more accurately and more rapidly, while breeders will be able to pyramid resistance genes more appropriately while using the discovered effect or repertoire for effect or-mediated R-gene discovery (43, 117).

## CONCLUSIONS

*Phytophthora* biology currently is an area of work for plant pathologists undergoing rapid discovery of new species and brimming with the hope of novel, genome-enabled translational discoveries. The *Phytophthora* community is hoping to sequence the genome of each known *Phytophthora* species in a concentred effort, which will provide a useful, novel platform for development of targeted disease management strategies and result in novel insights about the evolution and patterns of emergence of this important plant pathogen genus.

## REFERENCES


1. Abad ZG and Coffey M 2008 Development of a Morphological/Phylogenetic Lucid Key for the identification of Oomycetes *Phytophthora* 3rd International *Phytophthora* and *Pythium* Workshop *in* Integration of Traditional and Modern Approaches for Investigating the Taxonomy and Evolution of *Phytophthora, Pythium* and Related Genera, (http// wwwphytophthoradborg/pdf/ 3rd_Inter_workshoppdf)
2. Abad ZG, Abad JA, Coffey MD, Oudemans PV, Man in 't Veld WA, de Gruyter H, Cunnington J, and Louws FJ 2008 *Phytophthora bisheria* sp nov, a new species identified in isolates from the Rosaceous raspberry, rose and strawberry in three continents, *Mycologia* **100** 99–110
3. Abad ZG, Ivors KL, Gallup CA, Abad JA, and Shew DH 2011 Morphological and molecular characterization of *Phytophthora glovera* sp nov from tobacco in Brazil *Mycologia* **103** 341–350
4. Abad ZG, and Abad JA 2003 Advances in the integration of morphological and molecular characterization in the genus *Phytophthora* The case of *P niederhauseria* sp Nov, *Phytopathology* **93** S1
5. Aragaki M, and Uchida JY 2001 Morphological distinctions between *P capsici* and *P tropicalis* sp Nov, *Mycologia* **93** 137-145
6. Balci Y, Balci S, Blair JE, Park S, Kang S, and McDonald WL 2008 *Phytophthora quercetorum* sp nov, a novel species isolated from eastern and north-central US oak forest soils, *Mycological Research* **11** 2906–916
7. Belbahri L, Moralejo E, Calmin G, Oszako T, Garcı´a JA, Descals E, and Lefort F 2006 *Phytophthora polonica*, a new species isolated from declining *Alnus glutinosa* stands in Poland, *FEMS Microbiology Letters* **26** 1165–174
8. Blair JA, Coffey MD, Park S, Geiser DM, and Kang S 2008 A multilocus phylogeny for *Phytophthora* utilizing markers derived from complete genome sequences, *Fungal Genetics and Biology* **45** 266–277
9. Bonants PJM, Hagenaar-de Weerdt M, Man in 't Veld WA, and Baayen RP 2000 Molecular characterization of natural hybrids of *Phytophthora nicotianae* and *P cactorum, Phytopathology* **90** 867–874
10. Brasier C 2009 *Phytophthora* biodiversity how many

*Phytophthora* species are there? in EM Goheen, SJ Frankel (tech coords) *Phytophthoras* in Forests and Natural Ecosystems Proc, 4thMeeting of IUFRO Working Party 070209, Monterey, CA Gen Tech Rep PSW-GTR-221, Albany, CA, 101–115 pages

11. Brasier CM, Vettraino AM, Chang TT, and Vannini A 2010 *Phytophthora lateralis* discovered in an old growth Chamaecyparis forest in Taiwan, *Plant Pathology* **59** 595-603

12. Brasier C and Webber J 2010 Plant pathology Sudden larch death, *Nature* **466** 824-5

13. Brasier CM, Kirk SA, Delcan J, Cooke DEL, Jung T, and Man in 't Veld WA 2004 *Phytophthora alni* sp nov and its variants designation of emerging heteroploid hybrid pathogens spreading on Alnus trees, *Mycological Research* **108** 1172–1184

14. Brasier CM, Beales PA, Kirk SA, Denman S and Rose J 2005 *Phytophthora kernoviae* sp nov, an invasive pathogen causing bleeding stem lesions on forest trees and foliar necrosis of ornamentals in the UK, *Mycological Research* **109** 853–859

15. Brasier CM, Sanchez-Hernandez E, and Kirk SA 2003a *Phytophthora inundata* sp nov, a part heterothallic pathogen of trees and shrubs in wet or flooded soils, *Mycological Research* **107** 477–484

16. Brasier CM, Cooke DEL, and Duncan JM 1999 Origin of a new *Phytophthora* pathogen through interspecific hybridization, *PNAS,*USA, **96** 5878–5883

17. Brasier CM, Cooke DEL, Duncan JM, and Hansen EM 2003b Multiple new phenotypic taxa from trees and riparian ecosystems in *Phytophthora gonapodyides–P megasperma* ITS clade 6, which tend to be high-temperature tolerant and either inbreeding or sterile, *Mycological Research* **107** 277–290

18. Brown AV and Brasier CM 2007 Colonization of tree xylem by *Phytophthora ramorum, P kernoviae* and other *Phytophthora* species, *Phytopathology* **56** 227– 241

19. Burgess TI, Webster JL, Ciampini JA, White DW, Hardy GEStJ, and Stukely MJC 2009 Re-evaluation of *Phytophthora* species isolated during 30 years of vegetation health surveys in Western Australia using molecular techniques, *Plant Disease* **93** 215–223

20. Chmielarz M, Sobkowiak S, Lebecka R, and Śliwka J 2010 Chosen characteristics

of Polish *Phytophthora infestans* isolates PPO-Special Report no 14 39 – 44 Twelfth EuroBlight workshop Arras (France),

21. Chowdappa P, Brayford D, Smith J, and Flood J 2003 Molecular discrimination of *Phytophthora* isolates on cocoa and their relationship with coconut, black pepper and bell peper isolates based on rDNA repeat and AFLP fingerprints, *Current Science* **84** 1235-1238

22. Cline ET, Farr DF, and Rossman AY 2008 Synopsis of *Phytophthora* with accurate scientific names, host range, and geographic distribution Plant Health Progress doi101094/PHP-2008-0318-01-RS

23. Cooke DEL, Drenth A, Duncan JM, Wagels G, and Brasier CM 2000 A molecular phylogeny of *Phytophthora* and related Oomycetes, *Fungal Genetics and Biology* **30** 17–32

24. De Bary A 1873 The potato disease, *The Journal of the Quekett Microscopical Club* **31** 39-145

25. De Bary A 1876 Researches into the nature of the potato fungus, *Phytophthora infestans, Journal of the Royal Agricultural Society of England* Series **2** 12149-154

26. de Cock AWAM and Lévesque CA 2004 New species of *Pythium* and *Phytophthora, Studies in Mycology* **50** 481-487

27. Dick MW, Dobbie K, Cooke DEL, and Brasier CM 2006 *Phytophthora captiosa* sp nov and *P fallax* sp nov causing crown dieback of Eucalyptus in New Zealand, *Mycological Research* **110** 393 404

28. Dick MW 2001 Straminipilous Fungi Kluwer Academic Publishers, The Netherlands

29. Donahoo R, Blomquist CL, Thomas SL, Moulton JK, Cooke DEL, and Lamour KH 2006 *Phytophthora foliorum* sp nov, a new species causing leaf blight of azalea, *Mycological Research* **110** 1309–1322

30. Dorrance A and Grünwald NJ 2009 *Phytophthora sojae* Diversity Among and Within Populations in Oomycete genetics and genomics diversity, interactions and genomic tools (eds) L Kurt and Sophien Kamoun John Wiley & Sons, 197-212 pages

31. Durán A, Gryzenhout M, Slippers B, Ahumada R, Rotella A, Flores F, Wingfield BD, and Wingfield MJ 2008 *Phytophthora pinifolia* sp nov associated with a serious needle disease of *Pinus radiata* in Chile, *Plant Pathology* **57** 715–727

32. Durán A, Gryzenhout M, Drenth A, Slippers B, Ahumada R, Wingfield BD, and Wingfield MJ 2010 AFLP analysis reveals a clonal population of *Phytophthora pinifolia* in Chile, *Fungal Biology* **114** 746 –752

33. Erwin DC and Ribeiro OK 1996 *Phytophthora* Diseases, Worldwide APS Press

34. Flier WG, Grünwald NJ, Kroon LPNM, Sturbaum AK, van den Bosch GBM, Garay-Serrano E, Fry WE, and Turkensteen LJ 2003 The population structure of *Phytophthora infestans* from the Toluca Valley in central Mexico suggests genetic differentiation between populations from cultivated potato and wild *Solanum* species, *Phytopathology* **93** 382-390

35. Flier WG, Grünwald NJ, Kroon LPNM, van den Bosch BM, Garay-Serrano E, Lozoya-Saldana H, Bonants PJM, and Turkensteen LJ 2002 *Phytophthora ipomoeae* sp nov, a new homothallic species causing leaf blight on *Ipomoea longipedunculata* in the Toluca Valley of central Mexico, *Mycological Research* **106** 848–856

36. Fry WE, Grünwald NJ, Cooke DE, McLeod A, Forbes GA, and Cao K 2009 Population Genetics and Population Diversity of *Phytophthora infestans* in Oomycete genetics and genomics diversity, interactions and genomic tools (eds) L Kurt and Sophien Kamoun John Wiley & Sons, 139-164 pages

37. Gallegly ME and Hong CX 2008 *Phytophthora* Identifying species by morphology and DNA fingerprints St Paul, Minnesota APS Press 158 page

38. Goss EM, Cardenas MC, Myers K, Forbes GA, Fry WE, Restrepo S, and Grünwald NJ 2011 The plant pathogen *Phytophthora andina* emerged via hybridization of an unknown *Phytophthora* species and the Irish potato famine pathogen, *P infestans, PLoS One* (in press)

39. Goss EM, Larsen M, Vercauteren A, Werres S, Heungens K, and Grünwald NJ 2011 *Phytophthora ramorum* in Canada Evidence for migration within North America and from Europe, *Phytopathology* **101** 166-171

40. Goss EM, Carbone I, and Grünwald NJ 2009 Ancient isolation and independent evolution of the three clonal lineages of the exotic sudden oak death pathogen *Phytophthora ramorum, Molecular Ecology* **18** 1161–1174

41. Goss EM, Larsen M, Chastagner GA, Givens DR, and Grünwald NJ 2009 Population genetic analysis infers migration pathways of *Phytophthora ramorum* in US nurseries, *PLoS Pathogens* **5**,e1000583
42. Greslebin AG, Hansen EM, and Sutton W 2007 *Phytophthora austrocedrae* sp nov, a new species associated with *Austrocedrus chilensis* mortality in Patagonia (Argentina), *Mycological Research* **111** 308–316
43. Grünwald NJ 2012 Genome sequences of *Phytophthora* enable translational plant disease management and accelerate research, *Canadian Journal of Plant Pathology*
44. Grünwald NJ, Martin FN, Larsen MM, Sullivan CM, Press CM, Coffey MD, Hansen EM, and Parke JL 2011 Phytophthora-IDorg A sequence-based *Phytophthora* identification tool, *Plant Disease* **95** 337-342
45. Grünwald NJ, Werres S, Goss EM, Taylor CR, and Fieland VJ 2011 *Phytophthora obscura* sp nov, a new species that is a member of the novel *Phytophthora* subclade 8d, *Plant Pathology*
46. Grünwald NJ and Goss EM 2011 Evolutionary and Population Genetics of Exotic and Re-emerging Pathogens Traditional and Novel Tools and Approaches, *Annual Review of Phytopathology* **49** 249-267
47. Grünwald NJ 2009 Standardizing the nomenclature for clonal lineages of the sudden oak death pathogen, *Phytophthora ramorum Phytopathology* **99** 792-795
48. Grünwald NJ, Goss EM, and Press CM 2008 *Phytophthora ramorum* a pathogen with a remarkably wide host-range causing sudden oak death on oaks and ramorum blight on woody ornamentals, *Molecular Plant Pathology* **9** 729–740
49. Grünwald NJ and Flier WG 2005 The biology of *Phytophthora infestans* at its center of origin, *Annual Review of Phytopathology* **43** 171-190
50. Grünwald NJ, Flier WG, Sturbaum AK, Garay-Serrano E, van den Bosch TBM, Smart CD, Matuszak JM, Lozoya-Saldaña H, Turkensteen LJ, and Fry WE 2001 Population structure of *Phytophthora infestans* in the Toluca Valley region of Central Mexico, *Phytopathology* **91** 882-890

51. Guha Roy S 2008 Progress in *Phytophthora* research identification, species diversity and population diversity, *Journal of Mycopathological Research* **46** 163 - 184

52. Guha Roy S, Bhattacharyya S, Mukherjee SK, and Khatua DC 2009 Molecular Identification of *Phytophthora* spp Affecting some Economically Important Crops in Eastern India through ITS-RFLP and Sequencing of the ITS Region, *Journal of Phytopathology* **157** 666-674

53. Haas BJ 2009 Genome sequence and analysis of the Irish potato famine pathogen *Phytophthora infestans*, *Nature* **461** 393-398

54. Hansen EM, Reeser PW, Davidson JM, Garbelotto M, Ivors K, Douhan L, and Rizzo DM 2003 *Phytophthora nemorosa*, a new species causing cankers and leaf blight of forest trees in California and Oregon, USA, *Mycotaxon* **88** 129–138

55. Hansen EM, Wilcox WF, Reeser PW, and Sutton W 2009 *Phytophthora rosacearum* and *P sansomeana*, new species segregated from the *Phytophthora megasperma* "complex", *Mycologia* **101** 129–135

56. Hansen EM, Goheen DJ, Jules ES, and Ullian B 2000 Managing Port-Orford-Cedar and the introduced pathogen *Phytophthora lateralis*, *Plant Disease* **84** 4-10

57. Ho HH 2001 *Phytophthora oryzo-bladis*, a new name for *P fragariae* var *oryzo-bladis*, *Mycotaxon* **78** 17–21

58. Hong C, Gallegly ME, Richardson PA, Kong P, and Moorman GW 2008 *Phytophthora irrigata*, a new species isolated from irrigation reservoirs and rivers in eastern United States of America, *FEMS Microbiology Letters* **285** 203–211

59. Hong CX, Gallegly ME, Browne GT, Bhat RG, Richardson PA, and Kong P 2009 The avocado subgroup of *Phytophthora citricola* constitutes a distinct species, *Phytophthora mengei* sp Nov, *Mycologia* **101** 833–840

60. Hong CX, Gallegly ME, Richardson PA, and Kong P 2011 *Phytophthora pini* Leonian resurrected to distinct species status, *Mycologia* **103** 351–360

61. Hong CX, Gallegly ME, Richardson PA, Kong P, Moorman GW, Lea-Cox JD, and Ross DS 2010b *Phytophthora hydropathica*, a new pathogen identified from irrigation water, *Rhododendron catawbiense* and *Kalmia latifolia*, *Plant Pathology* **59** 913–921

62. Hong CX, Richardson PA, and Kong P 2008 Pathogenicity to ornamental plants of some existing species and new taxa of *Phytophthora* from irrigation water, *Plant Disease* **92** 1201–1207

63. Hong CX and Moorman GW 2005 Plant pathogens in irrigation water challenges and opportunities, *Critical Reviews in Plant Science* **24** 189–208

64. Hulvey J, Gobena D, Finley L, and Lamour K 2010 Co-occurrence and genotypic distribution of *Phytophthora* species recovered from watersheds and plant nurseries of eastern Tennessee, *Mycologia* **102** 1127–1133

65. Hwang J, Oak SW, and Jeffers SN 2009 Monitoring occurrence and distribution of *Phytophthora* species in forest streams in North Carolina using baiting and filtration methods In Goheen EM, Frankel SJ (tech coords), Proceedings of the fourth meeting of the International Union of Forest Research Organizations (IUFRO) Working Party S070209 Phytophthoras in forests and natural ecosystems Gen Tech Rep PSW-GTR-221 91– 95 US Department of Agriculture, Forest Service, Pacific Southwest Research Station, Albany, CA, USA

66. Jiang RHJ, Tripathy S, Govers F, and Tyler BM 2008 RXLR effector reservoir in two *Phytophthora* species is dominated by a single rapidly evolving superfamily with more than 700 members *PNAS* **105(12)** 4874-4879

67. Jung T and Burgess TI 2009 Re-evaluation of *Phytophthora citricola* isolates from multiple woody hosts in Europe and North America reveals a new species, *Phytophthora plurivora* sp Nov, *Persoonia* **22** 95–110

68. Jung T and Nechwatal J 2008 *Phytophthora gallica* sp nov, a new species from rhizosphere soil of declining oak and reed stands in France and Germany, *Mycological Research* **112** 1195–1205

69. Jung T, Stukely MJC, Hardy GESJ, White D, Paap T, Dunstan WA, and Burgess TI 2011 Multiple new *Phytophthora* species from ITS Clade 6 associated with natural ecosystems in Australia evolutionary and ecological implications, *Persoonia* **26** 13–39

70. Jung T 2009 Beech decline in Central Europe driven by the interaction between *Phytophthora* infections and

climatic extremes, *Forest Pathology* **39** 73-84

71. Jung T, Hansen EM, Winton L, Obwald W, and Delatour C 2002 Three new species of *Phytophthora* from European oak forests, *Mycological Research* **106** 397-411
72. Jung T, Nechwatal J, Cooke DEL, Hartmann GC, Blaschke M, Oswald WF, Duncan JM, and Delatour C 2003 *P pseudosyringae* sp nov, a new species causing root and collar rot of deciduous tree species in Europe, *Mycological Research* **107** 772-789
73. Kong P, Hong CX, Richardson PA, and Gallegly ME 2003 Single-strand-conformation polymorphism of ribosomal DNA for rapid species differentiation in genus *Phytophthora, Fungal Genetics and Biology* **39** 238–249
74. Kroon LP, Bakker FT, van den Bosch GB, Bonants PJ, and FlierWG 2004 Phylogenetic analysis of *Phytophthora* species based on mitochondrial and nuclear DNA sequences, *Fungal Genetics and Biology* **41** 766–782
75. Lamour K and Kamoun S 2009 Oomycete Genetics and Genomics Diversity, Interactions and Research Tools Wiley-Blackwell, 582 page
76. Lamour K 2009 *Phytophthora capsici* Sex, Selection, and the Wealth of Variation, in Oomycete genetics and genomics diversity, interactions and genomic tools (eds) L Kurt and Sophien Kamoun John Wiley & Sons, 165- 178 pages
77. Lamour KH, Daughtrey ML, BensonDM, Hwang J, and Hausbeck MK 2003 Etiology of *Phytophthora drechsleri* and *P nicotianae* (= *P parasitica*) diseases affecting floriculture crops, *Plant Disease* **87** 854-858
78. Lees AK, Wattier RAM, Shaw DS, Sullivan L, Williams NA, and Cooke DEL 2006 Novel microsatellite markers for the analysis of *Phytophthora infestans* populations, *Plant Pathology* **55** 311-319
79. Li Y, Govers F, Mendes O, Testa A, Jacobsen E, Huang SW, and van der lee TAJ 2010 A new set of highly informative ssr markers for *Phytophthora infestans* population analysis assembled into an efficient multiplex, *Molecular Ecology Resources* **10(6)** 1098-1105
80. Mammella MA, Cacciola SO, Martin F, and Schena L 2011 Genetic characterization of *Phytophthora nicotianae* by the analysis of polymorphic regions of the mitochondrial

DNA, *Fungal Biology* **115** 432-442

81. Man in 't Veld WA, de Cock AWAM, Ilieva E, and Levesque CA 2002 Gene flow analysis of *Phytophthora porri* reveals a new species *Phytophthora brassicae* sp Nov, *European Journal of Plant Pathology* **108** 51–62

82. Man in 't Veld WA 2001 First report of natural hybrids of *Phytophthora nicotianae* and *P cactorum* on loquat in Taiwan, *Plant Disease* **85** 98–98

83. Man in't Veld WA 2007 Gen flow analysis demonstrates that *Phytophthora fragariae* var *rubi* constitutes a distinct species, *Phytophthora rubi* comb Nov, *Mycologia* **99** 222–226

84. Man in'tVeld WA, de Cock AWAM, and Summerbell RC 2007 Natural hybrids of resident and introduced *Phytophthora* species proliferating on multiple new hosts, *European Journal of Plant Pathology* **117** 25–33

85. Man in't Veld WA, Rosendahl KCHM, Brouwer H and De Cock AWAM 2011 *Phytophthora gemini* sp nov, a new species isolated from the halophilic plant *Zostera marina* in the Netherlands, *Fungal Biology* doi101016/jfunbio201105006

86. Martin FN and Tooley PW 2003 Phylogenetic relationships among *Phytophthora* species inferred from sequence analysis of mitochondrially encoded cytochrome oxidaseI and II genes, *Mycologia* **95** 269–284

87. Mascheretti S, Croucher PJP, Vettraino A, Prospero S, and Garbelotto M 2008 Reconstruction of the Sudden Oak Death epidemic in California through microsatellite analysis of the pathogen *Phytophthora ramorum, Molecular Ecology* **172** 755-2768

88. Maseko B, Burgue´s TI, Coutinho TA, and Wingfield MJ 2007 Two new Phytophthora species from South African Eucalyptus plantations, *Mycological Research* **111** 1321–1338

89. Mayr E 1942 Systematics and the Origin of Species, New York Columbia University Press

90. Mirabolfathy M, Cooke DEL, Duncan JM, Williams NA, Ershad D, and Alizadeh A 2001 *Phytophthora pistaciae* sp nov and *P melonis* the principal causes of pistachio gummosis in Iran, *Mycological Research* **105** 1166–1175

91. Mostowfizadeh-Ghalamfarsa R, Cooke DE, and Banihashemi Z 2008 *Phytophthora parsiana* sp nov, a new high temperature tolerant species,

*Mycological Research* **112** 783–794

92. Myers K, Small I, Jensen S, Zuluaga P, Guha Roy S, and Fry WE 2010 Characterization of *Phytophthora infestans* isolates from potato/tomato in 2010, *Phytopathology* **101** 261-262
93. Naher M, Motohash K, Watanabe H, Chikuo Y, Senda M, Suga H, Brasier C, and Kageyama K 2010 *Phytophthora chrysanthemi* sp nov, a new species causing root rot of chrysanthemum in Japan, *Mycological Progress*
94. Nelson S and Abad ZG 2010 *Phytophthora morindae*, a new species causing black flag on Noni (*Morinda citrifolia* L) in Hawaii, *Mycologia* **102** 122–134
95. Newhook FJ, Waterhouse FM, and Stamps DJ 1978 Tabular key to the species of *Phytophthora* de Bary Commonwealth, *Mycol Inst Mycol* Pap No 3, 20 page
96. Nirenberg HI, Gerlach WF, and Gra¨fenhan T 2009 *Phytophthora* X *p elgrandis*, a new natural hybrid pathogenic to *Pelargonium grandiflorum* hort, *Mycologia* **101** 220–231
97. Oliva RF, Kroon LPNM, Chaco´n G, Flier WG, Ristaino JB, and Forbes GA 2010 *Phytophthora andina* sp nov, a newly identified heterothallic pathogen of solanaceous hosts in the Andean highlands, *Plant Pathology* **59** 613–625
98. Prospero S, Grünwald NJ, Winton LM, and Hansen EM 2009 Migration patterns of the emerging plant pathogen *Phytophthora ramorum* on the west coast of the United States of America, *Phytopathology* **99** 739-749
99. Prospero S, Hansen EM, Grünwald NJ, and Winton LM 2007 Population dynamics of the sudden oak death pathogen *Phytop-hthora ramorum* in Oregon from 2001 to 2004, *Molecular Ecology* **16** 2958–2973
100. Reeser PW, Hansen EM, and Sutton W 2007 *Phytophthora siskiyouensis*, a new species from soil, water, myrtlewood (*Umbellularia californica*) and tanoak (*Lithocarpus densiflorus*) in southwestern Oregon, *Mycologia* **99** 639–643
101. Reeser PW, Sutton W, Hansen EM, Remigi P, and Adams GC 2011 *Phytophthora* species in forest streams in Oregon and Alaska, *Mycologia* **103** 22–35
102. Ristaino JB 2011 Key for Identification of Common *Phytophthora* Species, APS Press, St Paul, MN
103. Rizzo DM, Garbelotto M, Davidson JM, Slaughter GW, and Koike ST 2002 *Phytophthora ramorum* as

the cause of extensive mortality of *Quercus* spp and *Lithocarpus densiflorus* in California, *Plant Disease* **86** 205-214

104. Rizzo DM, Garbelotto M, and Hansen EM 2005 *Phytophthora ramorum* Integrative research and management of an emerging pathogen in California and Oregon forests, *Annual Review of Phytopathology* **43** 309-335

105. Saude C, Hurtado-Gonzales O P, Lamour KH, and Hausbeck MK 2008 Occurrence of a *Phytophthora* sp pathogenic to asparagus *(Asparagus officinalis)* in Michigan, *Phytopathology* **98** 1075–1083

106. Savidor A, Donahoo RS, Hurtado-Gonzales O, Land ML, Shah MB, Lamour KH, and McDonald WH 2008 Cross-species Global Proteomics Reveals Conserved and Unique Processes in *Phytophthora sojae* and *Phytophthora ramorum*, *Molecular Cellular Proteomics* **78** 1501-1516

107. Schwingle BW and Blanchette RA 2008 Host range investigations of new, undescribed, and common *Phytophthora* spp isolated from ornamental nurseries in Minnesota, *Plant Disease* **92** 642-647

108. Scott PM, Burgess TI, Barber PA, Shearer BL, Stukely MJC, Hardy GEStJ, and Jung T 2009 *Phytophthora multivora* sp nov, a new species recovered from declining Eucalyptus, Banksia, Agonis and other plant species in Western Australia, *Persoonia* 221–13

109. Stamps DJ, Waterhouse GM, Newhook FJ, and Hall GS 1990 Revised tabular key to the species of *Phytophthora* Mycological Papers 162 CAB International Mycological Institute, Kew, Surrey

110. Sullivan, M J, Parks EJ, Cubeta MA, Gallup CA, Melton T A, Moyer JW, and Shew HD 2010 An assessment of the genetic diversity in a field population of *Phytophthora nicotianae* with a changing race structure, *Plant Disease* **94** 455-460

111. Tjosvold SA, Chambers DL, Koike ST, and Mori SR 2008 Disease on nursery stock as affected by environmental factors and seasonal inoculum levels of *Phytophthora ramorum* in stream water used for irrigation, *Plant Disease* **921** 566–1573

112. Truong NV, Liew ECY, and Burgess LW 2010 Characterisation of *Phytophthora capsici* isolates from black pepper in

Vietnam, *Fungal Biology* **114** **(2-3)** 160-170

113. Tyler BM 2007 *Phytophthora sojae* root rot pathogen of soybean and model oomycete, *Molecular Plant Pathology* **8** 1-8

114. Tyler BM 2006 *Phytophthora* genome sequences uncover evolutionary origins and mechanisms of pathogenesis, *Science* **313** 1261-1266

115. Vercauteren A, De Dobbelaere I, Grünwald NJ, Bonants P, Van Bockstaele E, Maes M, and Heungens K 2010 Clonal expansion of the Belgian *Phytophthora ramorum* populations based on new microsatellite markers, *Molecular Ecology* **19** 92-107

116. Vettraino AM, Brasier CM, Brown AV, and Vannini A 2011 *Phytophthora himalsilva* sp nov an unusually phenotypically variable species from a remote forest in Nepal, *Fungal Biology* **115** 275 -287

117. Vleeshouwers V, Rietman H, Krenek P, Champouret N, Young C, Oh SK, Wang MQ, Bouwmeester K, Vosman B, Visser RGF, Jacobsen E, Govers F, Kamoun S, and Van der Vossen EAG 2008 Effector genomics accelerates discovery and functional profiling of potato disease resistance and *Phytophthora infestans* avirulence genes, *PLoS ONE* **3** e2875

118. Wang Y, Meng Y, Zhang M, Tong X, Wang Q, Sun Y, Quan J, Govers F, and Shan W 2011 Infection of *Arabidopsis thaliana* by *Phytophthora parasitica* and identification of variation in host specificity, *Molecular Plant Pathology* **12** 187-201

119. Waterhouse GM 1963 Key to the species of *Phytophthora* de BaryMycol Pap, No 92, 1–22 CMI, Kew, UK

120. Werres S, Marwitz R, Man in 't Veld WA, de Cock AWAM, Bonants PJM, de Weerdt M, Themann K, Ilieva E, and Baayen RP 2001 *Phytophthora ramorum* sp nov, a new pathogen on *Rhododendron* and *Viburnum, Mycological Research* **105** 1155–1165

121. Werres S, Wagner S, Brand T, Kaminski K, and Seipp D 2007 Survival of *Phytophthora ramorum* in recirculating irrigation water and subsequent infection of *Rhododendron* and *Viburnum, Plant Disease* **91** 1034–1044

122. Wolfe MS and Caten CE 1987 Populations of Plant Pathogens Their Dynamics and Genetics (eds), Blackwell Scientific Publications, Oxford UK

123. Zeng HC, Ho HH, and Zheng FC 2009 A survey of *Phytophthora* species on

Hainan Island of South China, *Journal of Phytopathology* **157** 33 – 39

124. Zhang XG, Sun WX, Guo L, Yu JF, and Chang CJ 2003 Genetic and pathogenic variation among tobacco black shank strains of *Phytophthora parasitica* var *nicotianae* from the main tobacco growing in China, *Journal of Phytopathology* **151** 259-266.

*Rev. Plant Pathol. Vol. 6, 2014*
*Indian Society of Mycology and Plant Pathology*
*Scientific Publishers (India), Jodhpur pp. 413-436*

# A POLYPHASIC APPROACH BASED ON MORPHOLOGICAL, MOLECULAR AND SECONDARY METABOLITE PROFILES FOR RESOLVING TAXONOMIC COMPLEXITIES IN *ALTERNARIA*

***P. Chowdappa***

Indian Institute of Horticultural Research, Hessaraghatta Lake Post, Bangalore-560 089

**KEY WORDS :** Alternaria, Embellisia, Nimbya, Stemphylium, Ulocladium

## ABSTRACT

The genus *Alternaria* was erected with *Alternaria alternata* (former name: *A. tenuis*) as main species. It is ubiquitous, found worldwide on a wide variety of substrates. Many species are saprophytes, but most are known as potent pathogens. *Alternaria* have a wide host range as plant pathogens. As postharvest pathogens, *Alternaria* species contribute to extensive losses of our agricultural output due to spoilage. In addition, they are one of the most common airborne allergens. The taxonomy of the genus *Alternaria* is not well defined and has been extensively discussed. As a genus, *Alternaria* encompasses considerable morphological diversity and there have been a number of attempts to organize taxa into subgeneric groupings based on morphological characters. Fourteen morphological groups have been proposed to describe discrete clusters of morphospecies. Subsequent molecular studies have supported many of these groups as monophyletic lineages, termed "species-groups" but not separated species within group due to lack of sufficient genetic information. Secondary metabolite profiling has been shown to be of great value in the differentiation of certain plant pathogenic *Alternaria* spp although not extensively studied. To resolve complexities in *Alternaria* taxonomy, a combined approach based on a multi-gene phylogeny, with comparison made with type specimens, in conjunction with recognisable polyphasic characters, such as morphology, physiology, pathogenicity and secondary metabolites is proposed.

## INTRODUCTION

The genus *Alternaria* was erected in 1917 by Nees von Esenbeck with *Alternaria alternata* (former name: *A. tenuis*) as main species (73). The genus stands in $10^{th}$ position among nearly 2000 fungal genera listed based on the total number of plant host records. The information available in the USDA Fungal Plant Host Index, indicated that over 4000 Alternaria/host associations are recorded. The Index fungorum shows 491 records of *Alternaria* species (11). Now, the genus *Alternaria* includes nearly 100 species of dematiaceous hyphomycetes that occur worldwide in a variety of habitats (78). Several species are saprobes and can found on wide range of substrates such as sewage, leather, wood pulp, paper, textiles, building supplies, stone monuments, optical instruments, cosmetics, computer disks, jet fuel, soil and dead or dying plant tissue. As plant pathogens, they cause a range of economically important diseases on variety of crops such as cereals, ornamentals, oil-yielding, vegetables and fruit crops (73). They also cause contamination of food and feed by production of mycotoxins (31,73,103). *Alternaria* species are also emerging as important human pathogens, particularly in immunocompromised patients (4,72) and spores as potent airborne allergens (38,102). Numerous *Alternaria* species have been found to be associated with infections of the cornea, oral and sinus cavities, respiratory tract, nails, and skin of humans. *Alternaria* species are known to produce toxic secondary metabolites varying toxicity. Some of these toxic metabolites have been known to cause certain forms of cancer. The occurrence of *Alternaria* secondary metabolites in food materials such as grains, peanuts, tomato products, apple sauce, olive oil, and fresh fruits and vegetables is a major concern (58). *Alternaria* spp are widely distributed, extremely variable and have been described under different names and their identification becoming difficult. The identification and classification of *Alternaria* species were primarily based on morphological characteristics of conidia and conidiation patterns. Adopting these criteria, *Alternaria* species were placed into several species groups (78). Several researchers questioned the use of morphological characters as sole criteria (6,7,45,59,62,73) for distinguishing species due to presence of considerable variation in conidium size, shape and septation within a species and single colony (78). More over, the conidial characters over lap among the closely related species, making the identification process difficult and also hindering the creation of distinct species boundaries (73). A Precise and correct identification of *Alternaria* species is very crucial for taxonomic purposes and formulating effective control strategies. An integrated approach of morphological and molecular criteria as well as secondary metabolite profiles could be the key for a better discrimination and identification of species (47).

## ALTERNARIA TAXONOMY: HISTORICAL PERSPECTIVES

The genus *Alternaria* was described with *A. tenuis* as the type species (55). The production of chains of dark-coloured multi-celled conidia with longitudinal and transverse septa (phaeodictyospores) and a beak or tapering apical cells were the morphological identification markers of the genus. Subsequently, the genus was re-described to include additional new species, which did not produce conidia in chains and or have a conidium beak (24,36,54,84,104) but the production of dark-coloured conidia with longitudinal and transverse septa has remained a key taxonomic criterion of the genus. Fries (27) equated Nees' description of *A. tenuis* to *Torula alternata* and erected a new genus, *Macrosporium*, which included several species that had phaeodictyosporic characters with *Alternaria*. Because of this early taxonomic difference, subsequently identified phaeodictyosporic fungi were included in both *Alternaria* and *Macrosporium* and no clear consensus emerged for over 100 years. Later, fungal isolates that did not produce conidia in chains and/or have a conidium beak were placed in "atypical" *Alternaria* and *Macrosporium* species. Further, two additional genera, *Stemphylium* and *Ulocladium* were created. These also were characterized by the production of phaeodictyospores, further complicating the taxonomic position of this group of fungi. To accommodate a growing list of new species, these genera were re-described (24,39,74). Keissler (39) synonymized both *A. tenuis* of Nees and *T. alternata* of Fries with *Alternaria alternata* Kiessl. nov. nom to avoid ambiguity. Wiltshire (104) further revised *Alternaria* and *Stemphylium*. *Macrosporium* was placed on the list of nomina ambigua. Many of the "atypical" species were included in a subgenus, *Pseudostemphylium* under *Stemphylium*. *Ulocladium* was not formally recognized. After reviewing the genus *Alternaria* and related species, Joly (36) suggested that most of these "atypical" species should be classified as *Alternaria* rather than *Pseudostemphylium*. Simmons (84) reviewed *Alternaria*, *Stemphylium*, and *Ulocladium*, and concluded that many of the "atypical" *Alternaria* and *Stemphylium* species should be placed under *Ulocladium*.

Most *Alternaria* species do not have known teleomorphic affinities except *A. infectoria* and closely related species, However, a number of anamorphically-defined taxa within the *Pleosporaceae* have recognized teleomorphs. These teleomorphs are representative of nearly all major lineages within the *Pleosporaceae*. Many species of *Stemphylium*, the closely related group to *Alternaria*, have well established teleomorphs in the genus *Pleospora*. Based on detailed teleomorph and anamorph characters of *Pleospora* and *Stemphylium*, Simmons (82) created new genus *Lewia* for Pleospora-like fungi with *Alternaria* teleomorphs, which is morphologically distinct from *Pleospora* spp with *Stemphylium* anamorphs. Simmons (83,85) described two additional anamorphic based genera, *Embellisia* and *Nimbya*. Species in both genera contains characteristics of conidia that are morphologically reminiscent of both phragmosporic *Helminthosporium* species and dictyosporic *Alternaria* species. Moreover, numerous species in both genera are synonyms of earlier described species of *Helminthosporium*, *Alternaria* or other morphologically similar genera.

Simmons (85) erected the genus *Embellisia* originally to accommodate the atypical *Helminthosporium allii* Campanile, causal agent of garlic bulb canker based on the production of conidiophores as successive sympodial proliferations by *H. allii* compared to the erect growth of *Helminthosporium* conidiophores. This pattern was more similar to those produced by *Curvularia*,

*Bipolaris*, *Drechslera*, *Alternaria* and *Ulocladium* (2,30,76,84). Like *Curvularia* spp., conidia of *H. allii* variously can be curved and branched (12). But unlike *Curvularia*, the curvature is not generated by the inequilateral growth of a single central cell (85). Moreover, the production of significant numbers of dictyoconidia among a majority of phragmoconidia by *H. allii* was inconsistent with the strict production of phragmoconidia by *Curvularia*, *Bipolaris* and *Drechslera* and more resembled those produced by *Alternaria* and *Ulocladium*. Then, erection of genus *Embellisia* with *E.allii* (Campanile) as type species was justified (85) based on combination of several characters including thick, dark and rigid conidium septation in contrast to the external wall of the conidium. Later, 19 additional species of *Embellisia* have been described or transferred from other genera (19,21,22,53,81,85,86).

The genus *Nimbya* was erected with *N. scirpicola* (Fuckel) Simmons as type species to accommodate the atypical isolates of *Sporidesmium scirpicola* Fuckel, a pathogen of river bulrush (83). Earlier, this fungus has been described as the species as *Sporidesmium*, *Clasterosporium*, *Cercospora* or *Alternaria* (29,74,83,88,106). Simmons (1989) indicated that *Sporidesmium scirpicola* was the earliest taxa described with a combination of conidium characters suggestive of both *Alternaria* and *Drechslera*. Since its inception, 13 additional taxa have been added to *Nimbya*, several of which formerly were recognized in other genera (14,35,77,79,80,83).

Several *Embellisia* spp. have notable teleomorphs, including the former *A. eureka*. (a close relative of *A. infectoria*), and these teleomorphs are morphologically similar to *Lewia* spp. Therefore, to preserve exclusive relationships between teleomorph and anamorph, it was proposed that *Pleosporaceae* spp. with *Embellisia* anamorphs be considered members of a new genus *Allewia* (79).

## MORPHOLOGICAL TAXONOMY

The *Alternaria* and related genera and species within genera could be distinguished based on morphological traits that include size, length/breadth ratio, surface ornamentation of the conidium, presence and length of conidium beak, conidium septation and mode of catenulation (81,84,85)

***Alternaria:*** Classification and identification of *Alternaria* species have been based primarily on morphological characteristics of conidia and conidiation patterns. Based upon common conidium characteristics of length, width, and septation, the genus was organized into six groups, with each group designated by a typical species (24). Later, the genus was divided into 3 sections (54) based upon the formation of conidia in long chains (Longicatenatae), short chains (Brevicatenatae), or singly (Noncatenatae). Utilizing both the concepts (24,54), the genus was further organized into 14 species-groups based upon characteristics of conidia and catenulation (78). Additional species-groups, the alternata group, the tenuissima group and the infectoria group arborescens, brassicicola, porri and radicina groups were created (68,71,79,87).

***Embellisia:*** Generally, conidia are borne singly in most of the species even though the production of secondary conidia is recorded for several species. However, conidia often are borne in chains up to 10–12 conidia in length in *E. indefessa* (81) as a exception. The other combination of characters that are diagnostic for *Embellisia* such as large number of phragmoconidia with a low percentage of dictyoconidia, dominant population of straight-elliptical or oblong-elliptical conidia with variously swollen, smoothly curved or sigmoid conidia as exceptions, presence of conidium production pores at conidiophore geniculations, and hyphal coils in culture with intra hyphal proliferating chlamydospores (81,85). A key identifying character is the presence of conidial septation that distinctly contrasts with the external wall of the conidium.

***Nimbya*** : The species of *Nimbya* produce one to several conidia on closely geniculate conidiophores. Conidia generally are solitary or rarely catenate, long obclavate or short to long rostrate. Most conidia are porogenic; however, in some species conidia are holoblastic. Initially the conidia are distophragmoseptate becoming rarely, partially or completely euphragmoseptate, which is a key taxonomic character and clearly distinguishes *Nimbya* from *Drechslera*, *Alternaria* and other related genera (83). Conidia often are distinctly constricted at eusepta. Longitudinal or oblique septa occasionally are produced, similar to *Alternaria* and *Ulocladium*.

***Stemphylium*:** The genus *Stemphylium* was established with *S. botryosum* as the type species (101). The diagnostic criteria for species of *Stemphylium* include muriform, septate, usually pigmented conidia produced on a conidiophore that proliferates percurrently. The percurrently proliferating conidiophore is the principal morphological characteristic that distinguishes this genus from other genera with muriform conidia such as *Ulocladium* and *Alternaria* (84). The sexual state of *Stemphylium* is *Pleospora*.

***Ulocladium*:** Conidiophores are simple or branched or straight or flexuous, strongly geniculate, leading to a zigzag appearance and bear the conidia. Conidia are brown to black, round to oval in shape, smooth, verrucous and rough-walled. Conidia are most or less egg-shaped and have a narrower base compared to its apex. The conidia are typically muriform and divided into several cells by transverse and longitudinal walls. Most species of *Ulocladium* are typically solitary (*Ulocladium botrytis*), but secondary conidia on short secondary conidiophores can give rise to short chains (*Ulocladium chartarum*). *Ulocladium* differs from *Alternaria* by its strongly geniculate conidiophores, and the absence of beak-like tapered apex of conidia. It differs from *Bipolaris*, *Curvularia*, and *Drechslera* by producing muriform conidia. *Ulocladium* is differentiated from *Stemphylium* by having geniculate, sympodial conidiophores.

## MOLECULAR TAXONOMY

Based on morphological characters of conidia and conidial formation patterns the *Alternaria* species were placed into 14 species-groups (78). These species-groups can not resolve definite species boundaries within *Alternaria*. But, the species-group concept has provided an important framework for hypothesis-testing in advanced studies on *Alternaria* phylogeny. The use of morphological characters as the sole determinants for delimiting species has been debated (6,7,45,59,62,73) due to the presence of considerable variability in conidium size, shape, and septation within each species as well as within a single *Alternaria* colony (78). Further, conidium characteristics often overlap between closely related species, hindering the establishment of distinct and unambiguous species boundaries.

Small-spored species within the alternata species group are particularly challenging because few morphological characters are able to clearly differentiate taxa and these characters are strongly influenced by the environment In addition to morphology, variation exists in respect to host range. Examples of host-specific small-spored taxa include *A. mali* on apple, *A. gaisen* on pear, *A. longipes* on tobacco, *A. alternata* pathotype strawberry on strawberry, *A. citri* (pathotypes tangerine and rough lemon) on citrus and *A. alternata* f. sp. *lycopersici* on tomato. As with *A.tenuissima* and *A. infectoria*, many of the conidium characteristics of these taxa overlap those of *A. alternata*, and some researchers have suggested that many of the host-specific taxa should be referred to as pathotypes or as formae speciales of *A. alternata* (73).

With the advancement of molecular techniques, several studies have examined taxonomic relationships among small-spored catenulate *Alternaria* spp. using a variety of methods in an attempt to establish consensuses with contemporary morphological-based species. Analyses of DNA-DNA re-association kinetics revealed high DNA homology among several small-spored host-specific *Alternaria* spp. and *A. alternata*, which suggests these species represent intra-specific taxa (43). Supporting these results, restriction fragment length polymorphism (RFLP) analyses using the phage clone Alt1 as a hybridization probe were not able to differentiate small-spore host-specific *Alternaria* spp. from *A.alternata* (44). However, RFLP analyses using a fragment of the Alt1clone were able to segregate small-spored *Alternaria* isolates from pistachio into three distinct groups, although critical morphological characterization of these isolates was not performed (8). Sequence analysis of the nuclear r DNA internal transcribed spacer(ITS) region has revealed variation (1 to 4 nucleotides) among small-spored *Alternaria* spp., but failed to resolve these taxa as phylogenetically distinct from *A. alternata* (45). However, random amplified polymorphism DNA (RAPD) analyses of *A. alternata*, *A. tenuissima*, *A. infectoria*, and small-spored host specific-species revealed distinctive RAPD fragment patterns for all species, and cluster analysis did resolve these species into distinct clades, which suggests that these taxa constitute well-defined species. Based on morphological characteristics, the pistachio isolates were grouped as identical or very similar to either *A. alternata*, *A. tenuissima*, *A. arborescens*, or *A. infectoria* (69). Isolates from the *alternata*, *tenuissima* and *arborescens* species-groups were pathogenic to pistachio and no significant differences in pathogenicity were observed. Isolates from the infectoria species-group were only weakly pathogenic to pistachio. Based on cluster analysis of RAPD and PCR-RFLP data, three distinct clusters were evident; the *infectoria* cluster, the *arborescens* cluster, and a combined *alternata/tenuissima* cluster. Based on analysis of ITS sequence data, the infectoria species-group was phylogenetically distinct from the other species-groups. Isolates of the *alternata*, *tenuissima*, and *arborescens* species-groups comprised a monophyletic clade in which the three species-groups could not be further resolved.

The phylogenetic analyses based on nuclear internal transcribed spacer (ITS), mitochondrial small-subunit (mt SSU), glyceraldehyde 3-phosphate dehydrogenase (gpd), mitochondrial large-subunit (mt LSU), ß-tubulin, endo-polygalacturonase (endo-PG) and anonymous genomic regions (OPA1-3 and OPA2- 1) sequence have supported the organization of *Alternaria* species into distinct species-groups (15,23,31,61,65,66). Composition of species-groups was relatively well supported, but relationships among species within the same species group were often not well resolved because of insufficient genetic information contained within the loci chosen for genetic analysis. Based on ITS and mt SSU r DNA sequences, typical members of *Alternaria* and *Ulocladium* formed a large monophyletic group with *Stemphylium* as the sister group (66). Within the large *Alternaria/Ulocladium* clade, species were clustered into distinct species-groups; the porri clade with *A. porri*, *A. solani*, *A. dauci*, *A. crassa* and *A. macrospora* ;the brassicicola clade with *A. brassicicola*, *A. japonica*, and *U.alternariae* ; the *Ulocladium* clade with *U. atrum*, *U. botrytis*, *U.chartarum*, and *A. cheiranthi* ; the alternata clade with *A.alternata* and *A. tenuissima* ; the radicina clade with *A. radicina*,*A. carotiincultae*, *A. petroselini*, *A. selini* and *A. smyrnii* ; and the infectoria clade with *A. infectoria* and *P. infectoria,* which corresponded fairly well with previously described species-groups based on morphology (78,79).

Based on restriction site information from the intergenic spacer (IGS) region of nuclear rDNA, five species of the alternata species-group comprised a monophyletic group supported by high bootstrap value (31). Among them, *A. alternata, A. tenuissima* and *A. arborescens* comprised a monophyletic group leaving the other two species, *A. gaisen* and *A. longipes* as sister taxa. *A. brassicicola* and *A. mimicula* comprised a monophyletic group. *A. carotiincultae* and *A. radicina* comprised another monophyletic group. Although these two monophyletic groups were grouped as sister clades, bootstrap support for this relationship was very low. Two other species that shared all restriction sites, *A. petroselini* and *A. selini* comprised another monophyletic group. Relationships among other species except those mentioned above were not clear. Although *A. crassa* and *A.tagetica* of the porri species-group clustered together and *A. japonica* of the brassicicola species-group and *A. dauci* of the porri species-group clustered together, bootstrap support for these groupings were very low.

The four genera such as *Alternaria, Nimbya, Embellisia and Ulocladium* formed a large monophyletic clade with *Stemphylium* as the sister taxon based on analysis of ITS, mt SSU,and *gpd* sequences (15,23,65,66). It was also revealed that *Alternaria, Embellisia*, and *Ulocladium* are polyphyletic, and eight species-groups were proposed based on monophyletic relationships. Most of the species-groups were supported by high bootstrap values, especially the alternata, infectoria, porri, and radicina species-groups. Analyses of Alt a 1 gene and *gpd* exon sequences also strongly supported grouping of *Alternaria* spp. and related taxa into several species-groups described in previous studies, especially the infectoria, alternata, porri, brassicicola, and radicina species-groups and the *Embellisia* group. The sonchi species group was newly suggested (31).

Analysis of the relatively conserved nuclear 18S sequence has been used to describe family-genus relationships within the *Pleosporales*, and revealed a close phylogenetic relationship between *Pleospora* (anamorph *Stemphylium*) and *Alternaria* (9,52). Molecular phylogenies have supported the division between the alternata and porri species-groups, morphologically distinguished by small and large conidial sizes, respectively. Conidia produced by fungi in the alternata species-group are 20–50 mm long, while conidia produced by the porri species-group are generally more than 100 mm long (95).

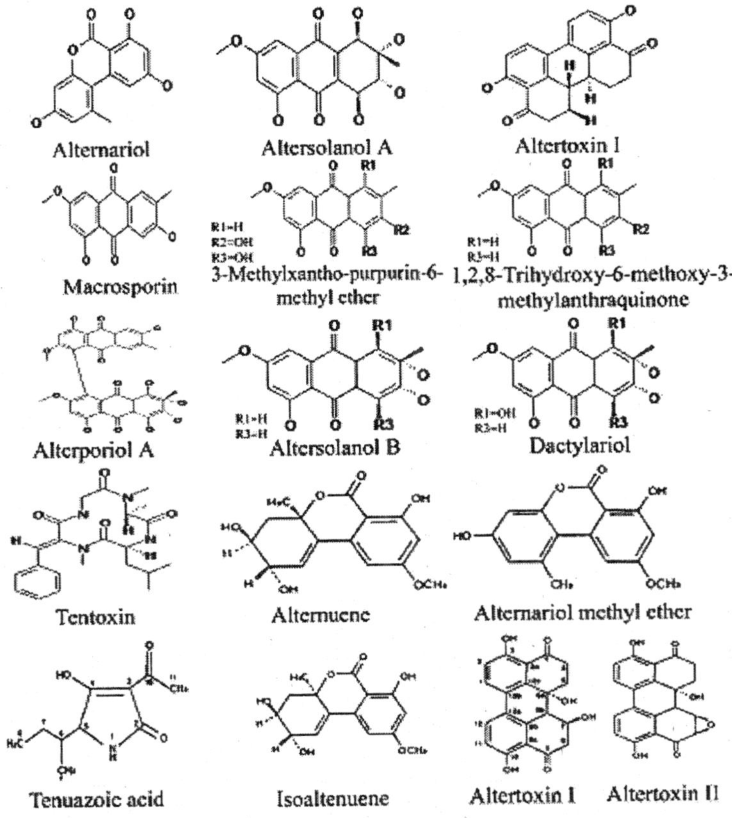


**Fig.1**. Chemical structures of the known secondary metabolites produced by *Alternaria* spp

## CHEMOTAXONOMY

Many plant pathogenic *Alternaria* spp are known producers of toxic secondary metabolites, mycotoxins (10,13,40,61), which are of low molecular weight, but are not necessary for growth or development (25). *Alternaria* species are known to produce a wide diversity of secondary metabolites belonging to different chemical groups including dibenzopyrones, tetramic acids, lactones, quinones and cyclic peptides. The *Alternaria* species are capable to produce more than 120 secondary metabolites (60). However, more than 30 toxic secondary metabolites have been isolated and characterized from *Alternaria* species (58).They are mostly host-specific or non-host specific phytotoxins, which play an important role in the pathogenesis of plants (17,34,46,48,49, 64,70,94,95, 96,97,98,100). Several reviews on the *Alternaria* mycotoxins toxins have been published (10,13,28,58,60,75,99). Among the *Alternaria* mycotoxins, alternariol (AOH), alternariol monomethyl ether (AME), altenuene (ALT), altertoxins I, II, III (ATX-I, -II, -III) and tenuazonic acid (TeA) are the most important ones, which belong to three structural classes (10), dibenzopyrone derivatives (AOH, AME, ALT), perylene derivatives (ATX-I, -II, -III) and tetramic acid derivatives (TeA). The chemical structures some of the known secondary metabolites of *Alternaria* spp are elucidated (Fig. 1).

The use of secondary metabolite profiling as chemotaxonomic markers, has been demonstrated in the classification and identification of species in many genera in Ascomycota (28). Even though a large amount of work has been done secondary metabolites of pathogenic *Alternaria* spp.(51,73), only few studies have employed metabolite profiles for taxonomic purposes (5,6). Production of secondary metabolites (mycotoxins) by *Alternaria* spp is shown in Table.1. A large number of small-spored *Alternaria* isolates have been misidentified as *A. alternata* based solely on conidial size (1,42,56,57,59,73) as squash mounts disrupts any evidence of sporulation pattern by breaking up the conidial chains. Several valid species, such as *A. longipes*, *A. gaisen* (syn. *A. kikuchiana*), *A. citri* and *A. mali*, have been referred to as special forms or pathotypes of *A.alternata* (56,59) because they have similar spore ranges. However, 3- dimensional sporulation patterns have been introduced for segregation and identification of small spored *Alternaria* isolates from lesions of Japanese pear (87). Later, it has been reported that RAPD-PCR could discriminate between different sporulation patterns when anlaysed using same isolates (71). Using RAPD-PCR, various species of *Alternaria* have also been separated including *A. longipes* and *A. alternata* (16). *A. alternata* has been reported to produce altenuene, alternariol, alternariol monomethyl ether, altertoxin I, tentoxin and tenuazonic acid (10). *A. gaisen* produces 3,6,8-trihydroxy-3-methyl-3,4-dihydroisocoumarin, alternariol, alternariol monomethyl ether, tentoxin and tenuazonic acid (10,37,51). *A. longipes* has only been reported to produce alternariol monomethyl ether, tenuazonic acid and tentoxin (10,51). *A. longipes* and *A.gaisen* can produce host-specific toxins,

AT-toxin and AK-toxin, respectively (42,57). Based on analyses of morphological, molecular and secondary metabolite profiles, that the correct names, *A. gaisen* and *A. longipes*, should be used instead of the misleading terms `*A. alternata* Japanese pear pathotype' and *A. alternata* tobacco pathotype (6). *A. tenuissima* has been reported to produce alternariol, alternariol monomethyl ether, altenuene, altertoxin I and tenuazonic acid (10,51). *A. infectoria* species-group is chemically very different from both the *A. arborescens* and the *A. tenuissima* species-groups with only a few metabolites in common (7). The *A. arborescens* was able to produce altenuene, alternariol, alternariol monomethyl ether; tenuazonic acid, altertoxin I, and tentoxin. *A. tenuissima* produce altenuene, alternariol, alternariol monomethyl ether; tenuazonic acid, altertoxin I, and. tentoxin. None of isolates belonging to *A.infectoria* were able to produce any of these secondary metabolites even though alternariol monomethyl ether and tenuazonic acid were reported (10).

**Table 1.** Production of secondary metabolites (mycotoxins) by *Alternaria* spp

| ***Alternaria* species** | **Secondary metabolite** | **Source** |
|---|---|---|
| *A. solani* Sorauer | Alternaric acid | 5,32,51,90,91,92,93 |
| | Altersolanols | |
| | Altertoxin I | |
| | Alterporriols | |
| | Macrosporin | |
| | Tentoxin | |
| | Alternariol | |
| | Alternariol monomethyl ether | |
| | Solanapyrones | |
| | Tenuazonic acid | |
| | Zinniol | |
| | Zinnolide | |
| *A. porri* (Ell.) Cif. | Alterporriols | 10,32,51,92,93 |
| | Altersolanols | |
| | Alternariol monomethyl ether, | |
| | Anthraquinones | |

| | | |
|---|---|---|
| | Erythroglaucin<br>Tentoxin<br>Tenuazonic acid<br>Zinniol | |
| *A. dauci* (Kühn) Groves & Skolko | Alternariol<br>Alternariol monomethyl ether<br>Zinniol | 10,26,32,51 |
| *A. tomatophila* | Altersolanol , Altertoxin I,<br>Macrosporin. | 5 |
| *A. brassicae* (Berk.) Sacc. | Alternariol,<br>Alternariol monomethyl ether | |
| *A. brassicicola* | Alternariol,<br>Alternariol 9-methyl ether, Altechromone A, Herbarin A,<br>Ccerevisterol,<br>3β,5α-dihydroxy-(22E,24R)-ergosta-7,22-dien-6-one,<br>3β-hydroxy-(22E,24R)-ergosta-5,8,22-trien-7-one | 30 |
| *A. alternata* (syn. *A. tenuis*) | Altenuene, Alternariol<br>Alternariol monomethyl ether, Altertoxin I, Tentoxin<br>Tenuazonic acid | 10 |
| *A. capsici-anui* Săvul. & Sandu | Alternariol,<br>Alternariol monomethyl ether, Tenuazonic acid | 10 |
| *A. cassiae* Jurair & A. Khan | Altertoxins -I, -II | 33 |
| *A. citri* Ell. & Pierce | Alternariol,<br>Alternariol monomethyl ether,<br>Tenuazonic acid | 41 |
| *A. cucumerina* (Ell. & Ev.) Elliott | Alternariol,<br>Alternariol monomethyl ether | 26 |
| *A. japonica* Yoshii | Tenuazonic acid | 41 |
| *A. gaisen* (syn.*A. kikuchiana* Tanaka | Altertoxin I, Alternariol,<br>Alternariol monomethyl ether,<br>Tenuazonic acid | 6,10,37,41,51,97 |

| | | |
|---|---|---|
| | 3,6,8-trihydroxy-3-methyl-3,4-dihydroisocoumarin,Tentoxin | |
| *A. longipes* (Ell. & Ev.) | Altertoxin I, Alternariol monomethyl ether, Tenuazonic acid | 10,50 |
| *A. mali* Roberts | Altertoxins-I, -II, -III, Tenuazonic acid | 41 |
| *A. oryzae* Hara | Tenuazonic acid | 41 |
| *A. radicina* Meier, Drechsler & Eddy | Altertoxins-I, -II, -III, Tenuazonic acid | 10,89 |
| *A. tenuissima* (Kunze) Wiltshire | Alternariol, Alternariol monomethyl ether Altertoxins -I, -III, Altenuene Tenuazonic acid | 10,20,63,105 |
| *A.infectoria* | Alternariol monomethyl ether Tenuazonic acid | 10 |

The large spored and long beaked Alternaria species namely *A. carthami, A. macrospora, A. porri, A. solani* and *A. tagetica* produces zinniol as their common metabolite where as other species lacking large spores and long beaks (*A. alternata, A. citri* and *A. raphani*) did not produce (18,32,51). Alterporriols, altersolanols, macrosporin, and tentoxin have been reported from *A. porri* and *A. solani* (51,91,92,93). Alternariol monomethyl ether has been reported from *A.dauci*, erythroglaucin and other anthraquinones from *A. porri*, and alternaric acid, alternariol, solanapyrones and zinnolide from *A. solani* (51). The presence of alternariol in *A. dauci* and *A. solani,* altersolanol A and macrosporin in *A. porri* and *A. solani*, tentoxin in *A. porri*, altertoxin in *A. solani* and altersolanol, A, altertoxin, and macrosporin in *A. tomatophila*. Alternariol was not detected in *A. porri* and altersolanol A, altertoxin, macrosporin, and tentoxin were not detected in *A. dauci*. Altenuene, erythroglaucin, and tenuazonic acid were not detected in any of the four species (5).

## THE FUTURE

The identification of *Alternaria* species should be simple and rapid as it has important implications in biosecurity, disease management and disease resistance programmes. A major effort is under way to establish species concepts and formalise names in current use. It is unlikely that morphology alone will resolve many problems in species separation, but a polyphasic approach can be used to establish species boundaries, for epitypification of existing names and for the introduction of new species. Ideally, a single gene needs to be identified that can be used for development barcodes for quick identification of species in simple way. A polyphasic approach to classification, integrating chemical, physiological and molecular data to traditional morphology gives a more precise means of identification. We have examined 12 *Alternaria* species associated with various crops in India based on multilocus gene phylogeny and secondary metabolite profiling using thin layer chromatography, high performance chromatography and MALDI-TOF MS in addition to morphology and the results are at various stages of publication.

## REFERENCES


1. Akamatsu H, Taga M, Kodama M, Johnson R, Otani H, and Kohmoto K 1999 Molecular karyotypes for *Alternaria* plant pathogens known to produce host-specific toxins, *Current Genetics* **35** 647-656
2. Alcorn JL 1988 The taxonomy of "*Helminthosporium*" species, *Annual Review of Phytopathology* **26** 37–56
3. Alcorn JL 1991 New combinations and synonymy in *Bipolaris*and *Curvularia*, and a new species of *Exserohilum*, *Mycotaxon* **41** 329–343
4. Anaissie E J, Bodey GP, and Rinaldi MG 1989 Emerging fungal pathogen, *European Journal of Clinical Microbiology and Infectious Diseases* **8** 323–330
5. Andersen B, Dongo A, and Pryor BM 2008 Secondary metabolite profiling of *Alternaria dauci, A porri, A solani*, and *A tomatophila, Mycological Research* **112** 241-50
6. Anderson B, Kroger E, and Roberts RG 2001 Chemical and morphological segregation of *Alternaria alternata, A. gaisen* and *A. longipes, Mycological Research* **105** 291-299
7. Anderson B, Kroger E, and Roberts RG 2002 Chemical and morphological segregation of *Alternaria arborescens, A. infectoria* and *A. tenuissima* species-groups, *Mycological Research* **106** 170-182

8. Aradhya MK, Chan HM, and Parfitt DE 2001 Genetic variability in the pistachio late blight fungus, *Alternaria alternate, Mycological Research* **105** 300-306

9. Berbee ML 1996 Loculoascomycete origins and evolution of filamentous Ascomycete morphology based on 18S gene sequence data, *Molecular Biology and Evolution* **13** 462–470

10. Bottalico A and Logrieco A 1998 Toxigenic *Alternaria* species of economic importance, in KK Sinha and D Bhatnagar (eds) Mycotoxins In agriculture and food safety Marcel Dekker, Inc, New York, USA, 65-108 pages

11. CABI 2004 Index fungorum Available at http// www. speciesfungorumorg Accessed 21-4-2008

12. Campanile G 1924 Su di una nuova malattia dell'aglio dovuta ad "*Helminthosporium allii*" nov sp Nuovi, *Ann dell'Agric* **4** 86–106

13. Chełkowski J and Visconti A 1992 *Alternaria* Biology, plant diseases and metabolites Elsevier, Amsterdam, the Netherlands,573 page

14. Chen WQ, Lin XF, and Zhang TY 1997 A new species of *Nimbya, Mycosystema* **2** 106–108

15. Chou HH and Wu WS 2002 Phylogenetic analysis of internal transcribed spacer regions of the genus *Alternaria*, and the significance of filament-beaked conidia, *Mycological Research* **106** 164-169

16. Cooke DEL, Forster JW, Jenkins PD, Jones DG, and Lewis DM 1998 Analysis of intraspecific and interspecific variation in the genus *Alternaria* by the use of RAPD-PCR, *Annals Applied Biology* **132** 197-209

17. Coombe RG, Jacobs JJ, and Watson TR 1970 Metabolites of some *Alternaria* species The structures of altenusin and dehydroaltenusin, *Australian Journal of Chemistry* **23** 2343-2351

18. Cotty PJ and Misgahi IJ 1984 Zinniol production by *Alternaria* species, *Phytopathology* **74** 785-788

19. David JC, Coles K, Fisher J, and Moss ST 2000 A new species of *Embellisia* from soil with high levels of heavy metals, *Mycoscience* **41** 533–537

20. Davis ND, Diener UL, and Morgan-Jones G 1977 Tenuazonic acid production by *Alternaria alternata* and *Alternaria tenuissima* isolated from cotton, *Applied and Environmental Microbiology* **34** 155-157

21. de Hoog GS and Muller PJ 1973 A new species of *Embellisia* associated with skin

disease of hyacinths, *Netherlands Journal of Plant Pathology* **79** 85–93

22. de Hoog GS, Seigle-Murandi F, Steinman R, and Eriksson KE 1985 A new species of *Embellisia* from the North Sea, *Journal of Microbiology* **51** 409–413

23. de Hoog GS and Horre R 2002 Molecular taxonomy of the *Alternaria* and *Ulocladium* species from humans and their identification in the routine laboratory, *Mycosis* **45** 259-276

24. Elliott and John A 1917 Taxonomic characters of the genera *Alternaria* and *Macrosporium, American Journal of Botany* **4** 439-476

25. Fox EM and Howlett BJ 2008 Secondary metabolism regulation and role in fungal biology, *Current Opinion in Microbiology* **11** 481-487

26. Freeman GG 1965 Isolation of alternariol and alternariol monomethyl ether from *Alternaria dauci* (Kukn) Groves and Skolko, *Phytochemistry* **5** 719-721

27. Fries E 1832 *Systema Mycologicum*, VolI Officiana Berlingianalund, 520 page

28. Frisvad JC, Andersen B, and Thrane U 2008 The use of secondary metabolite profiling in chemotaxonomy of filamentous fungi, *Mycological Research* **112** 231– 240

29. Fuckel L 1863 Fungi rhenani exsiccate, Fasc I, No 1–100

30. Gu W 2009 Bioactive metabolites from *Alternaria brassicicola* ML-P08,an endophytic fungus residing in *Malus halliana, World Journal of Microbiology and Biotechnology* **25** 1677–1683

31. Hong GS, Liu D, and Pryor, BM 2005 Restriction mapping of IGS regions from *Alternaria* spp and their taxonomic implication, *Mycological Research* **109** 87-95

32. Horiuchi M, Ohnishi K, Iwase N, Nakajima Y, Tounai K, Yamashita M, and Yamada Y 2003 A novel isoindoline, porritoxin sulfonic acid from *Alternaria porri* and the structure phytotoxicity correlation of its related compounds, *Bioscience Biotechnology and Biochemistry* **67** 1580–1583

33. Hradil CM, Hallock IF, Clardy J, Kenfield DS, and Strobek G 1989 Phytotoxins of *Alternaria cassiae, Phytochemistry* **28** 73-75

34. Janardhanan KK and Husein A 1975 Isolation of tenuazonic acid, a phytotoxin from *Alternaria alternata* causing leaf blight of *Datura innoxia* Mill, *Indian Journal of Experimental Biology* **13** 321-322

35. Johnson DA, Simmons EG, Miller JS, and Stewart EL 2002 Taxonomy and

pathology of *Macrospora/ Nimbya* on some North American bulrushes (*Scirpus* spp), *Mycotaxon* **84** 413–428

36. Joly P 1964 Le Genre *Alternaria Encyclope'die Mycol* **33** 1–250
37. Kameda K, Aoki H, Tanaka H, and Namiki M 1973 Studies on metabolites of *Alternaria kikuchiana Tanaka*, a phytopathogenic fungus of Japanese pear, *Agricultural and Biological Chemistry* **37** 2137-2146
38. Karlsson-Borga A, Jonsson P, and Rolfsen W 1989 Specific IgE antibodies to 16 widespread mold genera in patients with suspected mold allergy, *Annals of Allergy, Asthma & Immunology* **63** 521–526
39. Keissler KV 1912 Zun Kenntnis den Pilzflora Krains Beih, *Bot Zbl* **29** 395-400
40. King AD and Schade JE 1984 *Alternaria* toxins and their importance in food, *Journal of Food Protection* **47** 886-901
41. Kinoshita T, Renbutsu Y, Khan I, Kohmoto K, and Nishimura S 1972 Distribution of tenuazonic acid production in the genus *Alternaria* and its phytopathological evaluation, *Annals of the Phytopathological Society of Japan* **38** 397-404
42. Kohmoto K, Otani H, Cavanni P, and Bugiani R 1992 Occurrence of the Japanese pear pathotype of *Alternaria alternata* in Japanese pear orchards in Italy, *Phytopathologia Mediterranea* **31** 141-147
43. Kuninaga S and Yokusawa R 1987 Studies on the taxonomy of plant pathogenic fungi by a comparison of DNA homology Genetic relatedness among species in the genus *Alternaria* (In Japanese), *Annual Phytopathological Society of Japan* **53** 368-369
44. Kusaba M and Tsuge T 1994 Nuclear ribosomal DNA variation and pathogenic specialization in *Alternaria* fungi known to produce host-specific toxins, *Applied and Environmental Microbiology* **60** 3055-3062
45. Kusaba M and Tsuge T 1995 Phylogeny of *Alternaria* fungi known to produce host-specific toxins on the basis of variation in internal transcribed spacers of ribosomal DNA, *Current Genetics* **28** 491-498
46. Liebermann B, Ihn W, Baumann E, and Tresselt D 1988 Dihydrotentoxin related dipeptide produced by *Alternaria alternate, Phytochemistry* **27** 357-359
47. Logrieco A, Moretti A, and Solfrizzo M 2009 *Alternaria* toxins and plant diseases an overview of origin, occurrence and risks, *World Mycotoxin Journal* **2** 129–140

48. Logrieco A, Bottalico A, Solfrizzo M, and Mule G 1990 Incidence of *Alternaria* species in grains from Mediterranean countries and their ability to produce mycotoxins, *Mycologia* **82** 501-505
49. Maekawa N, Yamamoto M, Nishimura S, Kohmoto K, Kawada M, and Watanabe Y 1984 Studies on host-specific AF-toxins produced by *Alternaria alternata* strawberry pathotype causing *Alternaria* black spot of strawberry (1) Production of host- specific toxins and their biological activities, *Annals of the Phytopathological Society of Japan* **50** 600-609
50. Mikami Y, Nishijima T, Imura H, Sužuji A, and Tamura S 1971 Chemical studies on brown-spot disease of tobacco plants Part Itenuazonic acid as vivotoxin of *Alternaria longipes, Agricultural and Biological Chemistry* **35** 611-618
51. Montemurro N and Visconti A 1992 Alternaria metabolites – chemical and biological data, in Chelkovski J, Visconti A (eds), *Alternaria Biology, Plant Diseases and Metabolites* Elsevier, Amsterdam, 449–557 pages
52. Morales VM, Jasalavich CA, Pelcher LE, Petrie GA, and Taylor JL 1995 Phylogenetic relationship among several *Leptosphaeria* species based on their ribosomal DNA sequences, *Mycological Research* **99** 593–603
53. Muntanola-Cvetkovic M and Ristanovic B 1976 A new species of *Embellisia* isolated from seawater, *Mycologia* **68** 47–51
54. Neergaard P 1945 Danish species of *Alternaria* and *Stemphylium,* Copenhagen Einar Munksgaard
55. Nees von Esenbeck CG 1816 *Das System der Pilze und SchwaXmme* Stahelschen Buchhandlung, Wu$ rzburg
56. Nishimura S, Sugihara M, Kohmoto K, and Otani H 1978 Two different phases in pathogenicity of *Alternaria* pathogen causing black spot disease of Japanese pear, *Journal of the Faculty of Agriculture, Tottori University* **13** 1-10
57. Nishimura S 1980 Host-specific toxins from *Alternaria alternata* Problems and prospects, *Proceedings of the Japan Academy* **56** 362-366
58. Ostry V 2008 *Alternaria* mycotoxins an overview of chemical characterization, producers, toxicity, analysis and occurrence in foodstuffs, *World Mycotoxin Journal* **1** 175-188
59. Otani H and Kohmoto K 1992 Host-specific toxins of *Alternaria* species, in *Alternaria Biology, Plant Diseases and*

*Metabolites* J Chelkowski and A Visconti (ed), Elsevier, Amsterdam 123-156 pages

60. Panigrahi S 1997 *Alternaria* toxins In D'Mello JPF, (eds) Handbook of plant and fungal toxicants CRC Press, Boca Raton, Florida

61. Patriarca A, Azcarate MP, Terminiello L, and Fernandez Pinto V 2007 Mycotoxin production by *Alternaria* strains isolated from Argentinean wheat, *International Journal of Food Microbiology* **119** 219-222

62. Peever TL, Ibanez A, Akimitsu K, and Timmer LW 2002 Worldwide phylogeography of the citrus brown spot pathogen, *Alternaria alternata Phytopathology* **92** 794-802

63. Pero RW and Main CE 1969 Chlorosis of tobacco induced by alternariol monomethyl ether produced by *Alternaria tenuis, Phytopathology* **60** 1570-1573

64. Pero RW, Harvan D, and Blois MC 1973 Isolation of the toxin altenuisol, from the fungus, *Alternaria tenuis* Auct, *Tetrahedron Letters* **12** 945-948

65. Pryor BM and Bigelow DM 2003 Molecular characterization of *Embellisia* and *Nimbya* species and their relationship to *Alternaria, Ulocladium*, and *Stemphylium, Mycologia* **95** 1141-1154

66. Pryor BM and Gilbertson RL 2000 Phylogenetic relationships among *Alternaria* and related fungi based upon analysis of nuclear internal transcribed sequences and mitochondrial small subunit ribosomal DNA sequences, *Mycological Research* **104** 1312-1321

67. Pryor BM and Gilbertson RL 2001 A PCR-based assay for detection of *Alternaria radicina* from carrot seed, *Plant Disease* **85** 18-23

68. Pryor BM and Gilbertson RL 2002 Relationships and taxonomic status of *Alternaria radicina, A carotiincultae,* and *A petroselini* based upon morphological, biochemical, and molecular characteristics, *Mycologia* **94** 49-61

69. Pryor BM and Michailides TJ 2002 Morphological, molecular, and pathogenic characterizations of Alternaria isolates associated with *Alternaria* late blight of pistachio, *Phytopathology* **92** 406-416

70. Rizk AM, Hammouda FM, El-Missiry MM, Mayergi HA, Lashin, SM, and Nofal MA 1985 Mycotoxins of *Lolium* seeds in response to fungal infections, *Annals of Agricultural Science* **30** 607-615

71. Roberts RG, Reymond ST, and Anderson B 2000 RAPD fragment pattern analysis and morphological segregation of small-spored

*Alternaria* species and species group, *Mycological Research* **104** 151-160

72. Rossmann SN, Cernoch PL, and Davis JR 1996 Dematiaceous fungi are an increasing cause of human disease, *Clinical Infectious Diseases* **22** 73–80
73. Rotem J 1994 The Genus *Alternaria* biology, epidemiology and pathogenicity, *American Phytopathological Society Press*, St Paul, MN
74. Saccardo PA 1886 Sylloge Fungorum Vol IV Padova Publ by the author 807 page
75. Scott PM 2001 Analysis of agricultural commodities and foods for *Alternaria* mycotoxins, *Journal of AOAC International* **84** 1809-1817
76. Shoemaker RA 1959 Nomenclature of *Drechslera* and *Bipolaris*, grass parasites segregated from "*Helminthosporium*", *Canadian Journal of Botany* **37** 879–887
77. Simmons EG 2000 *Alternaria* themes and variations (244–286)species on Solanaceae, *Mycotaxon* **75**1–115
78. Simmons EG 1992 *Alternaria* taxonomy current status, viewpoint, challenge, in *Alternaria Biology, Plant Diseases and Metabolites,* J Chelkowski and A Visconti, (eds) Amsterdam, Netherlands Elsevier Science Publishers, 1-35 pages
79. Simmons EG 1995 *Alternaria* themes and variations (112–144), *Mycotaxon* **55** 55–163
80. Simmons EG 1997 *Alternaria* themes and variations (151–223), *Mycotaxon* **65** 1–91
81. Simmons EG 1983 An aggregation of *Embellisia* species, *Mycotaxon* **17** 216–241
82. Simmons EG 1986 *Alternaria* themes and variations (22–26), *Mycotaxon* **25** 287–308
83. Simmons EG 1989 *Macrospora* Fuckel (Pleosporales) and related anamorphs, *Sydowia* **41** 314–329
84. Simmons EG 1967 Typification of *Alternaria, Stemphylium*, and *Ulocladium, Mycologia* **59** 67–92
85. Simmons EG 1971 *Helminthosporium allii* as type of a new genus, *Mycologia* **63** 380–386
86. Simmons EG 1990 *Embellisia* and related teleomorphs, *Mycotaxon* **38** 251–265
87. Simmons EG and Roberts RG 1993 *Alternaria* themes and variations (73), *Mycotaxon* **48** 109-140
88. Sivanesan A 1984 The Bitunicate Ascomycetes Vaduz JKramer 701 page
89. Solfrizzo M, Girolamo AD, Vitti C, Tylkowska K, Grabarkiewicz- Szczesna J, Szopinska D, and Dorna H 2005 Toxigenic profile of *Alternaria alternata* and

*Alternaria radicina* occurring on umbelliferous plants, *Food Additives and Contaminants* **22** 302-308

90. Stoessl A 1969 Metabolites of *Alternaria solani, Canadian Journal of Chemistry* **47** 767-776
91. Suemitsu R, Horiuchi K, Horiuchi M, and Hanabata M 1992 Detection of tentoxin by *Alternaria solani, A cucumerina, A bataticola* and *A japonica, Bioscience Biotechnology & Biochemistry* **56** 139
92. Suemitsu R, Horiuchi K, Kubota M, and Okamatsu T 1990a Production of alterporriols, altersolanols and marcosporin by *Alternaria porri* and *A solani, Phytochemistry* **29** 1509–1511
93. Suemitsu R, Ohnishi K, Nobuhara T, Horiuchi M, and Horiuchi K,1990b Studies on the metabolic products of *Alternaria porri* 16Isolation and identification of tentoxin from *Alternaria-porri* (Ellis) Ciferri, *Agricultural & Biological Chemistry* **54** 2449–2450
94. Tadakazu N, Ueno T, Fukami H, TagaT, Masuda H, Osaki K, Otani H, Kohmoto K, and Nishimura S 1985 Isolation and structures of AK-toxins I and II, host-specific phytotoxic metabolites produced by *Alternaria alternata* Japanese pear pathotype, *Agricultural and Biological Chemistry* **49** 807-816
95. Templeton GE, Grable CI, Fulton ND, and Meyer WL 1967 Tentoxin from *Alternaria tenuis* its isolation and characterizationIn USDA (ed) Proceedings of the 1967 mycotoxin research seminar, Washington DC, USA, 27-29 pages
96. Thomma BPHJ 2003 *Alternaria* spp from general saprophyte to specific parasite, *Molecular Plant Pathology* **4** 225-236
97. Tirokata H, Ohkawa M, Sassa T, Yamada T, Ohkawa H, Tanaka H, and Aoki H 1969 Studies on resistance of Japanese pears to black spot disease fungus (*Alternaria kikuchiana*) VIII Alternariol and its monomethyl ether, *Annals of the Phytopathological Society of Japan* **35** 62-66
98. Ueno T, Nakashima T, Hayashi Y, and Fukami H 1975 Structures of AM-toxin I and II, host-specific phytotoxic metabolites produced by *Alternaria mali, Agricultural and Biological Chemistry* **39** 1115-1122
99. Visconti A and Sibilia A 1994 *Alternaria* toxins In Miller, JD and Trenholm, HL (eds) Mycotoxins in grain Eagan Press, St Paul,MN, USA, 315-336 pages

100. Visconti A, Bottalico A, Solfrizzo M, and Palmisano F 1988 Isoaltenuene, a new metabolite of *Alternaria alternate, Proceedings of the Japanese Association of Mycotoxicology* **1** 139-140

101. Wallroth FG 1833 Flora Cryptogamica Germaniae, pars post Nurenberg J L Schrag 923 page

102. Wilken-Jensen K and Gravesen S 1984 Atlas of Moulds in Europe Causing Respiratory Allergy Foundation for Allergy Research in Europe, ASK Publishing, Copenhagen, Denmark

103. Wilson CL and Wisniewski M E 1994 Biological Control of Post harvest Diseases theory and practice CRC Press, Boca Raton, FL

104. Wiltshire SP 1933 The foundation species of *Alternaria* and *Macrosporium, Transactions British Mycological Society* **18** 135-160

105. Young AB, Davis ND, and Diener UL 1980 The effect of temperature and moisture an tenuazonic acid production by *Alternaria tenuissima, Phytopathology* **70** 607-609

106. Zinderen Bakker van EM 1940 *Cercospora scirpicola* (Sacc) nov comb, *Revue de Mycology* **5** 64–69.

*Rev. Plant Pathol. Vol. 6, 2014*
*Indian Society of Mycology and Plant Pathology*
*Scientific Publishers (India), Jodhpur pp. 437-472*

# FOLIAR BLIGHT: THE MAJOR BIOTIC CONSTRAINT OF WHEAT IN RICE WHEAT SYSTEMS OF EASTERN GANGETIC PLAINS

***A.K. Chowdhury, P.M. Bhattacharya and Gyanendra Singh[1]***

Department of Plant Pathology, Uttar Banga Krishi Viswavidyalaya, Coochbehar, West Bengal, India;
[1]Directorate of Wheat Research, ICAR, Karnal 132001, Haryana, India

**KEY WORDS:** wheat, foliar blight, spot blotch, yield loss. Disease resistance, management

## ABSTRACT

Wheat, the winter cereal of Indo – Gangetic plains, is a major contributor to the agricultural economy of India, occupying nearly 29.5 x $10^6$ ha with total production around 85.93 x $10^9$ kg (2010-11 crop season). Although the production has increased considerably, yet the actual average yield continues to be low and regional disparity exists. One of the reasons of yield disparity is poor disease management, particularly the foliar blight of wheat, and a serious problem in eastern India. Among the different foliar blights, Spot blotch is gaining attention due to its fast spread and severity in the areas where it was not very important previously. Typically symptoms of *Bipolaris sorokiniana* occur in warm and humid environments but recently it has spread to moderate and temperate regions where irrigated rice – wheat cropping system area is on increase. The development of resistant varieties seems to be only viable option for effective management of spot blotch in

India. The disease resistance strategies have gained a key role in modern production systems.

## INTRODUCTION

Wheat is now cultivated on over 24 to 27 million hectares in India, depending on the moisture availability at the post monsoon phase. During the last decade an additional area of more than 1.5 million hectares was added to wheat in the Central Zone and North East Plain Zone. Area expansion also recorded in the Peninsular zone and North West Plain zone but was limited to 0.2 million hectares only. Out of the 25.0 million hectares, 9.0 million hectares is exposed to various types of leaf diseases other then rusts and powdery mildew. Of this, it is estimated that more than 6.0 million hectares is in the North East Plain zone or the eastern drain of the Indus – Ganges basin where rainfed upland rice, lowland or medium lowland rice precedes wheat (88). During the last two decades there has been a substantial increase in rice – wheat sequence (32). Apart from delayed wheat sowing under difficult field conditions where the crop stand is invariably poor, the high level of organic matter (rice stubble) and poor nutrient application results in spot blotch due to *Bipolris sorokiniana*. In the early days, when the disease intensity was lower, the soil become solarized during the summer, reducing the population of the organism as evidenced in solarization experiments conducted in Nepal (27). The solarized soil invariably promoted better rooting of wheat and decreased foliar blight severity. The increased blight severity in eastern India is, therefore, a second, generation problem arising from crop intensification. The disease is also prevalent in warmer growing areas of eastern plains of South Asia characterized by an average temperature in the coolest month above 17°C (26,29,36,59,110). At present, control measures are predominantly based on chemicals, however, a holistic approach is needed to combat this serious disease.

## YIELD LOSSES

Yield losses due to foliar blights are variable but are considered to be very significant. The disease affects all plant parts and cause up to 100 per cent yield loss (74). There is a growing body of evidence that highlights increased yield losses due to *Helminthosporium* leaf blight (HLB). While Duveiller and Gilchrist (36) reported that losses up to 20 per cent in farmer's fields in several areas of South-East Asia but same is up to 30 per cent in experimental studies. (27). Dubin and van Ginkel (29) conducted numerous trials and estimated the losses which may exceed 20 per cent, depending on cultivar susceptibility and environmental fluctuation. Older cultivated varieties released prior to 1970's and varieties susceptible to leaf rust tended to suffer greater losses.

At Poza Rica, in Mexico which is the hot spot for screening resistance to spot blotch, 49-90 per cent yield losses have been recorded despite the fact that, wheat is not grown as commercial crop (35). In the Mixteca region of Mexico, on-farm trials under severe natural infection by tan spot indicated plots under zero tillage suffer loss in yield around 37 per cent even with one spray of propiconazole (96). In Argentina, tan spot is recognized as major leaf blight and severity level beyond 50 per cent is a common feature with potential yield losses up to 10-20 per cent (4). In Paraguay, during 1972 losses due to the disease complex in the wet year reached as high as 70 per cent (101,141). In Bolivia, yield losses up to 57 per cent have been observed but are highly dependent on planting date, wheat variety and climatic conditions.

In farmers' fields in Bangladesh, the average losses due to these foliar blights were estimated to be 15 per cent (2). In Bangladesh yield losses due to spot blotch were about 20 per cent in Sonalika whereas 14 and 8 per cent losses have been reported in wheat varieties Akbar and Kanchan, respectively (106). Late sowing also dramatically increases yield loss. In Bangladesh, the late sowing of the two leaf rust susceptible cultivars Sonalika and Kanchan resulted in a yield loss of 71 per cent and 41 per cent, respectively (5). A study conducted at Bangladesh into the effect of *B. sorokiniana* on wheat production revealed that upon the artificial inoculation of plants at the flag leaf stage reduced the number of grains per ear head and 1000-grain weight by 7-100 per cent and 12-100 per cent, respectively compared with the control (50). Yield loss assessment studies in Nepal have shown losses were around 27 per cent (20).

The losses caused by *Alternaria triticina* and *Helminthosporium* spp. have been estimated independently in India. Flag leaf in almost all cereals has major impact on yield potential of the concerned crop. Chenulu and Singh (14) estimated losses due to *Alternaria triticina* to the extent of 99 per cent in a highly susceptible variety under artificial conditions of inoculations at the boot stage in pot experiment. Nema and Joshi (92) correlated reduction in grain weight to the number of lesions incited by *Helminthosporium sativum* and disease intensity per unit area of flag leaf in pot experiments. Prabhu and Singh (102) estimated the relative effects of *Alterneria triticina* and *Helminthosporium sativum,* independently and in combination on yield and magnitude of losses .They observed losses increased at the rates of 0.92, 0.52 and 0.36 per unit increase of the disease incited by *Helminthosporium sativam, Alternaria triticina* and combined infection of both the pathogens, respectively. Studies conducted at Rajendra Agricultural University, Pusa, India revealed that 20-25 per cent losses in grain yield are inflicted due to blights in different wheat cultivars under normal years, which may be more in epiphytotic years (44).

The yield losses due to foliar blight at Narendra Deva University of Agriculture and Technology, Faizabad, India during 1994-95 crop season, was estimated 20 per cent and 22 per cent in the wheat varieties UP 262 and HD 1633, respectively (99). Yield loss assessment conducted at Cooch Behar, West Bengal showed 42 per cent yield loss in highly susceptible varieties in experimental plots and on farm studies up to 21 per cent (85).

## THE PATHOGEN

### *Naming of the disease*

The working groups of CIMMYT recommended the name *B. sorokiniana* (Sacc. in Sorok) Shoem for the spot blotch pathogen. They recommended it as appropriate, given that the anamorph plays the main role in nature and as teleomorph is also effective during the disease cycle, *P. tritici-repentis*(Died.) Drechs., should be used for tan spot (31). The foliar blight diseases represent a complex and are collectively referred to as Helminthosporium leaf blight (36). Another leaf blight fungus reported from India is *Alternaria triticina* Prasada & Prabhu (102). As the foliar blight symptoms are associated with more than one pathogen and often difficult to differentiate in naked eye, it is considered as disease complex and its importance was highlighted previously in reports from India (91), Bangladesh (5) and Nepal (20).

### *Evolution of the nomenclature*

The functional linkage between anamorph and teleomorph has helped the taxonomists to review the status of many taxonomic groups. This name had been widely used for the tan spot and yellow leaf spot pathogen of wheat (49,145). Nevertheless, this created confusion from the phytopathological point of view because *Helminthosporium* on the various hosts are associated with very different diseases with regard to pathogenecity, ecological requirements, and cultivar resistance, and because *H. tritici-repentis* also has, *per se*, a wide host range (117).

The name given to the tan spot fungus has also evolved considerably. The nomenclature of leaf blight pathogens of wheat was much overlapping and confusing to a relatively younger reader. In literature, the names of *Helminthosporium sativum* and *Helminthosporium tritici-repentis* are still widely followed for fungi that cause spot blotch and tan spot of wheat, respectively. This is unfortunate in the literature despite the fact that Shoemaker (115) considered the *Helminthosporium* an illegitimate orthographic variant of *Helmisporium* Link ex S.F.Gray (meanwhile *nomen rejciendum*). He also proposed the anamorph names actually recommended for graminicolus *Helminthosporium* species associated with leaf blight of wheat. The problem is two fold. On one hand, Plant Pathologists have deep-rooted attachment and preference to the name *Helminthosporium* linked to important diseases referred in textbooks. On the other hand, they have the difficulty and reluctance to follow various discussions of systematists regarding taxonomy and nomenclature. In later years however, some of these changes in nomenclature have become widely now adopted. Now in order to favour mutual understanding, researchers should reach to a consensus on a name(s) for the same organism.

The graminicolous *Helminthosporium* was erected as type species *H. velutinum* Link:S.F.Gray. However, it differs fundamentally from the modern *Helminthosporium* spp. Alcorn (3) reviewed the mechanism of conidium production, conidiophore architechture and conidial structure of *Helminthosporium* spp. and finally opined that there remains sufficient justification for renaming of graminicolous *Helminthosporium* spp. Sivanesan (123) is of the opinion that the classification made by Shoemaker (115,117) is still tenable although characters like type of hilum, conidial wall and known or unknown teleomorph state were considered later on.

*Helminthosporium sativum* Pammel, King and Baker (97) was proposed without considering the earlier account of the earlier *H. sorokinianum* Sacc. During 1890. Shoemaker's (115) proposal for the generic name *Bipolaris* to replace *Helminthosporium* species with fusoid, straight, or curved conidia germinating by one germ tube from each end (bipolar germination). The spot blotch pathogen was thus renamed as *Bipolaris sorokiniana* (Sacc.) Shoem which is currently widely adopted. Various other synonyms like *H. acrothecioides* Lindfors, *H. californicum* Mackie and Paxton, and *Drechslera sorokiniana* (Sacc.) Subrum. & Jain can be traced in literature. The teleomorph state is rare in nature in many fungi. Credit goes to Ito and Kuribyasi who in the laboratory on natural media only in the presence of opposite mating types first developed and first observed the ascigerous state (teleomorph).They described the teleomorph state as *Ophiobolous sativus* Ito and Kurib. It was latter renamed *Cochliobolus sativus* (Ito and Kurib.) Drechsler ex Dastur (25) which is characterized by globose ascomata usually with a long cylindrical neck, obclavate-asci, and helically coiled filiform ascospores. This genus is associated with either *Bipolaris* or *Curvularia* anamorphs. Teleomorphs of *C. sativus* were only reported from Zambia (104).

Studying the link between *Helminthosporium* anamorphs and *Pleospora* teleomorphs on various grasses, *H .grmineum* Rab.ex Schlecht.f.sp.*Tritici repentis* on quake grass *Triticum repens*, an ancient name (*Agropyrum repens* (L.) Beauv.). The anamorph of *H .grmineum* was linked with the teleomorph *Pleospora trichostoma* (Fr.) Wint.f.sp. *Tritici repentis* and was first described, in 1902. Diedicke (22) changed the names of *P. trichostoma* (Fr.)Wint and *Pleospora Tritici repentis* based on pathogenicity (22). *Pleospora tritici repentis* was renamed *Pyrenophora tritici- repentis* (Died.) Drechsler (25) because of the presence of setae on the ascocarp. Many other species of *Pleospora* (anamorph of Helminthosporium ) erected by Diedicke (22) have not been accepted by Noak based on morphological similarities of the teleomorph(94).

Wehmeyer (144) also supported the concept f Diedicke and reintroduced *Pleospora trichostoma* (Fr.) Ces. and de Not. (144). The exclusion by of all forms having a *Helminthosporium* state from *Pleospora trichostoma*, the erection of the five binomials for the ascus state of the different species of *Helminthosporium*, and the description of Drechsler. Based on morphological similarities of the ascocarp, he and subsequently *Pyrenophora trichostoma* (Fr.) Fck. as the teleomorph of the various species of *Helminthosporium* obtained on *Bromus*, *Hordeum*, *Poa*, *Secale*, and *Triticum* (144).

In his description of *H. tritici-repentis*, Drechsler (25) described the subhyaline, straight-cylindrical conidia. Nisikado (93) associated leaf spots on wheat with *H. tritici-vulgaris* Nisikado, characterized by conidia with uninflated, long-conical basal cell, *H. tritici-repens* being considered to have inflated basal cells. Ito (54) proposed the generic name *Drechslera* Ito for *Helminthosporium* species with conidia germinating from every cell and renamed *H. tritici-vulgaris*(Nisikado) Ito as *D. tritici-repentis* (Nisikado) Ito. However, Ito made indepth taxonomic studies of this *Drechslera*, the hyphomycete genus was not taken up until Shoemaker (117) considered the differences between *H .tritici-repentis* and *H. tritici-vulgaris* as not consistent and proposed *Drechslera tritici-repentis* (Died.) Shoem as the anamorph of *Pyrenophora tritici-repentis*. By this simplification, *Pyrenophora tritici-vulgaris* Dickson became the synonym of *P. tritici-repentis*.

For different states of fungi with pleomorphic life-cycles allowed by the International Code of Botanical Nomenclature to be given Separate names; if a teleomorph is present, the name automatically refers to that morph even if the anamorph is also present (47). Thus correct names for Helminthosporium leaf blight of wheat are accordingly: *Cochliobolus sativus* (Ito and Kurib.) Drechsler ex Dastur, anamorph *Bipolaris sorokiniana* (Sacc.) Shoem., and *Pyrenophora tritici-repentis* (Died.) Drechsler, anamorph *Drechslera tritici-repentis* (Died.) Shoem.

### *Causal organisms of the foliar blight disease complex*

Spot blotch and tan spot in wheat are caused by the fungi *Bipolaris sorokiniana* (Sacc.) Shoem., in anamorph state and *Pyrenophora tritici- repentis* (Died.) Drechs. in teleomorph state respectively (110). Another leaf blight fungus reported from India is *Alternaria triticina* Prasada and Prabhu. A detail description of *Bipolaris sorokiniana* is available at International Mycological Institute (UK) formerly C.M.I. (124). A key for distinguishing species of *Bipolaris* was described by Subhramanian (129). The evolution of nomenclature of causal organisms have been reviewed in section 2.4. *B. sorokiniana* is characterized by thick-walled, elliptical conidia (60-120 μm ×12-20μm) with 5-9 cells. In axenic culture, the mycelium is composed of hyphae interwoven as a loose cottony mass and appears as white or light to dark grey depending on the isolates (66). These fungi are differentiated from the *Bipolaris* genus on the basis of morphological features of conidiophores and conidia. Tan spot (also known as yellow spot) is caused by *P. tritici-repentis*, which is the sexual stage of *Drechslera tritici-repentis* (Died.) Shoemaker (Syn. *H. tritici-repentis* Drechs). On the leaf, lesions are due to anamorph *Bipolaris sorokininana*, characterized by long multicellular spores, whereas the ascospores of *Cochliobolus sativus* are formed in pseudothecia developed on the wheat residue. The anamorph is also found on stems, glumes, seeds and stubble. Septate conidiophores measure 80-400 × 6-9 μm and bear a single sub-hyaline 1-9 septate conidium, which is 80-250 × 14-20 μm in size and easily recognized by its basal cell suggesting a snake head (112,147). On PDA, the pathogen grows as dense, fluffy, greenish-grey mycelium without sporulation (Schilder and Bergstrom, 1993). On V8 agar medium, the mycelium is white to light grey. Conidia are produced after exposure to 12-24 hours near-UV light followed by 12-24 hours of darkness (112). Immature pseudothecia are formed only in old cultures. The pseudothecia are black, erumpent, 0.2- 0.35 mm in diameter with dark spines surrounding the short beaks and produced on weathered stubbles or straw. At maturity, they contain asci with ascospores measuring 40-60 μm long × 18-25 μm wide (147). The mature ascospores are yellow to brown, oval shaped, having three transverse septa and 1 to 2 longitudinal septa. Ascopores and conidia may both serve as primary inoculum.

Another foliar blight disease, zonate-eyespot (C.O. *Drechslera gigantea* ) was reported by Chowdhury *et al.* (15) along with *B. sorokiniana* from infected wheat plants from Cooch Behar, West Bengal. The conidiophores are dark brown, 3-6 septate and 180-350 $\mu$m × 10 $\mu$m in size. Conidia are straight, cylindrical with rounded ends, sub-hyaline and measure 330-490 $\mu$m ×15-20 $\mu$m in size. They are 6-8 septate and the middle cells are larger than the terminal ones.

In Brazil, another fungi associated with foliar blight of wheat has been isolated from wheat seeds as *Bipolaris bicolor* (Mitra) Shoem. This fungus causes spot blotch with very minute lesions and is most deleterious for the producers (80). On PDA culture *Bipolaris sorokiniana* presented black colonies, whereas *Bipolaris bicolor* greyish, aerial, cotton-like colonies. The size of the conidia of the *B. bicolor* is small but differed in length and width from *B.sorokiniana.* The septa of *B. bicolor* are deep ones with dark colour bases and rarely with dark apex whereas *B. sorokiniana* has homologous colour.

Physiological specialization at the species level of *Bipolaris sorokiniana* was first described by Christensen (16), who showed that fungal isolates varied considerably in virulence to wheat and barley. There are several reports on the genetic variability of *B. sorokiniana* populations (40,79). Differential reaction of progenies of crosses between isolates that differed in pathogenicity to different grass species indicated complex nature of inheritance involving several genes (89,90). The field population shifted to more aggressive races with long-term continuous wheat cultivation when race-cultivar level is considered (38). Duveiller and Garcia Altamirano (34) suggested that *B. sorokiniana* forms a continuum of isolates varying in virulence and aggressiveness with specific and non specific interactions.

Pathogenicity studies on a global collections of *B.sorokiniana* monoclonal strains on a differential set of wheat entries showed that differences between strains existed without any clear host specialization (48). Researchers from all over the world are also fascinated to investigate and understand the strain differences of organisms based on molecular available tools. Similarly, Ruckstuhl (109) found differences between groups of strains from South Asia, Mexico and Bolivia based on Random Amplified Polymorphic DNA (RAPD) analysis but without any relationship to pathogenicity. Toxins produced by *B. sorokiniana* are non specific and not directly associated with aggressiveness (33).

### *Pathogenic variability*

*B. sorokiniana* is believed to be a variable fungus with many morphological (16,73,95,98,131,133,134,135) and physiological variants as observed by different researchers(16,135).There are sufficient evidence that *B .sorokiniana* populations are comprised of pathogenic races(16,79). In North Dakota, USA, Valjavec-Gratian and Steffenson, (136,137) identified three pathotypes as 0,1,2 out of 33 isolates from different geographical areas, tested on the basis of interaction phenotypes with three host genotypes. They used three to six rows barley differential. Pathotype 1 was found in the east central and north eastern region only, whereas pathotype 2 and 0 were distributed throughout the state. Fetch and Steffenson, (40) reported distinct variation of two isolates of *B. sorokiniana* prevailing in North Dakota, USA namely ND85F and ND90Pr on a set of barley genotypes.

Five arbitrary 10-mer primers were able to reveal DNA polymorphism within three Helminthosporia named *Drechslera teres*, *Drechslera graminea* and *B. sorokiniana*. The total DNA of 8 different isolates of *B. sorokiniana* were digested with several restriction enzymes and probed with XYL1cDNA of *C. carbonum* (47). Southern hybridization patterns were similar for all isolates except the one from Temeni wheat and were distinct from those obtained with DNA of the related species *C. heterostrophus*. Isozyme patterns of esterase, malate dehydrogenase and superoxide dismutase were analysed by ValimLabres *et al.* (135) among 27 isolates of *B. sorokiniana*, recovered from susceptible , moderately resistant and resistant cultivars of wheat. The esterase system was the most polymorphic, and the isolate recovered from the susceptible cultivar showed the highest variability.

Pathogenic variability among the isolates of *B. sorokiniana* and probable cause of its emergence was studied in India (12). All isolates were collected from infected seeds and leaves on PDA by monoconidial isolation using fluorescent staining. The presence of nucleus on mycelial cells was also recorded. Observing colony morphology, the isolates were classified into 5 categories. This classification was confirmed by Cluster analysis. In the nature, the maximum frequency (44.63%) of isolates was black suppressed type. This type was the most virulent and identified as the epidemic population. Minimum frequency (4.96%) was reported for white colony in which number of conidia were very few .These isolates when inoculated on resistant and susceptible host showed varied AUDPC range. The aggressive population showed cells with either two nuclei and in other categories cells with three or four nuclei were most frequent.

Pathogenicity of *Alternaria* spp were collected from different locations where leaf blight has been reported, compared with that of IMI reference isolates of *A. triticina* and *A. alternata.* Morphology of the culture filtrate as well as the DNA sequence analysis based on ribosomal DNA from the internal transcribed spacer (ITS) region (ITS15.8s rRNA ITS2) and toxicity bioassay were added. Leaf blight symptom was only observed when *A. triticina* isolates inoculated on durum wheat *cv.* Bansi, while the other isolates – *A. alternata, A. tenuissima* and *A. arborescens* were found to be nonpathogenic on the different wheat cultivars tested. It differed from the other isolates from sporulation pattern. Phylogenic analysis proved effective in separating them as pathogenic and non-pathogenic. In agarose gel electrophoresis, it was clearly visualized that *Alternaria triticina* yielded longer ITS sequences than *A. alternata, A. tenuissima* and *A. arborescens* (72).

### Pathogenesis mechanism

*B. sorokiniana* is a hemibiotrophic organism (65) meaning that the fungus exerts a biotrophic and a subsequent necrotrophic growth phase. Microscopic analysis of fungal development on leaves of subsequent barley host reveals clearly delimited growth phases characterized by (i) cuticle and cell wall penetration followed by development of infection hyphae within the invaded, living epidermal host cell (biotrophic phase) and (ii) spread of hyphae into mesophyll layer, accompanied by epidermal and mesophyll cell death (nectrotic phase). Host cell collapse in the second phase is probably brought about by toxin secretion since cell die without contact to fungal hyphae (66). A histochemical study using diaminobenzidine (DAB) to visualize the reactive oxygen intermediate $H_2O_2$ shows association of the necrotrophic phase with the oxidative burst in epidermal and mesophyll cells. A positive correlation between host susceptibility and the amount of $H_2O_2$ produced in leaf lesions suggests that the $H_2O_2$ is necessary for successful tissue infestation. This view is also supported by the finding that the $H_2O_2$ scavenger ascorbic acid, when infiltrated into the inoculated leaves, reduces both $H_2O_2$ accumulation and disease development (lesion diameter). Interestingly, the early biotrophic phase of fungal development is also associated with a $H_2O_2$ generation, though the burst is known to be the highly confined to cell wall appositions (papilla-like structures) and single cells at the site of attempted fungal penetration. The positive DAB staining of papilla structures correlates with penetration resistance. Whole cell DAB staining followed by rapid cell death is sometimes observed in penetrated epidermal cells. Rapid cell collapse of invaded cells results in a retardation of fungal growth. Hence, $H_2O_2$ accumulation at early infection stages reflects a defense mechanism of barley plant against *B. sorokiniana*. This indicates that $H_2O_2$ has an ambivalent functions in the host depending on its spatio-temporal accumulation profile (66). Papilla formation and hypersensitive cell (HR) are well known defense reactions against biotrophic pathogens. It is very important that at early interaction stages cereals exert the same effective defense features (papilla and cell death) against hemibiotrophic and biotrophic pathogens.

## SYMPTOMS

Typical symptoms caused by *B. sorokiniana, A. triticina, P. tritici-repentis* are found in regions where the spot blotch, leaf blight and tan spot diseases occur respectively, separately. However, it is often impossible, without the help of a microscope, to recognize the causal organism of a blight symptom in wheat-growing areas where all the pathogens occur together. The term *Helminthosporium* leaf blight is commonly used to refer to this disease complex.

*Bipolaris sorokiniana* is an aggressive pathogen that causes spot blotch, root and crown rots, node cankers, ear head and seedling blight (147). Lesions in leaves are small, chlorotic and oval shaped with dark centers. Lesions have reddish brown centers with yellow margins and tapered ends. They may reach several centimeters before coalescing and inducing the death of the leaf. If spikelets are affected, it can result in shriveled grain and black point, a dark staining of the embryo at the end of the seed (33).

The lesions of *Alternaria triticina* are irregular in shape and dark brown to grey. Initially, the disease appears as small, oval discoloured lesions, which coalesce with disease progression, resulting in leaf death (118).

Typically, the first foliar symptoms of tan spot (C.O. *Pyrenophora tritici-repentis*) appears as small, light brown blotches that develop into oval shaped, light brown, necrotic lesions bordered with a yellow halo (112). Lesions coalesce with age and favour the early senescence of the entire leaf. Necrosis often begins near the tip and progresses towards the base of the leaf.

## Management of Foliar blight disease

The management of foliar blight by different pathogens was the centre of interest along with the development of Plant Pathology. It becomes critical particularly under favourable environmental regime specially for fungal pathogens, as they are highly sporulating with shorter life cycle as well as easy dissemination. Generally multipronged strategy becomes essential and application of a single tactics results in obvious failure. Conventionally foliar spray has been a first choice to researchers, extension agents and to a practicing farmers. IPM with various stages of development in different countries has been attempted, however, not yet accepted in larger scale among the growers' community. Researches relevant to foliar blight management can be considered and categorized as component for developing IPM.

### *Seed treatment*

Seed treatment with different fungicide and various other substances including non-conventional chemical may provide variable protection against foliar blight incidence of wheat. Hait and Sinha (46) reported that phytoalexin inducers like Cupric chloride, Ferric chloride at $10^{-3}$ M conc. protected the wheat seedlings from foliar blight infection.The other effective fungicides includes captan, mancozeb, thiram, pentachloronitrobenzene, proline and tridemefon (77,126). The foliar pathogen can be controlled with seed treating fungicides like guazatine and guazatine+ imazalil (93).

Pavolva *et al.* (100) indicated that seed treatment with flutriafol + thiabendazole, carboxin + thiram, difenoconazole + cyproconazole, carbendazim, fludioxonil, triticonazole, tebuconazole and diniconazole provided protection against root rot caused by *Fusarium, Helminthosporium,Bipolaris* and *Rhizoctonia* spp.

Domanov (24) conducted field trials to test carbendazim + carboxin which gave good control of root rots and increased yield of wheat.

The efficacy of Raxil (tebuconazole) was studied on spring wheat cultivar and seed treatment with tebuconazole controlled root rots caused by *Bipolaris sorokiniana* (*Cochliobolus sativus* ) and *Alternaria* sp. (62).

Seed treatment with Vitavax 200 B and Bavistin increased seed germination by 43 pet cent and reduced seedling infection by *B.sorokiniana* in Nepal (115).

The seed treatment of a newly developed fungicidal formulation Vitavax 200 WS (carboxin + thiram 1:1) @ 2.0, 2.5 and 3.0g/kg gave good results in reducing seedling mortality, incidence of foliar diseases at multilocations of India including Uttar Banga Krishi Viswavidyalaya, West Bengal (119).

### *Cultural practices*

Various information from different countries on managing foliar blight through manipulation of agronomic practices suggest that different mineral nutrients may reduce foliar blight (52,63,81). Although soil moisture and soil nutrient stress occur together in the wheat fields of South Asia, little quantitative information is available on the effect of low soil moisture and poor soil fertility on foliar blight severity (107).

In long term trials with differing nutrient management in rice-wheat cropping system, Gami *et al.*(41) and Regmi *et al.*(107) reported that declining soil fertility affects foliar blight severity. Factors that may alter the severity of wheat foliar blight caused by nonspecific pathogens include change in seeding date (132) and mineral nutrients (52,63,107). Sharma and Duveiller (114) opined soil fertility stress has pronounced effect on foliar blight severity and/or disease levels than moisture stress. They observed that under low soil fertility, reduction in grain yield and Thousand Kernel Weight due to HLB were 52.1 per cent and 15.5 per cent respectively, compared to 26.1and 8.05 under optimum soil fertility whereas HLB caused yield reduction of 22.1 per cent under low soil moisture compared to 11.85 under optimum soil moisture.

Ledinghan (68) showed that a longer time gap between two crops in the same season with different susceptible hosts may cause lowering of the rate of common root rot. The effect of tillage and crop rotation may not be consistent all over the world. Duczek *et al.* (30) observed a low sporulation rate of *B. sorokiniana* in the soil when rotation is made with oil seed rape and red clover. As the frequent tillage of soil produces more infectivity of *B. sorokiniana* (6,108), lesser tillage or reduction in frequency of tillage is advocated to reduce infectivity of the soil.

Clearing or ploughing the stubble, grass weeds and volunteer cereals reduce inoculum similar to crop rotation (23). They further noted that eradicant fungicide treatment of the seed and crop rotation with non- host crops can control spot blotch. However, Cook and Veseth (19) contradicted the views of Diehl *et al.*(23) in reducing the blight incidence.

Hossain *et al.*(51) observed that nutrient elements, especially application of micro-elements Cu, B, and Mo in soil is an alternate method of leaf spot control with the least possible disturbance of the overall natural soil fertility balance. Boron application increased grain formation and wheat yields. An *in-vitro* study revealed that Boron has a strong inhibitory effect on *B. sorokiniana*.

Studies conducted in Nepal from 2000-2002, on experimental station and in farmers fields, comparing plots sprayed with fungicides with untreated check , exhibited that yield losses due to HLB was 17-34% when losses were recorded up to 14-19 per cent, the recommended level of NPK 120-60-30 kg/ha was used. Similarly losses due to disease increased from 14 per cent to more than 19 per cent when supplementary irrigation was not available and physiological stress was higher (114). In different locations and of north-eastern plains zones of India, application of higher doses of potash in combination with other nutrients (120:60:60 kg N, P, K/ha) reduced foliar blight incidences as compared to 120:60:40 kg N, P, K/ha (120). Application of farm-yard manure 210 t/ha, led to decline of AUDPC by 30 per cent, suggesting not only that FYM contains a significant amount of potassium, but that copious levels of organic matter impart beneficial role and reduce disease severity.

Crop rotation allows time for the decomposition of stubble on which pathogens carry over, and natural competitive organisms reduce the pathogens on the remaining residue while unrelated crops are being grown (19,64). Leaf spot disease severity can be greater in monoculture compared with following an alternative crop. When tan spot [*Drechslera tritici-repentis* (Died.) Shoemaker] and *Stagonospora nodorum* blotch [*Stagonospora nodorum* (Berk) Castellani and E.G. Germano] were the dominant diseases, Fernandez *et al.* (1998) observed that monoculture wheat had a higher disease severity than wheat grown after flax or lentil. It is interesting to note sometime that rotations have limited impact on wheat disease severity, pathogen populations reduced yet, resulting higher yields under, and higher yields diversified wheat based crop rotations compared to wheat in monoculture (7).

Mahto *et al.* (71) compared the foliar blight incidence on three different wheat varieties Rohini, BL 1473 and Bhrikuti under surface seeding conditions and conventional tillage. Surface seeding was superior to conventional tillage. The results indicate that the average foliar blight induced reductions in grain yield , thousand grain weight and Kernel per spike were 16, 12 and 155 respectively under conventional tillage whereas the corresponding reductions were 23, 13 and 195 under surface seeding. The same observation was noticed at Cooch Behar, West Bengal by using three different wheat varieties PBW 343, NW 2045 and NW 1012 (8).

### *Chemical Management*

Despite the harmful effect of fungicides, it had proved useful and economical in the control of tan spot and spot blotch (70).

In Brazil Mehta (78) observed that foliar fungicides is effective in controlling spot blotch. Non systemic and systemic foliar fungicides belonging to the dithiocarbamates (viz. mancozeb) and triazoles (viz. propiconazole, tebuconazole, flutriazol, procloraz, and triadimenol) and dicarboxymides (viz. iprodione) are known to be effective. The first spray should be applied soon after the onset of disease symptoms appear. In a non-endemic zone generally spraying may be done 45-55 days after sowing for the spring wheat cultivars (75). Foliar applications specially with systemic fungicides such as tebuconazole, epoxiconazole, flutriafol, cyproconazole, flusilazole, epoxiconazole, and metaconazole, applied between heading and grain filing stages, have been proved to be cost effective. Under severe disease levels, a second spray can result in a grain yield increase by 38-61 per cent (141).

In fungicide tests, manzate reduced disease severity by 25 per cent. There were no significant differences between two treatments reduced the average disease severity on combined flag leaf (F) and (F-1) leaves by more than 75 per cent. Only two sprays of tebuconazole (Folicur) caused a significantly larger reduction in disease severity (86%) (96).

In Argentina, Annone (4) observed positive response with foliar fungicides for reduce disease development under conditions of moderate to low inoculum pressure, but inconsistent results under high inoculum pressure. Among fungicides tested under field conditions, tebuconazole and propiconazole achieved the highest level of control, though results varied widely (30-80%).

Tewari and Wako (130) reported that the mixture of Tebuconazole (Folicur) and Metacid applied as foliar spray at the boot stage of wheat crop effectively suppressed all the foliar diseases (brown and yellow rusts, powdery mildew and leaf blight) showing curative property with no phytotoxic effect on the plant. They also observed that Sencor and Propiconazole (Tilt) was incompatible and highly phytotoxic. They further noticed that the mixture of Zinc Sulphate and Urea completely inhibited mycelial growth of *Bipolaris sorokiniana* and 78 per cent growth inhibition of *Alternaria triticina*.

Rashid *et al.* (105) reported propiconazole (tilt 250 EC) to be very effective against foliar blight of wheat (*Bipolaris sorokiniana*). The different fungicides like Flusilazole, Prochloraz, Propiconazole and Tebuconazole were effective against tan spot disease of wheat (17).

### *Induced resistance and Biological control*

Yield losses in wheat and barley leaf blight (C.O. *Bipolaris sorokiniana*) indicate the need to search for alternative strategies of disease control. On a global basis one of the promising strategies is induced resistance. In the broadest sense, induced resistance means the control of parasite and pest by a prior activation of the plants own defense system. Defense was activated by necrotizing pathogen as well as by chemicals mimicking factors of the natural defense systems, such as salicylic acid (110,118). Chemical induction of resistance to *Bipolaris sorokinian.* in barley by pre-treatment with inducers 2,6 – dichloroisonicotinic acid (DCINA), benzo (1,2,3) thiadiazole –7-carbothioic acid –s-methylester (BTH) or jasmonates leads to symptom reduction in the range of 10-20 per cent (Kumar and lbeagha personal communication). Hait and Sinha (46) reported that seed treatment with SAR chemicals (heavy metals) provide wheat seedlings from *Helminthosporium* infection.

Pre-treatment of wheat leaves with the inappropriate *Bipolaris oryzae* (inducer organism) reduced spot blotch lesion numbers and lesion size of *Bipolaris sorokiniana* on the same level (111). A preliminary study identified antifungal substance in apoplastic fluids of 'induced' plants were able to prevent spore germination of *Bipolaris sorokinina in vitro*. It was seen that barley seeds are shown to cause systemic induced resistance on the plants to subsequent infection with *B. sorokiniana* which was inoculared by fungi *I. bolleyi* (69) Reduced symptoms could also be achieved, although not to the full extent, after spraying with the bacterial biocontrol agent *Pseudomonas chlororaphis* strain MA 342. Treated seed could be stored dry for at least 2 years without losing the disease suppressing effect of the bacterial treatment (57). *Pseudomonas chlororaphis* (Cedomon) again used for seed treatment against the pathogen *Drechslera* sp. is available but it does not have sure effect on *B. sorokiniana* (95).

Plant growth-promoting fungal rhizosphere isolates of *Phoma spp.* from Zoysia grass (*Zoysia* sp.) suppressed *Bipolaris sorokiniana* due to the competitive root colonization. Successful antagonists against seed borne *Bipolaris sorokiniana* were *Chaetomium* sp., *Idriella bolleyi*, and *Gliocladium roseum* (61). The antagonistic potential of *Chaetomium globosum* agaist

*Drechslera sorokiniana* was confirmed by Biswas *et al.* (9). Suppressions of soil-borne fungi, including *Bipolaris sorokiniana*, was observed in the presence of isothiocyanates released into soil by *Brassica* species (60). Biological control of leaf blight of wheat with botanicals has been successfully done by many scientists (50,51,53). Islam *et al.* (53) reported reduction in disease severity in flag leaf stage, flowering stage , milking stage and hard dough stage which was achieved by seed treatment with *Allium sativum* clove extract and *Polygonum hydropiper* leaf and stem extract and yield enhancement to 29.74 per cent over untreated control. Keeping in mind Rahman, Aminuzzaman and Islam tried to use extracts of thirteen plant species as a means of controlling leaf blight of wheat (*B. sorokiniana*).Experiments were done in natural condition during the period from November, 2006 to March, 2007. Disease severity was best controlled by using seed treat treatment with *Curcuma longa,* rhizome extract and *Allium sativum* clove extract. Among all the treatments *Curcuma longa* rhizome extract treated seeds resulted the highest grain yield (3.60 t/ha) which was 39.44 higher over untreated control.

### *Breeding for Resistance*

The best approach to manage foliar blight disease is to develop resistant varieties. Inheritance of its resistance has been reported by many workers. In many countries highly susceptible cultivars are quickly replaced by less susceptible cultivars.

Sources of spot blotch resistance had been identified over the years and governed by one or more genes, and their origins can be differentiated into three categories: Latin America, China and wild relatives of wheat or alien species (138). The Latin American sources are mainly derived from Brazil, and may trace back to Italian ancestry. Older two resistant Brazilian varieties are BH 1146 and CNT 1 (Mehta 1985). The Chinese sources are mostly materials from Yangtze River Basin. Early Chinese sources of resistance used at CIMMYT include Shanghai# 4, Suzhoe# 8, and Yangmai# 6.

Alien resistance sources are many. Alien sources used at CIMMYT had centered around crosses involving *Thinopyrum curvifolium* (83,142). Using these alien sources in combination with Chinese resistance sources, outstanding lines such as Mayoor and the Chirya series were developed. It may sometimes be difficult to accurately quantify or determine the exact contribution of alien genes in such multi-parent crosses. van Ginkel and Rajaram(1998) indicated that there is a component of alien-derived resistance present in certain advanced materials developed by Dr. A. Mujeeb-Kazi at CIMMYT. van Ginkel and Rajaram (1998) found *Aegilops squarrosa* crosses had shown impressive resistance to spot blotch in Mexico.Some progress in transferring resistance genes from *Thinophylum curvifolium, Elymus curvifolius* and *Triticum tauschii* to wheat germplasm has been achieved (83,84) . Among 250 synthetic hexaploid(2N=6X=42) amphiploides of wheat some resistant stocks against *B.sorokiniana* have been identified (82,85).

Researches in Punjab Agricultural university in Punjab, India, had studied more than 14,000 entries of wheat and related wild species, representing 13 genera and 136 species, for spot blotch resistance (21,122).They identified several promising entries including *Ae. Triaristata, Triticum dicoccoides, Ae. cylindrical and T. boeoticum.*

Studies on the inheritance on spot blotch resistance with Latin American and Indian resistance sources had shown one or two genes to be involved (1,126). Velazquez- cruz (143) found six genes to be segregating in four moderately resistant to resistant lines (Gisuz, Cugap, Chirya 1, and Sabut) developed at CIMMYT, with two to three genes providing good resistant levels. Conner (18) and Gilchrist *et al.* (42,43) had shown that foliar and seed infection by *B. sorokiniana* may be determined by different genetic systems.

Heritabilities of resistance tend to be low and much influenced by the changing environment, as determined by temperature, humidity, inoculum pressure, planting date, observation date, etc. These environmental components and the known variability in the pathogen make breeding for resistance difficult (36,37).

In Zambia, spot blotch is the devastating disease of rain fed wheat production (87,88). Most resistance cultivars in Zambia contain brazilian ancestry. Brazilian sources of resistance were first introduced in 1979 from Passo Fundo (Raemaekers 1988 a and b). In 1984, the introduction PF 7748 was released as whydah, and in 1986 Hornbill (=IAS64/Aldan). PF 73339) Hahn, a CIMMYT cross, was released as Coucal in 1988 (Raemaekers 1988b). The release of these three varieties allowed yield potential to increase from 1.6-1.7 t ha- to 2.7 t $ha^{-1}$( 87).

More previous studies had reported the resistance to spot blotch in wheat and barley is inherited is governed by quantitative trait and no major genes are known to involve (74).Though, sources of higher levels of resistance had been achieved using conventional breeding methods, mainly because of the polygenic nature of this resistance. It is necessary to identify the quantitative trait loci (QTLs). It is expected that once the genes corresponding to QTLs are cloned, transformation method may offer new options for combating the disease in future( 134,135). Induced variation for higher levels of resistance is also expected through somatic embryogenesis (139,74,11). Valjavec-Gratin and Steffenson (1997) reported that both virulence in the pathogen and resistance in the host are under monogenic control in progenies of barley cross Bowman/ND 5883 and *B. sorokiniana* isolate ND90Pr. Resistance to *B. sorokiniana* in barley is reported to be conferred by a single gene at the seedling stage. This gene is mapped to the distal region of chromosome 1P on the basis of both qualitative and quantitative data analyses. Two QTLs are identified for spot blotch resistance at the adult plant stage: the largest QTL mapped to chromosome 5P and the other mapped to chromosome 1P near the seedling resistance locus (128). Baily and Wolfe (6) reported that cultivated barley conserved in situ in the middle east preserve resistance to *B. sorokiniana* in either single or multiple combinations in cultivated landrace population, but it was less effective in preserving the diversity for resistance than in wild barley. Baily and Wolfe (6) suggested that a breeding strategy for developing lines resistant to common root rot, while maintaining high yield should involve screening to eliminate the highly susceptible lines from crosses and then selecting the best yielding lines with the lowest disease. While analyzing resistance to common root rot and spot blotch in barley, the heritability to common root rot resistance of cross Fr926- 77 x Deuce was found to range from 56to 85%. Heritability for spot blotch resistance in this cross was 43-615.for another cross of Virden x Elice, the heritability to common root rot resistance ranged from 53-78%, while 73 to 78% for spot blotch. However, no association could be detected between heritability to common root rot and spot blotch resistance in barley (67). A high level of resistance to spot blotch was achieved by crossing Chinese hexaploid wheat with a moderately resistant commercial cultivar in Nepal by Sharma *et al.* (114). They made selections for low AUDPC of spot blotch in segregating generations.

*B. sorokiniana* is highly aggressive on barley with mutations in the Mlo gene.A central position in barley defense signaling is taken by the Mlo locus. Importantly, the functional MLO protein has an ambivalent role in interactions with biotrophic or hemibiotrophic fungal parasites that attack barley epidermal cells. MLO is required for establishment of a compatable interaction of barley and the biotrophic powdery mildew fungus, as barley mutants lacking MLO (mlo mutants) are highly resistant to this fungus (54). Resistance to powdery mildew is executed in the mutants at a very early, prehaustorial stage of the interaction (54), involving formation of papilla, highly localized $H_2O_2$ accumulation, phytoalexin synthesis and defense gene expression. On the other hand, mlo mutants are hyper-susceptible to *B. sorokiniana* (unpublished results), toxin containing cultural filtrate(62) as well as to the hemibiotrophic fungus *Magnaporthe grisea* (56). Compared to wild type barley, mlo mutants accumulate much more $H_2O_2$ and develop leaf necrosis in response to the toxic culture filtrate of *B. sorokiniana* in epidermal and mesophyll cells (65).

Chaurasia *et al.* (13) reported that somaclones ($R_2$, $R_3$ and $R_4$generations) regenerated from immature embryos of two spring wheat varieties, HUW-206 and HUW-234 displays improved earliness, enhanced resistance to spot blotch disease and increased yield over the parent. The superiority of variants for yield traits and disease resistance was established in $R_4$ generation, confirming the possibility of wheat improvement through somaclonal variation.

## CONCLUSIONS

Spot blotch is a disease of great concern especially in South Asia and Latin American countries, where warm-humid conditions persist during wheat cropping system. Nearly, 12 million hectares of land in these regions of world under cultivation are affected. Moreover, this disease is now expanding towards non-traditional cooler reasons likes North Western Plain Zones, India and European countries. Disease appears initially as brown eye shaped necrotic spots surrounded by yellow halo on leaves. Later such spots coalesce to make a larger area of leaf blighted and thus adversely affect photosynthesis. It takes a heavy toll through reducing annual yields by about 20%. The complex of symptoms caused by Spot blotch can be controlled, but only partially, by applying methods of soil/residue management and crop rotation. Although, a satisfactory control can be achieved by fungicidal treatments but it increases input cost beyond the capacity of Indian farmers and pollutes environment also. At this juncture, cultivation of resistant varieties is the only economically feasible and ecologically friendly approach of keeping disease losses to a minimum.

## REFERENCES

1. Adlakha KL, Wilcoxson RD, and Raychaudhury SP 1984 Resistance of wheat to leaf spot caused by *Bipolaris sorokiniana, Plant Disease* **68** 320-321
2. Alam KB, Banu SP, and Shaheed MA 1997 The occurrence and Significance of spot blotch disease in Bangladesh, Proc of the International workshop on helminthosporium Disease of wheat spot Blotch and Tan spot, (eds Duveiller, E, Dubin, H J, Reves, J, Mcnab, A) CIMMYT, El Batan, Mexico, 63-66.
3. Alcorn JL 1988 The taxonomy of "*Helminthosporium*" Species, *Annual Review of Phytopathology* **26** 37-56
4. Annone J 1997 Tan spot of wheat in Argentina Importance and prevailing disease management strategies, Proc of the International Workshop on Helminthosporium Diseases of wheat Spot Blotch and Tan Spot, (eds Duveiller, E, Dubin, HJ, Reves, J, McNab, A) CIMMYT, El Batan, Mexico, 339-345
5. Badaruddin M, Saunders DA, Siddique AB, Hossain MA, Ahmed MU, Rahman MM, and Parveen S 1994 Determining yield constrains for wheat Production in Bangladesh, In *Wheat in Heat -Stressed Environments Irrigated, Dry Areas and Rice-Wheat Systems* (eds DA Saunders and G P Hettle), 265-271
6. Bailey KL and Wolfe RI 1994 Genetic relationships between reaction to common root rot and yield in the progeny of a barley cross, *Canadian Journal of Plant Pathology* **16** 163-169

7. Bailey KL, Gossen BD, Lafond G R, Watson PR, and Derksen 2001 Effect of tillage and crop rotation on root and foliar diseases of wheat and pea in Saskatchewan from 1991-1998 Univariate and multivariate analyses, *Canadian Journal of Plant Science* **81** 789-803

8. Bandyopadhaya S, Chowdhury AK, Laha SK, and Singh G 2009 Response of wheat lines to foliar blight and biochemical changes associated with infection, *Journal of Mycopathological Research* **47** 119-128

9. Biswas SK, Srivastava KD, Aggarwal R, Dureja P, and Singh DV 2000 Antagonism of *Chaetomium globosum* to *Drechslera sorokiniana*, the spot blotch pathogen of wheat, *Indian Phytopathology* **53** 436-440

10. Bochow H, Kogel KH, Poehling H, von Alten H, and Wyss U 2001 Mechanisms of interactions in the system of plant, parasite and beneficial organisms, *Journal of Plant Disease and Protection* **108** 626–652

11. Bohorova NE, van Ginkel M, Rajaram S, and Hoisington DA 1995 Tissue culture response of CIMMYT elite bread wheat cultivars and evaluation of regenerated plants, *Cer Res Comm* **23** 243 – 249

12. Chand RK, Pandey SP, and Joshi AK 2003 Proceedings of Fourth International Wheat Tan Spot and Spot Blotch Workshop, Bemidji, Minnesota, USA

13. Chaurisa S, Joshi AK, Dhari R, and Chand R 1999 Resistance to foliar blight of wheat, *Genetic Resources and Crop Evolution* **46** 469 – 475

14. Chenulu VV and Singh A 1967 A not on estimation of losses due to leaf blight of wheat caused by *Alternaria triticina, Indian phytopathology* **17** 256 – 57

15. Chowdhury AK, Garain PK, Mukharjee S, Datta S, Bhattacharys PM, Singh D P, and Singh, G 2005 Zonate eyespot of wheat new report, *Journal of Mycopathological Research* **43** 139-140

16. Christensen JJ 1926 Physiological specialization and parasitism of *Helminthosporium sativum* Minn, *Agric Expt Sta Tech Bull* **37** 101

17. Colson ES, Platz GJ, and Usher T R 2003 Fungicidal control of *(Pyrenophora tritici repentis)* in wheat, *Australian Plant Pathology* **32** 241-246

18. Conner RL 1990 Interrelationship of cultivar reaction to common root rot, black point, and spot blotch in spring wheat, *Plant Disease* **74** 224-227

19. Cook RJ and Veseth RJ 1991 Wheat health management St Paul, MN, USA, APS Press 152 page

20. Devkota RN 1994 Wheat Breeding objectives in Nepal

The National Testing system and Recent progress Wheat in Heat stressed Environment Irrigated, dry area and rice wheat farming systems D A Saunders and G P Hettel (eds) Mexico CIMMYT pp 216 –223

21. Dhaliwal HS, Singh H, Gill KS, and Randhawa HS 1993 Evaluation and cataloguing of wheat germplasm for disease resistance and quality, in A B Dhamania (ed), Biodiversity and Wheat Improvement, New York John Wiley and sons, 123-140 pages
22. Diedicke H 1903 Ueber den Zusammenhang zwischen *Pleospora-* und *Helmintho-sporium-* Arten II Bacteriolo-gisches Centralblatt Abt II 1152-59
23. Diehl JA, Tinline RD, Kochhann RA, Shipton PJ and Rovira AD 1982 The effect of fallow periods on common root rot of wheat in Rio Grande do Sul, Brazil, *Phytopathology* **72** 1297-1301
24. Domanov NM 2002 Kolfugo Duplet for seed treatment of cereals *Zashchita Karantin Rasteni* 6 29 Moscow, Russia (Cab International)
25. Drechsler C 1923 Some gram-inicolous species of *Helmin-thosporium, Journal of Agricultural Research* **24** 641-740
26. Dubin HJ and Bimb HP 1991 Effects of soil and foliar treat-ments on yield and diseases of wheat in lowland Nepal, in Wheat in Heat – Stressed Environments Irrig-ated, Dry Areas and Rice – Wheat Farming Systems (eds D A Saunders and G P Hettel), 484-485
27. Dubin HJ and Bimb HP 1994 Studies of Soil borne diseases and foliar blights of wheat at the National Wheat Research Experiment Station, Bhaira-hawa, Nepal Wheat Special Report No 36, Mexico CIMMYT 30
28. Dubin HJ and Duveiller E 2000 Helminthosporium leaf blights of wheat integrated control and prospects for the future Proceedings of the International Conference on Integrated Plant Disease Management for Sustainable agriculture, Indian Phytopa-thological Society, New Delhi, India, 575-579
29. Dubin HJ and Van Ginkel M 1991 The status of wheat diseases and disease and disease research in the warmer areas, Proceedings of the Wheat for Non-traditional Warm Areas Conference, Foz de Iguazu, Brail, (eds D A Saunders), UNDP/CIMMYT, Mexico, 549
30. Duczek LJ, Sutherland KA, Reed SL, Bailey KL, and Lafond GP 1999 Survival of leaf spot pathogens on crop residues of wheat and barley in Saskatchewan, *Canadian Journal of Plant Pathology* **21** 165-173

31. Duveiller E 2002 Helminthosporium blight of wheat challenges and strategies for a better disease control, In Advances of wheat breeding in china China Science and Technology Press, Jinan, Shandong, Peoples Republic of China 57 – 66

32. Duveiller E 2004 Controlling foliar blights of wheat in te rice-wheat System of Asia, *Plant Disease* **88** 552-556

33. Duveiller E and Dubin HJ 2002 Helminthosporium Leaf blights Spot blotch and Tan Spot, in Bread Wheat Improvement & Production Series, No 30 (eds Curtis, BC, Rajaram, S, Gomez Macphercon, H), FAO, Rome

34. Duveiller E and Garcia Altamirano I 2000 Pathogenicity of *Bipolaris sorokiniana* isolates from wheat roots leaves and grains in Mexico, *Plant Pathology* **49** 235-242

35. Duveiller E, van Ginkel M and Dubin J 1997 Helminthosporium diseases of wheat Summary of Group Discussions and recommendations in Helminthosporium Blights of Wheat Spot Blotch and Tan Spot (eds, Duveiller, E, Dubin, HJ, Reeves, J and McNab, A), pp1-5CIMMYT/ UCL/BADC

36. Duveiller E and Gilchrist L 1994 Productions constraints due to *Bipolaris sorokiniana* in wheat Current situation and future prospects, in *Proceedings* of the Wheat in the Warmer Areas, Rice/ Wheat Systems, (eds D A Saunders and Hettel, E,) CIMMYT/UNDP, Nashipur, (Dinajpur), Bangladesh, 343-352 pages

37. Duveiller E, Van Ginkel M and Dubin HJ 1997 Helminthosporium diseases of wheat Summary of group discussions and recommendations, in Duveiller, E, Dubin, H J, Reeves, J, McNab, A (eds), *Helminthos-porium* Blights of Wheat Spot Blotch and Tan Spot CIMMYT, Mexico, DF, 1-5

38. El-Nashaar, and HM and Stack RW 1989 Effect of long-term continuous cropping of spring wheat on aggressiveness of *Cochliobolus sativus, Canadian Journal of Plant Science* **69** 395-400

39. Fernandez MR, Zentner BG, McConkey, and Campbel 1998 Effect of crop rotations and fertilizer management on leaf spotting diseases of spring wheat in southwestern Saskatchewan, *Canadian Journal of Plant Science* **78** 489-496

40. Fetch TG and Steffenson BJ 1994 Identification of *Cochliobolus sativus* isolates expressing differential virulence on two-row barley genotypes from North Dakota, *Canadian Journal of Plant Pathology* **16** 202-206

41. Gami SK, Ladha JK, Pathak H, Shah MP, Pasaquin E, Pandey SP, Hobbs PR, Joshy

D, and Misra R 2001 Longterm changes in yield and soil fertility status in a 20-year rice-wheat experiment in Nepal, *Biology and Fertility of Soils* **34** 73-78

42. Gilchrist LI, Pfeiffer WH, and Rajaram S 1991a Progress in developing bread wheats resistant to *Helminthosporium sativum,* in Wheat for the Nontraditional, Warm Areas, (ed DA Saunders), Mexico, DF CIMMYT, 469-472

43. Gilchrist LI, Pfeiffer WH, and Velaquez C 1991b Resistance to *Helminthosporium sativum* in bread wheat relationship among infected plant parts and associated with agronomic traits, in Developing Sustainable Wheat Production Systems. The 8th Regional Wheat Workshop for Eastern, Central and Southern Africa, (Ed DG Tanner), Addis Ababa, Ethiopia; CIMMYT, 176-181

44. Goel LB, Nagarajan S, Singh RV, Sinha VC, and Kumar, J 1999 Foliar blight of wheat Current states in India and Identification of donor lines for resistance through multilocational evaluation, *Indian Phytopathology* **52** 398-402

45. Hack E, Emami K, and Hassan AA 1997 Characterization of *Cochliobolus sativus* isolates from the UK and Yemen, in *Helminthosporium* Blights of Wheat Spot Blotch and Tan Spot pp 313 (eds, Duveiller, E, Dubin, HJ, Reeves, J and McNab, A), 376

46. Hait GN and Sinha AK 1986 Protection of wheat seedlings from *Helminthosporium* infection by seed treatment with chemicals, *Journal of Phytopathology* **115** 97-107

47. Hawksworth DL, Kirk PM, Sutton BC, and Pegler DN 1995 Ainsworth and Bisby's dictionary of the fungi Wallingford, UK CAB International

48. Hetzler J, Eyal Z, Mehta YR, Campos LA, Fehrmann H, Kushnir U, Zekaria – Oren J, and Cohen L 1991 Interaction between spot blotch and wheat cultivars, in Saunders, D A (ed), Wheat for the Nontraditional, Warm Areas CIMMYT, Mexico, DF, 146-164

49. Hosford RB 1982 Proceedings of the Tan Spot of wheat and Related Diseases workshop, North Dakota Agricultural Experiment Station, Fargo, USA

50. Hossain I and Schlosser E 1993 Control of *Bipolaris sorokiniana* in wheat with neem extracts, *Bangladesh Journal of Microbiology* **10** 39-42

51. Hossain MM, Khalequzzaaman FM, Mollah MRA, and Rehman GMM 2005 Effect of Plant Extract on the Incidence of Seed-Borne fungi of Wheat, *Journal of Agrilculture and Rural Develpment* **3** 39-43

52. Huber DM, Lee TS, Ross MA, and Abney TS 1987 Amelioration of tan spot infected wheat with nitrogen, *Plant Disease* **71** 49-50

53. Islam MA, Aminuzzaman FM, Islam MR, and Zamal MS 2006 Seed treatment with plant extract and Vitavax 200 in controlling leaf spot (*Bipolaris sorokiniana*) with increasing grain yield of wheat, International Journal of Sustainable Agrilcultural Technology **2(8)** 15-20

54. Ito S 1930 On some ascigerous stages of the species of *Helminthosporium* parasitic on cerealsProc Imp Acad, Tokyo 6352-355) Proc Int Workshop Helminthosporium Diseases of Wheat Spot Blotch and Tan Spot, (eds, Duveiller, E, Dubin, HJ, Reeves, J and McNab, A) CIMMYT, El Batan, Mexico, Mexico, DF, CIMMYT, 88-106

55. Jana S and Bailey KL 1995 Responses of wild and cultivated barley from west Asia to net blotch and spot blotch, *Crop Science* **35** 242-246

56. Jarosch B, Kogel KH, and Schaffrath U 1999 The ambivalence of the barley *Mlo* locusMutations conferring resistance against powdery mildew (*Blumeria graminis* fsp *hordei*) enhance susceptibility to the rice blast fungus *Magnaporthe grisea,* Molecular *Plant-Microbe Interactions* **12** 508-513

57. Johnsson L, Hokeberg M, and Gerhardson B 1998 Performance of the *Pseudomonas chlororaphis* biocontrol agent MA 342 against cereal seedborne diseases in field experiments, *European Journal of Plant Pathology* **104** 701–711

58. Jørgensen JH 1992 Discovery, characterization and exploitation of *Mlo* powdery mildew resistance in barley, *Euphytica* **63** 141-152

59. Joshi AK, Kumari M, Singh VP, Reddy CM, Kumar S, Rane J, and Chand R 2007 Stay green trait Variation, inheritance and its association with spot blotch resistance in spring wheat (*Triticum aestivum* L), *Euphytica* **153** 59- 71

60. Kirkegaard JA, Wong PTW, and Desmarchelier JM 1996 *In vitro* suppression of fungal root pathogens of cereals by *Brassica* tissues, *Plant Pathology* **45** 593–603

61. Knudsen IMB, Hockenhull J, and Jensen DF 1995 Biocontrol of seedling disease of barley and wheat caused by *Fusarium culmorum* and *Bipolaris sorokiniana* Effects of selected fungal antagonists on growth and yield components, *Plant Pathology* **44** 467–477

62. Korobov VA and Korobova LN 2002 Raxil in the south of West Siberia *Zashchita Karantin Rasteni* 6 29-30 novosivirec Agrarian University,

Siberia, Russia, Cab International

63. Krupinski JM and Tanaka DL 2001 Leaf spot diseases on winter wheat influenced by nitrogen, tillage and haying after a grass alfalfa mixture in conservation resource program, *Plant Disease* **85** 785-789

64. Krupinsky JM, Bailey KL, McMullen MP, Gossen BD, and Turkington 2002 Managing plant disease risk in diversified cropping systems, *Agronomy Journal* **94** 198-209

65. Kumar J, Hueckelhoven R, Beckhov U, Nagarajan S, and Kogel KH 2000 A compromised Mlo pathway affects the response of barley to the necrotrophic fungus *Bipolaris sorokiniana* (teleomorph *Cochliobolus sativus*), *Phytopathology* **91** 127 – 133

66. Kumar J, Schafer P, Huckelhoven R, Langen G, Baltruschat H, Stein E, Nagarajan S, and Kogel K 2002 *Bipolaris Sorokiniana,* a cereal pathogen of global concern Cytological and molecular approaches towards better control, *Molecular Plant Pathology* **3** 185-195

67. Kutcher HR, Bailey KL, Rossnagel BG, and Legge WG 1994 Heritability of common root rot and spot blotch resistance in barley, *Canadian Journal of Plant Pathology* **16** 287–294

68. Ledingham RJ 1961 Crop rotations and common root rot in wheat, *Canadian Journal of Plant Science* **41** 479-486

69. Liljeroth E, Franzon Almgren I, and Gunnarsson T 1993 Effect of pre – helminthosporol, a phytotoxin produced by *Bipolaris sorokiniana*, on barley roots, *Canadian Journal of Botany* **72** 558 – 563

70. Loughman R, Wilson RE, Roake JE, Platz GJ, Rees RG, and Ellison FW 1998 Crop Management and breeding for control of *Pyrenophora tritici repentis* Causing yellow spot of wheat in Australia Proceedings of (*Indian phytopathological Society Golden Jubilee)* IARI New Delhi, 571-579 pages

71. Mahto BN, Deveiller E, and Sharma RC 2006 Effect of surface seeding on foliar blight severity and wheat performance, *Field Crops Research* **97** 344-52

72. Marcedo VB, Renard ME, Deveiller E, and Maraite H 2006 Identification of Alternaria species on wheat by pathogenecity assay and sequencing, *Plant Pathology* **55** 485-93

73. Matsumura AT 1991 Variability intraespecifica quanto a patogenicidade, caracteristicas da cultura e padrao isoesterasico em populaces naturais de Bipolaris sorokiniana (*Helminthos-porium sativum*) Phd thesis,

Departmento de Genetica, Universidade Federal do Rio Grande do Sul, Porto Alegre, RS, Brazil

74. Mehta YR 1996 Variacao somaclonal nas plantas de trigo regeneradas atraves de embriogenese *Fitopatol Bras* 21 (abstract)

75. Mehta YR and Igarashi S 1985 Chemical control measures for major diseases of wheat with special attention to spot blotch, in Wheats for the More Tropical Environments A Proceedings of the International symposium, Mexico, DF CIMMYT, 196-200

76. Mehta YR 1985 Breeding wheats for resistance to *Helminthosporium* spot blotch, in Wheats for the More Tropical Environments A Proceedings of the International Symposium, Mexico, DF CIMMYT, 135-144

77. Mehta YR 1993 Manejo Integrado de Enfermedades de Trigo Santa Cruz, Bolivia CIAT/IAPAR 314

78. Mehta YR 1998 Constraints on the integrated management of spot blotch of wheat In *Helminthosporium Blights of Wheat Spot Blotch and Tan Spot* (Duveiller, E, Dubin, HJ, Reeves, J and McNab A, eds) Mexico, DF, Mexico CIMMYT, 18–27

79. Misra AP 1979 Variability, physiologic specialization and genetics of pathogenicity in graminicolous *Helminthosporia* affecting cereal crops, *Indian Phytopathology* **32** 1–22

80. Morejon KR , Moraes HMD, and Bach EE 2006 Identification of *Bipolaris bicolor* and *Bipolaris sorokiniana* on Wheat Seeds (*Triticum aestivum* L) in sBrazil, *Brazilian Journal of Microbiology* **37** 247-250

81. Morh RM, Flasten DN, Bernier CC, and Racz GJ 1987 Effect of chloride fertilization on Bedford barley and Katepwa wheat, *Canadian Journal of Soil Science* **75** 15-25

82. Mujeeb-Kazi A 1998 Alien genetic diversity for bread wheats resistant to *Helminthosporium sativum (Cochliobolus sativus)* current status and future projections In *Helminthosporium Blights of Wheat Spot Blotch and Tan Spot* (Duveiller, E, Dubin, HJ, Reeves, J and McNab A, eds) Mexico, DF, Mexico CIMMYT, 223-229

83. Mujeeb-Kazi A, Villareal RL, Gilchrist LI, and Rajaram S 1996a Registration of five wheat germplasm lines resistant to Helminthosporium leaf blight, *Crop Science* **36** 216–217

84. Mujeeb-Kazi A, Rosas V, and Roldan S 1996b Conservation of the genetic variation of *Triticum tauschii* (Coss) Schmalh *(Aegilops squarrosa osa auct non* L) in synthetic hexaploid wheats *(T*

*turgidum* L s lat *T tauschii*= 42, AABBDD) and. □; 2n = 6 its potential utilization for wheat improvement, *Genetic Resources and Crop Evolution* **43** 129–134

85. Mujeeb-Kazi A, Villareal RL, Gilchrist LI, and Rajaram S 1998 Registration of five wheat germplasm lines resistant to *Helminthosporium* leaf blight, *Crop Science* **38** 116–131

86. Mukherjee S, Chowdhury AK, Bhattacharya PM, and Singh G 2011 Incidence of Helminthosporium leaf blight of wheat and biochemical background of disease resistance in the Eastern Gangetic Plains, *Journal of Wheat Research* **3** 26-28

87. Mukwavi MV 1995 Breeding for Disease Resistance with emphasis on Durability Proceedings of a Regional Workshop for Eastern, Central and Southern Africa, 143-145

88. Muyanga S 1994 wheat production and research in Zambia constraint and sustainability In Developing Sustainable Wheat Production System The 8th Regional Workshop for Eastern, Central and Southern Africa pp 57-64 Addis Ababa Ethiopia CIMMYT

89. Nagarajan S and Kumar J 1998 Foliar blights of wheat in India Germplasm improvement and Future Challenges for Sustainable, High Yielding Wheat Production In *Helminthos-porium Blights of Wheat Spot Blotch and Tan Spot* (Duveiller, E, Dubin, HJ, Reeves, J and McNab A, eds) Mexico, DF, Mexico CIMMYT, 52-58

90. Nelson RR 1960 Evolution of sexuality and pathogenicity Interspecificcrosses in the genus *Helminthosporium*, *Phytopathology* **50** 357–377

91. Nelson RR 1961 Evidence of gene pools for pathogenecity in species of *Helminthosporium*, *Phytopathology* **51** 736–737

92. Nema KG and Joshi LM 1973 Spot blotch disease of wheat in relation to host age, temperature and moisture, *Indian Phytopathology* **26** 41-48

93. Nema KG and Joshi LM 1971 The spot blotch disease of wheat caused by *Hilminthosporium sativum*, *Proceedings of Second International symposium of Plant Pathology IARI*, New Delhi, 42

94. Nisikado Y 1928 Studies on the *Helminthosporium* diseases of Gramineae in Japan Ann Phytopath Soc Japan 296 Special Report, Ohara Institute Agricultural Research 4 iv, 384, 11

95. Noak F 1905 *Helminthosporium gramineum* Rabenh Und *Pleospora trichostoma* Wint Z Pflanzenkrankh **15** 193-205

96. Olliviera AMR, Matsumura ATS, Prestes AM, Matos GS, and Van Der Sand ST 1998

Variabilidade patogenica e morfologica em isolados de *Bipolaris sorokiniana*, *Fytopatol Gras* **23** 437-442

97. Osorio L, Garcia I, Lopez F, and Duveiller E 1997 Improving the control of tan spot caused *Pyrenophora tritici-repentis* in the Mixteca Alta of Oaxaca, Mexico In *Helminthosporium Blights of Wheat Spot Blotch and Tan Spot* (Duveiller, E, Dubin, HJ, Reeves, J and McNab A, eds) Mexico, DF, Mexico CIMMYT, 142-145

98. Pammel LH, King CM, and Baker AL 1910 Two barley blights with comparison of species of *Helminthosporium* upon cereals, *Bull Iowa Agric Exp Stn* **116** 178-190

99. Pandey SP, Sharma S, Chand R, Sahi P, and Joshi AK 2008 Clonal variability in the spot blotch pathogen *Bipolaris sorokiniana* of wheat and its relevance in generation of new pathotypes, *Current Microbiology* **56** 33- 41

100. Parashar M, Nagarajan S, Goel LB and Kumar J 1995 Report of the coordinated Experiments 1994 –95 Crop protection (pathology) AICWIP, Directorate of wheat Research, karnal, 206

101. Pavlova VV, Dorofeeva LL, and Kozhukhovskaya VA 2002 Effectiveness of seed treatments against root rots of cereals No 8, 21-23 Moscow Russia; Izdatel stvokolos [Ru] All Russian Research Institute of Phytopathology, Russia (Cab International)

102. Pedretti R and Viedma 1988 Estrategia de la DIEAF-PARAGUAY para el Mejoramiento para Resiatencia a Enfermedades In Seminario sobre Mejoramiento Genetico sobre Resiatencia a Enfermedades de Trigo Passo Fundo, PR, Brasil IICA/PROCISUR/BID

103. Prabhu AS and Singh A 1974 Appraisal of yield losses in wheat due to foliar disease by *Alternaria triticina* and *Helminthosporium sativum*, *Indian phytopathology* **27** 632 – 834

104. Raemaekers RH 1988 *Helminthosporium sativum* Disease complex on wheat and sources of resistance in Zambia In Wheat Production Constraints in Tropical Environments (ed) Klatt, AR, Mexico, DFCIMMYT, 175-185

105. Raemaekers R 1991 Contribution to the epidemiology of *Bipolaris sorokiniana* diseases and the development of rain fed wheat, a new crop in Zambia Dissertationes de Agricultura, Feculteit der Landbouwwetenschappen, Katholieke Universiteit te Leuven, Belgium

106. Rashid AQMB, Sarker K, and Khalequzzaman KM 2001 Control of *Bipolaris* leaf blight of wheat with foliar spray of Tilt 250 Ec, *Bangladesh Journal of plant pathology* **17** 45-47

107. Razzaque MA and Hossain AS 1991 The wheat development programme in Bangladesh In Wheat for the Non Traditional warm areas, D A Saunders (ed), Mexico, D F CIMMYT, 44 –45

108. Regmi AP, Ladha JK, Pasuquin EM, Pathak H, Hobbs PR, Srestha LL, Gharti DB, and Duveiller E 2002 The role of potassium in sustaining yield in long-term rice-wheat experiment in the Indo-Gangetic plains of Nepal, *Biology and Fertility of Soils* **36** 240-247

109. Reis EM 1990 Integrated Disease Management the changing concept of controlling head blight and spot blotch, In: DA Saunders (ed), Wheats for the nontraditional Warm Areas, Bangkok, Thailand CIMMYT, 165-177 pages

110. Ruckstuhl M 1998 *Bipolaris sorokiniana* Studies on its epidemiology and population structure in the rice-wheat cropping pattern of Nepal, in *Helminthosporium* Blights of Wheat Spot blotch and Tan spot (eds E Duveiller, HJ Dubin, J Reeves and A McNab), CIMMYT, DF Mexico, 88-105

111. Saari EE 1998 Leaf blight diseases and associated soil borne pathogens of wheat in south and South East Asia, in *Helminthosporium* Blights of Wheat Spot blotch and Tan spot (eds E Duveiller, H J Dubin, J Reeves and A McNab), CIMMYT, DF Mexico, 37-51

112. Sarhan ART, Kiraly Z, Sziraki I, and Smedegaard-Peterson V 1991 Increased levels of cytotoxins in barley leaves having the systemic acquired resistance to *Bipolaris sorokiniana* (Sacc) Shoemaker, *Journal of Phytopathology* **131** 101-108

113. Schilder and Bergstrom G 1993 Tan spot In SB Mathur and BM Cunfer, (eds) Seed-borne diseases and seed health testing of wheat, Copenhagen, Denmark, Jordburgsforlaget, 113-122 pages

114. Sharma RC, Dubin HJ, Devkota RN, and Bhatta MR 1997 Heritability estimates of field resistance to spot blotch in four spring wheat crosses, *Plant Breeding* **116** 64-68

115. Sharma RC and Duveiller E 2004 Effect of Helminthosporium leaf blight on performance of timely and late-seeded wheat under optimal and stressed levels of soil fertility and moisture , *Field Crops Research* **89** 205-218

116. Sharma-Poudyal D, Duveiller E, and Sharma RC 2005 Effects of seed treatment and foliar fungicides on *Helminthosporium* leaf blight and on performance of wheat in warmer growing conditions, *Journal of Phytopathology* **153** 401-408

117. Shoemaker RA 1959 Nomencleture of *Drechslera*

and *Bipolaris*, grass parasites segregated from *'Helminthosporium'*, *Canadian Journal of Botany* **378** 79-887

118. Shoemaker RA 1962 *Drechslera* Ito, *Canadian Journal of Botany* **40** 809-836

119. Singh DV and Srivastava KD 1997 Foliar blights and Fusarium scab of wheat present status and strategies for Management In Management of Threatening plant Disease of National Importance, New Delhi Malhotra publishing House, 1-16

120. Singh DP, Chowdhury AK, and Pankaj K 2007 Management of losses due to seedborne infection of *Bipolaris sorokiniana* and *Alternaria triticina* in wheat (*Triticum aestivum*) using seed treatment with Vitavax 200WS, *The Indian Journal of Agricultural Sciences* **77(2)** 101103

121. Singh DP, Sharma AK, Kumar P, Chowdhury AK, Singh KP, Mann SK, Singh RN, Singh AK, Kalappanvar IK, and Tewari AN 2008 Management of leaf blight complex of wheat (*Triticum aestivum*) caused by *Bipolaris sorokiniana* and *Alternaria triticina* in different agroclimatic zones using an integrated approach, *Indian Journal of Agricultural Sciences* **78** 513-517

122. Singh PJ and Dhaliwal HS 1993 Resistance to foliar blight of wheat in *Aegilops* and wild *Triticum* Species, *Indian Phytopathology* **46(3)** 941-943

123. Singh RV, Singh AK, Singh D, Singh RK, and Singh D 1997 Inheritance studies of the foliar blight of wheat caused by *Helminthosporium sativum*, **20** 37-39

124. Sivanesan A 1987 Graminicolous species of *Bipolaris*, *Curvularia*, *Drechslera*, *Exerohilum* and their teleomorphs *Mycological Papers*, No 158 Wallingford, UK CAB International Mycological Institute

125. Sivanesan A and Holliday P 1981 *CMI Description of Pathogenic Fungi and Bacteria Sheet-No701* Kew CAB-International Mycological Institute

126. Srivastava OP, Luthra JK, and Narula PN 1971 Inheritance of seedling resistance to leaf blight of wheat, *Indian Journal of Genetics and Plant Breeding* **31** 209-211

127. Stack RW and McMullen M 1988 Root and crown rots of small grains NDSU extension service PP-785 Fargo, ND, USA 8

128. Steffenson B J Hayes PM, and Kleinhofs A 1996 Genetics of seedling and adult plant resistance to net blotch (*Pyrenophora teres f teres*) and spot blotch (*Cochliobolus sativus*) in barley, *Theoretical and Applied Genetics* **92** 552-558

129. Sticher L, Mauch-Mani B, and Métraux JP 1997 Systemic acquired resistance, *Annual Review of Phytopathology* **35** 235–270

130. Subramanian CV 1971 The phialide, in B Kendrick, *Taxonomy of fungi Imperfecti*, Univ Toronto Press

131. Tewari AN and Wako K 2003 Effect of some agrochemicals on foliar disease of wheat, *Plant Disease Research* **18** 39-43

132. Tinlin RD 1960 Pathogenic and cultural variation in *Cochliobolus sativus, Transaction of British Mycolological Society* **43** 696-697

133. Toledo BJ and Guzman AE 1997 Importance of spot blotch caused by *Bipolaris sorokiniana* in Bolivia, in *Helminthosporium* Blights of Wheat Spot Blotch and Tan Spot pp 146-149 (Eds, Duveiller, E, Dubin, HJ, Reeves, J and McNab, A), 376 CIMMYT, 376 page

134. Kumar U, Kumar S, Joshi AK, Chand R, and Roder RS 2009 Mapping of resistance to spot blotch disease caused by *Bipolaris sorokiniana* in spring wheat, *Theoretical Applied Genetics* **118** 783-792

135. Kumar U, Joshi AK, Kumar S, Chand R, and Roder RS 2010 Quantitative trait loci for resistance to spot blotch caused by *Bipolaris sorokinina* in wheat (*T estivum*) line Ning 8201 and Chirya 3, *Molecular Breeding* **26** 477-491

136. Valim-Labres, M E 1995 Variabilidade intra-especifica em *Bipolaris sorokiniana* analise morfologica, isoenzimatica e de patogenecidade M Sc Thesis, Universidade Federal do Rio Grande do Sul, Porto Alegre, RS, Brazil Addy, SK and Goodman, R N (1972) Polyphenoloxidase activity in apple leaves inoculated with a virulent or avirulent strain for *Erwinia amylovora, Indian Phytopathology* **25** 575-579

137. Valim-Labres ME, Porto MD, and Matsumura ATS 1997 Effects of host resistance on the isozymatic pattern of *Bipolaria sorokiniana* (Dematiaceae, moniliales), *Brazilian Journal of Genetics* **20** 541-545

138. Valjavec-Gratian M and Steffenson BJ 1997 Pathotypes of *Cochliobolus sativus* on barley in North Dakota, *Plant Disease* **8** 11275-1278

139. Van Ginkel M and Rajaram S 1998 Breeding for resistance to spot blotch in wheat global perspective, in *Helminthosporium* Blights of Wheat Spot Blotch and Tan Spot (Eds Duveiller, E, Dubin, HJ, Reeves, J and McNab A,) Mexico, DF, Mexico CIMMYT, 162–170 pages

140. Vasil I and Vasil V 1986 Regeneration in cereal and other grass species, in I Vasil and V Vasil (eds), *Cell*

*Culture and Somatic Cell Genetics of Plants* **3** 121 – 150

141. Velazquez Cruz C 1994 Genetica de la resistencia a *Bipolaris sorokiniana* en trigos harineros MSc thesis, Coligo de postgraduados, Montecillo, Mexico, 84 page
142. Viedma LQ 1989 Importancia y distribucion de la fusariosis del trigo en el Paraguay In dialogo XIII, Royas de Cerales de invierno, PROCISUR, 153-164 pages
143. Villareal RL, Rajaram S, Gilchrist L, and Mujeeb-Kazi A 1992 Advanced lines derived from wheat (*Triticum aestivum* L) and *Thinopyrum curvifolium* resistant to *Helminthosporium sativum*In Absracts International Crop Science Congress, Ames, IA, USA Crop Science Society of America/Iowa State University
144. Wehmeyer LE 1953 On the status of the generic names *Pyrenophora* and *Pleospora, Mycologia* **45** 562-571
145. Wehmeyer LE 1961 A world monograph of the genus *Pleospora* and its segregates Ann Arbor, USA University of Michigan Press
146. Wiese MV 1977 Compendium of Wheat Diseases St Paul, Minnesota APS
147. Wisniewska H, Wakulinski W, and Chelkowski J 1998 Susceptibility of barleys to *Bipolaris sorokiniana* seedling blight determined by disease scoring and electrolyte leakage, *Journal of Phytopathology – Phytopath-ologische - Zeitschrift* **146** 563-566
148. Zillinisky FJ 1983 Common Diseases of small Grain Cereals A Guide to Identification Mexico, D F CIMMYT 141 page.

*Rev. Plant Pathol. Vol. 6, 2014*
*Indian Society of Mycology and Plant Pathology*
*Scientific Publishers (India), Jodhpur pp. 473-499*

# EPIDEMIOLOGY AND MANAGEMENT OF LEAF BLIGHT AND FRUIT ROT OF BELL PEPPER

***S.K. Gupta*[1] *and Kumud Jarial*[2]**

[1]Dr. Y.S. Parmar University of Horticulture and Forestry, Horticultural Research Station/ Krishi Vigyan Kendra, Kandaghat, Solan-173 215 (H.P.),
[2]Dr. Y.S. Parmar University of Horticulture and Forestry, Institute of Biotechnology and Environmental Sciences, Neri-1, Khaggal, Hamirpur-177 001 (HP)

**KEY WORDS :** Leaf Blight, Fruit Rot, Capsicum annum, Phytophthora capsici, Phytophthora nicotianae

## ABSTRACT

Bell pepper, an important crop cultivated throughout the world, suffers significant losses due to leaf blight and fruit rot diseases. The initial symptoms appear as water- soaked to dull or faded green areas differing slightly from the colour of the normal healthy leaf tissue which under humid conditions rapidly enlarge covering entire leaf, resulting in premature leaf fall. In case of fruits, the disease first appears in the form of dark green water- soaked spots in contrast to light green colour of the healthy fruit tissue which enlarge rapidly, ultimately covering the entire fruit. Two species of *Phytophthora* i.e. *P. capsici* and *P. nicotianae* var. *nicotianae* have been found to be associated with this disease. Both pathogens survive in the form of oospores in the soil as well as in infected seed. Both fungi remain viable for more than one year under ordinary field conditions. *P. nicotianae* var. *nicotianae* is generally associated with the symptom produced on leaves and fruits whereas *P. capsici* with the roots and stem at collar region. Combined use of fungicides, resistant varieties and modified cultural practices such as drainage with high ridges, crop rotation, intercropping and mixed cropping may help in achieving effective management of this disease.

## INTRODUCTION

Bell pepper (*Capsicum annum* L.) popularly known as 'Shimla Mirch', is an important solanaceous crop grown throughout the world. The fruits are a good source of vitamins A and C. It is a native of Mexico with secondary centre of origin in Gautemala (68) and Portuguese explorers introduced it to Asia. In India, it is commonly grown in the states of Tamil Nadu, Karnataka, Himachal Pradesh, at places having mild climate with humid environmental conditions, thus, the mid hill tracts of Himachal Pradesh are more suitable for its cultivation during summer months. In mid hills, the crop is transplanted in April-May and fruits are available during June to September. Recently, this vegetable has been adopted by the farmers for cultivation at higher elevations because of very high returns. They plant it in June-July and the crop is ready for harvest during September to November. Thus, the bell pepper fruits from Himachal Pradesh are available from June to November in the markets located in the plains of Northern India.

Since the active growth period of the crop coincides with the advent of monsoon rains, which favours the development of various fungal and bacterial diseases, of which leaf blight and fruit rot [*Phytophthora capsici* Leon. and *Phytophthora nicotianae* (Breda de Haan) var. *nicotianae* Watershouse] is the predominant and most destructive diseases in the mid hills. This disease is highly destructive to both leaves as well as fruits, resulting in complete defoliation of the crop in most cases (67). This review aims to deal with the information available on different aspects of this disease world over with special reference to the work done under mid hill conditions of Himachal Pradesh.

## LOSSES

Losses from leaf blight and fruit rot disease are greater than any other diseases of bell pepper. These results from:

(i) defoliation and mortality of young plants at post transplanting stage,

(ii) premature defoliation of leaves and fruit rot,

(iii) reduction in fruit production due to flower blight and reduction in net photosynthetic area and

(iv) increased expenses due to spraying operations.

The prevalence of warm-humid conditions during rainy season (July-September) may result in heavy blight development on leaves and rotting of fruits, necrotic lesions on stems and branches, and thereby resulting in rapid multiplication of inoculums in the soil, thus reducing the potential yields, drastically. Another burden, which this disease places on the growers, is the increased cost of production due to expenditure incurred on repeated sprays of fungicides at short intervals (7-10 days).

Leonian (1922) reported that the-

Reduction in yield has been reported (41) mainly due to premature defoliation of plants and rotting of fruits. The yield loss of infected plants ranged from 43-100 per cent (42). In India, the 58 to 100 per cent incidence of fruit rot under Himachal Pradesh conditions (67), while leaf blight severity and fruit rot incidence during crop season varied between 35.0 to 62.7 and 30.4 to 68.7 per cent, respectively (6).

## GEOGRAPHICAL DISTRIBUTION:

The disease is prevalent world over wherever warm- humid conditions exist during the growing season. Leonian was the first to report (41) its occurrence from Mexico. Subsequently, it was reported from Puerto Rico (74), Argentina (43), Florida (79), Greece (61), Russia (52), Venezuela (44), Japan (33), France (14) and Pakistan (27).

In India, the disease caused by *Phytophthora capsici* Leon was first time reported from the Solan area of Himachal Pradesh (71), Later on, it was reported (65) the disease to be caused by *Phytopthora nicotianae* Breda de Haan var. *nicotianae* Waterhouse.

## SYMPTOMS

Both the pathogens have been reported to attack all the aerial plant parts such as leaves, stem, branches, growing shoots, collar region and fruits (25) and are described in detail as below:

## Leaves

The initial symptoms appear as water- soaked to dull or faded green areas differing slightly from the colour of the normal healthy leaf tissue (Fig 1 A). Under humid conditions, the diseased areas rapidly enlarge covering entire leaf, resulting in premature leaf fall. However, with the decrease in relative humidity and intervening clear sky, further development of the disease is restricted and the spots get bleached to light tan colour with brown coloured margins.

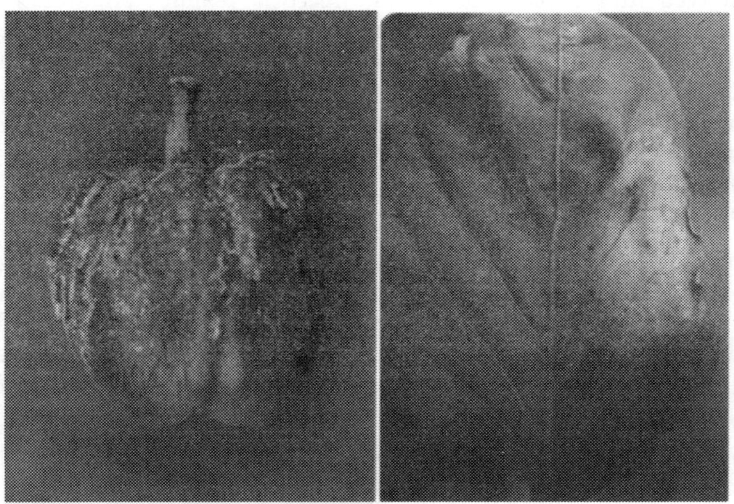

Fruit rot Leaf Blight

**Fig 1.** A, B and C

### Stems and shoots

Symptoms on stems appear first as dark green water soaked areas, irregular in shape, turning dark brown to black with the passage of time, giving a blighted appearance (Fig 1 B). The infection of growing shoots results in the rapid killing of the growing tips as well as flower buds progressing downwards towards the base. At such a stage, complete defoliation is often observed.

### Collar region

Symptoms on collar regions of adult plants appear as water soaked areas engirdling the collar region and the point of contact of the soil line with whitish growth of mycelium. The rot often progresses downward to the roots in the affected plants and there is sudden drooping of leaves of the whole plant, giving the appearance of sudden wilt.

### Fruits

The fruits are vulnerable to infection irrespective of their size and age except the fruits nearing maturity with the change of fruit colour from light green to red. The disease first appears in the form of dark green water- soaked spots in contrast to light green colour of the healthy fruit tissue. The spots enlarge rapidly, ultimately covering the entire fruit (Fig 1 C). The fruits showing initial symptoms when cut open exhibits profuse mycelial growth inside the fruit cavity, even covering entire surface of the pericarop as well as immature seeds. Completely rotten fruits may fall down on the ground, often invaded by saprophytic fungi and bacteria causing soft rots.

## THE PATHOGEN(S)

Two species of *Phytophthora* i.e. *P. capsici* Leon and *P. nicotianae* Breda de Haan var. *nicotianae* Waterhouse have been found to be associated with this disease (65,70). However, *P. capsici* is of worldwide occurrence. Morphological characters (24) of both the species are being described in detail as below:

***Phytopthora capsici***

The fungus shows very good growth on oatmeal agar medium. The colonies are fluffy and cottony white in appearance. Mycelium is hyaline, branched 3.3-8.6 μm wide and non-septate but few septa are found in case of old hyphae (Fig. 2). Typically hyphal branches arise at right angles (Fig. 2) often variously swollen and tuberous and of abnormal diameter. Sporangia are rarely produced or almost absent in culture but are abundantly produced in tap water or $KNO_3$ (0.1 M) solution. The sporangia are hyaline, ovoid to pyriform or sometimes round to lemon shaped, non-pedicillate with a predonminant, and hemispherical papilla at the apex measuring 29.7-59.8 x 33.0-72.5 μm. The mean size of the sporangia is 38.4 x 45.5 μm. Sporangia germinate readily in tap water at room temperature (17-23°C) by releasing 15-20 zoospores. The zoospores are reniform to oval, biflagellate and remain motile for 20-30 minutes then shed flagella and become encysted. The encysted zoospores germinate by means of germtubes. Oospores are formed abundantly in aerial as well as submerged mycelia in the medium. The oogonia are hyaline, circular to spherical. The oogonia are hyaline, circular to spherical. The oogonial stalk is encompassed by persistent, amphygynous antheridium. The oospores are circular to spherical, 16.5 to 26.4 μm, averaging 21.28 μm. Single oosporic cultures of the pathogen exhibited the phenomenon of homothalism. The oospores are found to germinate either by germ tubes or by the rupture of oosporic walls or disintegration of oospores or by stalked or sessile germ sporangia. The fungus does not produce chlamydospores.

***Phytophthora nicotianae* var. *nicotianae***

The colonies of the fungus on oatmeal are cottony white and fluffy in appearance. The hyaline branched and non-septate mycelium is 3.5 – 8.5 μm in width with branches at right angles (Fig. 3). The sporangia are not produced in culture but are abundantly produced in tap water or $KNO_3$ (0.1 M) solution. The sporangia are hyaline, non-pedicellate, globose to subglobose with a prominent beak like papilla at the apex at least 4 μm deep and measured 15.0 – 42.0 x 17.0 -47.0 μm. The length: breath ratio is generally 1.2:1, however, in certain cases it raises to 1.4:1. The sporangia germinate by production of 15 to 20 zoospores. Oospores are formed abundantly on aerial and submerged mycelium in culture. The oogonia are hyaline, globose and measured 23.0 x 21.0 $\mu$m. The oogonical stalk was encompassed by a persistent amphigynous antheridium. The oospores are thick walled golden brown in colour and measured 19.0 -37.0 $\mu$m.

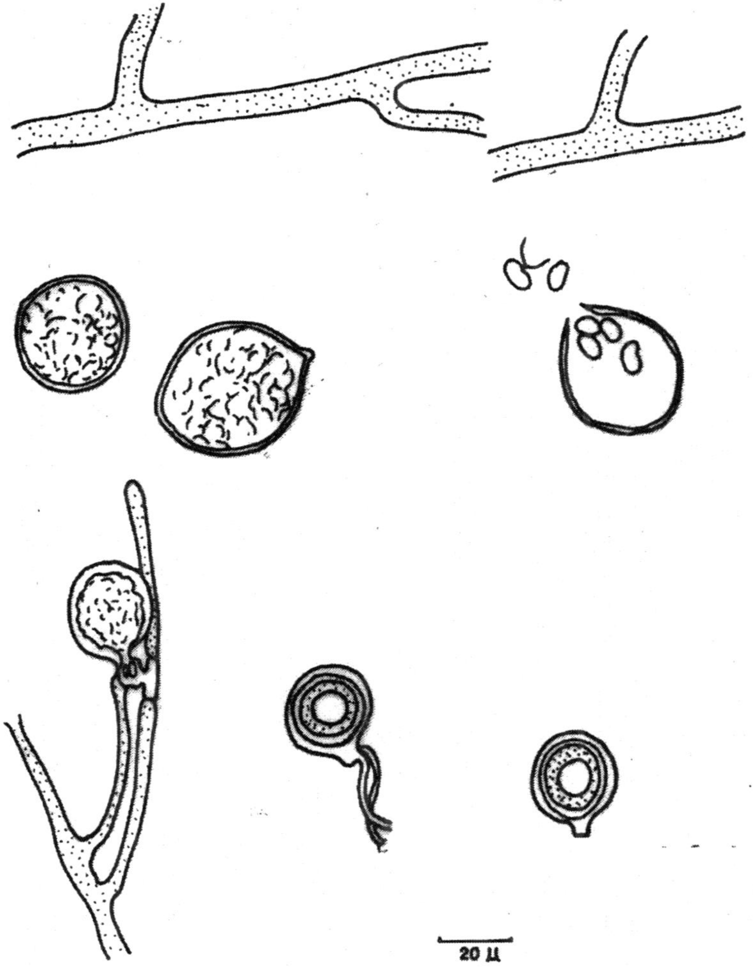


Hyphae, Sporangia, Release of zoospores, oogonium, fertilizing oogonium, mature oospore

**Fig. 2.** Morphological structures of *Phytophthora capsici*

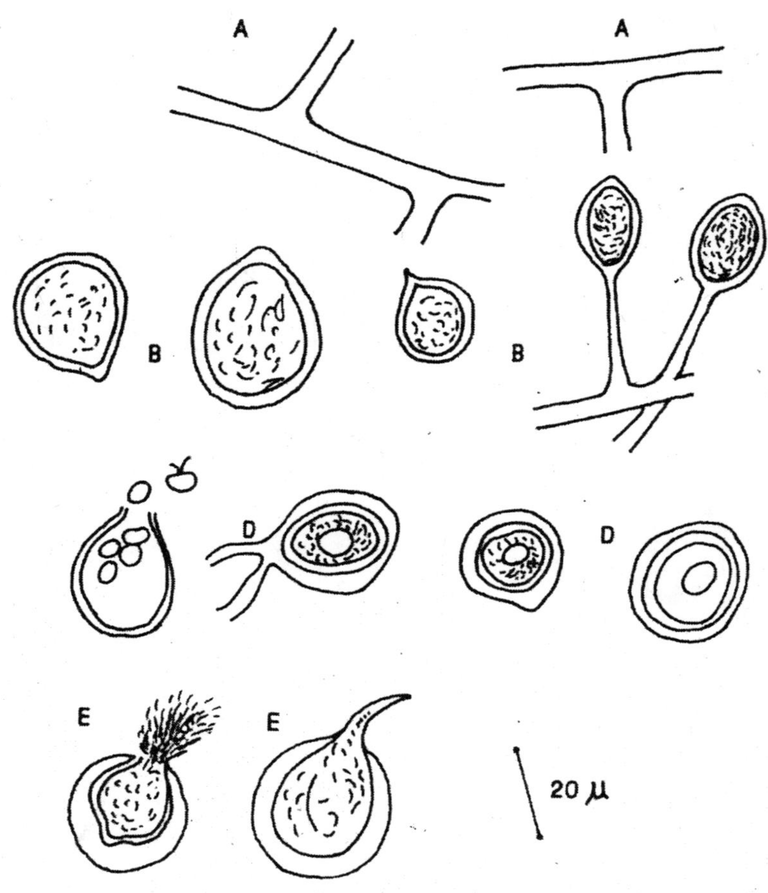


A. MYCELIUM B. SPORANGIA AND SPORANGIOPHORES
C. ZOOSPORES RELEASE FROM SPORANGIUM
D. OOSPORES E. GERMINATING OOSPORES

**Fig. 3.** Morphological structures of *Phytophthora nicotianae* var. *nicotianae*

## Variability

*Phytophthora nicotianae* var *nicotianae* and *P. capsici*are important oomycetous fungi which occur as aggregate of strains varying in cultural appearance, morphological characteristics, virulence and physiology which in turn is manifested by variation in genotypes of different strains. Variation among different species of *Phytophthora* is inherent as seen from the various reports on the subject as also from the review (5). The existence of pathogenic variation among bell pepper isolates of *P. nicotianae* var. *nicotianae* collected from different parts of Himachal Pradesh have been reported. Traditional markers used to study the variability in plant pathogens are based on the differential hosts, cultural characters, morphological markers and biochemical tests. Cultural, morphological and genetic variability among 23 isolates of *P. nicotianae* var. *nicotianae* revealed that the colour of mycelium varied from white to creamy or dirty white and the appearance of the mycelium was fluffy or suppressed and dry or slimy. Morphological studies in relation to sporangial and oospore size revealed that the sporangial size ranged from 7.21 x 6.18 μm (PN 4) to 59.62 x 34.65 μm (PN 21). The shape of the sporangia ranged from round to oblong and they were papillate. Oospores were recorded only in ten isolates and were round in shape, brown in colour with thick wall. The oospore size ranged from10.30 x 9.27 μm to 45.32 x 45.32 μm (23).

These markers are influenced by host age, inoculum quality and environmental conditions. Moreover, these techniques are time consuming and laborious. In such cases, molecular markers are used for studying genetic variability in plant pathogens. Using PCR, very closely related strains of pathogen can be distinguished without prior *knowledge* of the nature of polymorphic regions by use of RAPD (Randomly Amplified Polymorphic DNA). RAPD offers very promising, versatile and informative molecular tool to detect genetic variation within populations of plant pathogens. Isolates of *P. nicotianae* var. *nicotianae* have high degree of phenotypic variability (5). The genetic variation by using twenty four decamer oligonucleotide RAPD primers have been studied, out of which 14 primers yielded informative, strong and reproducible DNA amplicons of *P. nicotianae*var. *nicotianae*. The dendrogram (Fig. 4) produced from computerized cluster analysis of the DNA fingerprints revealed a great deal of heterogeneity among isolates as expected after DNA banding pattern analysis of gel image (Fig. 5).

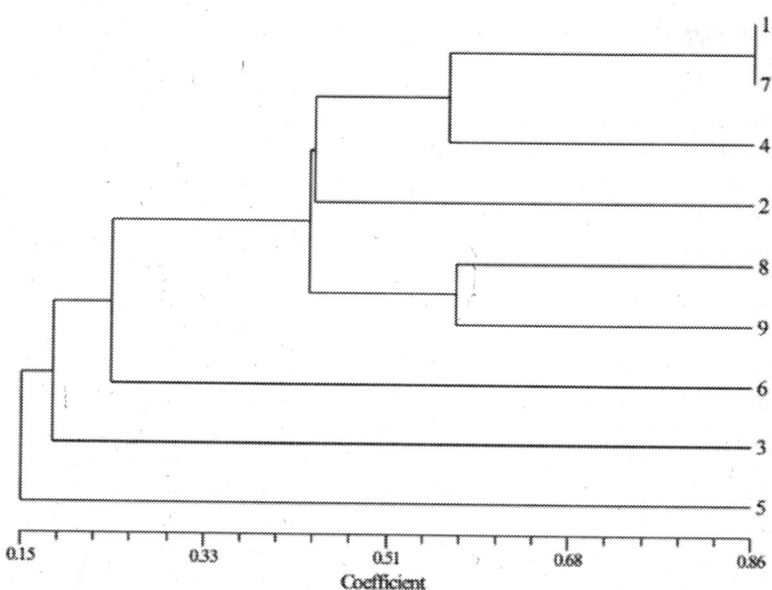


**Fig. 4.** Dendogram derived from banding patterns of RAPD analysis

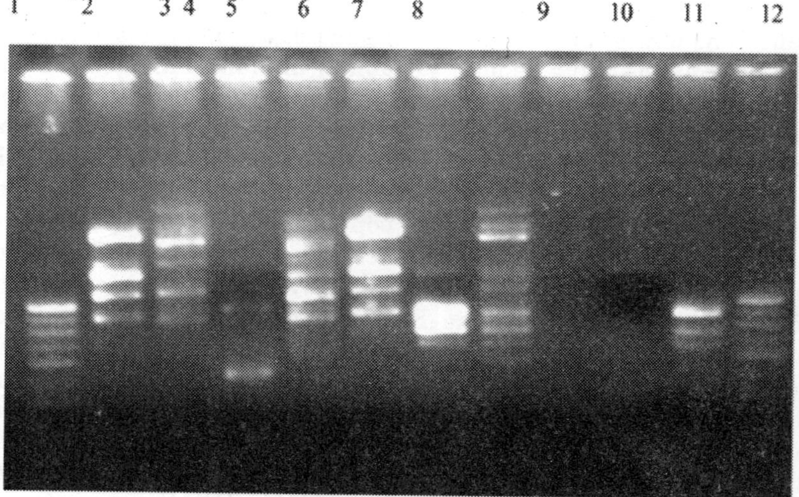


**Fig. 5.** Amplification products of *Phytophthora nicotiane* var. *nicotiane* isolates with primer OPV 14. Lane 1 &12 : 100 bp (molecular weight marker), Isolates 1, 2, 3, 7, 16, 19, 20, 22, 8, 11 in lanes 2-11, respectively.

## HOST RANGE

The host range of the *P. capsici* as reported by various workers includes pepper, brinjal, tomato, cucumber, pumpkin, watermelon, bottle gourd, squash melon, tobacco, petunia (4,5,33,62,67). Ornamental plants like paparnomia, Saintpaulia, Gloxima and carnation have been reported to be infected by this fungus (57) while also reported it on *Hibiscus* (66).

## DISEASE CYCLE

In areas, where the crop is grown regularly in the same field, the fungus over winter in the form of oospores in soil (67) and they remain viably during unfavourable weather conditions in affected plant debris or on the infected seeds from rotten fruits which are thrown in the field during the crop season. *P. capsici* could survive in moist soil for at least 15 months even in the absence of host while oospores remain viable for the least 7-8 months in the soil. Infected seed constitute an important source of dissemination of the fungus (32,41). The fungus has been reported to be both externally and internally seed borne (67). With the onset of monsoon rains in the second fortnight of June, in the presence of high soil moisture and moderate temperature (20-25°C), the oospores germinate either by germ tube or by stalked or sessile germ sporangia. The sporangia in the presence of free water standing on the soil surface during rainy period produce biflagellate zoospores (Fig. 6), which after swimming for some time in a film of water lose flagella and become encysted. Such encysted zoospores constitute the primary inoculums, which bring about infection on leaves and fruits through rain drop splashes from the soil. Zoospores of *P. capsici* encysted on the surface of Capsicum leaves and penetrate either directly or by a germ tube or by an infection peg from an appressorium into an epidermal cell, occasionally through stomata (35). Penetration is complete in 4 h. The fungus invades intracellularly, downward as well as radially. The hyphae invading intracellularly are smooth with granular protoplasm, branching profusely in the walled parenchymatous cells of the host. These hyphae penetrate deep into the tissue. The symptoms develop on different parts after 24-72 h. The mycelium in the epidermal layer of the affected portion of the fruit, agglomerate, push through the epidermis and cuticle by rupturing it and grow on the fruit surface under humid conditions, producing a crop of sporangia on simple branched sporangiophores. No sporangia are produced on the infected leaf tissue. These sporangia from fruits spread rapidly during the rainy season and cause secondary infection. When the conditions become unfavorable, the fungus enters in sexual reproduction phase and produce oospores.

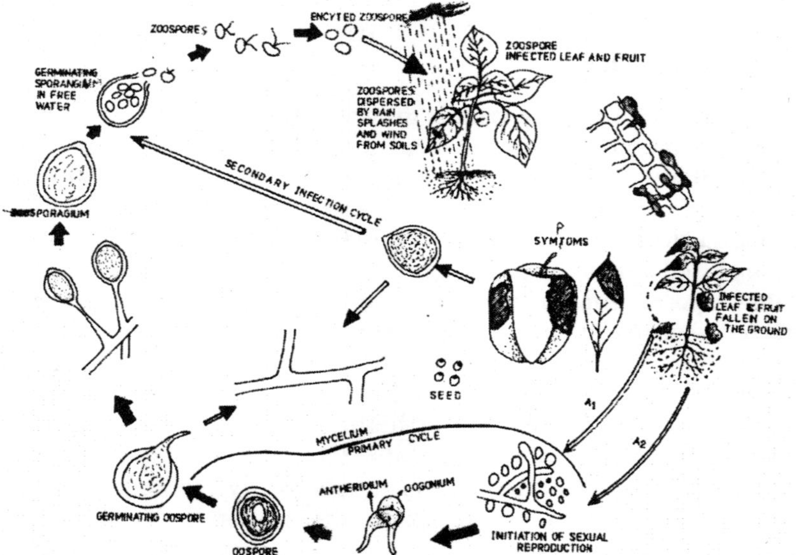


**Fig. 6.** Disease cycle of leaf blight and fruit rot (*Phytophthora nicotianae* var. *nicotianae*) of bell pepper

## EPIDEMIOLOGY

Both pathogens survive in the form of oospores in the soil as well as in infected seed (6,67). Both fungi remain viable for more than one year under ordinary field conditions. *P. nicotianae* var. *nicotianae* is generally associated with the symptom produced on leaves and fruits whereas *P. capsici* could be isolated from the roots and stem at collar region. Both the species were found pathogenic on leaves, fruits, stems and roots with shortest incubation period on leave followed by fruits (Table 1). *P. nicotianae* var. *nicotianae* had shorter incubation periods on leaves (26 h) and fruits (28h) than *P. capsici* while the reverse was true in *P. capsici*having shorter incubation period on roots (42h) and stems (52h). The fruits situated on the lower branches were first to be infected (41). Mechanism of dispersal of primary inoculums of *P. capsici* in soil was studied (71) and found that the incidence was highest when sporangia and mycelia were buried directly in the soil. Presence of abundant rainfall, high RH and warm weather is essential for initiation of this disease while it is completely absent in dry weather. The incidence of Phytophthora blight was more during 1995 as compared to 1996 because the rain fall was 44.3 cm in 1995 as compared to 23.1 cm in 1996 (71). The inoculums surviving in soil is splashed from the soil with splattering rains and after infection; sporangia are produced on the fruit surface and are carried by wind to the adjoining fruits. When infected fruits fall down on the ground, the sporangia and zoospores get mixed with rain water in soil and they are either splashed to the foliage and also move in soil and cause root infection.

**Table 1.** Pathogenecity of *P. nicotianae* var. *nicotianae* and *P. capsici* on different plant parts of bell pepper (78)

| Plant parts | Incubation period (h) | |
|---|---|---|
| | ***P.n.* var. *nicotianae*** | ***P. capsici*** |
| Leaves | 26 | 30 |
| Fruits | 28 | 32 |
| Stems | 60 | 52 |
| Roots | 85 | 42 |

Various workers have reported warm temperature ranging between 20 to 30°C in combination with high relative humidity (>70%) and rainfall as essential parameters (Fig. 4) for maximum disease development on fruits and leaves (18,31,48,67). A positive correlation between disease development and high relative humidity, rainfall and temperature has been reported (8). The distribution of the disease in the field has been reported to follow field drainage pattern. Splashing rain and run off water have been observed to be the main source of secondary spread of the disease.

Sporangia of *P. capsici* liberated zoospores only in the presence of free water (34). Zoospore penetrates the host either through stomata or directly through epidermal cells within 60-120 minutes. The incidence of *P. capsici* is more in clay loan soil in comparison to sandy loam soil (81).

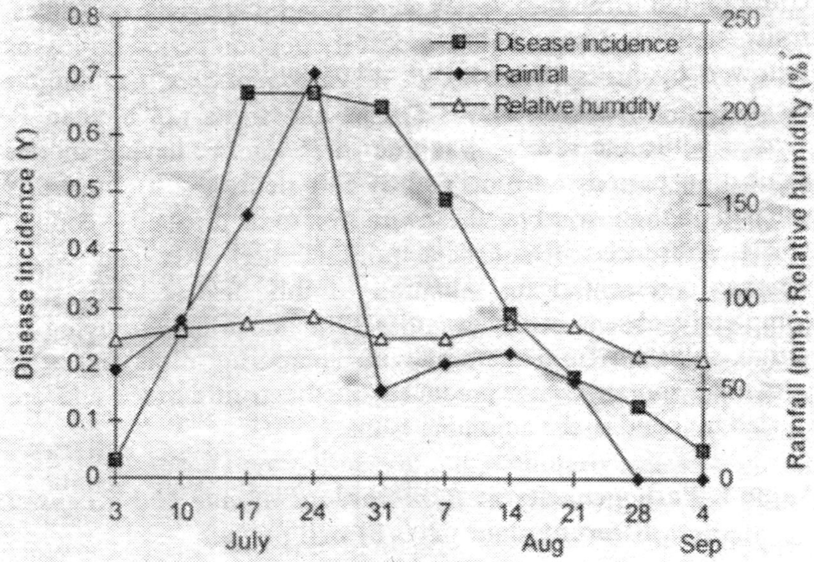


**Fig. 7** Relationship between disease incidence and meteorological factors

## MANAGEMENT

Since, the casual fungi persist either in the soil or in infected seed, therefore, integration of various cultural biological, host resistance and chemical methods is expected to enhance substantially the efficacy of the disease management.

### a) Host resistance

Growing of resistant plants is the most practical effective and economical approach for suppression of soil borne plant diseases. It not only reduces the crop losses but also lessen the current expenditure on the applications of chemical control measures besides also being an eco friendly approach. Several workers evaluated the available cultivars and germplasm lines to identify source of resistance against this disease. Cvs./lines like PI-188376,PI-202132, PI-201234 (37), Waxy Globe (64), PI-123469, PI-163192, PI-201232, PI-201234, PI-224445, P 51, AC 2258,

Riogrande and Gosung (10), Sweet peppers like Smith 5, Sweet pepper 469, Stal Cristal and L 5279 (15), Kiraaseong (13), Kingkun (29), Calafyuco INTA (17), Serrano Criolla de Morelos (72), Szechwan (26), PM-702 (76), Santaka Pimento Naciional (60), PI-59236 (9), Tian Jiao-2 (49), PBC-408 (59), PI-201232 and SCM-334 (50) have been reported to be resistant towards *P. capsici*.

None of the cvs./lines reported to exhibit resistance against Phytophthora blight (*P. n.* var. *nicotianae*) except PBG-631 (38). PBG-631 is horticulturally superior as fruits are non pungent and large in size (15-16 x 3-4cm), with nearly appropriate shape. Similarly, cv. Kandaghat Selection was also found resistant both under laboratory and field conditions (77). They further reported higher degree of horizontal resistance in the moderately resistant cvs./lines as reflected by longer incubation periods and lower rates of lesion development while apparent infection rate and AUDPC were inversely proportion to the degree of resistance. Out of 39 germplasm lines screened, lines K 1046, HC-202 and K 1028 were found resistant to leaf blight and fruit rot and remained completely disease free (63). Out of 28 cultivars / lines evaluated for resistance to this disease under field conditions, only one line DG (Dark Green) was found resistant while rest were moderately resistant to highly susceptible (22). Higher degree of horizontal resistance was present in moderately resistant lines 'HC 201" and 'Feroz' as reflected by slowest and moderate apparent infection rate, respectively under field conditions besides comparatively longer incubation periods under laboratory conditions.

The sources of resistance against this disease mostly belongs to pungent peppers, however, a few sweet pepper genotypes were also reported to be resistant (15) which are of tremendous use because many generation of backcrossing are needed to remove undesirable traits of pungent peppers introduced together with resistance. The most important problem in transferring resistance from pungent peppers into bell peppers is the removal of pungency and attaining of proper size, shape and colour of fruits because these traits are controlled by polygenes (19). The inheritance of resistance to *P capsici* has been reported to be monogenic dominant (64), digenic (58), trigenic (51), polygenic (53) and epistatic (3).

While studying the inheritance of resistance, Waxy Globe appeared to be controlled by one or several major dominant genes with some modifier (64), while resistance in line LS 279 was conferred by a single incompletely dominant gene (80). Resistance in pungent types of *C. annuum* has been observed to be controlled by two recessive genes and a dominant modifier (46). When one of the resistance genes is present in the heterozygous conditions, resistance occurs if a modifier is present. When both resistance genes are present in heterozygous conditions, the modifier must be homozygous.

Seven bell pepper germplasm lines possessing different degree of resistance/ susceptibility were also screened against 23 isolates of *P. nicotianae* var. *nicotianae,* out of which, DG (Dark Green) was found to be tolerant to maximum 16 isolates while Feroz and Danish Wonder were tolerant to 14 isolates (23). Genetic variation among germplasm lines including resistant and susceptible ones by employing molecular markers revealed a great deal of variation among the lines. In all, 23 RAPD primers were used out of which 10 were recorded to be polymorphic and only one (D-17) was monomorphic (Fig. 8). The dendrogram thus prepared revealed that the similarity coefficient values ranged from 0.59 to 0.77 indicating a fair range of variability in the genotypes studied. Analysis of dendrogramme thus obtained exhibited lines Feroz, Danish Wonder and Dark Green to be individual groups at 59, 65 and 71 per cent values of genetic similarity (Fig. 9), respectively. These lines can thus be employed in breeding programmes for developing resistant varieties/hybrids having resistance against broader range of pathogen isolates.

**b) Cultural practices**

Various cultural practices like selection of healthy seed (41), rouging of diseased plants/parts, mulching with straw (52), transplanting of the plants on top of furrow increased to 30 cm (16), soil solarization with transparent polythene sheet along with application of fumigants (11) and cultivation of Capsicum with non-host plants like groundnut sesame, onion, garlic, pea and spinach (39) have been reported effective in reducing the severity of disease. Different mulches and soil amendments have been evaluated for the management of this disease and found application of pine needle mulch before the onset of monsoon rains as effective against this disease.

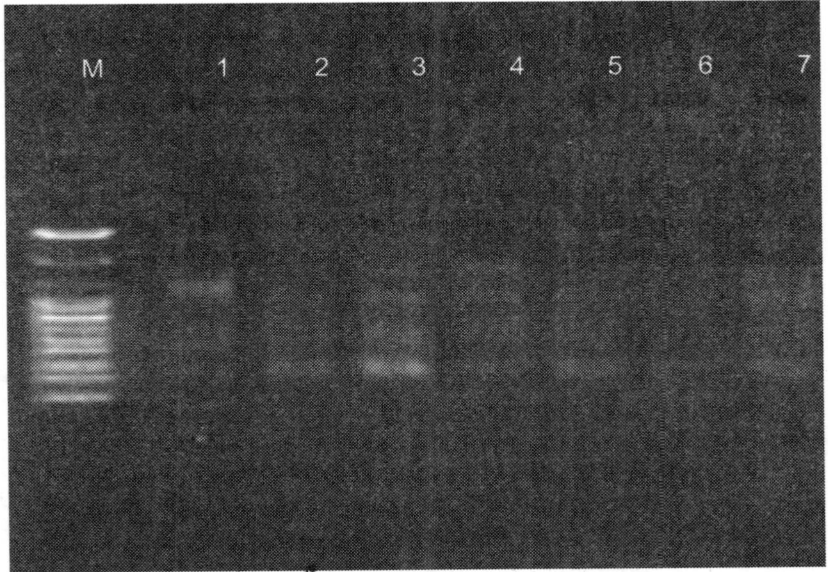


**Fig. 8.** Amplification products of bell pepper germplasm lines with primer 34MP10A16. Lane M: 100 bp (molecular weight marker), Lane 1: Sel-64-1-4, Lane 2: SP 701, Lane3: Dark Green (DG), lane 4: California Wonder (CW), Lane 5: Sel 22-1, Lane 6: Danish Wonder (DW), Lane 7: Feroz

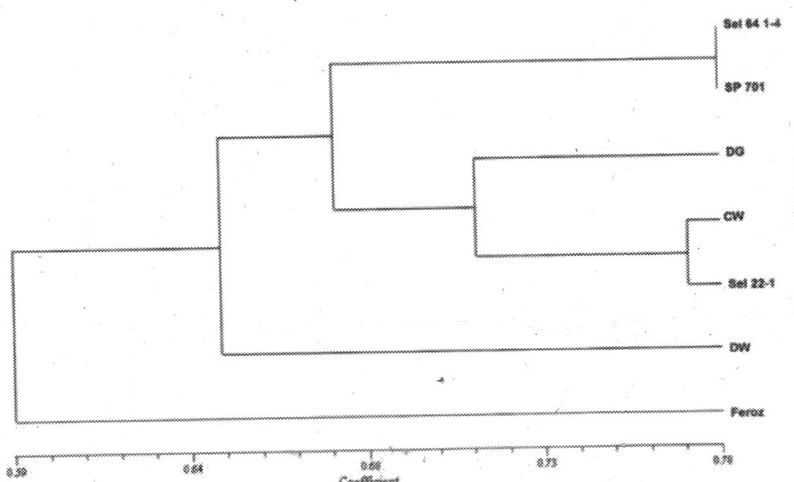


**Fig. 9.** Dendogram derived from banding patterns of RAPD analysis with primers

### c) Biological Control

Several fungal and bacterial antagonists viz., *Trichoderma Pseudomonas cepacia, Bacillus polymyxa and Streptomyces parvullus* (30,40,55) inhibited the growth *of P. capsici and P. nicotianae* var *necotianae* under *in vitro* conditions. Immersion of seed or dipping of roots and lower stems of seedlings into the homkogenate of C/2-9 (new race of *Streptomyces ochraceis-cleroticus*) before planting in natural, inoculated soil gave 73.3 per cent protection of *C. annuum* against *P. capsici* (75). Promising results were also obtained when antagonists like *Pseudomonas cepacia, Bacillus polymyxa* and *Bacillus* sp. were applied with suitable organic soil amendments and calcium fertilizers (12). Application of cultural filtrates of actinomycetes could control *Phytophthora* blight markedly (2). *Trichoderma* spp. and *Streptomyces californicus* (MITCC No. 8771) caused >50 per cent mycelial growth inhibition of the pathogen (*P. nicotianae* var *necotianae*). These antagonists were further evaluated for their plant growth promoting activity like siderophore production and phosphate solublizing activity. *Streptomyces californicus* showed phosphate solublizing activity (Plate 2). Under field conditions, soil and foliar application of *Streptomyces californicus* reduced the fruit rot incidence to 11.70 per cent as compared to 48.26 per cent in control and increased yield by > 100 per cent (73).

### d) Chemical Control

Various systemic and non-systemic fungicides have been reported to be quite efficacious in the management of this disease both under laboratory and field conditions. However, the frequent rains during the month of July-August, which are essential for good growth of bell pepper plants, also promote rapid multiplication of inoculums of the pathogen and spread of the disease. Sometimes, the delay in the routine applications of fungicides often results in heavy disease outbreak. Proper selection of fungicide and timing of initial sprays is of prime importance to keep the disease under check.

Fungicides like captafol, prothiocarb, metalaxyl, Fosetyl-Al, copper sulphate and captan are reported effective under *in vitro* conditions against *P. capsici* (1,14,54,62) while Dichlone, captafol, pyroxychlor, Blitox-50, Dithane Z-78, Antracol and ziram were effective against *P. nicotianae* var. *niocotianae* (6,47). Efficacy of potassium phosphonate and monoethylphosphonate against isolates of Phytophthora species has also been reported (56).

Bordeux mixture, Kirti copper, Zincop, zineb and captafol have been reported effective in controlling the disease on bell pepper caused by *P. capsici* under field conditions (14,41,44,47, 54, 69). Blitox-50 (0.3%) was found to be the most effective in checking the infection of *P. nicotianae* var *necotianae* var. *nicotianae* (7) and increased fruit yield, followed by ziram (0.2) and Bordeaux mixture (0.8%). Soil drenching with metalaxyl was the most effective chemical treatment and gave 40 to 60 per cent control of the disease (36), while seed treatment with captan and Ridomil and spraying with metalaxylwas also effective for control of this disease (27). Activities of sodium tetrathiocarbonate (STTC) and metalaxyl on the viability of *P. capsici* within colonized *C. annum* tissue was studied and found that pathogen was not recovered from colonized tissue buried in soil that was drenched with 4900 mg/ml of STPC while recovery from metalaxyl (10 µg/ml) drenched tissue did not differ from that obtained from water drench only (45).

Proper timing of initial sprays is of prime importance to achieve desirable level of disease control. Systemic and non-systemic fungicides against this disease including two strobilurin fungicides both under *in vitro* and *in vivo* conditions were evaluated (21). Under *in vitro* evaluation, metalaxyl +mancozeb, mancozeb, cymoxanil + mancozeb, carbendazim + mancozeb and copper oxychloride completely inhibited the mycelial growth of the fungus while, kresoxim methyl was least effective. When evaluated by detached leaf method under laboratory conditions, metalaxyl + mancozeb, mancozeb, cymoxanil + mancozeb, copper oxychloride and azoxystrobin exhibited 100 per cent disease control while carbendazim + mancozeb treated leaves developed 66.67 per cent disease severity. In protective spray programme, four sprays of metalaxyl + mancozeb (0.25%) at fortnightly intervals were found highly effective followed by copper oxychloride and cymoxanil + mancozeb while captan was least effective. Maximum fruit yield was also recorded in cymoxanil + mancozeb followed by metalaxyl + mancozeb which were statistically at par with each other. Though the efficacy of strobilurin fungicides was found slightly less as compared to other combi – fungicides like metalaxyl + mancozeb and cymoxanil + mancozeb but being ecofriendly with least residual toxicity these can be applied in integrated management of this disease at critical stages of the crop even after fruit formation for obtaining produce with minimum residues. Therefore, to check the buildup of inoculums of this disease effectively, the different activities of fungicides can be utilized in planning different spray schedule.

## INTEGRATED DISEASE MANAGEMENT

Combined use of chemicals, resistant varieties and modified cultural practices may help in achieving effective management of this disease. Use of integrated disease management practices like cultural methods (drainage with high ridges, crop rotation, intercropping and mixed cropping) resistant cultivars alongwith appropriate fungicides (manocozeb and metalaxyl) have been suggested (28) for the effective management of this disease. Application of pine needle mulch on the field floor alone and in combination with the sprays of metalaxyl + mancozeb (0.25%) on mulch and foliage resulted in significantly lower disease levels followed by sprays of copper oxychloride (0.3%) both on mulch and foliage and increased fruit yield (20).

## INTEGRATED MANAGEMENT PRACTICES FOR IMPORTANT DISEASES OF BELPEPPER AND CHILLIES

| SNo. | CROP STAGE / DISEASE | MANAGEMENT PRACTICES |
|---|---|---|
| 1.<br>2.<br>3.<br>4.<br>i)<br>ii) | Pre-sowing treatments Pre-sowing treatment of nursery beds for reducing the population of soil borne pathogens.<br>At nursery state<br>After seedling emergence for the control of damping-off<br>At transplanting stage<br>Post transplanting stage<br>With the onset of pre-monsoon showers and during rainy season for the management of fruit rot and bacterial leaf spot | Mix the soil well with FYM and prepare raised beds. The surface of beds should not be concave.<br>Mix captafol or captan (6-8 g/m·) in upper5 cm soil layer and level the surface or treat the soil with formalin (5%).<br>Drench the beds with metalaxyl + mancozeb (0.25%) at 8-10 days interval.<br>Give light irrigations and ensure drainage in case of heavy soils.<br>Select healthy and vigorous seedlings. |

| | |
|---|---|
| During dry and cool nights with dew at fruit maturity initiation stage | Ensure proper drainage during rainy season and collect diseased leaves and |
| | fruits regularly to destroy them. |
| | Spray the crop with the combination of metalaxyl + mancozeb (0.25%) followed by sprays of copper oxychloride (0.25%) at 8-10 day interval. |
| | Spray crop with copper oxychloride (0.25%) or carbendazim (0.1%) to protect fruits from anthracnose at 10-15 days interval. |

## Recommended brand names of fungicides

| | |
|---|---|
| Captafol- | Difolatan/Foltaf |
| Captan- | Captaf/Hexcap |
| Metalaxyl+mancozeb- | Ridomil MZ |
| Copper oxychloride- | Blitox/Fytolan |
| Carbendazim- | Bavistin/Bengard/JK Stein/Derosal. |

## REFERENCES

1. Abdellaouni-Maane S, Seng JM, Saindrenan P, and Bompeix G 1988 Fosetyl-AI is effective against mutants of *Phytophthora capsici* resistant to metalaxyl, *Cryptogamie Mycologie* **9** 47-56
2. Ahn SJ and Hwang BK 1992 Isolation of antibiotic-producing actinomycetes anatagonistic to *Phytophthora capsici* from pepper growing soil, *Korean Journal of Mycology* **20(3)** 259-268
3. Bartual R, Lacasa A,Marsal JI, and Tello JC 1994 Epistas is in the resistance of pepper to *Phytophthora* stem blight (*Phytophthora capsici* Leon) and its significance in the production of double cross performances, *Euphytica* **72 (1/2)** 149-152
4. Bell DK and Haasis FA 1967 Etiology and epiphytology of root rot, stem necrosis and foliage blight of Box –wood caused *by Phytophthora parasitica* Dast. *Bulletin of North Carolina Agricultural Experiment Station* **177** 47
5. Bhardwaj SS 1978 Pathogenic variability in *Phytophthora nicotianae* var. *nicotianae* causing fruit rotof bell pepper (*Capsicum annuum* L.),

*Pantnagar Journal of Research* **3** 140-141

6. Bhardwaj SS 1983 Studies on fruit rot and leaf blight of bell pepper (*Capsicum annum* L.) caused by *Phytophthora nicotianae* B. de hann var. nicotianae Waterhouse. Ph.D. thesis, HPKVV, Solan 114 page
7. Bhardwaj SS and Sharma S L 1985 Chemical control of fruit rot and leaf blight of bell pepper in India, *Agriculture, Ecosystems & Environments* **14** 151-154
8. Bhardwaj SS, Sharma SL, and Dohroo NP 1985 Role of meteorological factors in development of bell pepper fruit rot (*Phtophthora nicotianae* var. *nicotianae* Waterhouse), *Indian Journal of Mycology and Plantpathology* **15** 328-329
9. Boiteux LS, Cupertino FP, and Reifschneider FJB 1983 Capsicum chinensis PI 159236: a source of resistance to *Phytophthora capsici* and tomato spotted wilt virus (ISWV), *Capsicum and Eggplant Newsletters* **12** 76
10. Byung SK 1988 Characteristics of bacterial spot resistant *Phytophthora* blight resistant lines of *Capsicum* pepper, *Journal of Korean Society for Horticultural Science* **29(4)** 247-252
11. Cartia G, Cipriano T, and Greco N 1989 Effect of solarization and fumigants on soil borne pathogens of pepper in green house, *Acta Horticulturae* **255** 111-116
12. Cho EK 1987 Strategies for biological control of soil borne disease in economic crop in Korea, *Korean Journal of Plant Pathology* **3(4)** 313-317
13. Choi JK, Park DY, Woo YH, and Sung JM 1990 Studies on the resistance of red pepper varieties to *Phytophthora* blight and anthracnose, *Res Rep Rural Dev Admn Hort* **32(2)** 1-9
14. Clerjeau M and Beyries A 1978 Comparative study of the preventive action and systematic capacity of some new fungicides (phosphates, prothiocarb, pyroxychlor) against *Phytophthora capsici* (Leon) on pepper, *Station de Pathologie Vegetable Centre de Researches d' Avignon*, INRA, Mntafavet, France 73-83 pages
15. Fernandez MC 1988 Evaluation of sweet and chilli pepper (*Capsicum annuum*) genotypes for Resistance to *Phytophthora capsici* Leonian, *Agricultural Tecnica* **48(4)** 359-362
16. Ferreyra ER, Tosso TJ, and Fernandez MC 1984 Effect of irrigation water management on *Phytophthora capsici* Leonian, casual agent of *Capsicum annuum* blight, *Agricultura Tecnica* **44** 319-324
17. Galmarini A, Sentiner A, and Galmarini H 1991 Breeding pepper (*Capsicum annuum* L.) for Resistance to *Phytop-*

*hthora capsici* in Argentina: Calfayuco INTA' a new cultivar, *Capsicum Newsletters* 10-61

18. Godoy EF 1940 The mildew or blight of chilli produced by *Phytophthora capsici* in Argentina Republic Ev, *Fac Agron La Plata* **24** 235-280
19. Greenleaf WH 1986 Pepper breeding. In: Breeding Vegetable Crops, (ed) M.J. Basett, Avi Publishing Co, 67-134 pages
20. Gupta SK and Kumud J 2008 Management of fruit rot (*Phytophthora nictianae* var *nicotianae*) of bell pepper (*Capsicum annuum*) with mulch and fungicide application, *Journal of Mycology and Plant Pathology* **38 (3)** 689 – 690
21. Gupta SK and Kumud J 2010. Comparative efficacy of two strobilurins in management of leaf blight and fruit rot (*Phytophthora nicotianae* var. *nicotianae*) of bell pepper, *Indian Phytopathology* **63(2)** 216-218
22. Gupta SK, Kumud J, Sandeep K, Thakur MC, Vikram A, and Seema M 2007 Evaluation of bell pepper (*Casicum annuum*) germplasm for resistance to leaf blight and fruit rot caused by *Phytophthora nicitianae* var *nicotianae*, *Indian Journal of Agricultural Science* **77(4)** 267-269
23. Gupta SK, Kumud J, Saharan MS, Kaur R, and Bhawana S 2011 Morphological, genetic and pathological variation among *Phytophthora nicotianae* var. *nicotianae* isolates incitant of fruit rot of bell pepper and genetic variability among host genotypes, *Indian Phytopathology* (In Press)
24. Gupta SK, Shyam KR, and Bhardwaj SS 2002 Leaf blight and fruit rot of bell pepper. Research Information Bulletin No. 4. Department of Mycology and Plant Pathology, Dr. Y.S. Parmar University of Horticulture and Forestry, Nauni, Solan, 24 page
25. Gupta SK and Thind TS 2006 Diseases of bell pepper and chillies, in *Disease Problems in Vegetable Production*. Scientific Publishers (India), Jodhpur
26. Hartman GL and Hawang TC 1992 *Phytophthora* blight of pepper: screening for disease resistance, *Tropical Pest Management* **38(3)** 319-322
27. Hussain A, Ahmad MN and Akhtar AS 1990 Protect chillies crop from *Phytophthora* new disease In Pakistan, *Capsicum Newsletter* **59** 8-9
28. Hawang BK and Kim CH 1995 *Phythphthora* blight of pepper and its control in Korea, *Plant Disease* **79** 221-227
29. Hawang BK and Kim YJ 1990 Capsidiol production in pepper plants associated with age-related resistance to *Phytophthora capsici*, *Korean*

*Journal of Plant Pathology* **6 (2)** 193-200

30. Jee HJ and Kim CH 1988 Studies on biological control of *Phytophthora* blight of red pepper, isolation of antagonists and evaluation of antagonistic activity in vitro and in green house, *Korean Journal of Plant Path*ology **4(4)** 305-312

31. Jia JS 1992 On chilly pepper *Phytophthora*blight and its control in Xinjiang, *Acta Phytopathologica Sinica* **22** 257-263

32. Jindal KK, Gupta SK, and Shyam KR 1994 Studies on the germination, vigour and microflora of bell pepper seeds from healthy and diseased fruit, *Indian Journal of Mycology and Plant Patho*logy **24(3)** 227-228

33. Katsura K and Tokura R 1954 Studies on *Phytophthora* diseases of economic plants (VII) "A brown rot of watermelon caused by *Phytophthora capsici* Leon, *Sci Rep Fac Agric Saiko Uni*versity **6** 38-48

34. Katsura K 1956 Studies on the physiology of sporangial germination of *Phyto phthora capsici* Leon, *Sci Rep Fac Agric Saiko University* **8** 97-106

35. Katsura K and Miyazaki 1960 Leaf penetration by *Phytophthora capsici* Leonian, *Sci Rep Kyota University Agriculture* **12** 65-69

36. Kim CH 1989 Soil borne diseases: *Phytophthora* blight and other diseases or red pepper in Korea, *Extension Bull.- ASPAC, Food and Fertilizer Technology Centre* **302** 10-17

37. Kimble KA and Grogan RG 1960 Resistance to *Phytophthora* root rot of pepper, *Plant Disease Reporter* **44** 872-873

38. Kohli UK 1997 Breeding for resistance against *Phytophthora* blight of bell pepper. In: *Breeding for resistance to biotic and abiotic stresses in vegetable crops*, CAS Horticulture (Vegetables), Dr Y.S. Parmar University of Horticulture and Forestry, Solan, 3-8 pages

39. Lee HU, Kim CD, Lee YS, and Yoo CT 1993 Effects of soil chemical properties and soil Microflora by anti plants *on Phytophthora* blight incidence of red pepper, *RDA Journal of Agricultural Science and Soil Fertility* **35(2)** 305-310

40. Lee HU, Kim CD, Lee YS, and Yoo ID 1990 Antifungal antibiotic against fruit rot disease of red pepper from *Streptomyces parvulus, Korean Journal of Applied Mycology and Biotech*nology **18 (2)** 142-147

41. Leonian LH 1922 Stem and fruit blight of pepper caused by *Phytophthora capsici* sp. nov, *Phytopathology* **12** 401-408

42. Liang Y, Lu JD, and Ji RZ 1992 Study on the occurrence and yield loss of pepper *Phytophthora* blight in Shaanxi, *Plant Protection* **1** 14-15

43. Lindquest JC 1932 On the presence of *Phytophthora capsici* in Argentina Republic, *Physis* **11** 170-174

44. Malaguti G and Pontis VRE 1950 Blight or mildew of Peimento pepper in Venezuela, *Agric Venz* **14** 4-5

45. Matheron ME and Matejka JC 1995 Comparative studies of sodium tetrathiocarbonate and metalaxyl *on Phytophthora capsici* and root and crown rot on chilli pepper, *Plant Dis*ease **79** 56-59

46. Matsuoka K, Venetti CA, Costa H, and Pinto CMF 1996 Diseases caused by fungi in Capsicum and pepper, *Informe Agropecuario (Belo Horizonte)* **18 (184)** 64-66

47. McIntyre JC and Lukens RJ 1977 Screening chemicals to control *Phytophthora parasitica* var. *nicotianae, Plant Disease Reporter* **61** 366-370

48. Nakayama RM 1960 *Verticillium* wilt and *Phytophthora* blight of chilli pepper, Disease Abst **20** 4484-4485

49. Niu E, Shuquin Q, Niu EZ, and Qu EZ 1994 New released pungent *Capsicum* pepper cultivar, Tian Jiao 2, *China Veg*, **6** 6-7

50. Ortega RG, Espanol CP, and Zueco JC 1995 Interactions in the pepper-*Phytophthora capsici* system, *Plant breeding* **114** 74-77

51. Ortega RG, Palazon EC, and Cuartero ZT 1992 Genetic relationships among four pepper genotypes resistant to *Phytophthora capsici, Plant Breeding* **108(2)** 118-125

52. Osnitzkaya NEA 1949 *Phytophthora* of tomatoes in protected soil, *Orchard and Kitchen Gard* **2** 62-64

53. Palloix A, Abak K, Daubeze AM, Guldur M, and Selassie KG 1994 Survey of pepper diseases affecting the main production regions of Turkey with special interest in viruses and poty virus pathotypes, *Capsicum & Egg plant Newsletter*s **13** 78-81

54. Papavizas CC and Bower JH 1981 Comparative fungitoxicity of captafol and metalaxyl to *Phytophthora capsici, Phytopathology* **71** 123-128

55. Puri A, Agarwal A, and Mehrotra RS 1994 Antagonism of *Trichoderma* against fungal pathogens andtheir enzyme production, Advances in *Plant Science* **7(1)** 147-153

56. Quinette DG and Coffey MD 1989 Comparative antifungal activity of for uphosphonate compounds against isolates of nine *Phytophthora* species, *Phytopathology* **79** 761-767

57. Rattink H 1981 Characteristics and pathogenecity of six isolates from pot plants, *Netherlands Journal of Plant Patholo*gy **87** 83-80

58. Reifschneider FJB, Boiteux LS, Vechia, Della PT, Populos JM, Kuroda N, and Della Vechia PT 1992 Inheritance of adult plant resistance to *Phytophthora capsici* in pepper, *Euphytica* **62(1)** 45-49

59. Riley MK and Bosland PW 1995 Additional sources of resistance to *Verticullium* wilt and *Phytophthora* root rot of *Capsicum*, *Capsicum and Eggplant Newsletters* **14** 65-67

60. Saborio MM and Pacheco LH 1993 Reactions of progenitors and hybrids of Capsicum to artificial inoculation with *Phytophthora capsici* under field conditions, *Manejo Integradode Plages* **30** 11-14

61. Sarejanni JA 1936 Collar rot of cultivated solanaceae and the classification of the genus *Phytophthora, Ann Inst Phytopathology Benaki, Greece* **1** 35-52

62. Satour MM and Butler ED 1968 Comparative morphology and physiological studies on progenies from interspecific matings of *Phytophthora capsici*, *Phytopathology* **58** 183-193

63. Sharma HR, Gupta SK, Thakur AK, and Kumud J 2006 Evaluation of bell pepper germplasm for resistance to bacterial spot, leaf blight and fruit rot diseases, *Journal of Mycology and Plant Pathology* **36 (2)** 219-223

64. Sharma PP and Saini SS 1979 Inheritance of resistance to *Phytophthora capsici* Leon in *Capsicum annuum* L, *Haryana Journal of Horticultural Science* **8** 51-57

65. Sharma SL and Bhardwaj SS 1976 A fruit rot and leaf blight of bell pepper caused by *Phytophthora nicotianae* var. *nicotianae*, *Indian Phytopathology* **29** 93-94

66. Shukla RS, Sattar A, Kumar S, Singh KP, and Hussain A 1981 *Phytophthora* leaf blight of *Hibiscus moschatus* - new record from India, *Indian Phytopathology* **34** 539

67. Shyam KR 1969 The fruit rot of bell pepper (*Capsicum annuum* L.) caused by *Phytophthora capsici* Leon, M Sc Thesis, Punjab University, Chandigarh, 80 page

68. Smith PG and Heiser CB 1957 Breeding behavior of cultivated peppers, *Proceedings of American Society of Horticultural Science* **10** 286-290

69. Sohi HS, Shyam K R, and Sharma SL 1976 Fruit rot of bell pepper caused by *Phytophthora capsici* Leonian, *Indian Journal of Horticulture* **33** 266-273

70. Sohi HS, Shyam KR, and Sharma SL 1971 Fruit rot of *Capsicum annuum* caused by *Phytophthora capsici* Leon, *International Symposium of Plant Pathology*, Bangalore, 31 page (Abstr)

71. Sujkowaski LS, Para GR, Gumpertz MI, and Ristanio J B 2000 Temporal dyanamics of Phytophthora blight on bell

pepper in relation to the mechanisms of dispersal of primary inoculums of *Phytophthora capsici* in soil, *Phytopathology* **90** 148-156

72. Tamietti G, Nervo G, and Restaino F 1992 Screening of Capsicum genotypes for resistance to *Phytophthora capsici* Leon. and *Verticillium dahlia* Kelb, *Capsicum Newsletter, Spl Issue on EUCARPIA meeting, Rome*, 194-195 pages
73. Thakur MC, Gupta SK, Kansal S, and Vikram A 2009 Developing superior varieties/ hybrids of bell pepper resistant to fruit rots and their epidemiology and management. Final Progress Report of the ICAR Adhoc research project. Dept. of Vegetable Science, Dr YS Parmar University of Horticulture and Forestry, Solan, 113
74. Tucker CM 1928 Report of Plant pathologists, *PR Agricultural Experiment Station Rep* 25-27
75. Turhan G 1981 A new race of *Streptomyces ochraceisclerot-icus* in the biological control of some soil borne plant pathogens, in vitro studies on the possibilities of using CI 2-9 against some important diseases, *Zeitschrift fur Pflancenkrankheiten and Pflancenschut* **88(7)** 422-434
76. Ustun AS 1993 Effect of inoculums concentrations on root rot (*Phytophthora capsici* Leon) in peppers, *Doga, Turk Tarimve Ormancilik Dergist* **17(3)** 683-693
77. Verma S, Shyam KR, Gupta SK, and Sharma HR 2001 Evaluation of bell pepper (*Capsicum annuum*) germplasm for resistance against leaf blight and fruit rot (*Phytophthora nicotianae var nicotianae*), *Indian Journal of Agricultural Science* **71(3)** 219-221
78. Verma S1997 Epidemiology and management of leaf blight and fruit rot of bell pepper, M.Sc Thesis, Dr YS Parmar University of Hort & Forestry, Solan, 82 page
79. Weber GF 1932 Blights of pepper in Florida, *Florida Agricultural Experimental Station Bull*etin **244-46**
80. Yamakawa K, Mochizuk T, and Yasui H 1979 Screening cultivated and wild red peppers for *Phytophthora capsici* resistance and its inheritance, *Bull Veg Ornamental Crop Research Station* **6** 29-37
81. Yang SS, Kim CH, Cho EK, and Lee EJ 1991 Distribution and characteristics of suppressive soil to *Phytophthora* blight of red pepper in Korea, *Res Rep Rural Dev Adm Crop Protection* **33(3)** 18-22
82. Zhang PR, Chen HS, and Gu HF 1992 On the control effect of fourteen fungicides against bell pepper *Phytophthora* rot fungus, *Plant Protection* **18** 25-26.

*Rev. Plant Pathol. Vol. 6, 2014*
*Indian Society of Mycology and Plant Pathology*
*Scientific Publishers (India), Jodhpur pp. 501-518.*

# NON-TARGET EFFECTS OF FUNGICIDES

***Sanjay Goswami, Anju Bala, T.S. Thind, Sandeep Raheja, Ratinderbir Kaur and Mandeep Hunjan***

Department of Plant Pathology, Punjab Agricultural University, Ludhiana 141004

**KEY WORDS** : Fungicide, Plant disease control, Health hazard, Fungicide residue, Fungicide resistance

## ABSTRACT

Fungicides serve as important tools for managing plant diseases. Rational and recommended use of fungicides is very important to avoid non-target effects of fungicides on environment. Residue and resistance development are major problems due to non judicious use of these fungicides. Fungicides also affect the useful organisms. Few fungicides have been reported to be toxic to human being as well eg. Ethylene bis dithio carbamate (EDBC). Some other fungicides are tested amongst the cancer risk pesticides by National Research Council (NRC) are captafol and chlorothalonil which are widely used fungicides in agriculture. So, judicious use of fungicide is important to avoid non-target and harmful effects of fungicides. Keeping, these points into consideration an attempt has been made to open the chapter of non-target affects of fungicides

## INTRODUCTION

Fungicides serve as an important tool in plant disease management. These have been in use since ancient times for the control of plant pathogens and find a prominent place among important methods of disease control. In some cases these fungicides are the only means of defense in the absence of suitable host resistance and other control measures. Although they are successful in controlling plant diseases but their excessive, irrational and indiscriminate uses pose problems pertaining to the safety of the consumers as well as threaten our environment. The major problems include residue accumulation, resistance build up and non-target impact on other micro-flora.

The residue problems are serious especially when the fungicides are applied at the maturity stage and the minimum waiting period is not followed. Some fungicides get leached down into the lower soil layers/ strata contaminating the soil and ultimately ground water table.

Some pathogens also develop resistance in response to the non-judicious use of these fungicides thus resulting in the development of more virulent strains. Also these fungicides affect the non-target beneficial organisms such as antagonistic organism and predators or parasites of noxious insects.

First public awareness in relation to toxicity and non-target effects of fungicides came into light with publications of Rachel Karson's "Silent Spring" in 1962 that led to ban on DDT, which was believed to be responsible for the near extinction of such spp. as Bald eagles and California candors.

## ROLE OF FUNGICIDES IN CONTROL OF PLANT DISEASES

Although some plant diseases may be managed through resistant varieties and alteration of cultural practices, several diseases are only managed acceptably with the application of a suitable fungicide. Unlike insecticides and some herbicides, which kill established insects or weeds, fungicides are most commonly applied to protect healthy plants from infection. To be effective, fungicides are normally applied before infections become established and in a sufficient spray volume to achieve thorough coverage of the crop. About 150 different chemicals belonging to different classes, costing 4-5 billion US dollars, are used as fungicides in various countries (13).

## ADVERSE EFFECTS ON CROPS

Phytotoxic effects of fungicides are rarely noticed in arable crops, but can be a problem in orchard and glass house crops (71). Sometime these effects are only temporary and are overcome during the growing season. In some instances all cultivar of a specific crop may be effected, for example, inhibitors of sterol C14 demethylation (DMIs), which reduce growth of barley and wheat when applied at high rates on seeds. This effect is mainly related to inhibition of gibberellin biosynthesis (29).

Benomyl has negative effect on mitotic divisions in onion root tip cells as it causes abnormalities in mitotic cell divisions and the mitotic frequency in the onion root tip cells (19). The formulated product of famoxadone and cymoxanil causes anomalies in pollen meiosis and hence negatively affects pollen fertility (14).

Carbendazim, besides its direct antibiotic action against pathogens, could be phytotoxic at higher dosages to sensitive plants like tobacco. (34). Root nodulation and subsequent biological nitrogen fixing in soybean is alse greatly reduced with the application of carbendazim (44). Benomyl and carbendazim can have an iatrogenic effect on barley net blotch whereby, the fungicides increases the disease development (70).

## ADVERSE EFFECTS ON BENEFICIAL MICROFLORA

Several fungicides reduce saprophyte populations of microorganisms in the phyllosphere of the crops. If fungicides reduce such populations (e.g. yeasts) and have little or no effect on certain minor pathogens, infection by these pathogens may increase when fungicides treatments are applied. One example is the role of phyllosphere yeasts on the incidence of *Septoria nodorum* in wheat (23). A reduction of the population of yeast may also effect the fermentation of grape juice. This result has been observed when multisite fungicides were applied to control *Botrytis cinerea* (e.g. captan, dichlofluanid, thiram). Consequently these fungicides can not be applied just before harvest (4). Recent investigations conducted with DMIs, applied against *Uncinula necator*, indicate that they can disturb fermentation by acting directly on the yeast or by modifying the sterol content of the grape (50). The attack of oomycetes fungi has been found to be more serious where applications of benzimidazoles are made (21).

Fungicides may also affect antagonistic interactions between pathogens and soil microflora (10). Many systemic fungicides reduce competition for nutrients by common soil fungi like *Penicillium* and *Trichoderma* spp, thereby increasing pathogens with a natural insensitivity to these compounds. Such effects may also be based on reduction of antibiosis or parasitism. Differential sensitivity of pathogens may sometimes result in disease trading, which is a change of the dominant pathogen. For instance, use of benzimidazole fungicides to control common eyespot caused by *Pseudocercosporella herpotrichoides* resulted in higher incidence of sharp eyespot caused by *Rhizoctonia solani* (75). The use of benzimidazoles also causes reduction in population of *Entomophthora* spp. Which are natural biocontrol agent of aphids, and hence, the population of latter may build up to damaging level (31). Overall, side effects of fungicides on soil microflora are economically important in only a limited number of pesticide-crop combinations (10).

The fungicides triadimefon and propiconazole when applied repeatedly could affect the soil microbial ecology over a longer period of time (43). Soil microbial populations including various fungi, denitrifying bacteria and aerobic diazotrophs are also greatly affected by the application of mancozeb (58). Various fungicides like Benomyl, fenpropimorh, fenhexamid, mepronil, furalaxyl, thiram also affect symbiotic relationship between the plant and mycorhhizal activity (27, 56, 17, 24). Four fungicides namely thirum, benomyl, captan and carboxin used commonly are shown to be inhibitory to the axenic cultures of several species of cyanobacteria. Thus, fungicides may limit the beneficial effects of these microorganisms on soil nutrition (15).

## ADVERSE EFFECTS ON BENEFICIAL ARTHROPODS

Development of integrated crop protection requires fungicides with few or no adverse effects on parasites and predators of noxious insects and mites. Several old fungicides such as sulphur, pyrazophos, chinomethionate, which are suitable against powdery mildew of fruits, are not recommended because they are more or less toxic to various Phytoseiidae and coccinellidae (71). Consequently the only fungicide available against mildew of apple are DMIs. However, intensive use of this type of fungicide is not recommended because of the risk of resistance. Problems encountered from the adverse effects of fungicides are much less severe than those associated with the application of insecticides (71). Elemental Suphur and benomyl is toxic to non-target parasitic and predatory arthropods such as *Galendromus occidentalis* (59, 11, 3). However, other fungicides namely fenbuconazole, myclobutanil propiconazole, boscalid, fenhexamid and pyraclostrobin have been revealed to be harmless to adult *G. occidentalis* (11).

The fungus *Nomuraea rileyi,* is one of the most important natural enemies of soybean caterpillars. Outbreaks of the fungus *Microsphaera diffusa* have necessitated of difenoconazole and benomyl application on soybeans, which could result in outbreaks of noctuid populations by reduction of the natural inocula of *N. rileyi* (65). A transient negative effect of propiconazole and triadimenol on the overall abundance of higher collembolan taxa such as springtails (30).

## HEALTH HAZARDS

Ethylene bis dithio carbamate (EBDC) fungicides are the most important and widely used group of organic sulphur fungicides. These chemicals have been reported to be toxic to human beings when consumed orally. The toxic effects are expressed as inflammation of the eyes and the upper respiratory tract and allergic skin reactions (47, 79, 80). After alcohol consumption the sensitivity against dithiocarbamate is increased.

EBDC fungicides decompose under high temperature and oxidative conditions and yield ethylenethiourea (ETU) as one of their degradation products. There are reports that ETU is carcinogenic and goitrogenic (37, 38, 41, and 74). Graham et al (36) found ETU to be thyroid carcinogen in rats when supplied for two years in their diet at 250 and 500 ppm and a thyroid tumorigen at 125 ppm. At 5 and 25 ppm ETU was not harmful. Larson et al (49) found that high doses of mancozeb, and especially of maneb were teratogenic in rats but not in mice. Doses of 730 and 770 mg of mancozeb and maneb per kg of body weight, respectively, induced haemorrhages or malformatiuon in fetuses of pregnant rats. In mammals, cytotoxicity of various fungicides viz. mancozeb (61, 16), methyl thiophanate (72), have been reported recently.

Vinclozolin is a common dicarboximide fungicides used in vineyards. It is a known endocrine disruptor and leads to alterations in DNA. Effects of Vinclozolin are anti-androgenic which affects male secondary sex characters and reproductive success (7). Mutagenic and embryotoxic effects expressed by still birth, decreased survival rate, deformities of other skull bones, shortening or absence of tail with dithiocarbamates are also reported (57).

In laboratory studies made by National Research Council (NRC), these chemicals shown to pose serious cancer risk especially to children from exposure to chemical residues on processed food. These fungicides become more dangerous when exposed to heat because of production of ETU. It is also produced when fruits or vegetables carrying residues of these fungicides are cooked. This conversion may represent about 20 % of the parent residues on pre-cooked produce (53). However, results of another agency, Food and Drug Administration (FDA) over the past two decades persistently show that the residues in excess of established tolerance level are encountered infrequently (<1%). Therefore NRC's report levels to distortions in risk rankings of these chemicals. When applied suitably, the residues occurring on plant foods do not represent an acute danger for man as well as animals. Keeping in view the clearance given by Environmental Protection Agency (EPA), the use of mancozeb belonging to EBDC fungicides has cleared by Central Insecticide Board (CIB) in India.

Some other fungicides tested amongst the cancer risk pesticides by NRC are captafol and chlorothalonil. The Ministry for Environment and Energy in Denmark has recommended the ban of carbendazim and thiophanate methyl also in addition to chlorothalonil because they either threaten the environment or human health (6). The use of captafol has been banned by the Directorate of Plant Protection, Quarantine and Storage, Government of India. Captafol is reported to cause skin irritation and rash on people working with this compound. As for as environmental safety is concerned, chlorothalonil has been designated as relatively non-toxic compound. Due to low water solubility of the active ingredient, the product is tenacious and has rainfastedness. Although, this fungicide is listed amongst the cancer risk pesticides by NRC, however, the comparative estimates by FDA have found about 300 times less residue of this fungicide, thus indicating the large difference between the NRC's and FDA's interpretation on the estimated cancer risk possibilities.

Environmental hazards due to fungicides may also be there due to direct poisoning of human beings. This kind of hazard arise moré due to ignorance, carelessness, negligence rather than from inherently toxic nature of the pesticide (1). Among the poisonous fungicides still in use are mercury compounds used as seed dressers for control of seed borne diseases. The mercurials whether organic or inorganic are extremely persistent toxicants in that they undergo changes in the soil to release mercury which presents the continuing danger of toxicity inherent in the mercury ions. Based on the classification by WHO, these fungicides fall under highly hazards category and cause irritation to skin and mucous membrane, and my cause demititis and allergy (40). The use of these compounds has now been banned in several countries because of high rate of persistence of mercury ions and the risk of their getting into the food chain. These compounds are also known to produce permanent injury to nervous system.

Poisoning may also occur due to inadequate protection in some formulation manufacturing units or while applying pesticides in field. Apart from these, there are reports where the treated seeds to ensure better germination and growth, has been consumed for eating purposes resulting in poisoning of human beings. A well known example of this sort is reported from Iraq where hundreds of deaths occurred and thousands of men and women and children were permanently disabled after consuming grams dressed with a methyl mercury fungicide (18).

When sulphur is used as dust, it creates pollution hazards due to inhaling of the fungicides. The use of elemental sulphur against powdery mildew on grapes is a common practice as it adds color to the berries also. This practice amounts to direct consumption of the fungicides along with the berries. Thiram reported to have moderate acute oral toxicity to mammals cause reduction in egg laying when led to hens at 35 ppm in the diet (76). Human exposure to mancozeb can occur during treatment of mandarin fields (52). Komareka *et al* (46) reviewed high level of contamination of agricultural soils with inorganic copper based fungicides with especial reference to vineyard soils.

Pollution may also arise due to inhalation of pesticides during spray. The main rout of exposure is the skin and not the respiratory tract. Reports regarding harmful effects to human health due to direct exposure to fungicides are also available. Maneb and streptocyclin cause dermatitis and some people working with captaf or in fields treated with it showed symptoms of skin irritation and rashes (63). Residues of captan were even reported from the hands beneath cotton gloves in field workers working in peach orchards sprayed with captan (28)

Due to toxicity of soil from accumulation of some chemicals, reports of pollution of ground water are also available. Neil and Williams (55) reported detectable levels of ETU with a maximum value of 23 ppb in 45 % well sampled in potato fields. Fushiwaki et al (32) also reported high river water contamination of pentachloronitrobenzen near the cabbage fields where it was applied. A concern over the possibility of ground water contamination by chlorothalonil applications stimulated the study in the pine barren area in New Jersy, USA. It was that there is a little risk of chlorothalonil contamination (78). Soil toxicity from the accumulation of copper- based fungicides used in banana rendered the soil useless for growing any crop in Costa Rica (69). Due to change in the soil chemistry of the surface organic horizon following the oxidation of elemental sulphur, Kennedy et al (45) reported dramatic decline of the vegetation of a lodgepole pine stand.

## FUNGICIDE RESIDUE

Work done on residues of fungicides in India is meagre although a number of fungicides are being used at present. The residue levels in the soil or edible parts vary with the dose of the fungicides used and with total number of sprays done (73, 54). If the dose use is high and it is applied at the improper time and the total number of sprays exceeded than the recommended ones, there is every chance that the residues left in the crops at harvest time are high than the tolerance limit prescribed.

Beever et al (8) reported that the residues of captan when used as spray against *Botrytis* storage rot of Kiwi fruit were within the acceptable limits when used as per recommended dose. However the increased number of sprays resulted in more than acceptable limit of residue level. Tripathi et al (73) reported no residue of carboxin on grain and straw of crops when raised from treated seeds at concentration of less than 0.25%. However, at higher doses (0.4%) residues of its degraded product was reported.

Dubey and Nath (25) reported that the residues of EBDC fungicides exceed the MRL of 3 mg per kg $CS_2$ in 5% apple and 12% tomato samples showed non-detectable residues of EBDC.

Dose and number of sprays of Bavistin are reported to have considerable influence on carbendazim residues in grapes (48). They reported that two field applications of Bavistin @ 600 mg a.i. the residues remained higher than the recommended tolerance limit of 10 mg per kg throughout storage. However, two sprays with lower dose resulted in residues closer to the tolerance limit by the end of storage under ambient and refrigerated conditions. With one application at lower dose, the residues at harvest were within the acceptable limits. Residues of carbendazim on grapes may be reduced if these are washed with water.

Similarly, residues of carbendazim on cauliflower curd after 15 days of application were reported to be more with higher dose of the fungicide used. The residues were 1.85 ppm and 0.71 ppm with 500 g/ha and 250 g/ha doses used respectively (2). On betelvine also, with 0.05 % conc. the residues of carbendazim was 6.35 ppm as compared to 12.07 ppm residues with 0.1 % conc. (39). Residue of metalaxyl MZ on betelvine are also reported to be 0.1825 mg/kg and 0.4461 mg/kg with 18 kg a.i. /ha and 36 kg a.i. /ha doses respectively at 1 day after application. However, no residues of metalaxyl MZ were detected at 10 days after application (60).

The residues of thiophanate methyl and MBC on apple also were more with high dose used as compared with the less dose used. Following dipping in 250 ppm and 500 ppm aqueous suspensions of these fungicides, apples contained 0.76 ppm and 1.20 ppm thiophanate methyl and 0.52 and 0.90 ppm MBC residue respectively. The residues in both the cases persisted beyond 30 days (42).

In apple, the amount of residues of carbendazim on treated fruits vary with the dose of the fungicide used as well as contact time with the fungicide. Residues gradually declined with time. If the time for dipping treatment was increased from 4 min. to 8 min.

the deposits of carbendazim increased from 4.87 to 6.24 ppm at 500 ppm conc. of the fungicide and 6.58 ppm to 7.67 ppm at 1000 ppm conc. More than 80 % residues remained on the peel and more than 50 % residues in pulp were lost during the last days of storage. Following 4 min. dipping of fruits in 500 ppm carbendazim solution, the residues were below the MRL value of 5 ppm as given by the WHO (62).

Residues of thiophanate methyl are reported to vary on curds and leaves of cauliflower at the same dose applied to these parts. The residues reported were 0.40 ppm and 0.80 ppm on leaves and curds respectively with 250 g/ha dose applied whereas the residues was 0.55 ppm and 1.25 ppm on leaves and curds respectively with 500 g/ha dose applied (2).

## FUNGICIDE RESISTANCE

Under certain circumstances a fungicide might fail to control disease development. Poor disease control with fungicides can result from several causes, including insufficient dose, inherently low fungicide effectiveness, improper timing of application and longer interval between applications, expired product, excessive wind or wash off by rain or extremely heavy disease pressure. Field resistance to a fungicide occurs when reduced efficacy is due to the presence of resistant strains of the target fungus that were initially absent (5). This phenomenon affects most fungicide groups including some multisite compounds (e.g. mercurials, dodine). However, most prominent cases are encountered with systemic fungicides, which often posses a single biochemical target site within fungal cells (22, 35, 51, 64). Field resistance can occur more or less rapidly after a fungicide is introduced, depending upon genetic and epidemiological factor of the pathogen, the type of compound and mode of action (20). Resistance to fungicides has become a challenging problem in the management of crop diseases and has threatened the performance of some highly potent commercial fungicides (13). Worldwide, resistance in pathogen populations to more than 100 different active ingredients has been reported.

The first case on development of resistance in pathogens to fungicide in practice were reported soon after the registration and wide spread use of the systemic fungicide benomyl (benzimidazole) and dimethrimol (amino pyrimidine) in the early 1970's (9, 26, 77). Majority of the fungicides developed and registered since the introduction of benzimidazoles have site-specific mode of actions and carry risk of resistance development. While resistance risk of many of these fungicides, except phenylamides and recently introduced strobilurins, may not be as great as benomyl, strategies to manage the resistance risk should be developed and implemented to avoid unexpected disease control failures and sustain their usefulness for longer periods. More novel compounds with different mode of action (notably strobilurins, phenylpyrroles, anilinopyrimidines, spiroxamines, quinolines and phenylpyridylanines) have been developed and they posses more potent action against diverse fungal pathogens at much lower rates compared to most of the earlier fungicides. However, strobilurins have been observed to carry a high risk of fungicide resistance. (68). The pathogen populations can remain resistant for many years even if the fungicide is withdrawn. However, a gradual recovery of sensitivity can occur after the at risk compound is withdrawn as is the case of metalaxyl resistance in *Phytophthora infestans*. But if the fungicide is put to regular use again, the resistance rapidly reappears (68).

The intensive use of at-risk fungicides in different chemical groups without following resistance management principles can result in the development of resistance to multiple fungicides. For example, cucurbit powdery mildew strains have been detected with resistance to as many as four classes of fungicides including the strobilurins (QoIs), benzimidazoles, and DMI's only after two years of intensive use in Japan and *Cercospora beticola* on sugarbeet in the Red River Valley has shown resistance to fungicides in at least two classes. Because of the ease of resistance developing to the benzimidazoles and strobilurins, resistance in the same pathogen to these two classes of fungicides is highly possible with sugarbeet powdery mildew (33).

Most of the research on fungicide resistance by Indian workers has been based on laboratory studies. Although no alarming situation arising out of disease control failure due to resistance development has been noticed so for, a few cases of resistance build up after a few years of fungicide applications have been reported in some pathogens from different parts of the country, such as metalaxyl resistance in *Plasmopara viticola* (grape) and *Phytophthora infestans* (Potato), carbendazim resistance in *Gloeosporium ampelophagum* (grape) and *Venturia inaequalis* (apple), edifenphos resistance in *Drechslera oryzae* and *Pyricularia oryzae* (rice), triadimefon resistance in *Uncinula necator* (grape). These reports have been discussed in detail in a recent review bulletin on fungicide resistance situation in India (66, 67).

## CHALLENGES FOR FUTURE

Fungicides, besides having some drawbacks, are likely to continue as the second line of defense in order to save our crops from the ravages of plant diseases. However, regular and irrational use of these chemicals often leads to problems of residues in edible plant parts and development of resistant strains of the pathogens, apart from environmental pollution. In order to cope up with these problems, there is a strong need to rationalize the use of fungicides and make it need-based. The study of disease epidemiology can be useful in initiating fungicide applications through development of disease prediction systems. In country like India, where most of the farmers have small land holdings, it may not be economically feasible to recommend regular fungicide applications as spray programmes. However, seed treatment in some important cereal and grain crops can be practiced to control early infections due to seed or soil-borne pathogens that would eventually help to maintain the required crop stand. Presently, not much work on fungicide residues is being done in India and most of the laboratories are engaged in work on insecticide residues. It is now imperative to produce residue free crops for exporting to other countries and it warrants setting up of fungicide residue laboratories in different states of India so that systematic work on this important work could be taken up. Technologies need to be developed for integration of fungicides with bio-control agents to minimize their use. Research needs to be strengthened to explore the potential of plant-based chemicals for the control of plant diseases as India is a storehouse of diverse plant species and extracts/oils of many species are known to possess antifungal and anti-bacterial properties. Some agrochemical companies have recently introduced eco-friendly compounds derived from microbial species and these can be tried in India as well in order to replace existing more toxic compounds.

## REFERENCES


1. Adam AV 1978 The importance of pesticides in developing countries, in DL Gunn and JGR Stevens (eds) *Pesticides and Human Welfare*, Oxford University Press, Oxford, UK, 115-130 pages
2. Agnihotri NP, Jain HK, and Gupta RK 1988 Persistence of thiophnate methyl and MBC residues on cauliflower, *Pesticides* **22 (8)** 22-23
3. Alston DG and Thomson SV 2004 Effects of Fungicide Residues on the Survival, Fecundity, and Predation of the Mites *Tetranychus urticae* (Acari: Tetranychidae) and *Galendromus occidentalis* (Acari: Phytoseiidae), *Journal of Economic Entomology* **97** 950-956
4. Anonymous 1987 Les actions secondaire des produits phytosanitaire, *Annual ANPP* Paris Assoc Prot Plantes 123 page
5. Anonymous 1988 Fungicide resistance definition and use of terms, *EPPO Bull*etin **18** 559-64
6. Anonymous 1995 Denmark to ban cancer linked pesticides, *Pesticides Research Journal* **7 (2)** 95
7. Bayley M, Larsen PF, Bækgaard H, and Baatrup E 2003 The effects of vinclozolin, an anti-androgenic fungicide, on male guppy secondary sex characters and reproductive success, *Biology of Reproduction* **69** 1951–1956
8. Beever DJ, McGrath HJW, Clarke DL, and Todd M 1984 Field application and residue of fungicide for control of Botrytis storage rot of Kiwifruit, *New Zealand Journal of Experimental Agriculture* **12 (4)** 339-346
9. Bent KJ, Cole AM, Turner JAW, and Woolner M 1971 Resistance of cucumber downy mildew to dimethrimol, in *Proceedings of 6th Brit Insecticides and Fungicides conference* **1** 274-282
10. Bollen GJ 1993 Mechanism involved in non-target effects of pesticides on the incidence of soil borne pathogens. In *Pesticide Interactions in Crop Production: Beneficial and Deleterios Effects*, (ed) J Altman, Boca Raton, FL CRC Press, In press
11. Bostanian NJ, Thistlewood HMA, Hardman JM, and Racette G 2009 Toxicity of six novel fungicides and sulphur to *Galendromus occidentalis* (Acari: Phytoseiidae), *Experimental and Applied Acarology* 47 63-69
12. Brent KJ 1995 Fungicide Resistance in Crop Pathogens: How it can be managed? *FRAC Monograph No. 1*, Global Crop Protection Federation, Brussels, 48 page
13. Brent KJ and Hollomon DW 2000 Fungicide resistance

management, *Plant Disease Research* **15 (1)** 1-13

14. Cali IO 2008 Effects of fungicide on meiosis of tomato (*Lycopersicon esculentum* Mill.), *Bangladesh Journal of Botany* **37** 121-125
15. Cameron HJ and Julian GR 1984 The effects of four commonly used fungicides on the growth of Cyanobacteria, *Plant and Soil* **78** 409-415
16. Cecconi S, Paro R, Rossi G, and Macchiarelli G 2007 The effects of the endocrine disruptors dithiocarbamates on the mammalian ovary with particular regard to mancozeb, *Current pharmaceutical design* **13** 2989-3004
17. Chiocchio V, Venedikian N, Mafrtinez AE, Menendez A, Ocampo JA, and Alicia Godeas A 2000 Effect of the fungicide benomyl on spore germination and hyphal length of the arbuscular mycorrhizal fungus *Glomus mosseae, International Microbiology* **3** 173-175
18. Damluji SF and Tikriti S 1972 Mercury poisoning from wheat, *British Medical Journal* **1** 804
19. Dane F and Dalgiç O 2005 The effects of fungicide benomyl (benlate) on growth and mitosis in onion (*Allium cepa* L.) root apical meristem, *Acta Biologica Hungarica* 56 119-128
20. De Waard MA 1993 Management of resistance to pesticides, *Modern Crop Protection in Europe*, Wageningen udoc,
21. Dekker J 1973 *Chemotherapy*, in "Plant Disease An Advance Treatise" (ed) JG Horsfall and AE Dimond Vol I, Academic Press, NY, 307-325 pages
22. Dekker J 1985 The development of resistance to fungicides, *Prog Pesticide Biochemistry and Toxicology* **4** 165-218
23. Dik A 1990 Population dynamics of phyllosphere yeasts: influence of yeasts on aphid damage, diseases and fungicide activity in wheat, Ph D thesis, Wageningen Agricultural University 108
24. Domenico Zoccoa D, Joel Fontaineb J, Lozanovaa E, Renarda L, Bivortc C, Durandb R, Grandmougin-Ferjanib A, and Declercka S 2008 Effects of two sterol biosynthesis inhibitor fungicides (fenpropimorph and fenhexamid) on the development of an arbuscular mycorrhizal fungus, *Mycological Research* **112** 592-601
25. Dubey JK and Nath A 1991 Monitoring of ethylene –bis-dithiocarbamate residues in apple and tomato, *Pesticide Research Journal* **3 (2)** 167-168
26. Ehrenhardt H, Eichhorn KW, and Thate R 1973 Zur frage der resistanzbildung von *Botrytis cinerea* gegnuber systemischen fungizidin, *Nachrichtenbl Deutsh Pflanzenschutz (Braunschw)* **25** 49-50

27. Estelle Campagnaca E, Fontainea J, Sahraouia ALH, Laruellea F, Duranda R, and Grandmougin-Ferjani A 2008 Differential effects of fenpropimorph and fenhexamid, two sterol biosynthesis inhibitor fungicides, on arbuscular mycorrhizal development and sterol metabolism in carrot roots, *Phytochemistry* **69** 2912-2919

28. Fenske RA, Birnhavm SG, Metner MM, and Soto R 1989 Methods for assessing field worker hand exposure to pesticides during each harvesting, *Bulletin of Environmental Contamination and Toxicology* **43** 805-813

29. Fletcher RA 1987 Plant growth regulating properties of sterol inhibiting fungicides, *Advances in Agricultural Biotechnology* **21** 103-13

30. Frampton GK and Wratten SD 2000 Effects of Benzimidazole and Triazole Fungicide Use of Epigeic Species of Collembola in wheat, *Exotoxicology and Environmental Safety* **46** 64-72

31. Fry WE 1982 Principles of Plant Disease Management, Academic Press, Inc, NY

32. Fushiwaki Y, Tase N, Saeki A, and Urano K 1990 Pollution by the fungicide pentachloronitrobenzene in an intensive farming area in Japan, *Science of the Total Environment* 92 55-67

33. Gallian JJ, Notle P, and Miller JS 2004 Managing Fungicide Resistance (Website:http//www.uidaho.edu/sugarbeet/Managing fungresist.htm.pp5)

34. García PC, Ruiz JM, Rivero RM, López-Lefebre LR, Sánchez E, and Romero L 2002 Is the application of carbendazim harmful to healthy plants? Evidence of weak phytotoxicity in tobacco, *Journal of Agricultural Food Chemistry* **50** 279-83

35. Georgopoulos SG and Skylakakis G 1986 Genetic variability in the fungi and the problem of fungicide resistance, *Crop protection* **5** 299-305

36. Graham SL, Davis KJ, and Hansen WH 1975 Effects of prolonged ethylenethiourea ingestion on the thyroid of the rat, *Food and Cosmetics Toxicology* **113** 493-499

37. Graham SL and Hansen WH 1972 Effects of short term administration of ethylenethiourea upon the thyroid of the rat, *Bulletin of Environmental Contamination and Toxicology* **7** 19-25

38. Graham SL, Hansen WH, Davis KJ, and Perry CH 1973 Effects of one year administration of ethylenethiourea upon the thyroid of the rat, *Journal of Agricultural Food Chemistry* **21** 324-329

39. Guha PK, Halder P, Dasgupta B, Sengupta K, and Bhatacharaya Anjan 1990 Studies of the dissipation of the carbendazim residues on

betelvine leaves, *Pestology* **1** **(11)** 23-26

40. Gujral AS 1987 Safe handling of pesticides-Toxicity, poisoning, first aid and treatment, labels and leaflets, *Pesticides Information* **13** **(2)** 15-21

41. Innes JRM, Ulland BB , Valerio MG, Petrucelli L, Fishbein L, Hart ER, Pallota AJ, Bates RR, Falk HC, Cart JJ, Klein M, Mitchell I, and Peter J 1969 Bioassay of pesticides and industrial chemicals for tumorigenicity in mice, *Journal; of the National Cancer Institute* **42** 1101-1114

42. Jain HK and Agnihotri NP 1987 Presence of thiophanate meyhyl and MBC residues in apple following post harvest dip, *Pesticides* **21** **(4)** 31-32

43. Jui-Hung Y, JinShu C, PinJui H, and YeiShung W 2009 Effects of fungicides triadimefon and propiconazole on soil bacterial communities, *Journal of Environmental Science and Health* **44** 681-689

44. Kaur C, Maini P, and Shukla NP 2007 Effect of Captan and Carbendazim Fungicides on Nodulation and Biological Nitrogen Fixation in Soybean, *Asian Journal of Experimental Sci*ences **21** 385-388

45. Kennedy KA, Addisson PA, and Maynard DG 1988 Effect of elemental sulphur on the vegetation of a lodgepole pine stand, *Environmental Pollution* **51** **(2)** 121-130

46. Komáreka M, Cadkováa E, Chrastnýc V, Bordasb F, and Bollinger JC 2010 Environment International **36** 138-151

47. Laborie FR and DEdieu EH 1964 Allergie aux fungicides de lagamme du manebe et du zinebe, *Archives des Maladies professionnelles* (Paris) **25** 419

48. Ladania, Millind S, Dhillon BS, and Randhawa SK 1987 Residues of MBC and captan in stored perlette grapes, *Professional Hortticulture* **19** **(3-4)** 301-305

49. Larson KS, Arnander C, Cekanova E, and Kjelberg M 1976 Studies of teratogenic effects of the dirhiocarbamates maneb, mancozeb and propineb, *Teratology* **14** 171-184

50. Larue F 1991 IBS: des fungicides qui perturbent les fermentations, *Viti* **159** 106-8

51. Leroux P 1991 Resistance des champignous phytopathogens aux fungicides, *Phytoma Def cult* **34** 20-26

52. Liu KH, Kim CS, and Kim JH 2003 Human exposure assessment to mancozeb during treatment of mandarin fields, *Bulletin of Environ-mental Contamination and Toxicology* **70** 336–342

53. McEven FL and Stephenson GR 1979 The use and significance of pesticides in the environment, John-Wiley and Sons, New York

54. Mithyantha MS, Agnihothrudu V, and Kulkarni DS

1977 Determination of captafol in its formulations and residues on crops, *Current Science* **46** 788-789

55. Neil CD and Williams JS 1988 Ethylenethiourea in Mine Ground Water. *In Proc. FOCUS Cont. on Eastern Reg. Ground water Issue*, Stamford, Connecticut, USA, 161-167 pages
56. Perrin R and Plenchette 1993 Effect of some fungicides applied as soil drenches on the mycorrhizal infectivity of two cultivated soils and their receptiveness to *Glomus intraradices*, *Crop Protection* **12** 127-133
57. Pilinskaia MA 1970 Khromosomnye aberratsii u lits, kon taktiuiuschchikhs tsiramon v prolzvodstvennykh usioviiakh, *Genetics* **6 (7)** 177
58. Pozo C, Rodelas B, Salmeron V, Martinez-Toledo MV, Vela GR, and Gonzalez-Lopez J 1994 Effects of fungicides maneb and mancozeb on soil microbial populations, *Toxicological & Environmental Chemistry* 43 123 – 132
59. Prischmann DA, James DG, Wright LC, Teneyck RD, and Snyder WE 2005 Effects of chlorpyrifos and sulfur on spider mites (Acari: Tetranychidae) and their natural enemies, *Biological Control* **33** 324-334
60. Reddy N, Rao N, and Sultan MA 1994 Residues of metalaxyl MZ and mancozeb in betelvine, *Pesticide Research Journal* **6 (1)** 84-86
61. Rossi G, Palmerini MG, Macchiarelli G, Buccione R, and Cecconi S 2006 Mancozeb adversely affects meiotic spindle organization and fertilization in mouse oocytes, *Reproductive Toxicology* **22** 51-55
62. Sharma ID, Dohroo NP, Dubey JK, and Korla BN 1991 Monitoring mancozeb and carbendazim residues in ginger (*Zinger officinale* Roscoe) following post harvest dip, *Plant Disease Research* **7 (1)** 43-47
63. Sharma VK and Kaur S 1990 Contact sensitization by pesticides in farmers, *Contact Dermatitis* **23 (2)** 77-80
64. Sisler HD 1988 Fungicidal action and fungal resistance mechanism, *Fungicide Resistance in North America* St Paul, APS Press, 6-8 pages
65. Sosa-Gómez DR, Delpin KE, Moscardi F, and Nozaki MH 2003 The impact of fungicides on *Nomuraea rileyi* (Farlow) Samson pizootics and on populations of *Anticarsia gemmatalis* Hubner (Lepidoptera: Noctuidae), on soybean, *Neotropical Entomology* **32** 287-291
66. Thind TS 2002a *Fungicide Resistance in India*, Indian Crop Protection Association, Mumbai, 50 page
67. Thind TS 2002b Recent advances in the development of fungicides for plant disease control and the situation in

India, in *Proceedings of Asian Congress of Mycology and Plant Pathology, Mysore,* 150 page

68. Thind TS 2008 Fungicide Resistance: A Perpetual Challenge in Disease Control, *Journal of Mycology and Plant Pathology* **38 (3)** 407-18
69. Thrupp LA 1991 Long term losses from accumulation of pesticide residues: a case of copper toxicity in soils of Costa Rica, *Geoforum* **22** 1-15
70. Toubia-Rahme H, Ali-Haimoud DE, Barrault G, and Albertini L 1995 Effect of Four Fungicides on Barley Net Blotch Caused by *Drechslera teres, Journal of Phytopathology* **143** 335 – 339
71. Touzeau J 1981 Les effets secondaire des produits fungicides utilizes pour la protection des plantes. *In Trois Colloq Eff Non intent Fongic* Paris Soc Fr Phytiatr Phytopharm 132 page
72. Traina ME, Fazzi P, Macri C, Ricciardi C, Stazi AV, Urbani E, and Mantovani A 1998 In vivo studies on possible adverse effects on reproduction of the fungicide methyl thiophanate, *Journal of applied toxicology* **18** 241-248
73. Tripathi RK, Singh A, and Singh SL 1976 Residue analysis of wheat grains and straw produced from vitavax treated seeds, *Pesticides* **10 (8)** 33-34
74. Ulland BM, Weisburger JH, Weisburger EK, Rice JM, and Cypher R 1972 Thyroid cancer in rats from ethylene-thiourea intake, *Journal of National Cancer Institute* **49** 583-584
75. Van der Hoeven EP and Bollen GJ 1980 Effect of benomyl on soil fungi assoc-iated with rye. I. effect on the incidence of sharp eye spot caused by *Rhizoctonia cer-ealis, Acta Bot Neerl* **21** 107-12
76. Waible PE, P Omerary BS, and Johanson EL 1995 Effect of Arasan treated corn on laying hens, *Science* **121** 401-402
77. Wicks T 1974 Tolerance of the apple scab fungus to benzimidazole fungicide, *Plant Disease Reporter* **58** 886-889
78. Winnet G, Marucci P, Redukar S, and Uchrin CG 1990 The fate of chlorothalo nil in ground water in commercial in canberry culture in the New Jersey Pine Barrens, *Journal of Environmental Science and Engineering* **25** 587- 595
79. Wolf F and Jung MD 1970 Akute kontaktdermatiden nach umgang nut Nemtin (Vapam), *Z ges Hyg Grenzgeb* **16** 423
80. Zorin PM 1970 Ahergicheskiy dermatit voznikshiy v. rezultate kontakta tsinebom, *Vestn Dermatol Venerol* **44 (2)** 65.

*Rev. Plant Pathol. Vol. 6, 2014*
*Indian Society of Mycology and Plant Pathology*
*Scientific Publishers (India), Jodhpur pp. 519-539*

# MYCOTOXINS AND MYCOTOXICOSES - AN OVERVIEW

***A.K. Roy[1], N.L. Mandal[2] and Sweta Roy[1]***

[1]University Department of Botany, T. M. Bhagalpur University, Bhagalpur – 812007
[2]Department of Botany, Visva-Bharati, Santiniketan – 713235

**KEY WORDS :** Mycotoxins, Mycotoxicoses, Aflatoxins

## ABSTRACT

Mycotoxins are structurally diverse and chemically complex group of fungal metabolites and many of which have been implicated as significant health hazards on worldwide scale. However, a sizeable population in under developed countries is exposed to great health risks due to consumption of mouldy foods. Frequent outbreak of mycotoxicoses in pet animal, poultry birds and among the down trodden is a matter of great concern. There is a critical need for practical and effective solution in this problem. Relatively little attention has been paid towards the control of mycotoxin posing health risks in man and animals. The development of essential components for prevention, detection and detoxification of these important substances along with optimal methods of food and feed production, harvesting, storage and processing will foster an integrated approach to the control of mycotoxins.

## INTRODUCTION

The word toxin means a poisonous substance produced by a microbe (pathogenic or nonpathogenic), a plant, an animal or any living organisms. Amongst microbes, fungi are well known for elaborating several hundred toxic metabolites which plays crucial role in various biological and biochemical interactions (11,37). Many definitions and terminologies were made with regard to toxins till 1960 but after this a new groups of toxic fungal compounds have gained prominence when it was discovered that many of these have highly deleterious effects on human and animal systems. The effect could be embryotoxic, teratogenic, carcinogenic, mutagenic and estrogenic etc. several vital organs like leaver, kidney, lungs, brain etc. get affected. These toxic compounds have created a separate niche for them in the literature under the banner 'Mycotoxin'. Most of the definitions of mycotoxins approach towards their origin and target of toxicity in the living systems. Uraguchi and Yamazaki (40) defined mycotoxins as "secondary fungal metabolites capable of causing pathological changes or physiological abnormalities in man and warm blooded animals". The definition falls short of justification because the coldblooded animals are also susceptible to such toxins. Moreau (21) defined mycotoxins as "extra cellular zootoxic metabolites produced by moulds in food consumed by man and animals". This definition also does not seem to be appropriate because mycotoxins are produced practically on all organic substrates. On the other hand, a number of mycotoxins induce a variety of physio-pathological and cytological abnormalities in plants also. International Society for Human and Animal Mycology (ISHAM) at its meeting held in New Zealand in 1982 defined mycotoxins as "a metabolite of micro fungus which when is ingested by a natural rout in sufficient concentration will cause ill effects in animals (including birds and men)". This definition includes both primary and secondary metabolites but does not include 1 toxin produced by mushrooms and insect, 2) compounds produced by fungal pathogen in an animal.

In the present context, the term "mycotoxins" may be defined as "a group of chemically unrelated secondary fungal metabolites which are detrimental to living organisms and cause illness and death in human and animals and in certain cases are also harmful to green plants". Certain toxigenic moulds produce these chemical compounds during their synthesis in food and feed materials. The toxic syndrome induced by ingestion of mycotoxins contaminated food and feeds are referred to mycotoxicoses.

## MYCOTOXINS: PAST AND PRESENT DIMENSIONS

The association of moulds with feed, food and other agricultural commodities is known since the Biblical time, however, their role in initiating diseases in human being was experienced only after wide spread outbreak of Ergotism in Europe, Alimentary Toxic Aleukia (ATA) in Russia during World War II due to consumption of mouldy grains and Red Mould disease in Japan and in other rice growing countries. Till the middle of this century mycotoxins and mycotoxicoses were the subject of attention by only a few mycologists particularly in USSR and Japan where outbreak of mycotoxicoses were more frequent. But the situation changed dramatically in 1960, when a disease of unknown etiology, tentatively termed as "Turkey-X Disease" appeared in England Sudden death of more than 100,000 young turkeys in the course of few months in the poultry farm of United Kingdom attracted immediate and serious attention towards this problem (6). The mortality was not confined to Turkeys alone, thousands of young ducklings and pheasants also perished in nearby farms of Kenya and Uganda (2). These reports initiated the interests in the scientific community and created need of intensive investigations with multidisciplinary approach by the veterinarians, biologists, microbiologists, toxicologists and nutritionalist.

Blount (6) and Sargeant et al. (34) recorded contamination of Brazilian groundnuts meal and on analysis it was revealed that there was heavy infestation of green mould, which was later identified as *Aspergillus flavus* Link ex Fries. The active ingredient was given the name aflatoxin on the basis of producing fungus, where 'a' represents *Aspergillus*, 'fla', represents *flavus* and 'toxin' as poison (35).

With the increasing input from large number of microbiologists and biochemists, the science of Mycotoxicology has made a multidimensional expansion during the last more than three decades and some active research centers have emerged in our country too. Various universities viz. Pantnagar, Ludhiana, Bhagalpur have been working on the problems of important crops and also on the herbal drugs with regard to mycotoxins contamination. NIN (Hyderabad), CFTRI (Mysore), BARC (Trombay) and AIMS (New Delhi) have also contributed facts in the area of mycotoxins research in relation to human health.

## MYCOTOXIGENIC FUNGI

Mycotoxins are produced by a large number of fungi, viz; *Aspergillus, Penicillium, Fusarium, Claviceps, Alternaria, Pithomyces, Stachybotrys, Phoma* and *Diplodia* of these, the species of *Aspergillus, Fusarium* and *Penicillium* are main producer of mycotoxins (39). Out of about 300 mycotoxins described so far only few have been found to occur naturally. Some of the important mycotoxins producing fungi and their toxic metabolites implicated in the outbreak of mycotoxicoses are depicted in Table 1.

**Table 1**: Toxic metabolites of some fungi and their biological effects

| Mycotoxin | Mycotoxin producing fungi | Biological effects |
|---|---|---|
| Aflatoxins ($B_1$, $B_2$, $G_1$, $G_2$) | *Aspergillus flavus, A. parasiticus* | Hepatotoxic, carcinogenic |
| Ochratoxin | *Ochraceus, A. melleus Penicillium viridicatum, P. cyclopium, P. communi* | Nephrotoxic |
| Citrinin | *P. viridicatum, P. implicatum P. citrinum* | Nephrotoxic |
| Citreoviridin | *P.citreo-viride* | Cardiovascular disorder |
| Cyclopiazonic acid (CPA) | *P. cyclopium* | Human intoxication |
| Patulin | *P. patulum* | Death of cattle |
| Trichothecenes | *Fusarium tricinctum* | Dermal necrosis |
| T-2 toxin | *F. solani* | Hemorrhage |
| Zearalenone | *F. graminearum, F. roseum* | Vulvovaginitis |
| Fumonisins | *F. moniliforme* | Abortion, carcinogenic |

## CHEMISTRY AND TOXIGENIC PROFILES OF MYCOTOXINS

Mycotoxins are secondary metabolites of some fungi, which are formed by consecutive series of enzyme-catalysed reaction from a few biochemically simple intermediates of primary metabolisms, e.g., acetate, mevalonate, malonate and certain amino acids. The main biosynthetic reactions include condensation, oxidation/reduction, alkylation and halogenations steps, which create a remarkable range of secondary compounds. The principal pathways involved in the formation of mycotoxins are the polyketide route (aflatoxin), the terpene route (tricothecenes), the amino acid route (gliotoxin) and the tricarboxylic acid route (rubratoxin). Some mycotoxins e.g., cyclopiazonic acid are formed by the combination of two or more of the principal pathways.

*Aflatoxins* (Afl.): Among all mycotoxins, aflatoxins are well known as well as widely investigated mycotoxins and attracted worldwide attention since 1960 when more than one lakh Turkey birds and fourteen thousand ducklings were died in the poultry farms of England. Scientists from different disciplines started investigation to isolate the pathogens, however, no bacteria, viruses, insecticides or pesticides were detected and therefore, this disease was named as "Turkey-X-disease". Later on after examination of feeds sources, the imported Brazillian groundnut meal was found to be heavily contaminated with *A. flavus* Link ex Fries and when extracted, a highly toxic chemical compound was identified and named as aflatoxins (35). Aflatoxins (*Aspergillus flavus* toxins) are naturally occurring mycotoxins that are produced by many species of *Aspergillus*; but most notably *A. flavus* and *A. parasiticus* (36). Presently three main species of *Aspergillus* viz. *A. flavus*, *A. parasiticus* and *A. nomius* are reported having the aflatoxin production potentiality. Some other species like *A. oryzae* and *A. niger* are also reported to produce aflatoxins but authentic confirmation is still awaited. Aflatoxins are toxic and carcinogenic to animals, including humans. The exposure of the Aflatoxins may be acute or chronic depending on the test systems, doses and frequency of exposures. Sensitivity towards these varies with species, age and sex of the organisms as well as diet and route of ingestion (9). Aflatoxins often occur in crops in the field prior to harvest. Post-harvest contamination can occur if crop drying is delayed and during storage if water exceeds the critical value for mold growth (36). The aflatoxin is a generic term that refers to one or more (four) principal metabolites produced as $B_1$, $B_2$, $G_1$ and $G_2$. Chemically aflatoxins constitute an unique group of highly oxygenated coumarin derivative heterocyclic compounds which consists of more than 18 closely related compounds having their characteristic emission of blue green fluorescence under UV light (27). On the basis of blue green fluorescence and Rf value, aflatoxin is categorized mainly into 2 major groups i.e. $B_1$ and $B_2$ for blue fluorescence and $G_1$ and $G_2$ for green fluorescence (1 and 2 stands for their Rf). The Rf value of $B_1$, $B_2$, $G_1$ and $G_2$ is in decreasing order from top to base on TLC plates. Their molecular formulae as established from elementary analyses and mass spectrometric determinations are: $AFB_1$- $C_{17}H_{12}O_6$; $AFB_2$- $C_{17}H_{14}O_6$; $AFG_1$- $C_{17}H_{12}O_7$; and $AFG_2$- $C_{17}H_{14}O_7$. $AFB_2$ $AFG_2$ established as the dihydroxyl derivatives of $AFB_1$and $AFG_1$, respectively. Whereas $AFM_1$ is 4-hydroxyl $AFB_1$and $AFM_2$ is 4-dihydroxyl $AFB_2$Aflatoxin $B_1$ ($AFB_1$) (Fig. 1a) is considered the most toxic and is produced by both *A. flavus* and *A. parasiticus*. Aflatoxin $B_2$ ($AFB_2$) (Fig. 1b) is less toxic that $AFB_1$ and is also produced by both *A. flavus* and *A. parasiticus*. Aflatoxin $G_1$ ($AFG_1$) (Fig. 2a) and aflatoxin $G_2$ ($AFG_2$) (Fig. 2b) are produced exclusively by *A. parasiticus* (27). Aflatoxin $M_1$ ($AFM_1$) (Fig. 3a) and aflatoxin $M_2$ ($AFM_2$) (Fig. 3b) were first isolated from milk of lactating animals fed aflatoxin preparations. Among all categories of aflatoxins, $B_1$ is the most toxic followed by $G_1$, $B_2$ and $G_2$ in order of their decreasing potency which is known to be potent hepato-carcinogenic teratogenic and immune-suppressive too. Naturally occurring different combinations of aflatoxin have been classified as class I human carcinogenic. While the presence of *Aspergillus* in food products does not always indicate harmful levels of aflatoxins, it does imply a significant risk in consumption of that product. After entering the body, aflatoxins are metabolized by the liver to an intermediate reactive aflatoxin $M_1$.

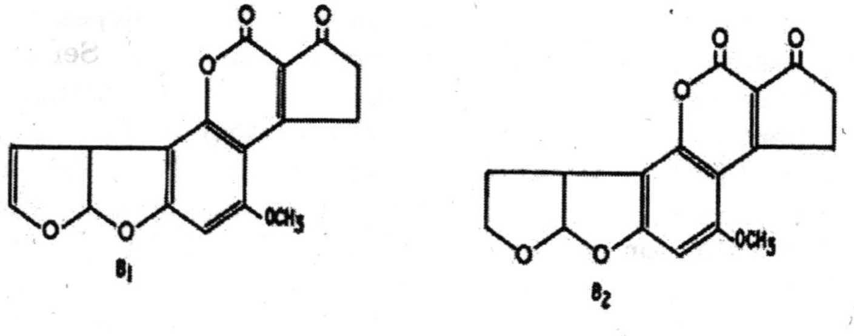

**Fig. 1a & 1b**: Aflatoxin $B_1$ and aflatoxin $B_2$ structures

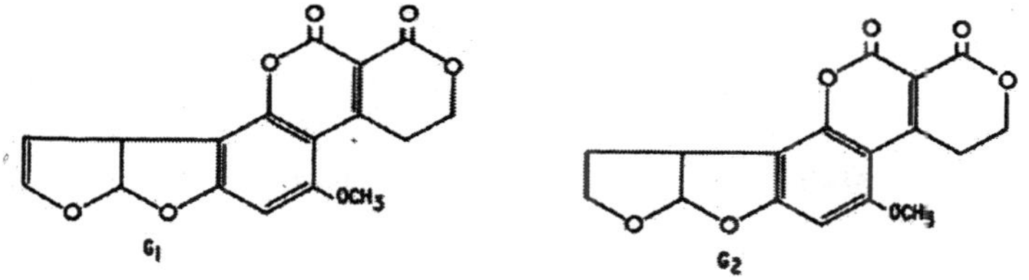

**Fig. 2a & 2b**: Aflatoxin $G_1$ and aflatoxin $G_2$ structures

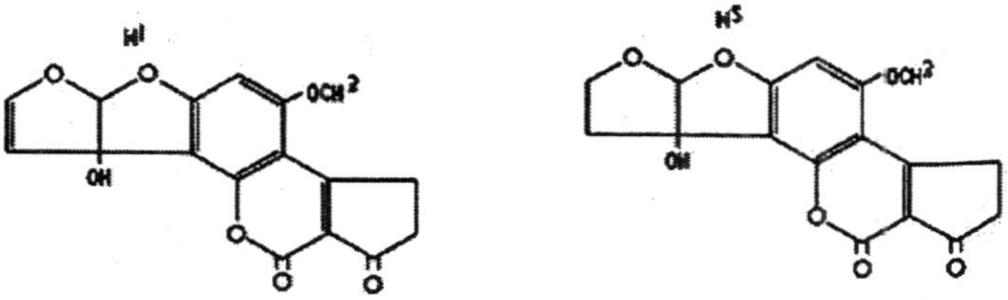

**Fig. 3a & 3b**: aflatoxin $M_1$ and aflatoxin $M_2$ structures

## Sterigmatocystin (Str.)

It is produced by *Aspergillus spp., Penicillium lateum* and *Biploaris* spp. Chemically Str. resembles the aflatoxins and it is a precursor in the biosynthesis of aflatoxins.

## Trichothecenes (Tri.)

It is produced by several fungal genera (*Fusarium, Cephalosporium, Myrothecium, Stachybotrys* and *Trichoderma*). There are several naturally occurring Tri. In food and feed elaborated by *Fusarium* spp. including T-2 mycotoxin and deoxynivalenol. T-2 mycotoxin has been studied most extensively and is a natural contaminant of food and feed.

## Citrinin (Cit.)

A methylated hetrocyclic compound produced mainly by the toxigenic stains of *Penicillium citrinum*. Several other species of *Penicillium* and *Aspergillus* are also able to produce these mycotoxins. The toxic effect of Cit. has been studied in various laboratory animals'. The lethal doses (10-70 mg/kg B.W.) in rabbits and guinea pigs cause swelling in the kidney and acute tubular necrosis. Besides renal damage, its exposure also causes depression in respiration and the cardiovascular system.

### Citriviridin (Ctv.)

Several species of *Penicillium* (e.g. *citro-viride*) and also a species of *Aspergillus* have been reported to produce ctv. in foods and feed stuffs (40). It causes paralysis, cardiomuscular disturbances and loss of eyesight in experimental animals(38).

### Cyclopiazonic acid (CPA)

It was mycotoxins isolated from the culture of *Penicillium cyclopium*. These mycotoxins have been shown to occur naturally in corn, peanuts and in Kodo millet. There are evidences that CPA might have been involved along with the aflatoxin in the "Turkey X" syndrome in England in 1960.

### Ochratoxin (Ochr.)

It constitutes a group of hazardous mycotoxins (Ochr A, B and C) which received considerable attention world-wide due to their carcinogenic, teratogenic, genotoxic and immune suppressive potency in man and animals. Porcine Nephropathy and Balkan Endemic Nephropathy have been established in man due to consumption of Ochr. Ochr. A and its congeners are produced mainly by the toxigenic strains of *Aspergillus ochraceus, A. sulphures, Penicillium viridicatum* and *P. cyclopium*. Chemically Ochr. A is dihydro-iso-coumarin derivative linked through a carboxy group to L – phenylalanine by an amide group. Alterations in DNA, RNA and hepatopoitic tissues are common phenomena due to feed of Ochr. A contaminated food. Ochr. B is a dechloratd derivative of Ochr. A and having comparatively lower toxic potency, while Ochr. C is more or less equally toxic as Ochr. A.

### Zearalenone (Zear)

It is naturally occurring mycotoxins, produced by *Fusarium moniliforme, F. roseum, F. oxysporum, F. tricintum and F. graminearum*, which results in estrogenic syndromes involving swelling of volva rectal, and vaginal prolapsed, enlargement of uterus, edematous tortoise, atropy of ovaries and abortion in animals. It has been found frequently in commercial swine feed. The low or moderate temperature and higher humidity favour growth of Zear producing fungi along with the natural elaboration of Zear. Estrogenism in swine is a common feature due to intake of Zear, contaminated feed degeneration of genital epithelium and sperms by 75% have also been recorded in bulls. However, poultry birds, especially chickens are less sensitive to Zear. Its 800 ppm concentration does not show any marked changes in body weight of chickens, whereas, 100 ppm Zear exposure in turkey birds caused 20% reduction in egg production. Chemically Zear is a 6-(10-hydroxy-6-oxotrans-1-undecenyl) b-resorcyclic acid lactone ($C_{18}H_{22}O_5$). It is a white crystal with a molecular wt. of 318 and is soluble in aqueous alkali, chloroform, acetonitrile, alcohols, methyl chloride and ether benzene. Zear is also slightly soluble in petroleum, either, and insoluble in water, carbon disulfide and carbon tetrachloride.

### Fumonisins (Fum.)

These are newly discovered mycotoxins isolated from *Fusarium monilifome,* and are found to cause diseases in horses. It has also been reported to have cancer promoting activity in Rat.

## NATURAL OCCURRENCE OF MYCOTOXINS

Natural contamination of food, feed and other agricultural commodities by mycotoxins is a worldwide problem more so in tropics than in temperate zones. Earlier reports (3,14,17,19, 22,38) confirm the association of moulds (toxigenic) with food and feed. In the first conference of mycotoxins held at Nairobi in 1977 a review was presented which clearly indicates that practically every edible materials including fruits, vegetables, cereals, pulses, spices and plant drugs are suitable substrates for mycotoxins elaboration (Table 2).

India being a tropical country has great diversity in agro climatic conditions due to the unseasonal rains high temperature and humidity, which adversely affect standing crops and stored food materials by favouring the growth of moulds. Aflatoxins are the most significant class of mycotoxins to be encountered as natural contaminants is maize under stored as well as field conditions(17,20). Aflatoxins are also spotted in rice grains, pulses, oil seeds, wheat, sorghum, pearl millet, dry fruits and spices (1,3,13,20,26,34). Aflatoxin in milk and milk products has been reported (10,41). Recently Roy and Associates (29,30,31,32) have also reported the natural occurrence of mycotoxins in large number of plant samples i.e. seeds, fruits, barks, woods and leaves used as crude herbal drugs beyond the tolerance level fixed by W.H.O. i.e. ppb in Indian context.

**Table 2**. Physical factors influencing mycotoxins production

| Fungus | Mycotoxin | Nature of substrate | Optimum range of production Moisture | Mycotoxins Temperature |
|---|---|---|---|---|
| *Aspergillus flavus* | Aflatoxin | Cereals | 17-24% | 25-30$^{0}$C |
| *A.ochraceus* | Ochratoxin A | Maize, wheat | 18.5% | 25$^{0}$C |
| *Fusarium moniliforme* | Zearalenone | Maize | 20-30% | 24-27$^{0}$C |
| *F. roseum* | Zearalenone | Oat, Barley | 17-24% | 24$^{0}$C |
| *F.graminearum* | Zearalenone | Wheat | 18-20% | 24$^{0}$C |
| *Penicillium citrinum* | Citrinin | In nutrient culture | - | 25$^{0}$C |

## FACTORS AFFECTING MYCOTOXINS PRODUCTION

Mycotoxin production in natural substrate or in nutrient culture depends on various factors, which can be categorized under three broad headings:

1. Physical factors – temperature, humidity, rainfall, moisture content of the substrate and duration of storage.
2. Chemical factors – Level of $CO_2$ and $O_2$, chemical nature of substrate, mineral nutrient and chemical treatment.
3. Biological factors – Plant varietal differences, spore load, strain differences, vectors and microbial interact.

Although all the above stated factors are important, however, oxygen level, temperature, moisture content of substrate and storage period etc. are determining factors for the moulds growth as well as mycotoxins elaborated by them. The optimum range of factors for mycotoxins production is given in Table 2.

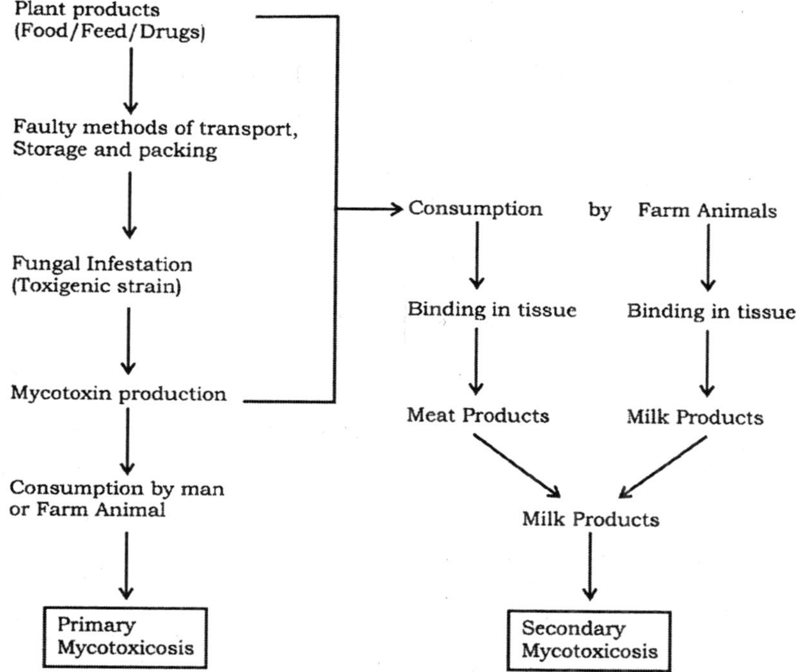


**Fig.** 4. Possible route of mycotoxin contamination and mycotoxicoses

## POSSIBLE ROUTES FOR MYCOTOXINS CONTAMINATION

Mycotoxins enter generally the human and animal food chains by direct or indirect contamination. In direct contamination the food materials are subjected to growth of the toxigenic fungi. Almost all food materials will be susceptible to growth of toxigenic fungi at some stage during their production, processing, transport and storage. In direct combination, the food materials are contaminated with mycotoxins. The possible route for mycotoxins entry into human and animal foods are : (a) toxigenic fungi damaged foodstuffs e.g. cereals, oilseeds, fruits, vegetables etc. (agricultural produced; consumer foods and compounded animal feeds (secondary infection); (b) residues in animal tissues and animal products e.g., milk, dairy produce, meat etc. (c) fungi – ripened foods, e.g., cheeses, fermented meat products, oriental fermentations, etc; and (d) fermented products, e.g., microbial proteins, enzymes, food additives, such as vitamins (Fig. 4).

## MYCOTOXIN- ANIMAL AND HUMAN HEALTH RISKS

Consumption of mycotoxin contaminated food/feeds by human or animal results in its accumulation and exerts a wide range of adverse effects on different vital organs. Some of the very noticeable adverse effects in several vital organs of animal are discussed below:

**Skin :** Tricothecenes are capable of inducing dermal irritation inflammation and desquamation. Aflatoxin contaminated meals cause irritation of tumor in mouse (16), loss of hairs and appearance of red tained papillae in guinea pigs (5) and papilloma in rabbits (29).

**Liver:** It is most target organs of animals for aflatoxins and several others mycotoxins. The diagnostic symptoms include hepatocellular necrosis, fibrorisis, cirrhosis and various types of lesions frequently leading to death of animals (23,24).

**Kidney:** Citrinin and Ochratoxin A are well known to effect kidney. Roy and Kumar (31) reported degeneration of epithilial cells of renal sinus and crystallization in the Bowman's capsule in rabbit exposed to aflatoxin $B_1$. Bilgrami et al. (4) observed degeneration of 'glomerulis' uriniferous tubules and enlargement of Bowman's capsule of swiss mice with citrinin.

**Heart:** Fatty infiltration of heart muscle and sub-endocardial hemorrhage has been recorded in the monkeys, dogs and cats after application of Aflatoxin $B_1$ and Rubratoxin (7,23).

**Lungs:** Aflatoxin induces petechial hemorrhage in the lungs of guenia pigs and some other mycotoxin like patulin and tricothecenes cause pulmonary congestion in intoxicated animals.

**Digestive Tract:** Hemorrhage and necrosis are common all along the digestive tract and in the mucous membrane of the affected animals; Saito and Ohtsubo (33) have demonstrated the toxic effect of some tricothecenes. Degeneration of columnar epithelial cells as well as numerous nacroytic patches in mucous layer of aflatoxin $B_1$ treated rabbits was observed (29).

**Reproductive Organs:** Zearlenone causes precocious sexual development in young female swine and inhibits normal development of testes in maleand also induces enlargement of vulvas, nipples and prolapse of vaginain swine (12). In rabbits degenerated spermatogonial cells, fibrosis of semniferous tubules, hypertrophy and hyperplasia in male and degenerated follicles and ovary in females have been observed by Aflatoxin $B_1$. All these changes in reproductive organs contribute to infertility.

**Blood Composition:** Decrease in total RBC and WBC counts include hemoglobin content in guinea pigs and rabbits by Aflatoxin $B_1$ contaminated foods.

## MYCOTOXINS HAZARDS ON PLANT SYSTEM

The death of plant cells or derangement of plant physiology by fungal toxins (phytotoxins, vivotoxins, pathotoxin or host specific toxins is known long back. Comparatively little work has so far been done with regard to effect of mycotoxins on plant systems. However, available reports reflect that some mycotoxins viz. aflatoxins, rubratoxins, patulin, zearelenone etc. induce a variety of physiological and biochemical disorders in various plants (4,8,15,18,25). The effect of various mycotoxins on plant systems based on experimental findings is summarized as below:

**Aflatoxin $B_1$:** Inhibition of seed germination in lettuce, cow pea, sorghum, maize and mung; Afla-root disease of groundnut; inhibition of chlorophyll synthesis in bhindi, maize and cucumber; reduction of respiratory rate in germinating maize seeds; suppression of synthesis of RNA, DNA and protein in germinating seeds and pollen grains; cytological abnormalities such as increase in number of anaphase chromosomes, inhibition of: mitosis chromosome bridge, c-mitosis and reduction of mitotic index; inhibition of fern spore, inhibition of growth of chlorella and several moulds; inhibition of nucleic acid synthesis in bacteria.

**T-2 toxin:** Inhibition of seed germination of *Pisum sativum* and mung; inhibition of root and shoot elongation; wilting; necrosis; and reduction in weight of seedlings; change in cell membrane permeability in tomato leaf, inhibition in elongation of wheat coleoptile.

**Trichothecenes:** - Leakage of electrolytes in tomato leaves.

**Zearelenone:** Inhibition of seed germination, loss in total chlorophyll, inhibition of protein and nucleic acid synthesis in mung and maize.

**Patulin:** Chlorosis, necrosis and narrowing of leaves in wheat.

## MYCOTOXICOSES

Mycotoxicoses known diseases of human beings develop by the consumption of mycotoxins contaminated foods. Some of the important diseases like ergotism, alimentary toxic aleukia (ATA) of Russia, Acute cardiac Beriberi and aflatoxicoses (liver cancer and Reye's syndrome) are discussed.

**Ergotism:** Ergotism is the oldest mycotoxicoses. Syndrome: sensation of cold in hands and feet along-with subsequent intense burning sensation. In advanced cases the extremities became gangrenous and necrotic. The occurrence of ergotism reached epidemic proportions in the Middle Ages, and it was thought long ago in Europe that a trip to the distant shrine of St. Anthony would bring about relief from the intense burning sensations; hence the disease was known popularly as St. Anthony's fire.

**Source:** The disease is associated with the consumption of bread made from flours of rye and other grains overgrown with toxigenic strains of the fungi Claviceps purpurea and Cpaspali. The fungi develop only in the female sex organs of the grasses, producing ergots, black or dark purple compact masses of hardened mycelium. As early in seventeenth century it was recognized that alkaloids produced by the ergot were responsible for the disease, even though alkaloids had long been known as powerful oxytocies.

**Chemistry and mode of action:** The biologically active ergot alkaloids are derivatives of d-lysergic acid. These compounds produce a-adrenergic blockade, inhibiting certain responses to adrenergic nerve activity, to epinephrine, and to 5-hydroxytryptamine. They bring about marked peripheral vasoconstriction, which if not corrected can result in gangrene. The compound is also highly active in direct stimulation of smooth muscles and has been used as oxytocies to increase the force and/or frequency of uterine contractions. Ergot alkaloids also have effects on the central nervous system, which include stimulation and hypothalamus and other sympathetic portion of the midbrain and depression of the vasomotor centre. These compounds are also centrally acting emetics. Several of the alkaloids are now widely used to treat human diseases.

**Alimentary Toxic Aleukia (ATA):** The outbreaks of ATA have been found to be associated with ingestion or over wintered grains infected with toxigenic fungi, *Fusarium sporotrichiodes* and *F. poae* The disease was first reported from Russia (60% mortalities) which was found to affect the hematopoetic system resulting in decrease of red and white blood cells and platelets. In Russia harvested crops left in the field all winter and under conditions of repeated freezing and thawing, fungi produce several toxic metabolites. Recently, it has been found that these toxic compounds contain sufficient amount of T-2 toxin i.e. epoxy tricothecenes. Which is assumed to be the causal agent of ATA?

**The outbreak of the clinical diagnosis:** a) a mild inflammation from the mouth of the stomach occur within hours of eating contaminated bread and which develops into acute gastroenteritis; this local effects continues several days and ends spontaneously; b) a period of several weeks follows in which there are few gross symptoms but the bone marrow undergoes destruction; after extensive damage of bone marrow hemorrhagic spots appear on the skin and there is a sharp decrease in the number of leukocytes; disturbances of the central and autonomic nervous system are apparent, especially weakness vertigo, headache, palpitation, slight asthma and lowered blood pressure; c) severe atrophy of the bone marrow with hemorrhagic diathesis, necrotic angina, sepsis, and marked decrease in the numbers of leukocytes, granulocytes, erythrocytes, and platelets; and d) either death or a period of recovery.

**Acute Cardiac Baribari :** In the nineteenth century epidemics of an acute heart disease broke out in rural areas of Japan. After a detailed retrospective analysis of the disease it was termed as acute cardiac beriberi (shoshin-kakke). It is not an avitominosis but was termed so because of its association with the consumption of polished rice. Syndrome: pericardial distress with palpitation and tachypnea, later on the victim suffers with severe anguish, pain and restlessness and at times can become violently maniacal. The right heart is dilated and the heart sounds are abnormal. In the last stage, as the dyspnea increase, the extremities become cold and cyanotic, the pupils dilate and the person loses consciousness. The neurotoxic mycotoxins of *Penicillium citroviride* is now known to be related to acute cardiac beriberi.

**Aflatoxicoses:** Aflatoxins is a widespread contamination in man's food and foodstuffs. Aflatoxins $B_1$ is known as highly carcinogenic compound and believed to be a causative factor of human liver cancer. Hepato-cellular carcinoma is unusually high in Sub- Sahara Africa and in Southeast Asia. After studying the evidences for mycotoxin contamination of man's food supply and the A geographical survey of liver cancer in the world associated with the consumption of mouldy foods reveals that aflatoxin is the main cause of this disease. The studies have been conducted in Uganda, Swaziland, Kenya, Thailand, Taiwan and India. There is reasonably strong epidemiological evidence indicating aflatoxin $B_1$ playing an active role in hepato-carcinoma in man at least in Africa, Kenya, India and Thailand.

## Management of mycotoxin contamination

The Mycotoxigenic fungi survive in plant residues and produce abundant spores. The spores are carried by the wind and infect food crops, usually through insect wounds. Prevention of natural occurrence of mycotoxin in various food grains by mycotoxigenic fungi and their detoxification are cumbersome process, therefore, 100% safe food and feed supply is not effective. However, it may be possible either by preventing moulding of substrates or detoxifying the mycotoxin contaminated foods. Prevention of mycotoxins contamination is thus prerequisite and their detoxification is secondary factor of its management. Decontamination and detoxification are two main strategies to control mycotoxin elaboration in foods, feeds and plant drugs. Once a product is contaminated a) the mycotoxin to be removed and b) the mycotoxin to be degraded into less toxic or nontoxic compounds for making product ready for human or animal consumption. The degradation of mycotoxin could be done by physical methods ie., irradiation (UV) and heating at 250°C; and chemical treatment with acids, alkalis, aldehydes, oxidizing agents and specifically with NaOH, Ca(OH)2 NH3 etc. In order to manage the mycotoxin problems in various foods and feeds substrate, different following methods are adopted in recent past.

i. **Cultural** practices help to prevention for the entry of mycotoxins into human and animal food chain due to consumption of contaminated agriculture commodities. This can be achieved by preventing fungal infection during cultivation, harvesting, transportation and storage.

ii. **Physical** processes help to reduce mycotoxin elaboration through heating, pasteurization, sterilization, cooling, vacuum packing, canning, drying and irradiation

iii. **Chemicals** applied to reduce mycotoxin elaboration by using Propionic acid, Ascorbic acid and Acetic acid singly or in combination.

iv. **Biological** controls the most promising and safest approach where living microbes such as bacteria, fungi, actinomycetes and yeasts are used to control these toxigenic strains.

The concept of Biological control based on the microbial antagonism to the toxigenic strains has opened a new vista under Control Program. Its application in the suppression or elimination of toxigenic microbes has been found significantly useful. Strategy of biological control is based on antibiosis (static or cidal effect of antagonists by producing toxic metabolites), competition (competition for survivability) or parasitism (where antagonist parasites on toxigenic strains).

## Conclusion and Future Prospects

The survival of mycotoxigenic strain and their natural contamination in food commodities are very common. The discovery of antagonist microbes such as bacteria, fungi and other has major role to manage the mycotoxin contamination, however, followings facts are to be considered to achieve the target:

- Identification and selection of effective strain of natural antagonists.
- To understand the biology, ecology, physiology and genetic behavior of bio control agents.
- To identify substrate-symbionts strains of antagonists.
- To develop mass culture technique of efficient antagonists.

- To evaluate the beneficial effect of antagonists under field conditions.
- To educate the public people for effective utilization and application of antagonists.
- Develop methods of seed bacterization/seed coating with antagonists to reduce the incidence of mycotoxigenic fungi in soil.
- Making transgenic plants against mycotoxigenic fungi.
- Making a gene library of antagonists and generate the data base of inhibitor and detoxification gene.

Fortunately the expansion of integrated pest management strategy has encouraged management approach with integrated multiple control practice, one of which may be the biological control. Research workers should pay due attention on the mycotoxin management using potential biocontrol agent in a concerned ecosystem.

## REFERENCES


1. Abbas HK, Mirocha CJ, and Tuite J 1986 Natural occurrence of deoxynivalenol, 15-acetyl-deoxynivalenol, and zearalenone in refusal factor maize stored since 1972, *Applied Environmental Microbiology* **51** 841–843
2. Asplin FD and Carnaghan RBA 1961 The toxicity of certain groundnut meals for poultry with special reference to their effect on ducklings and chickens, *Veterinary Record* **73** 1215–1219
3. Bhat RV, Nagarajan V, and Tulpule PG 1978 Health Hazards of Mycotoxin in India, ICMR, New Delhi, 58 page
4. Bilgrami KS and Sinha KK 1992 Mycotoxin incidence and effects, in *Mycotoxin in foods and feeds* (eds) KS Bilgrami and KK Sinha, CBS Pub, New Delhi, 73-82 pages
5. Bilgrami KS, Ranjan KS, and Sinha SP 1986 Aflatoxin-induced histopathology of skin in guinea pigs, *Proceedings of National Academy of Science* **352** 280-283
6. Blount WP 1961 Turkey "X" disease, *Turkeys* **9** 52-57
7. Busby WFJr and Wogan GN 1979 Food-borne mycotoxins and alimentary mycotoxicoses, in *Food-borne infections and intoxication*, (eds) H Reimann and F Bryan,

Academic Press, New York, NY, 519-610 pages

8. Chohan JS 1983 Effect of aflatoxin on plant system, in *Mycotoxin in Food and Feed* (eds) KS Bilgrami, T Prasad, and KK Sinha, Allied Press, Bhagalpur, 147 -157 pages
9. Coker RD, Jones BD, and Nagler MJ 1984 Mycotoxin Training Manual, *Tropical Development and Research Institute*, London, 291 page
10. Gopal T, Zaki S, Narayanaswamy M, and Premlata S 1968 Aflatoxicosis in dairy cattle, *Indian Veteri-nary Journal* **45** 707-712
11. Keeler F and Tu AT 1983 Plant and fungal toxins, Marcel Dekker, Inc New York and Basel, 934 page
12. Kumar S and Roy AK 2001 Inhibition of growth and aflatoxin production of *Aspergillus parasiticus* Speare by co-existing fungi *in vitro*, *Journal of Indian Botanical Society* **80** 149-151
13. Lalitha Kumari and Govindaswamy CV 1970 Role of aflatoxin in groundnut seed spoilage, *Current Science* **34** 306-309
14. Lee US, Jang HS, Tanaka T, Toyasaki N, Sugiura Y, Oh YJ, Cho CM, and Ueno Y 1986 Mycological survey of Korean cereals and production of mycotoxins by *Fusarium* isolates, *Applied Environmental Microbiology* **52** 1258-1260
15. Lillehoj EB 1980 Aflatoxin contamination of preharvest corn: role of *Aspergillus flavus* inoculum and insect damage, *Cereal Chemistry* **57** 255–257
16. Lindenfelser LA, Lilehoj EB, and Burmeister HR 1974 Aflatoxin and triehothecene toxins: skin tumor induction and synergistic acute toxicity in white miee, *Journal of National Cancer lnstitute* **52** 113-116
17. Mall OP, Ranawat KK, and Chauhan SK 1983 Mould flora and aflatoxin contamination in maize kernels, in *Mycotixin in food and foods* (eds) KS Bilgrami, T Prasad and KK Sinha, Allied Press, Bhagalpur, 37-44 pages
18. Marasas WFO, Wehner FC, van Rensburg SJ, and van Schalkwyk DJ 1981 Mycoflora of corn produced in human esophageal cancer areas in Transkei, southern Africa, *Phytopathology* **71** 792-796
19. Mishra RS 1977 Aflatoxin contamination of some agricultural commodities in IARI and biochemical effects of aflatoxin B1 on Maize seeds, PhD thesis, GB Pant University of Agriculture and Technology, Pantnagar, 134 page
20. Mishra NK and Daradhiyar SK 1991 Mould flora and aflatoxin contamination of

stored and cooked samples of Pearl millet in the paharia tribal belt of Santhal Pargana, Bihar, India, *Applied Environmental Microbiology* **57** 1223-1226

21. Moreau C 1979 Moulds, Toxins and Food. John Willey & Sons, New York, 477 page
22. Narita N, Suzuki M, Udagawa S, Sekita S, Harada M, Aoki N, Tanaka T, Hasegawa A, Yamamoto S, Toyzaka N, and Matsuda Y 1988 Aflatoxin potential of *Aspergillus flavus* isolates from Indonesian Herbal drugs, *Proceedings of Japan Association of Mycotoxin Col* **27** 21-26
23. Newberne PM and Butler WH 1969 Acute and chronic effects of aflatoxin on the liver of domestic and laboratory animals: a review, *Cancer Research* **1** 236–250
24. Pier AC 1987 Aflatoxicosis and immunosuppression in mammalian animals, in *Aflatoxin in Maize* (eds) MS Zuber, E B Lillehoj and BL Renfro, CIMMYT, Mexico, 58–65 pages
25. Rambo GW and Caldwell RW 1974 *A. flavus* and aflatoxin in pre-harvest corn from India in 1971 and 1972, *Cereals Chemistry* **51** 848-853
26. Rao KS, Madhawan TV, and Tulpule PG 1965 Incidence of toxigenic strains of *Aspergillus flavus* affecting ground nut crop in certain coastal districts of India, *Indian Journal of Medical Research* **53** 1169-1202
27. Reddy SV and Waliyar F 2012 Properties of aflatoxin and its producing fungi, *Aflatoxins* 20-32
28. Rosic JL, Skrinjar M, and Markor S 1991 Decrease of aflatoxin in yoghurt and acidified milks, *Mycopathologia* **113** 117-119
29. Roy AK and Chaurasia HK 1990 Aflatoxin production on *piper longum* fruits under different temperatures, *International Journal of Crude Drug Research* **28 (3)** 233-235
30. Roy AK 1989 Threat to medicinal plants and plant drugs by fungi, *Journal of Indian Botanical Society* **68** 149-153
31. Roy AK and Chourasia HK 1993 Inhibition of aflatoxins production by microbial interaction, *Journal of General and Applied Microbiology* **36** 59-62
32. Roy AK, Sinha KK, and Chourasia HK 1988 Aflatoxin contamination of some common Drug Plants, *Applied and Environmental Microbiology* **54(3)** 842-843
33. Saito M, Tsuruta O, Siriacha P, Kawasugi S, Manabe M, and Buangsuwon D 1986 Distribution and aflatoxin productivity of the atypical strains of *Aspergillus flavus*

isolates from soil in Thailand, *Proceedings of the Japan Association of Mycotoxicology* **24** 41-46

34. Sargeant I, Sheridan A, Kelly JO, and Carnatharn RBA 1961 Toxicity associated with certain samples of groundnuts, *Nature* **192** 1096 - 1097
35. Sinha AK and Ranjan KS 1989 Pre harvest aflatoxin problems in linseed (*Linum usitatissimum*) in Bihar, *Journal of Indian Botanical Society* **68** 19-20
36. Trail F, Mahanti N, and Linz J 1995 Molecular biology of aflatoxin biosynthesis, *Microbiology* **141** 755-765
37. Turner WB 1971 Fungal Metabolites, Academic Press, London, New York, 238 page
38. Ueno Y 1987 Trichothecenes in food, in *Mycotoxin in Food* (ed) P Krogh, Academic Press, London 123-147 pages
39. Ueno Y 1984 Toxicological features of T-2 toxin and related trichothecenes, *Applied Toxicology* **4** 124–132
40. Urguchi K and Yamazaki M 1978 Toxicology Biochemistry and Pathology of Mycotoxins, John Willey and Sons, New York, 246 page
41. Wicklow DT, Horn BW, and Shotwell OL 1988 Aflatoxin formation in preharvest maize ears co-inoculated with *Aspergillus flavus* and *Aspergillus niger*, *Mycologia* **79** 679-682
42. Wogan GN 1966 Chemical nature and biological effects of the aflatoxins, *Bacteriological Reviews* **30(2)** 460–470
43. Yadgiri B and Tulpule PG 1974 Aflatoxin in buffalo milk, *Indian Journal of Dairy Science* **27** 293-297.

*Rev. Plant Pathol. Vol. 6, 2014*
*Indian Society of Mycology and Plant Pathology*
*Scientific Publishers (India), Jodhpur pp. 541-576*

# FOSSIL EPIPHYLLOUS FUNGI AS POTENTIAL PALAEOENVIRONMENT INDICATOR: A REVIEW

***Subir Bera and Arkamitra Mandal***

Centre of Advanced Study, Department of Botany,
University of Calcutta, Kolkata 700019

**KEY WORDS:** Epiphyllous fungi, Fossil fungi

## ABSTRACT

The origin of fungi dates back to Precambrian but the earliest fungal fossil record is yet to be ascertained properly. A complete understanding of the occurrence and diversity of fungi in the fossil record helps in proper interpretation of the past environment. Fossil fungi of various groups reported so far are mostly represented as dispersed spores, fruit bodies and hyphal fragments from different sedimentary strata of the world. Among these, epiphyllous fungi are proved to be an important tool for phylogenetic studies, palaeoecological reconstruction and interpret-tion of environmental interactions and in solving stratigraphic problems. In the light of current palaeomycological research, the diversity in fossil epiphyllous fungal assamblages along with its application in interpretation of palaeoenvironment through the geological ages is presented in this review.

## INTRODUCTION

Palaeomycology is defined as a branch of microfloristic research dealing with the study of fossilized fungal remains (4).

Although consideration of fossil fungi has been started since the beginning of palaeobotanical investigations, records of fossil fungi are comparatively less utilized source of information concerning diversity, structure, evolution and activities of fungi in the geological past. However, interest in fossil fungi has been greatly enhanced in recent years due to the recognition of the utility of fungi as they have dynamic interactions with plants, animals and the surrounding environment. So palaeomycological data can be used for phylogenetic studies, palaeoecological reconstruction, and even solving stratigraphic problems (3,49,51,52,54,56). Evidence of fossil fungi helps in establishing intricate relationship of fungi with land plants, parasitic relationship with plants and animals, degradation of lignin in early land forests and finally their role in all important ancient ecosystems. These data may also help to interpret variable interactions of fungi with their hosts, including mutualism, cryptic commensalisms, and latent and virulent pathogenicity (66) in the past. Therefore, a complete underst-nding of the occurrence and diversity of fungi in the fossil record prove pivotal in assessing how fungi may have functioned in shaping ecosystems of the past (40).

The origin of fungi dates back to Precambrian but the earliest fungal fossil record is not so easy to detect (9), however, some early reviews of fossil fungi are available (53,68). Many earlier recorded forms proved later to be of non fungal nature. Several structures resembling fungi from ~1 Ga old sediments of the Lakhanda series of Siberia (Neoprtorezoic- Mesoproterozoic) have been reported (30). Though the recorded form is hypothesized as zygomycetes, its exact status is still controversial. Among the proven record of fossil fungi the earliest record comes from the Early Silurian of Virginia, USA as small fragmented hyphae and some conidia (55). The most important record though comes from Early Devonian bed of Rhynie chert, Scotland. The fungal body includes petrified hyphal fragments, sporocarps and numerous spores. Subsequently fossil fungi were reported as dispersed spores, fruit bodies and hyphal fragments from different parts of the world i.e. Devonian Zygomycetes from North America, Palaeozoic Basidiomycetes (54), Devonian arbuscular mycorrhiza (57), Lower Carboniferous Chytridiomycota from France (40) and numerous other fungi from Indian Tertiaries have been reported by many workers as dispersed specimens (33).

Fungi act as an important component of the ecosystem and play a vital role in the forestry and its management. In a certain forest ecosystem they act as scavengers by degrading litter, break down dead wood as wood rot fungi or may infect leaves as parasites. Parasitic fungi that infect leaf or grows on leaf i.e. epiphyllous or foliicolous fungi may be pathogenic or biotropic. Pathogenic fungi cause diseases by producing symptoms such as leaf spots, hypertrophy, canker, die-back, etc. To understand much about these microbes, their occurrence, interaction with the plants and other related information are urgently needed in all the biodiversity rich forest areas (31). It is important to determine the significance of microfungi of the leaf surface or phylloplane in the ecosystem as determination of epiphyllous fungal productivity can help in estimating the rate of energy flow between the host leaf and the epiphyllous fungi. Epiphyllous fungi create a unique physiological interaction in respect of enzymatic and signaling molecules with their host leaf and can also sometimes influence their host by production of growth substances. So the study of epiphyllous fungi and their host is an important tool in understanding certain forest ecology and its environment.

As the fossil fungus with proper host leaves helps in better understanding of the past, the importance of epiphyllous fungi is gaining importance day by day and subsequently *in situ* occurrence of fossil epiphyllous fungi on leaves proves to be an important indicator of palaeoclimate (3,16,73). This data can also be utilized in reconstructing micro-ecological condition and palaeovegetational pattern of the past. In most of the cases fossil fungi recovered as dispersed mycelia, spores or fruitbodies do not have any organic connection with host leaf tissues and thus it is difficult to establish any relationship with their extant comparable taxa. Epiphyllous fungi also provide a unique opportunity to assess the critical interaction with their host in certain forest ecology. So study of fossil epiphyllous fungi helps in proper interpretations of ancient palaeoenvironment, forest type, macro and micro-ecological conditions and host parasite interaction.

## FOSSIL EPIPHYLLOUS FUNGAL DIVERSITY

Fungi which occur on leaves in Tertiary assemblages are among the best known fossil epiphyllous fungi. These epiphyllous forms are commonly found in the areas where conditions of exceptional preservation permit recovery of leaves with intact cuticle often with the cellular structures (70). A large number of epiphyllous fungi have been recoded (74) in *'Genera et species Plantarum Fossilium'*. As previously mentioned, that the fungal origin of a large number of 'fungi' from earlier descriptions are in question, so some reported forms mentioned in those review may not be actually epiphyllous form. In the present review the epiphllous fungal assemblages are reported from different geological ages. In fossil fungal nomenclature due to presence of a large number of obligate synonyms many earlier reported forms may be mere repetition of the previous records. In this review the valid names with more recent author citations are described to avoid confusion. The classification of fossil fungi always posses a greater amount of challenge as classification and naming of extant fungi is rather confusing sometimes due to absence of perfect stages in large number of fungi which leads to a large number of synonyms. In fossil record it is almost impossible to assign fossil fungi in proper genera as in most of the cases teleomorph and anamorph connections are not found preserved. In most of the cases, for classiation of fossil fungal spores, Saccardo's system of Classifica-ion for Deuteromycetes is proposed where the classification is based on the conidial character and not related with any extant family. Classification of fossil mycelia is rather difficult as fungal myceliam doesn't possess much taxonomic importance in absence of spore and conidia. So absence of complete stage of life cycle makes fossil fungi impossible to compare with any extant fungal taxon. In the present review the taxonomic position of fossil fungi in most cases is followed as per Kalgutkar and Jansonius (34). Some older fossil records of extant forms are classified (16) and the systematic positions are mentioned in Table 1, which also provides some information about mode of interaction of fossil fungi with their host and possible Palaeo-climate. In some cases the mode of interaction was based on the basis of the modern counterpart of the fossil fungi as direct record of evidence was not found in fossil records. To cast away the confusion of classification we prefer reviewing the epiphyllous forms based on their age of occurrence rather than their systematic position.

## PALAEOZOIC EPIPHYLLOUS FUNGI

Palaeozoic epiphyllous record is meager with most of the early records are of putative epiphyllous forms.

### *Devonian*

The earliest record of fossil epiphyllous fungi comes from Devonian sediments of Germany. The reported form is *Hysterites opegraphoides* (24) on the leaf of *Neuropteris* sp. (Medullosaceae) and *Podozamites* sp. (Podocarpaceae). Another probable epiphyllous record from Devonian is *Melanosphaerites devonicus* (26) from Bear Island, Norway.

### *Carboniferous*

From Carboniferous there are two records of epiphyllous fungi, both are known from France. The reported forms are *Excipulites* sp. (24) on the leaf of *Hymenophyllites* sp. (Sphenopterideae/ cf. Hymenophyllaceae), *Callipteris* sp. (Peltaspermales), *Pecopteris* sp. (Psaroniaceae) and *Macrostachya* sp. and *Urophlyctites oliveriana* (42) on the leaf of *Alethopteris* sp. (Medullosaceae). Among the other records, the occurrence of *Sphaerites* sp. (75) on the leaf of *Abies* sp. (Pinaceae) and *Amygdalus* sp. (Rosaceae), *Hysterium* sp. on the leaf of *Zostera* sp. (Zosteraceae), *Pteridium* sp. (Dennstaedtiaceae) and *Acer* sp. (Sapindaceae) (28), *Macrosporites* sp. (58), *Mycocarpon pachyderma* (Williamson) (32) and *Sporocarpon compactum* Williamson (76) have been reported.

### *Permian*

Some unidentified epiphyllous fungal thyrothecia on dispersed leaf cuticle of *Glossopteris* (Glossopteridaceae) flora was reported (2) from Early Permian sediments of Saharjuri Outlier, Deogarh Coalfield, Bihar, India.

## MESOZOIC EPIPHYLLOUS FUNGI

Mesozoic assemblage is chiefly comprised of Cretaceous epiphyllous fungi recovered mostly from Gymnosperm and Pteridophyte leaves with a marked absence of epiphyllous forms in the Triassic and Jurassic.

### *Cretaceous*

From the Early Cretaceous strata there are a number of epiphyllous fungal records. The earliest records are *Puccinites lanceolatus* from Austria and Nebraska; *Phacidium rimosum* (41) from Bohemia, Germany on *Pyrus* sp. (Rosaceae), *Eugenia* sp. (Myrtaceae) and *Populus* sp. (Salicaceae) and *Ovularites barbourii* from Nebraska. Occurrence of *Excipula* sp. on the leaf of *Pecopteris* sp. (Psaroniaceae) (64), while *Sphaeropsis* sp. on a large number of leaf i.e. *Betula* sp. (Betulaceae), *Bumelia* sp. (Sapotaceae), *Carya* sp. (Juglandaceae) etc. from England have been reported (23). A large number of fungi have been reported (39) from Southern Primrosk, Russia preserved on a number of Gymnosperm and Pteridophyte leaves. The fungal forms are *Asterothyrites dictyozamiticola* (Krassilov) (34) on *Dictyozamites grossinervis* (Bennettitales); *Asterothyrites podocarpi* (Krassilov) on *Podocarpus harissii* (Podocarpaceae); *Perisporicites zamiophylli*, (39) on *Zamiophyllum buchianum* (Bennettitales) and *Subzamites borealis*; *Phragmothyrites doratophylli* (Krassilov) on *Doratophyllum sulcatum* (Cycadales/ Bennettitales); *Phragmothyrites nilssoniopterícola* (Krassilov) on *Nilssoniopteris rhitidorachis* (Bennettitales); *Phragmothyrites otozamiticola* (Krassilov) on *Otozamites* sp. (Bennettitales); *Trichopeltinites nilssoniopterícola*, on *Nilssoniopteris* sp. (Bennettitales) and *Cladophlebis frigid* (Osmundaceae); *Ussurithyrites araucariodendri*, on *Araucariodendron heterophyllum* (Coniferales) and *Microthyriacites cephalotaxi* (Krassilov) on *Cephalotaxus ussuriensis* (Cephalotaxaceae). Two epiphyllous from Santa Cruz province, Bajo Grande, Argentina viz. *Brefeldiellites argentinus* on *Podocarpus dubius* (Podocarpaceae), *Dictyozamites* sp. (Bennettitales), *Ruflorinia sierra* (Probable Seed Fern, Ruflorinaceae) and on an unknown conifer and *Microthyriacites baqueroensis* have been reported (45). The other Early Cretaceous record include *Chlamydosporites gramineum* from Mohgaon Kalan, Madhya Pradesh, India on Grass cuticle (Poaceae); *Molinea asterinoides* from Boyaca Basin, Colombia on dicotyledonous leaf and *Stomiopeltites cretacea* (1) from Compton Bay, Isle of Wight on Conifer leaf.

From the Late Cretaceous strata the records are comparatively less. Two epiphyllous fungi from Aachen region, Germany on *Dryophyllum* sp. (Fagaceae) i.e. *Aecidites stellatus*, and *Himantia* sp. (15) have been reported. The other reported forms include *Pleosporites shirainus* (72) from Hokkaido, Japan on *Cryptomeriopsis* sp. (Coniferales) and *Diplodites mohogaoensis* (Barlinge and Paradkar) from Mohgaon Kalan, Madhya Pradesh, India on Monocot leaf (36).

Three fossil records of modern day plant leaf pathogens i.e. *Colletotrichum*, *Erysiphe* and *Uncinula* recovered from the coprolite of *Isisaurus* (*Titanosaurus*) belonging to sauropods from the Lameta Formation have been recorded (69). The presence of these fungi in the coprolites indicates that the said dinosaur ate the leaves.

## COENOZOIC EPIPHYLLOUS FUNGI

### Tertiary Assemblages

Tertiary shows a wide array of diversity in epiphyllous fungal form. *Hysterites labyrinthiformis* (74) is one of the most early epiphyllous fungi in literature. The form was reported from a fossilized leaf fragment of Croatia. But according to Meschinelli the form may not be fungal. From Prussia, *Eurotium* sp. and *Oidium* sp., while *Cenangium* sp. on the leaf of *Typha* sp. (Typhaceae) have been reported (41). From the Oligocene-Miocene sediments of Germany, France, England, Greenland and Colorado, USA on the leaf of *Cinnamomum* Sp. (Lauraceae), *Populus* sp. (Salicaceae), *Cyperus* sp. (Cyperaceae), *Celastrus* sp. (Celastraceae) and *Gingko* sp. (Gingkoaceae) have been recorded. Among other early records worth mentioning are *Dothidea borealis* (29) from Mio-Pliocene sediments of Iceland and Switzerland on leaf of *Acer* sp. (Sapindaceae), *Andromeda* sp. (Ericaceae), *Sterculia* sp. (Sterculiaceae), *Myrica* sp. (Myricaceae), *Betula* sp. (Betulaceae) and *Cyperus* sp. (Cyperaceae); *Sclerotium acericola* (29) from Palaeo-Eocene sediments of Switzerland and Italy on the leaf of *Acer* sp. (Sapindaceae), *Rhus* sp. (Anacardeaceae), *Laurus* sp. (Lauraceae) and *Betula* sp. (Betulaceae); *Depazea acericola* Saporta (63) from Palaeo-Miocene sediments of France and Germany on the leaf of *Acer* sp. (Sapindaceae), *Salix* sp. (Salicaceae) and *Andromeda* sp. (Ericaceae) and *Leptosphaerites lemoinii* Richon (59) from France on a Monocot leaf fragment. Geyler (23) reported two form

*Hysteriopsis subopegraphoides* from the Greenland on the leaf of *Salix* sp. (Salicaceae) and *Sapotacites* sp. and *Phacidiopsis* sp. from Switzerland on the leaf of *Salix* sp. (Salicaceae), *Fagus* sp. (Fagaceae) and *Quercus* sp. (Fagaceae). Another Tertiary record was *Phyllerites* sp. from Prussia, Bohemia and France on the leaf of *Laurus* sp. (Lauraceae), *Myrica* sp. (Myricaceae), *Callicoma* sp. (Cunoniaceae), *Alnus* sp. (Betulaceae), *Platanus* sp. (Platanaceae) and *Ficus* sp. (Moraceae). *Palaeosordaria lagena* Sahni and Rao (61) is another probable Tertiary epiphyllous form from Chindwara District, Madhya Pradesh, India. Frantz (20) reported a number of microthyraceous germling on *Sapindus* sp. (Sapindaceae), *Pityophyllum* sp. and on numerous unidentified leaves from the Tertiary sediments of Germany, Hungary, England and Colorado and Tennessee of USA. Among Tertiary Assemblages, Eocene shows maximum diversity in epiphyllous fungal form.

### *Palaeocene*

Occurrence of *Brachysporium* sp. from Montana, a petrified epiphyllous form *Polyhyphaethyrites giganticus* from Deccan Intertrappean beds of Mohgaon Kalan, Madhya Pradesh, India have been reported.

### *Eocene*

Eocene represents quite a wide fossil epiphyllous fungal assemblage. Among the fossil records of extant form Berkeley (5) reported *Penicillium* sp. and *Dendryphium* sp. from Germany and Prussia; Braun (8) reported *Stegilla* sp. from Switzerland on the leaf of *Graminia* sp. (Cf. Poaceae) and *Cyperitum* sp. (Cf. Cyperaceae); Schmalhausen (65) reported *Erysiphe* sp. on the leaf of *Ficus* sp. (Moraceae) from Sicily; Caspary (10) reported *Stilbum* sp. and *Ramularia* sp. from Prussia. Some probable epiphyllous fungi from Transcaucasia i.e. *Perisporiacites larundae*, *Haplographites xylophagus* and *Haplographites cateniger* have been recorded (19). *Morosporium elongatum* Renault and Roche (58) is another probable epiphyllous form from France. Berry (7) reported *Pestalozzites* sp. on the leaf of *Thrinax* sp. (Arecaceae) and *Sabalites* sp. (Arecaceae) from Georgia, Florida and Louisiana of USA. The other reports include *Fungites apoidensis* (27) from

Germany; *Streptotrichites spiralis* and *Acremonites succineus* (53) from Prussia. Fritel (21) recorded some significant epiphyllous form from Early Eocene sediments of Cessoy, France. The forms are *Graphiolites sabaleos* on the leaf of *Sabalites sucssionensis* (Arecaceae); *Sphaerites nervisequus* on leaf of Lauraceae origin and *Phomites myricae* on a leaf of Myricaceae affinity. Berry (7) recovered some species of *Caenomyces* from the Early Eocene sediments of USA. The recorded species are *C. annulatus* and *C. cassia* on leaf of *Cassia emarginata* (Fabaceae) from Wilcox beds, Fayette County, Texas and *C. laurineus* on the leaf of *Nectandra lowii* (Lauraceae); *C. sapotae* on the leaf of *Sideroxylon premastichodendron* (Sapotaceae) and *C. pestalozzites* on the leaf of *Sabalites grayanusom* (Arecaceae) from Holly Springs sand; Oxford Gully. Another important epiphyllous fungal record from Eocene is *Phragmothyrites eocaenicus* (17) from a Conifer leaf from Scotland. The most illustrative documentation of epiphyllous fungi from Early Eocene is so far by Dilcher (16), who reported a wide number of fungi from two well preserved angiosperm leaf from Western Tennessee, USA. Among the reported form *Asterina eocenica* (16), *Asterothyrites spinulatus* (34), *Brefeldiellites fructiflab*ellus, *Haplopeltis mucoris*, *Meliolinites spinksii* (67), *Patoullardiella imbricate*, *Pelicothallos villosus*, *Sporidesmium henryense*, *Dictyotopileos yalensis*, *Haplopeltis mucoris* (16) and *Vizella memorabilis* (67) was recovered from leaf of *Chrysobalanus* sp. (Chrysobalanaceae) while *Asterina nodosaria*, *Asterothyrites tennesseensis*, *Callimothallus pertusus*, *Meliolinites anfractus* (34), *Microthyriella fungosa*, *Phragmothyrites lutosus*, *Plochmopeltidella antiqua*, *Stomiopeltites plectilis* (34), *Trichopeltina exporrecta*, *Trichopeltinites nilssoniopterícola*, *Parasterina plectopelta* and *Vizella memorabilis* (67) from the leaf of *Sapindus* sp. (Sapindaceae). Another species of *Meliolinites* was reported, i.e. *M. dilcheri* on unknown leaf fragment from Early Eocene sediments of a clay pit from Texas, USA. A comparatively recent record from Middle Eocene is *Palaeoserenomyces allenbyensis* (13) from British Colombia, Canada.

### *Oligocene*

The record is rather disputed. The earliest record of epiphyllous fungi from Oligocene probably is *Sphaerophorus moniliformis* (47) recovered from fossilized leaf of *Thuja occidentalis* (Cupressaceae) from the Baltic area and *Diplosporium* sp. from France. Some genus from the Baltic area of Poland, viz. *Acremonium succineum* (10) from the leaf of *Widdringtonites oblongifolius* G. var *longifolius* (coniferales); *Torula mengeana* on the leaf of *Thuites* sp. (Cupressaceae) and three species of *Fungites* e.g. *F. capillaries* on leaf of *Thuites succineus* (Cupressaceae); *F. pullus* on the leaf of *Chamaecyparis* sp. (Cupressaceae), *Sequoia* sp. (Cupressaceae) and *Libocedrites* and *F. hirtus* on a leaf of unknown affinity. The occurrence of *Didymosphaeria* sp. on the leaf of *Typha* sp. (Typhaceae) from Colorado, USA have been recorded (11). Five epiphyllous fungal taxa of which three thyrothecia genus viz. *Asterothyrites minutes* , *Euthythyrites oleinites* (12), *Trichothyrites ostiolatus* (34) and two species of *Phragmothyrites* e.g. *P. delicatissimus* and *P. sinuatus* (67), and *Trichopeltinites pulcher* (12), recovered from the leaf of *Oleinites willissi* of unknown affinity from Australia, New Zealand and *Plochmopeltinites masonii* (12) from New South Wales, Victoria.

### *Miocene*

Among the early record of Miocene epiphyllous fungi *Polysigmites priscus* (46) from Italy, *Sphaerites disciformis* reported from Parschlug, Styria, Austria on the leaf of *Quercus* sp. (Fagaceae) and *Vaccinum* sp. (Ericaceae) and *Linosporoidea* sp. (38) from Switzerland. Some probable epiphyllous form *Uncinulites baccarinii*, *Perisporites setosus*, *Perisporites hirsutus*, *Chaetomites intricatus*, *Melanosporites stephanii*, *Microthyrites disodilis* and *Monilites albida* (50) reported from Sicily. Other Miocene records are *Leptothyriomyces radiates* from Sumatra; *Cercosporites* sp. (62) from Italy, *Haplopeltis neyveliensis* (56) on some unknown leaf from Tamilnadu, India. Some epiphyllous fungi *Meliolinites nivalis* on the leaf of Myrtaceae and *Phragmothyrites kiandrensis* (67) on the leaf of Myrtaceae and Lauraceae origin were reported from the Early Miocene sediments of Kiandra, New South Wales, Australia. A number of unidentified fungi i.e fragmented hyphae, fruitbody and spore from Northern Idaho, USA on the leaf of *Betula* sp. (Betulaceae), *Smilax* sp. (Smilacaceae), *Lithocarpus* sp. (Fagaceae) and leaf of Lauraceae origin (70), while *Phragmothyrites concentricus* (52) preserved on the leaf of *Castanea* sp. (Fagaceae), *Betula* sp. (Betulaceae), *Quercus* sp. (Fagaceae), *Persea* sp. (Lauraceae), *Magnolia* sp.

(Magnoliaceae), *Lithocarpus* sp. (Fagaceae), *Pseudofagus* sp. (Fagaceae) and on *Pinus* sp. (Pinaceae) and *Stomiopeltites amorphos* on leaf of *Betula* sp. (Betulaceae), *Smilax* sp. (Smilacaceae) and on *Zizyphoides* sp. (Trochodendraceae) were recorded. Another Eocene form *Entopeltacites remberi* was reported on *Persea* sp. (Lauraceae) leaf from Clarkia, Idaho, USA.

There are several recent Indian reports from Middle Miocene sediments from Darjeeling foothills, West Bengal, India. Two forms, *Palaeocolletotrichum graminoides* on a monocot fossil leaf and *Palaeocercospora siwalikensis* on an angiospermic leaf (48), another fossil form *Palaeoasterina siwalika* on angiospermic leaf (49)were recorded. There are two reports of fossil epiphyllous fungi from the Middle to Upper Miocene sediments of Darjeeling Foothills, India, *i.e. Phomites* (21) on leaf of *Anthocephalus* sp. (Rubiaceae) and a new species of *Meliolinites* i.e. *M. siwalika* from a leaf of Lauraceae origin (43,44).

### *Pliocene*

*Laestadites nathostii* from the late Pliocene sediments of Japan and *Depazites* sp. on the leaf of *Cinnamomum* sp. (Lauraceae), *Alnus* sp. (Betulaceae), *Juglans* sp. (Juglandaceae), *Smilax* sp. (Smilacaceae) and *Myrica* sp. (Myricaceae) from Italy and Greenland are probably the earliest record of Pliocene epiphyllous fungi. Occurrence of *Asterina* sp. on the leaf of *Ilex* sp. (Aquifoliaceae) and *Sequoia* sp. (Cupressaceae), from Pliocene sediments of Germany (18), while occurrence of *Palaeoasterina siwalika* (49) on leaf of *Chonemorpha* sp. (Apocynaceae) was reported from the Upper Pliocene to Lower Pleistocene sediments of Arunachal Pradesh, India.

## QUATERNARY ASSEMBLAGES

The earliest record of Quaternary epiphyllous fungi is *Phacidites* sp. from Italy and Borneo from the leaf of *Buxus* sp. (Buxaceae), *Juglans* sp. (Juglandaceae), *Smilax* sp. (Smilacaceae) and *Palaeocassia* sp. (Fabaceae).

### *Pleistocene*

*Trichothyrites pleistocaenicus* (60) on fossilized leaf of *Picea mariana* (Pinaceae) and occurrence of *Microthyrium* sp. (25) from the Early Pleistocene sediments of Minnesota, USA were reported.

### *Holocene*

Some epiphyllous form Bengal basin, India on leaf of mangrove affinity. The recorded forms are *Callimothallus* sp. and two species of *Meliolinites* viz. *M. anfractus* (34) and *M. Spinksii* (67). Hyperparasite *Isthmospora* on *Meliola ellisii* in a Holocene intermediate was also reported (22).

## EPIPHYLLOUS FUNGI AS AN INDICATOR OF PALAEOENVIRONMENT

Previously vegetational changes caused by environmental changes are interpreted using fungal remains (77,78,79). In recent years edaphic consideration of the interaction between fungi and geological soil has also been done (34). Since fossil fungal spores are most abundantly found fungal remain in the sediments, several palaeoenvironmental assessments are made using those data. The prevalence of *Palaeocirrenalia* was correlated with brackish to marine waters from the Neogene sediments of Kerala, India, while *Helicoönites* and *Helicosporaiates* from Eocene sediments were correlated with pools of open water in a warm, humid reconstructed palaeoenvironment of Arctic Canadian region (35) and relative abundance of *Exesiporites* was correlated with warmer period and fluctuating cyclic climate during Pliocene of Mexico.

Since epiphyllous fossil fungi are preserved in direct connection with living tissue it is easy for them to correlate with certain environment as *in situ* association of fungi with mega-fossil remain provides more information in co-ordinated palaeobotanical research. Very few epiphyllous fungi have been recorded so far. The most striking among this is that a large collection of epiphyllous fungi has been reported in excellent condition from several well preserved angiospermic leaves from the Eocene sediments of Tennessee, USA (16). This kind of study has a great potential to detect palaeoenvironment. In accordance with the recovered host leaves and epiphyllous fungi, Dilcher reconstructed warm temperate or subtropical vegetation at the Eocene Tennessee. Using the epiphyllous fungal data a warm humid tropical climate with high rate of precipitation in the ancient forest of Darjeeling Foothill, India during Miocene have been reconstructed (43,44,48,49). Host parasite relation and palaeoenvironment of several well preserved fungi recovered from some petrified plant samples has been described from the Deccan Intertrappean bed of Madhya Pradesh, India. A comparatively cooler temperate climate in Clarkia, Idaho had been predicted (70) during Early Miocene. Emphasis was put a light on ecological significance of the fossil floras suggesting that palaeoenvironment could be deduced by identifying fossils and comparing with their nearest living relatives (16). Epiphyllous fungi are quite useful regarding determining their nearest living relatives as they are found preserved within definite host leaf tissues. The comparative study of extant fungal form and their ecology along with sufficient data of fossil epiphyllous fungi can successfully apply as proxies for interpretation of the past. Such an interesting investigation comes from the evidence of hyperparasitism of *Isthmospora spinosa* on *Meliola ellisii* on the leaf of *Calluna vulgaris* (22) which is a new aspect in the microecological condition of both the fungi and an unique host parasite interaction in the Holocene forest as extant *Meliola ellisii* grows on *Vaccinium* sp. and not on *Calluna vulgaris*. Similar record of extinct host parasite interaction that is not found in present day (14) from the Upper Pliocene to Lower Pleistocene sediments of Arunachal Pradesh, India between *Palaeoasterina siwalika* and *Chonemorpha* sp. and from the Miocene sediments from Darjeeling Foothills, India between *Phomites* sp. and *Anthocephalus* sp.

## FUTURE SCOPE IN FOSSIL EPIPHYLLOUS FUNGAL STUDY

The success of further work depends upon both the widespread recognition of the value of fossil fungi and the active and directed search for additional material (71). Since records of epiphyllous fungi are very few, it is still in juvenile state for using the data as palaeoenvironment indicator. For proper interpretation of palaeoenvironment analysis of compressed leaves and the discovery of well preserved fossil fungi is quite crucial.

Recovery of every stage of lifecycle of certain fungi is a huge challenge which can be recovered by the study of bio-molecular analysis. Due to the absence of anamorphic and teleomorphic connection it is mostly difficult to assign a fossil fungus in to its comparable extant form and this can be overcome by the study of activity of metabolically active fungal gene families eg., estimation and quantification of fungal chitin may help in estimating the rate of evolution of chitinase gene family. The fungal chitinase gene family performs the biological roles including growth, development and stress-responsive functions. Genome sequence data was used to analyze the size of the chitinase gene family in different modern fungal taxa to infer their phylogenetic relationships and action of adaptive natural selection (37). Similar studies can be made using fossil epiphyllous fungal specimen if the amounts of tissue are in sufficient amount and in minimal degraded state. Fossil epiphyllous fungal DNA analysis may also help in recognizing its nearest living relative or may help in proper identification of unidentified fossil fungal form. In another similar study (6) DNA from a 7000 year old Neolithic wood inhabiting polypore *Daedaleopsis tricolor* has been recovered and sequenced. Therefore, the cases where fossil epiphyllous fungi found in organic connection with host tissue with preserved DNA, determination of the rate of evolution of such fungi is possible. Since the origin of fungi and their evolutionary history is rather sketchy, more and more fossil epiphyllous fungal data may help to solve numerous evolutionary paradox of fungal origin. Analysis of nucleic acid variation in the dataset of 50 genes for 25 selected fungi, shows that the estimated age of the Ascomycota and Basidiomycota split was 1 808 000 000 y ago. While connected with host tissues, epiphyllous fungi provide unique opportunity to study their mode of interaction. Study of epiphyllous fungi thus can provide a precise understanding of ancient environment and interaction of fungi with its surrounding. We have earlier mentioned that there is a quite a good number of fossil epiphyllous fungal morphs, that couldn't be assigned into their respective taxonomic groups which can be resolved by restudying those morphs using recently emerged sophisticated techniques and more fossil findings.

**Table 1.** Fossil epiphyllous fungal diversity through ages with systematic position, host, location, probable palaeo-nviroment and mode of interaction

| Age | Epiphyllous fossil fungi | Systematic Position | Host | Loca-tion | Palaeo-enviro-nment | Pro-able Mode of Inter-ction |
|---|---|---|---|---|---|---|
| **1. Palaeozoic** | | | | | | |
| Devonian | *Hysterites opegraphoides* (24) | Ascomyetes? or Pyrenoycetes or Hysteriales | *Neuroperis* sp. *Podozaites* sp. | Germany | † | † |
| | **Melanosphaerites devonicus* (26) | Deuteromycetes-Phomales | — | Bear Island, Norway | † | † |
| Carbo-iferous | *Excipulites* sp. (24) | Ascoycetes Pezizales | *Hymeno-hyllites* sp. *Callipteris* sp. *Pecopteris* sp. *Macrost-chya* sp. | France | † | † |
| | *Urophlyctites oliveriana* (42) | Phycoycetes Protococales | *Alethoteris* sp. | France | † | † |
| | *Hysterium* sp. (28) | Ascomycota, Pezizomyotina, Dothideomycetes, Hysteriales, Hysteriaeae | *Zostera* sp. *Pteridium* sp. *Acer* sp. | Germany | † | † |
| | **Macrosporites* sp. | Deuteroycetes Moniliales | — | Germany | † | † |
| | **Mycocarpon pachydermum* (32) | Ascomycetes Eurotiales | — | England Kansas | † | † |
| | **Sporocarpon compactum* | Ascomycetes Eurotiales | — | England Kansas | † | † |
| Permian | *Rosellinia* sp. | Ascomycota, | — | Germany | † | † |

| | | | | | | |
|---|---|---|---|---|---|---|
| | | Pezizoycotina, Sordaioycetes, Xylariomyetidae, Xylariales, Xylariaceae | | | | |
| | Unidentified form (2) | — | *Glossopteris* sp. | Bihar, India | Tropical to sub-tropical | Probable Parasite |
| **2. Mesozoic** | | | | | | |
| Triassic | | | | | | |
| Cretaceous | **Puccinites lanceolatus* | Ascomycetes Pezizales | — | Austria Nebr-aska | ✝ | ✝ |
| | *Phacidium rimosum* (41) | Ascomycota, Pezizomycotina, Leotiomycetes, Leotiomycetidae, Helotiales, Phacidiaceae | *Pyrus* sp. *Eugenia* sp. *Populus* sp. | Bohemia, Germany | ✝ | Probable path-ogen |
| | **Ovularites barbourii* | Deuteromycetes Moniliales | — | Nebr-aska | ✝ | ✝ |
| | *Excipula* sp. (64) | Ascomycota, Pezizomycotina, Leotiomycetes, Leotiomycetidae, Helotiales, Dermateaceae | *Pecopteris* sp. *Macro-stachya* sp. | England | ✝ | ✝ |
| | *Sphaeropsis* sp. (23) | Anamorphic fungi | *Betula* sp. *Bumelia* sp. *Carya* sp. | England | ✝ | Probable path-ogen |
| | *Asterothyrites dictyozamiticola* (34) | Ascomycetes Microth-yriales | *Dictyoza-mites grossinervis* | Southern Primrosk Russia | ✝ | Ectopar asite |
| | *Asterothyrites podocarpi* (34) | Asco-mycetes Microthyriales | *Podocarpus harissii* | Southern Primrosk Russia | ✝ | Ectopar asite |
| | *Perisporicites zamiophylli* (39) | Ascomycetes Dothidiales | *Zamio-phyllum buchianum* | Southern Primrosk, Russia | ✝ | ✝ |

| | | | | | |
|---|---|---|---|---|---|
| | | *Subzamites borealis* | | | |
| *Phragmothyrites doratophylli* (34) | Ascomycetes Microthyriales | *Dorato-phyyllum sulcatum* | Southern Primrosk Russia | † | Probable patho-gen |
| *Phragmothyrites nilssoniopterícola* (34) | Ascomycetes Microthyriales | *Nilssoni-pteris rhitido-rachis* | Southern Primrosk Russia | † | Probable patho-gen |
| *Phragmothyrites otozamiticola* (34) | Ascomycetes Microthyriales | *Otozamites* sp. | Southern Primrosk Russia | † | Probable patho-gen |
| *Trichopeltinites nilssoniopterícola* (39) | Ascomycetes Microthyriales | *Nilssoni-pteris* sp. *Cladophl-ebis srigida* | Southern Primrosk Russia | † | Probable sapro-phyte |
| *Ussurithyrites araucariodendri* (39) | Ascomycetes Microthyriales | *Araucari-odendron hetero-phyllum* | Southern Primrosk Russia | † | † |
| *Microthyriacites cephalotaxi* (34) | Ascomycetes Microthyriales | *Cephalo-taxus ussuriensis* | Southern Primrosk Russia | † | Ectopar asite |
| *Brefeldiellites argentinus* (45) | ▫Ascomycetes Microthyriales | *Podocarpus dubius*, *Dictyoza-mites* sp., *Ruflorina sierra* Conifer leaf | Santa Cruz province, Bajo Grande, Argen-tina | † | † |
| *Microthyriacites baqueroensis* (34) | Ascomycetes Microthyriales | *Dictyoza-mites* sp. | Santa Cruz province, Bajo Grande, Argen-tina | † | Ectopar asite |
| *Chlamydosporites gramineum* | Basidiomycetes Ustilaginales | Grass cuticle | Madhya Pradesh, India | † | Parasite |

| | | | | | | |
|---|---|---|---|---|---|---|
| | *Molinea asterinoides* | Ascomycetes Myrangiales or Microthyriales | Dicot leaf | Boyaca Basin, Colombia | Warm moist climate | Parasite |
| | *Stomiopeltites cretacea* (1) | Ascomycetes Microthyriales | Conifer leaf | Compton Bay, Isle of Wight | † | Parasite / Probable Pathogen |
| | *Aecidites stellatus* (15) | Bacidiomycetes Uridinales | *Dryophyllum* sp. | Aachen region, Germany | † | Probable Pathogen |
| | *Himantia* sp. (15) | Anamorphic fungi | *Dryophyllum* sp. | Aachen region, Germany | † | † |
| | *Torula* sp. | Fungi imperfecti Phragmosporae | — | Prussia | † | Probable Saprophyte |
| | *Pleosporites shirainus* (72) | Ascomycetes Sphaeriales | *Cryptomeriopsis* sp. | Hokkaido, Japan | † | † |
| | *Diplodites mohogaoensis* (36) | Fungi imperfecti Sphaeropsidales | Monocot leaf | Madhya Pradesh, India | † | Probable Parasite |
| **3. Coenozoic** | | | | | | |
| A. Tertiary | *Hysterites labyrinthiformis* (74) | Ascomycetes? Pyrenomycetes Hysteriales | Unidentified Leaf | Croatia | † | † |
| | **Eurotium* sp. | Ascomycota, Pezizomycotina, Eurotiomycetes, Eurotiomycetidae, Eurotiales, Trichocomaceae | — | Prussia | † | † |
| | **Oidium* sp. | Anamorphic fungi | — | Prussia | † | Probable Pathogen/ Parasite |
| | *Cenangium* sp. (41) | Ascomycota, Pezizomycotina, Leotiomycetes, | *Typha* sp. | Prussia | † | Probable Pathogen |

| | | | | | |
|---|---|---|---|---|---|
| | Leotiomycetidae, Helotiales, Helotiaceae | | | | |
| **Fusidium* sp. | Anamorphic fungi | — | Prussia | † | Probable Pathogen/ Parasite |
| *Sclerotites* sp. | Deuteromycetes Moniliales | *Cinnamomum* Sp. *Populus* sp. *Cyperus* sp. *Celastrus* sp. *Gingko* sp. | Germany France England Greenland Colorado, USA | † | † |
| *Dothidea borealis*, (29) | Ascomycota, Pezizomycotina, Dothideomycetes Dothideomycetidae, Dothideales, Dothideaceae | *Acer* sp. *Andromeda* sp. *Sterculia* sp. *Myrica* sp. *Betula* sp. *Cyperus* sp. | Iceland Switzerland | † | Probable Pathogen |
| *Sclerotium acericola* (29) | Basidiomycota, Agaricomycotina Agaricomycetes, Agaricomycetidae, Agaricales, Typhulaceae | *Acer* sp. *Rhus* sp. *Laurus* sp. *Betula* sp. | Switzerland Italy | † | Probable Pathogen |
| *Depazea acericola* (63) | Anamorphic fungi | *Acer* sp. *Salix* sp. *Andromeda* sp. | France Germany | † | † |
| *Leptosphaerites lemoinii* (59) | Ascomycetes Dothidiales | Monocot leaf | France | † | Probable Pathogen |
| *Hysteriopsis subopegraphoides* (23) | Ascomycetes Phacidales | *Salix* sp. *Sapotacites* sp. | Greenland | † | † |
| *Phacidiopsis* | Ascomycetes | *Salix* sp. | Switzerla | † | Probable Pathoge |

| | | | | | | |
|---|---|---|---|---|---|---|
| | sp. (23) | Phacidales | *Fagus* sp. *Quercus* sp. | nd | | n |
| | *Phyllerites* sp. | Deuteromycetes Moniliales | *Laurus* sp. *Myrica* sp. *Callicoma* sp. *Alnus* sp. *Platanus* sp. *Ficus* sp. | Prussia Bohemia France | † | † |
| | **Palaeosordaria lagena* (61) | Ascomycetes Sphaeriales | — | Madhya Pradesh, India | † | † |
| | Microthyraceous germling (20) | Ascomycetes Microthyriales | *Sapindus* sp. *Pityophyllum* sp. Unidentified Leaf | Germany Hungary England Colorado and Tennessee, USA | † | Probable Pathogen |
| Palaeocene | **Brachysporium* (5) | Anamorphic fungi | — | Montana | † | Probable Pathogen/ Parasite |
| | **Polyhyphaethyrites giganticus* | — | — | Madhya Pradesh, India | † | † |
| Eocene | **Penicillium* sp. (5) | Anamorphic fungi | — | Germany | † | Parasite/ Saprophyte |
| | **Dendryphium* sp. (5) | Deuteromycetes Moniliales | — | Prussia | † | † |
| | *Stegilla* sp. (8) | Ascomycetes Phacidales | *Graminia* sp. *Cyperitum* sp. | Switzerland | † | † |
| | *Erysiphe* sp. (65) | Ascomycota, Pezizomycotina, Leotiomycetes, Leotiomycetidae, | *Ficus* sp. | Sicily | † | Pathogen |

| | | | | | |
|---|---|---|---|---|---|
| | Erysiphales, Erysiphaceae | | | | |
| **Stilbum* sp. | •Anamorphic fungi | — | Prussia | † | Probable Patho-gen |
| **Ramularia* sp. (10) | •Anamorphic fungi | — | Prussia | † | Patho-gen |
| **Perisporiacites larundae* (19) | Ascomycetes Dothidiales | — | Transcau casia | | |
| **Haplographites xylophagus* (19) | Fungi imperfecti Amerosporae | — | Transcau casia | † | Probable Sapro-phyte |
| **Haplographites cateniger* (19) | Fungi imperfecti Amerosporae | — | Transcau casia | † | Probable Sapro-phyte |
| **Morosporium elongatum* (58) | Deuteromycetes Moniliales | — | France | † | † |
| *Pestalozzites* sp. (7) | Deuteromycetes Melanconiales | *Thrinax* sp. *Sabalites* sp. | Georgia, Florida & Louisi-ana of USA | † | † |
| **Fungites apoidensis* (27) | Fungi imperfecti Mycelia sterilia | — | Germany | † | † |
| **Streptotrichites spiralis* | Deuteromycetes Moniliales | — | Prussia | † | † |
| **Acremonites succineus* (53) | Deuteromycetes Moniliales | — | Prussia | † | Probable Endophy te |
| *Graphiolites sabaleos* (21) | Basidiomycetes Ustilaginales or Fungi imperfecti Amerosporae | *Sabalites sucssion-ensis* | Cessoy, France | † | Pathoge n |
| *Sphaerites nervisequus* (21) | Ascomycetes Xylariales? | Lauraceae | Cessoy, France | † | Probable Patho-gen |

| | | | | | |
|---|---|---|---|---|---|
| *Phomites myricae* (21) | Fungi imperfecti Sphaeropsidales | Myricaceae | Cessoy, France | † | Pathogen |
| *Caenomyces annulatus* (7) | Ascomycetes? Pyrenomycetes? | *Cassia emarginata* | Wilcox beds, Fayette County, Texas | † | Probable Parasite |
| *Caenomyces cassia* (7) | Ascomycetes? Pyrenomycetes? | *Cassia emarginata* | Wilcox beds, Fayette County, Texas | † | Probable Parasite |
| *Caenomyces laurineus* (7) | Ascomycetes? Pyrenomycetes? | *Nectandra lowii* | Holly Springs sand; Oxford Gully | † | Probable Parasite |
| *Caenomyces sapotae* (7) | Ascomycetes? Pyrenomycetes? | *Sideroxylon premastichodendron* | Holly Springs sand; Oxford Gully | † | Probable Parasite |
| *Caenomyces pestalozzites* (7) | Ascomycetes? Pyrenomycetes? | *Sabalites graya-nusom* | Holly Springs sand; Oxford Gully | † | Probable Parasite |
| *Phragmothyrites eocaenicus* (17) | Ascomycetes Microthyriales | Conifer leaf | Scotland | † | Probable Pathogen |
| *Asterina eocenica* (16) | Ascomycetes Microthyriales | *Chrysobalanus* sp. | Western Tennessee, USA | Warm temperate subtropical | Pathogen |
| *Asterothyrites spinulatus* (34) | Ascomycetes Microthyriales | *Chrysobalanus* sp. | Western Tennessee, USA | Warm temperate subtropical | Ectoparasite |

| | | | | | |
|---|---|---|---|---|---|
| *Brefeldiellites fructiflabellus* (16) | Ascomycetes Microthyriales | *Chryso-balanus* sp. | Western Tenne-ssee, USA | Warm tem-perate sub-tropical | † |
| *Haplopeltis mucoris* (16) | Ascomycetes Microthyriales | *Chryso-balanus* sp. | Western Tenn-essee, USA | Warm tem-perate sub-tropical | † |
| *Meliolinites spinksii* (67) | Ascomycetes Erysiphales | *Chryso-balanus* sp. | Western Tenne-ssee, USA | Warm tem-perate sub-tropical | Probable Pathoge n |
| *Patoullardiella imbricate* (16) | Fungi imperfecti Moniliales | *Chryso-balanus* sp. | Western Tenn-essee, USA | Warm tem-perate sub-tropical | Probable Parasite |
| *Pelicothallos villosus* (16) | Ascomycetes Microthyriales | *Chryso-balanus* sp. | Western Tenn-essee, USA | Warm tem-perate sub-tropical | No evidence of direct parasitis m |
| *Sporidesmium henryense* (16) | Fungi imperfecti Phragmosporae | *Chryso-balanus* sp. | Western Tenne-ssee, USA | Warm tem-perate sub-tropical | No evidence of direct parasitis m |
| *Dictyotopileos yalensis* (16) | Ascomycetes Microthyriales | *Chryso-balanus* sp. | Western Tenne-ssee, USA | Warm tem-perate sub-tropical | Parasite |
| *Vizella memorabilis* (67) | Ascomycetes Microthyriales | *Chryso-balanus* sp. | Western Tenn-essee, USA | Warm tem-perate sub-tropical | Probable Pathoge n |

| | | | | | |
|---|---|---|---|---|---|
| *Asterina nodosaria* (16) | Ascomycetes Microthyriales | *Sapindus* sp. | Western Tenn-essee, USA | Warm tem-perate sub-tropical | Pathoge n |
| *Asterothyrites tennesseensis* (16) | Ascomycetes Microthyriales | *Sapindus* sp. | Western Tenne-ssee, USA | Warm tem-perate sub-tropical | Ectopara site |
| *Callimothallus pertusus* (16) | Ascomycetes Microthyriales | *Sapindus* sp. | Western Tenn-essee, USA | Warm tem-perate sub-tropical | No evidence of direct parasitis m |
| *Meliolinites anfractus* (34) | Ascomycetes Erysiphales | *Sapindus* sp. | Western Tenne-ssee, USA | Warm tem-perate sub-tropical | Probable Pathoge n |
| *Microthyriella fungosa* (16) | Ascomycetes Microthyriales | *Sapindus* sp. | Western Tenn-essee, USA | Warm tem-perate sub-tropical | Probable Pathoge n |
| *Phragmothyrites lutosus* (16) | Ascomycetes microthyriales | *Sapindus* sp. | Western Tenn-essee, USA | Warm tem-perate sub-tropical | Probable Pathoge n |
| *Plochmopeltidella antiqua* (16) | Ascomycetes Microthyriales | *Sapindus* sp. | Western Tenne-ssee, USA | Warm tem-perate sub-tropical | Probable Pathoge n |
| *Stomiopeltites plectilis* (34) | Ascomycetes Microthyriales | *Sapindus* sp. | Western Tenne-ssee, USA | Warm tem-perate sub-tropical | Parasite |

| | | | | | | |
|---|---|---|---|---|---|---|
| | *Trichopeltina exporrecta* (16) | Ascomycetes Microthyriales | *Sapindus* sp. | Western Tenn-essee, USA | Warm temp-erate sub-tropical | No evidence of direct para-sitism |
| | *Trichopeltinites nilssoniopterícola* (16) | Ascomycetes Microthyriales | *Sapindus* sp. | Western Ten-nessee, USA | Warm tem-perate sub-tropical | No evidence of direct para-sitism |
| | *Parasterina plectopelta* (16) | Ascomycetes Microthyriales | *Sapindus* sp. | Western Ten-nessee, USA | Warm tem-perate sub-tropical | Probable Patho-gen |
| | *Meliolinites dilcheri* | Ascomycetes Erysiphales | Unidentified Leaf | Texas, USA | † | Probable Patho-gen |
| | *Palaeoserenomyces allenbyensis* (13) | Ascomycetes Phyllachorales | Extinct Palm Leaf | British Colombia Canada | † | Parasite |
| Oligocene | *Sphaerophorus moniliformis* (47) | Fungi imperfecti Phragmosporae | *Thuja occidentalis* | Baltic area | † | Pathoge n |
| | **Diplosporium* sp. | ▫Deuteromycetes Moniliales | — | France | † | Probable Sapro-phyte |
| | *Acremonium succineum* (10) | Fungi Imperfecti Amerosporae | *Widdringtonites oblongifolius* G. var *longifolius* | Baltic area of Poland | † | Probable Sapro-phyte |
| | *Torula mengeana* | Fungi imperfecti Phragmosporae | *Thuites* sp. | Baltic area of Poland | † | Probable Sapro-phyte |
| | *Fungites capillaries* | Fungi imperfecti Mycelia sterilia | *Thuites succineus* | Baltic area of Poland | † | † |
| | *Fungites pullus* | Fungi imperfecti Mycelia sterilia | *Chamaecyparis* sp. *Sequoia* sp. | Baltic area of Poland | † | † |

| | | | | | |
|---|---|---|---|---|---|
| | | *Libocedrites* sp. | | | |
| *Fungites hirtus* | Fungi imperfecti Mycelia sterilia | Unidentified Leaf | Baltic area of Poland | † | † |
| *Didymosphaeria* sp. (11) | Ascomycota, Pezizomycotina, Dothideomycetes Pleosporomycetidae, Pleosporales, Didymosphaeriaceae | *Typha* sp. | Colorado, USA | † | Pathogen |
| *Asterothyrites minutes* (12) | Ascomycetes Microthyriales | *Oleinites willissi* | Australia New-zealand | Warmer and highly humid | Ecto-parasite |
| *Euthythyrites oleinites* (12) | Ascomycetes Microthyriales | *Oleinites willissi* | Australia New-zealand | Warmer and highly humid | Ecto-parasite |
| *Trichothyrites ostiolatus* (34) | Ascomycetes Microthyriales | *Oleinites willissi* | Australia New-zealand | Warmer and highly humid | Probable pathogen |
| *Phragmothyrites delicatissimus* (67) | Ascomycetes Microthyriales | *Oleinites willissi* | Australia New-zealand | Warmer and highly humid | Probable pathogen |
| *Phragmothyrites sinuatus* (67) | Ascomycetes Microthyriales | *Oleinites willissi* | Australia New-zealand | Warmer and highly humid | Probable pathogen |
| *Trichopeltinites pulcher* (12) | Ascomycetes Microthyriales | *Oleinites willissi* | Australia New-zealand | Warmer and highly humid | Probable saprophyte |
| **Plochmopeltinites masonii* (12) | Ascomycetes Microthyriales | — | New South Wales, Victoria | Warmer and highly humid | Probable pathogen |

| | | | | | | |
|---|---|---|---|---|---|---|
| Miocene | ******Polystigmites priscus*** (46) | Ascomycetes Sphaeriales | *Ficus* sp. | Italy | † | † |
| | ***Sphaerites disciformis*** | Ascomycetes Xylariales? | *Quercus* sp. *Vaccinus* sp. | Pars-chlug, Styria, Austria | † | Probable patho-gen |
| | **Linosporoidea* sp. (38) | ▫Ascomycetes Sphaeriales | — | Switz-erland | † | † |
| | **Uncinulites baccarinii* (50) | Fungi imperfecti? Amerosporae? | — | "Disodile" beds, Sicily | † | † |
| | **Perisporites setosus* (50) | Ascomycetes Erysiphales | — | "Disodile" beds, Sicily | † | † |
| | **Perisporites hirsutus* (50) | Ascomycetes Erysiphales | — | "Disodile" beds, Sicily | † | † |
| | *Chaetomites intricatus (50) | Ascomycetes Xylariales | — | "Disodile" beds, Sicily | † | † |
| | **Melanosporites stefanii* (50) | Ascomycetes Diaporthales | — | "Disodile" beds, Sicily | † | † |
| | **Microthyrites disodilis* (50) | Ascomycetes Microthyriales | — | "Disodile" beds, Sicily | † | † |
| | **Monilites albidus* (50) | Fungi imperfecti Phragmosporae? | — | "Disodile" beds, Sicily | † | † |
| | **Leptothyriomyces radiates* (50) | Ascomycetes Microthyriales | — | Sumatra | † | † |
| | **Cercosporites* sp. (62) | Fungi imperfecti Phragmosporae | — | Italy | † | Probable weak parasite or sapro-phyte |
| | *Haplopeltis neyveliensis* (56) | Ascomycetes Microthyriales | Unidentified Leaf | Tamil-nadu, India | Warm humid with | † |

| | | | | | |
|---|---|---|---|---|---|
| | | | | high precipitation | |
| *Meliolinites nivalis* (67) | Ascomycetes Erysiphales | Myrtaceae | Kiandra, New South Wales, Australia | † | Probable pathogen |
| *Phragmothyrites kiandrensis* (67) | Ascomycetes Microthyriales | Myrtaceae Lauraceae | Kiandra, New South Wales, Australia | † | Probable pathogen |
| Unidentified form (70) | — | *Betula* sp. *Smilax* sp. *Lithocarpus* sp. Lauraceae | Northern Idaho, USA | Temperate | † |
| *Phragmothyrites concentricus* (52) | Ascomycetes Microthyriales | *Castanea* sp. *Betula* sp. *Quercus* sp. *Persea* sp. *Magnolia* sp. *Lithocarpus* sp. *Pseudofagus* sp. *Pinus* sp. | Clarkia, Idaho, USA | Warm temperate | Probable pathogen |
| *Stomiopeltites amorphos* (52) | Ascomycetes Microthyriales | *Betula* sp. *Smilax* sp. *Zizyphoides* sp. | Clarkia, Idaho, USA | Warm temperate | Probable pathogen |
| *Entopeltacites remberi* (52) | #Dothideomycetes Dothideales | *Persea* sp. | Clarkia, Idaho, USA | Warm temperate | Probable host specific parasite |
| *Palaeocolletotrichum graminoides* (48) | #Coleomycetes Melanconiales | Monocot leaf | Darjeeling foothills, West | Moist tropical | Pathogen |

| | | | | | | |
|---|---|---|---|---|---|---|
| | | | | Bengal, India | | |
| | *Palaeocercospora siwalikensis* (48) | # Hyphomycetes Moniliales | Angiospermic leaf | Darjeeling foothills, West Bengal, India | Moist tropical | Pathogen |
| | *Palaeoasterina siwalika* (49) | #Ascomycetes Asterinales | Angiospermic leaf | Darjeeling foothills, West Bengal, India | Moist tropical | Exoparasite |
| | *Phomites* (44) | Fungi imperfecti Sphaeropsidales | *Anthocephalus* sp. | Darjeeling foothills, West Bengal, India | Moist tropical-evergreen | Pathogen |
| | *Meliolinites siwalika* (44) | Ascomycetes Erysiphales | Lauraceae | Darjeeling foothills, West Bengal, India | Moist tropical-deciduous | Probable Pathogen |
| Pliocene | **Laestadites nathostii* | Ascomycetes Sphaeriales | — | Japan | † | † |
| | *Depazites* sp. | Deuteromycetes Phomales | *Cinnamomum* sp. *Alnus* sp. *Juglans* sp. *Smilax* sp. *Myrica* sp. | Italy Greenland | Probable moist humid climate | Probable Pathogen |
| | *Asterina* sp. | Ascomycota, Pezizomycotina, Dothideomycetes Dothideomycetidae, Asterinaceae | *Ilex* sp. *Sequoia* sp. | Germany | Moist tropical-deciduous | Pathogen |

| | | | | | | |
|---|---|---|---|---|---|---|
| | *Palaeoasterina siwalika* (49) | #Ascomycetes Asterinales | *Chonemorpha* sp. | Arunachal Pradesh, India | Moist tropical-deciduous | Exo-parasite |
| B. Quaternary | *Phacidites* sp. | Ascomycetes Phacidales | *Buxus* sp. *Juglans* sp. *Smilax* sp. *Palaeocassia* sp | Italy Borneo | Temparate | Pathogen |
| Pleistocene | *Trichothyrites pleistocaenicus* (60) | Ascomycetes Microthyriales | *Picea mariana* | Minnesota, USA | Warm – sub tropical | Exo-parasite |
| | **Microthyrium* sp. (25) | Ascomycota, Pezizomycotina, Dothideomycetes Microthyriales, Microthyriaceae | — | England | Glaciation - wide climatic range | Pathogen |
| Holocene | *Callimothallus* sp. | Ascomycetes Microthyriales | Leaf of mangrove affinity | Bengal basin, India | Tropical-Subtropical with high precipitation | Exo-parasite |
| | *Meliolinites anfractus* (34) | Ascomycetes Erysiphales | Leaf of mangrove affinity | Bengal basin, India | Tropical-Sub-tropical with high precipitation | Parasite / Pathogen |
| | *Meliolinites spinksii* (67) | Ascomycetes Erysiphales | Leaf of mangrove affinity | Bengal basin, India | Tropical-Sub-tropical with high precipitation | Parasite / Pathogen |
| | *Isthmospora* spinosa (22) | Anamorphic fungi | *Calluna vulgaris* | Northern England | Probable high precipitation | Hyperparasite |

| | | | | | |
|---|---|---|---|---|---|
| *Meliola ellisii* Roum. (22) | Ascomycota, Pezizomycotina, Sordariomycetes, Meliolales, Meliolaceae | *Calluna vulgaris* | Northern England | Probable high preci-pitation | Patho-gen |

* Probable epiphyllous fungi

▫ Systematic position followed by Dilcher, 1965

• Systematic position followed by Mycobank.org.

# Systematic position followed by original describing author

✝ Not known

## REFERENCES


1. Alvin KA and Muir MD 1970 An epiphyllous fungus from the Lower Cretaceous, *Biological Journal of Linnean Society* **2** 55-59
2. Bajpai U and Maheshwari HK 1987 Epiphyllous fungi from the Gondwana, *The Palaeobotanist* **36** 210-213
3. Banerjee M 1995 Palaeobiology of Neogene sediments of Bhutan, Eastern Himalaya and environment of deposition Birbal Sahni Cent Vol Proc Int Conf Global Environment and Diversification of plants through geological time Soc Ind Plant Taxonomists, Allahabad 41-57
4. Benes K and Kraussova J 1962 Carboniferous fossil fungi from the Upper Silesian basin (Ostrava- Karvina Coal District), *Paleontologie* **4** 65-87
5. Berkeley M J 1848 On three species of mould detected by Dr Thomas in the amber of east Prussia, *The Annals and Magazine of Natural History* **2** 380-383
6. Barnicchia A, Fugazzola M A, Gemelli V, Mantovani B, lucchetti A, Cesari M, and Speroni E 2006 DNA recovered and sequenced from an almost 7000 y-old Neolithic polypore, *Daedaleopsis tricolor Mycological Research* **110** 14 – 17
7. Berry EW 1916 The lower Eocene floras of the South Eastern part of North America *US, Geol Surv Prof Paper* **91** 1-481
8. Braun A 1854 Einige Beiträge zur Flora der Tertiar, *Zeit N Jb* 138-147
9. Brunel M, Chaye D, Albissin M, and Locquin, M 1984 Détermination d'un âge cambrien des series carbonates magnésite situées sous le Grand Chevauchement Himalayan (Main Central Thrust- Népal Oriental), *par*

*la découverte de paléobasidiospores Géobios* **17** **(3)** 595-602

10. Caspary R 1907 Die Flora des Bernsteins und anderer fossiler Harze des ostpreussischen Tertiärs [Nach dem Nachlasse des Verstorbenen bearbeitet von Richard Klebs, in Königsberg] Abhandlungen der Königlish Preussischen Geologischen Landesanstalt, *Neue Folge* **4** 181
11. Cockerell TDA 1908 The fossil flora of Florissant, Colorado, *Bulletin of the American Museum of Natural History* **24** 71-110
12. Cookson IC 1947 Fossil fungi from Tertiary deposits in the Southern Hemisphere Part I, *Proceedings of the Linnean Society NSW* **72** 207-214
13. Currah RS, Stockey RA, and Lepage BA 1998 An Eocene tar spot on a fossil palm and its fungal hyperparasite, *Mycologia* **90** 667-673
14. Das P, Khan MA, De B, Samajpati N, and Bera S 2007 Evidence of exoparasitic relationship between *Asterina* (Asterinaceae) and *Chonemorpha* (Apocynaceae) from the Upper Siwalik (Kimin Formation) sediments of Arunachal subhimalaya, Indian *Journal of Mycopathological Research* **45** **(2)** 225-230
15. Debey MH and Ettingshausen A 1857 Die urweltlichen Thallophyten des Kreidegebirgs von Aachen und Maestricht, *Denkschriften der Akademie der Wissenschaften in Wien* **16** 131-214
16. Dilcher DL 1965 Epiphyllous fungi from Eocene deposits in Western Tennessee, USA, *Palaeontographica* **116** 1-54
17. Edwards W N 1922 An Eocene microthyracious fungus from Mull, Scotland, *Transaction in British Mycological Society* **8** 66-72
18. Engelhardt H and Kinkelin F 1908 Oberpliocäne Flora und Fauna des Untermaintales, insbesondere des Frankfurter Klärbeckens *Senckenberg naturforsch Ges Abh* **29** 150-306
19. Felix J 1894 Untersuchungen über fossile Hölzer, Stück 4, IV, Hölzer aus dem Kaukasus *Dt Geol Ges Z* **46** 79-110
20. Frantz U 1959 Die Pollenflora der Braunkohle von Lohsa/ Niederlausitz Inauguraldissertation zur Erlangung der Doktor- würde der Math, *Naturv Fakultät der Freien Univ Berlin* 46 page
21. Fritel PH 1910 Étude sur les vegetaux fossils de l'étage sparnacien du Bassin de Paris, Mémoires de la Société géologique de France *Palaéontologie* **40** 5-37
22. Geel BV, Aptroot A and Mauquoy D 2006 Sub-fossil

evidence for fungal hyperparasitism (*Isthmospora spinosa* on *Meliola ellisii*, on *Calluna vulgaris*) in a Holocene intermediate ombrotrophic bog in northern-England, *Review of Palaeobotany and Palynology* **141** 121-126

23. Geyler HT 1887 Ober fossile Pflanzen von Labuan *Vega-Exped Vetenskapliga Arb* **4** 475-507

24. Goepert HR 1836 Die fossilen Farrenkräuter (Systema filicum fossilium) *Nova Acta Acad Leop Carol* **17** 1-486

25. Godwin H and Andrew R 1951 A fungal fruitbody common in postglacial peat deposits, *New Phytologist* **50** 179-183

26. Grüss J 1928 Zur Biologie Devonischer Thallophyten, *Palaeobiologica* **1** 487-517

27. Hallier E 1865 Untersuchung fossiler Zellenpflanzen, *Botany Journal* **23** 189-191

28. Heer O 1855 Flora Tertiaria Helvetiae Vol I Winterthur 117 page

29. Heer O 1859 Flora Tertiaria Helvetiae Vol III Winterthur 377 page

30. Hermann TN and Podkovyrov VN 2006 Fungal remains from the Late Riphean, *Palaeontological Journal* **40** 307-214

31. Hosagoudar VB, Robin PJ, and Shivaraju B 2010 Foliicolous fungi from the Achankovil forests in Kollam District of Kerala State, India, *Journal of Threatened Taxa* **2 (3)** 760-761

32. Hutchinson SA 1955 A review of the genus *Sporocarpon* Williamson, *Annals of Botany II* **69** 425-437

33. Jain KP and Kar RK 1979 Palynology of Neogene sediments around quilon and Varkala, Kerala coast, South India-I Fungal remains, *Palaeobotanist* **26 (2)** 105-118

34. Kalgutkar RM and Jansonius J 2000 Synopsis of fossil fungal spores, mycelia and fructifications Published by American Association of Stratigraphic Palynologists Foundation, Dallas, Texas, USA

35. Kalgutkar RM and Mcintyre DJ 1991 Helicosporous fungi and early Eocene pollen, Eureka Sound Group, Axel Heiberd Islnad, Northwest Territories, *Canadian Journal of Earth Science* **28** 364-371

36. Kalgutkar RM, Nambudiri EMV and Tidwell WD 1993 *Diplodites sweetie* sp nov from the Late Cretaceous (Maastrichtian) Deccan Intertrappean beds of India, *Review of Palaeobotany and Palynology* **77** 107-118

37. Karlsson M and Stenlid J 2008 Comparative Evolutionary Histories of the Fungal

Chitinase Gene Family Reveal Non-Random Size Expansions and Contractions due to Adaptive Natural Selection, *Evolutionary Bioinformatics* **4** 47–60

38. Keller R 1895 Beiträge zur Tertiärflora des Kantons Gallen *St Gallenischen naturwiss Ges* **1894** 305-330

39. Krassilov V A 1967 Early cretaceous flora of the southern Primrosk and its significance for the stratigraphy (in Russia) Academy of Sciences of the USSR Siberian branch, Far East Geological Institute, Moscow 264 page

40. Krings M, Dotzler N, Galtier J, and Taylor TN 2009 Microfungi from the upper Visean (Mississippian) of central France Chytridiomycota and chytrid-like remains of uncertain affinity, *Review of Palaeobotany and Palynology* **156 (3-4)** 319-328

41. Ludwig R 1859 Fossile Pflanzen aus der altesten Abteilung der Rheinisch-Wetterauer Tertiär- Formation, *Palaeontographica* **8** 39-154

42. Magnus P 1903 Ein von F W Oliver nachgewiesener fossiler parasitischer Pilz *Dt Bot Ges Z* **12** 248-250

43. Mandal A, Samajpati N, and Bera S 2009 *In situ* occurrence of epiphyllous fungus *Phomites* Fritel from the Lower Siwalik sediments of Darjeeling Foothills, *Journal of Botanical Society of Bengal* **63 (1)** 37-40

44. Mandal A, Samajpati N, and Bera S 2011 A new species of Meliolinites (fossil Meliolales) from the Neogene sediments of sub-Himalayan West Bengal, India, *Nova Hedwigia* **92 (3-4)** 435-440

45. Martínez A 1968 Fossil Mycrothiriales (fungi, Ascomycetes) from the lower Cretaceous of Santa Cruz Province, Argentina *Revista de la Associaciòn Paleontològica Argentina, Aneghiniana* **5** 257-263

46. Massalongo A 1858 Synopsis florae fossilis Senogalliensis Verona 136 page

47. Menge A 1858 Beiträge zur Bernsteinflora Schriftn der naturforschendn Gesellschaft in Danzig **6** 3-18

48. Mitra S and Banerjee M 2000 On the occurrence of epiphyllous Deuteromycetous fossil fungi *Palaeocercospora siwalikensis* gen et sp nov and *Palaeocolletotrichum graminioides* gen et sp nov from Neogene sediments of Darjeeling foot hills, Eastern Himalaya, *Journal of Mycopathological Research* **37 (2)** 7-11

49. Mitra S, Bera S, and Banerjee M 2002 On a new epiphyllous fungus *Palaeoasterina siwalika* gen et sp nov from the Siwalik (Middle Miocene)

sediments of Darjeeling foothills, India with remarks on environment, *Phytomorphology* **52 (4)** 285-292

50. Pampaloni L 1902 I resti organic nel disodile di Melilli in Sicilia, *Palaeontographica Italica* **8** 121-130
51. Parsons MG and Norris G 1999 Palaeogene fungi from Caribou Hills, Mackenzie Delta, Northern Canada *Palaenographica*, Abt B 250 77-167
52. Phipps CJ and Rember WC 2004 Epiphyllous fungi from the Miocene of Clarica, Idaho reproductive structures, *Review of Palaeobotany and Palynology* **129** 67-79
53. Pia J von 1927 Thallophyta In M Hirmer, (ed), Handbuch der Paläobotanik 31-136 Verlag R Oldenbourg, Berlin
54. Pirozynski KA and Hodges Jr CS 1973 New hyphomycetes from South Carolina, *Canadian Journal of Botany* **51** 157-173
55. Pratt LM, Philips TL and Dennison JM 1978 Evidence of non vascular land plants from Early Silurian (Llandoverian) of Virginia, USA, *Review of Palaeobotany and Palynology* **25** 121-149
56. Reddy PR, Ramanujam CGK, and Srisailam M K 1982 Fungal fructifications from Neyveli lignite, Tamil Nadu – their stratigraphic and palaeoclimatic significance, *Rec geological Survey of India* **114 (5)** 112-122
57. Remy W, Taylor T N, and Hass H 1994 Early Devonian fungi A blastocladalean fungus with sexual reproduction, *American Journal of Botany* **81** 690-702
58. Renault B and Roche A 1898 Etude sur la constitution des lignites et les organisms qu'ils renferment, *Bulletin de la Société d'Histoire naturelle d'Autun* **11 (1)** 201-239
59. Richon C 1885 Notice sur quelques Sphaeriacès nouvelles, *Society bot France Bulletin* **32** supplvii-xii
60. Rosendahl C O 1943 Some fossil fungi from Minnesota, *Bull Torrey Bot Club* **17** 126-138
61. Sahni B and Rao H S 1943 A silicified flora from the Intertrappean cherts round Sausar in the Deccan, *Natlional Academy of Science India Proceedings* **13** 36-75
62. Salmon E S 1903 *Cercosporites* sp, a new fossil fungus, *Journal of Botany* **41** 127-130
63. Saporta G 1868 Études sur la végétation du Sut- Est de la France, *Annual Science and Nature Botany* , **9** 5-62
64. Schimper W T 1869 Traite de palaeontologic vegetale Vol I Baelliere et Fils, Paris 740 page

65. Schmelhausen J 1883 Veiträge zur Tertiärflora Süd-West Rußlands Palaeont Abh herausg von W Dames und E Kaiser *Bd I H 4* 285-335

66. Schulz B and Boyle C 2005 Review-The endophytic continuum, *Mycological Research* **109 (6)** 661–686

67. Selkirk DR 1975 Tertiary fossil fungi from Kiandra, New South Wales, *Proceedings of the Linnean Society NSW* **100** 70-94

68. Seward A C 1898 Fossil Plants Volume I Cambridge University Press, Cambridge 452 page

69. Sharma N, Kar RK, Agarwal A, and Kar R 2005 Fungi in dinosaurian (Isisaurus) coprolites from the Lameta Formation (Maastrichtian) and its reflection on food habit and environment, *Micropaleontology* **51 (1)** 73-82

70. Sherwood-Pike M and Gray J 1988 Fossil leaf inhibiting fungi from Northern Idaho and their ecological significance, *Mycologia* **80 (1)** 14-22

71. Stubblefield SP and Taylor TN 1988 Recent advances in palaeomycology, *New Phytologist* **108** 3-25

72. Suzuki Y 1910 On the structure and affinities of two new conifers and a new fungus from the Upper Cretaceous of Hokkaido (Yezo), *Botanical Magazine* **24** 181-196

73. Taylor T N, Galtier J, and Axsmith B J 1994 Fungi from the Lower Carboniferous of central France, *Review of Palaeobotany and Palynology* **83** 253-260

74. Unger F 1841 Beiträge zur Flora der Vorwelt *Chloria Protogaea* Teil I 1-16

75. Unger F 1850 Genera et species plantarum fossilium Vienna P 627

76. Williamson WC 1887 On the organization of fossil plants of the Coal Measures IX Philos, *Trans Royal Society of London B* **169** 319-364

77. Wolf FA 1966 Fungus spores in East African lake sediments, *Bulletin of Torrey Botanical Club* **93** 104-113

78. Wolf FA 1967 Fungus spores in East African lake sediments II, *Journal of Elisha Mitchell Scientific Society* **82** 57-61

79. Wolf FA 1968 Fungus spores in Lake Singletary sediments, *Journal of Elisha Mitchell Scientific Society* **84** 227-232.

*Rev. Plant Pathol. Vol. 6, 2014*
*Indian Society of Mycology and Plant Pathology*
*Scientific Publishers (India), Jodhpur pp. 577-608*

# PARASITIC ANGIOSPERMS AND BIOLOGY OF *CUSCUTA* SPECIES– AN OVERVIEW

***Yash Pal Sharma and Vijayata Kapoor***

Department of Botany, University of Jammu, Jammu -180006, India

**KEY WORDS :** Cuscuta reflexa, Cuscuta campestris, Cuscuta australis Parasitic plants

## ABSTRACT

Parasitic plants represent a taxonomically heterogeneous assemblage of angiosperms that rely partially or completely on host plants for organic, inorganic and water resources. They exist in a complex and dynamic relationship between host, environment, edaphic and biotic factors and are common in many natural and semi-natural ecosystems from tropical rain forests to the high Arctic. Of the known parasitic angiosperms, about 25 genera negatively affect host plants cultivated by humans and out of these *Striga, Orobanche, Loranthus, Dendrophthoe, Viscum* and *Cuscuta* are considered agronomically important and much research has been undertaken on these parasitic genera. Among these, *Cuscuta* species (dodders) are the best known .The parasitic relationship of *Cuscuta* with its hosts has been the subject of numerous investigations. Host morphology and phenology are considered as the two primary factors which are responsible for selection of different hosts by *Cuscuta* species. Representatives of the genus *Cuscuta* are widespread and comprise of about 180 species distributed throughout the temperate to the tropical regions of the globe (80) with the greatest species diversity in subtropical and tropical regions. They are known to cause direct and indirect damage to plantations through infestation of the aerial parts of many wild and cultivated plants. In India, among the various reported species of this holoparasite, *Cuscuta reflexa* has been recorded as leading destructive parasite on most economically important hosts from different regions. However, plants of *Cuscuta* have versatility in their use, in both traditional as well as modern medicine.

## INTRODUCTION

Parasitic plants are among the most exotic representatives of the plant kingdom with extraordinary appearance and unique mode of life. They represent a taxonomically heterogeneous assemblage of angiosperms that rely partially or completely on host plants for organic, inorganic and water resources. They exist in a complex and dynamic relationship between host, environment, edaphic and biotic factors and are common in many natural and semi-natural ecosystems from tropical rain forests to the high Arctic (12). Parasitic plants are prudent predators that are strongly adapted to the life cycle of their hosts. Each of the many individual stages through which these parasitic organisms pass, contain a compliment of stimuli with which the organism must contend. They can also alter the physical environments around their hosts including soil, water and nutrients, atmospheric carbon dioxide and temperature and thus can be considered as ecosystem engineers.

## HISTORY OF PLANT PARASITES

Theophrastus, who reported dodder like plant from Babylon and was aware of mistletoes, *Viscum album* and *Loranthus europaeus* (70), possibly carried out initial studies on parasitic plants. The herbalist, Culpeper (19), discussed about the medicinal potential of dodder species and recommended it to "to clean black or burnt choler". He established that dodder plucked-off thyme is the most efficacious, suggesting that parasites' medicinal properties are determined in part by its host. The *Cuscuta* species (dodders) are now among the best known and the most curious parasitic plants among the scientists and workers. In the ancient time, there have been misconceptions about the *Cuscuta* being growing on the grapes as "Uva barbata". Bauhin erected it as a monstrosity of the grape vine. Precise nature of the bearded grape (34), dodder branches on the grapevine (38) have been illustrated. Darwin (20) was moved to compare "the harlot-nymphs, the fair Cuscutas" with mythological monsters and described them as the highly evolved and greatest opportunists. Behaviour of *Cuscuta* was compared with an animal soul and the body was like parasitic plants , while Dioscorides was the first writer who clearly speaks of *Cuscuta* in the Materia Medica (79). Later, Yuncker (130) in his monograph of the genus *Cuscuta* also included the *C. reflexa* var. *anguina* in section Callianche of subgenus *Monogyna*.

## DIVERSITY OF PARASITIC PLANTS

Angiosperm parasites are predominantly dicotyledonous and have evolved independently in 20 plant families, comprising 270 genera and about 4000 species (83, 94) accounting for 1-2% of 2,58,650 angiosperm species (101, 95, 12, 52). Of the known parasitic angiosperms, about 25 genera negatively affect host plants cultivated by humans and out of these *Striga, Orobanche, Loranthus, Dendrophthoe, Viscum* and *Cuscuta* are considered agronomically important and much research has been undertaken on these parasitic genera. These parasitic plants vary greatly with respect to host range, and parasitize a wide range of plant families such as Asteraceae, Fabaceae, Solanaceae, Moraceae, Cucurbitaceae, and Rutaceae. The parasitic plant families, which are considered agronomically important (70, 22, 93), are:

| | |
|---|---|
| Lauraceae | *Cassytha filiformis.* |
| Orobanchaceae | *Aeginetia indica, Alectra vogellii, Christisonia wightii, Orobanche crenata, O. minor, O. ramosa, Rhamphicarpa fistulosa, Seymeria cassioides, Striga asiatica, S. gesnerioides, S. hermonthica.* |
| Loranthaceae | *Agelanthus spp., Amyema spp., Dendrophthoe spp., Phthirusa spp., Psittacanthus spp., Struthanthus spp. Tapinanthus bangwensis,* |

| | |
|---|---|
| Santalaceae | *Acanthosyris pauloalvimii, Exocarpos spp., Osyris alba, Pyrularia pubera, Thesium spp.* |
| Viscaceae | *Arceuthobium spp. Dendrophthora poeppigii, Phoradendron spp. and Viscum spp.* |
| Balanophoraceae | *Balanophora, Dactylanthus.* |
| Hydnoraceae | *Hydnora, Prosopanche.* |
| Lennoaceae | *Lennoa, Pholisma.* |
| Rafflesiaceae | *Rafflesia, Pilostyles.* |
| Scrophulariaceae | *Gerardia, Hyobanche.* |
| Cuscutaceae | *Cuscuta reflexa, C. campestris, C. australis, C chinensis, C.japonica* |

## FORMS OF PLANT PARASITES

Angiosperm parasites can be categorized as annual and perennial herbs, vines, shrubs and trees on the basis their life forms. They are further subdivided as either root or shoot parasites with respect to the site of their attachment to the host, whereas the presence or absence of functional chloroplasts defines the parasite as being either hemiparasitic or holoparasitic respectively (84). The holoparasites are obligate parasites with little independent capacity to assimilate carbon and inorganic nitrogen and depend entirely on the host for nutrients, minerals, and water e.g. *Cuscuta* and *Orobanche* species. All of these acquire a little or all of their water, carbon and nutrients via the vascular tissue of the host's root or shoots. Hemi-parasites, on the other hand, may be facultative or obligate, contain chlorophyll, thus can fix carbon dioxide and procure water and minerals from the host, e.g. *Striga* species and mistletoes. Invasion of the hosts by the parasitic plants varies in different hosts species. Some invade the roots with no noticeable parasitism, as is the case with *Castilleja coccinea* (Indian paintbrush), *Agalimis purpurea* (gerardia), *Striga asiatica* (witchweed), and *Orobanche cumana* (broomrape). The only distinguishing feature in such parasites is the subterranean attachment to the host plant, whereas others, such as *Cuscuta* spp. invade the aerial parts of the plant (87).

The characteristic feature of all these vascular parasites is a natural and specialized physiological bridge, the haustorium, which is a novel shoot oriented adventitious root organ that enables the parasitic plant in attachment, penetration of the host epidermis to establish connection and ultimately solute transfer from the host vascular system to the parasite (71, 116). The acquisition of host resources can exert strong effects on host growth, reproduction and physiology (95). Generally, parasitism reduces host productivity and reproductive effort (37). In most instances, reduction in host performance is considerable, and in the extreme cases, parasitism may result in host's death.

## *CUSCUTA*- FORM AND DIVERSITY

*Cuscuta* spp. (dodders), belonging to Cuscutaceae, are rootless annual or perennial, non-host specific, twining, above ground stem holoparasites with leafless, thread-like, orange, red or yellow stems that twine over other plants. They flower in the early summer, or later, depending upon the species. Dodder flowers are bisexual, self-fertile, range in colour from white, pink, yellow to cream and pollinated by insects. Water (91) and animals (82) facilitate seed dispersal in them over longer distances.

The parasitic dodder is known by many common names such as goldthread, helluvine, hairweed, devil's hair, love-vine, strangle-vine, pulldown etc. in different regions of the world and as Amarbel, Akasbel, Amarvalli, Swarnlata etc. in India. Many of these names reflect the agricultural and socio-cultural importance of the *Cuscuta*.

## *CUSCUTA* – CHLOROPHYLLOUS OR NON-CHLOROPHYLLOUS?

*Cuscuta* was earlier thought to be completely devoid of chlorophyll. While some species have retained a functional photosynthetic apparatus and are capable of performing photosynthetic carbon fixation up to the compensation point (46, 110, 120), others have adopted to a parasitic lifestyle by the loss of thylakoids and of chlorophylls (120). In green species, the plastid ultrastructure resembles that of chloroplasts of non-parasitic higher plants, whereas plastids of a chlorophyllous species have an amoeboid proplastid shape and are filled with vesicular structures. In several species, their low chlorophyll content is masked by high amounts of carotenoids, accounting for a characteristic yellow colour.

## LIFE CYCLE OF *CUSCUTA*

Life cycle of *Cuscuta* passes through various interesting stages (Fig 1 & 2). It begins with the germination of the seeds of the parasite near to the host and detection of xenognosins that the parasite recognizes as originating from a potential host for the development of structures by the parasite for its attachment to the host. It is followed by the formation of haustoria capable of penetrating the tissue of the host and overcoming host resistance mechanisms, if they exist. After haustoria formation, the invasion of the host and association with the vascular bundle of the host occurs and consequently, the growth and reproductive maturity of the parasite at the expense of the host is ensued. Finally, the dispersion of vegetative fragments and seeds takes place (102, 10).

In a successful contact between the dodder and its host, haustorial development goes through the following phases: initiation of a disc-like meristem next to the vascular bundles of the dodder shoot from which the pre-haustorium develops. The pre-haustorium adheres to the host and is transformed into a haustorium, which penetrates the host tissue (Fig. 3 & 4). The haustorium stops at a certain depth within the host tissue and branches into "searching hyphae" which then grows through the host tissue towards the host's vascular tissue establishing vascular contact (28, 22). Phloem contact develops from searching hyphae of the dodder, which changes to absorbing hyphae when touching a sieve tube. The common surface between the absorbing hyphae and the host sieve tube develops wall invaginations outside in close contact with the sieve element of the host (27, 28). Although dodder penetrates the host but it remains extra-cellular thus minimizing stress to the host because it relies totally on the host for both fixed carbon and water throughout its life cycle. Through these vascular connections, water and nitrogen compounds assimilate and even proteins or plant viruses are transferred into the parasitic plant (78, 56, 46, 46). When the connection with the host is established firmly, the original root of the vine in the soil shrivels and dies and the parasite becomes fully dependent on its host. The dodder can now simultaneously parasitize and compete with co-existing multiple host plants and can grow more or less continuously reaching high into the canopy of shrubs and trees.

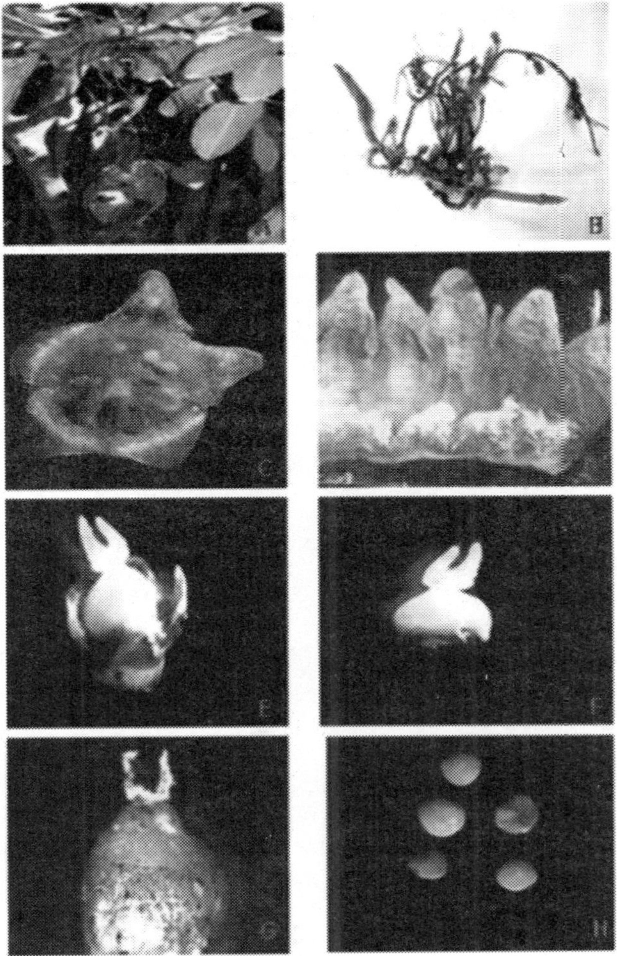

**Fig. 1.** Vegetative and reproductive phases of parasitic *Cuscuta gigantea* Griff. (Kapoor, 2007).
**A :** Parasite entwining host plant (*Psidium guajava);* **B :** Flowering branch with white flowers; **C :** A single flower; **D :** Corolla cut open with sessile stamens; **E :** Ovary with calyx; **F :** Ovary with well developed style; **G :** Fruit: a capsule; **H :** Seeds

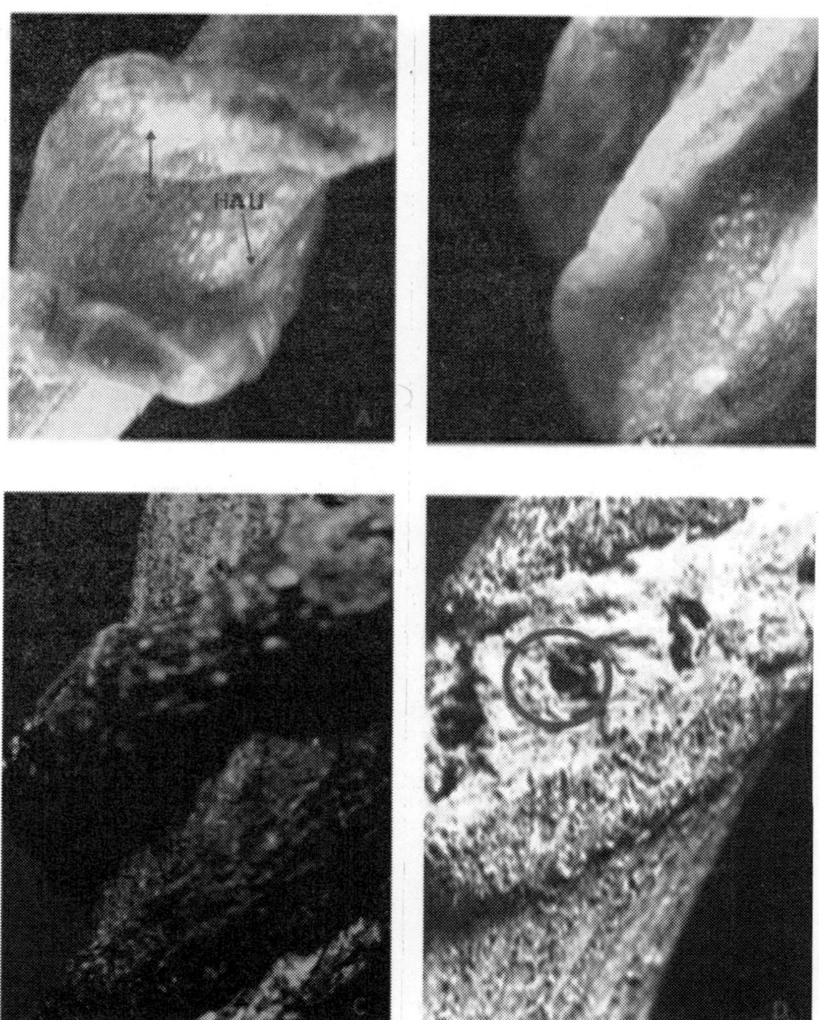


**Fig. 2. A & B :** Yellowish- brown *Cuscuta gigantea* vine entwined around its host stem. (attached concave surface of the parasite stem has increased its surface area; arrow indicates point of haustorial emergence); **C :** Host (*Olea ferruginea*) stem with dried and shriveled *Cuscuta* vine; **D :** Enlarged view of the host (*Olea ferruginea*) stem after removal of *Cuscuta* vine showing hole formation due to haustorial penetration.(Kapoor, 2007)

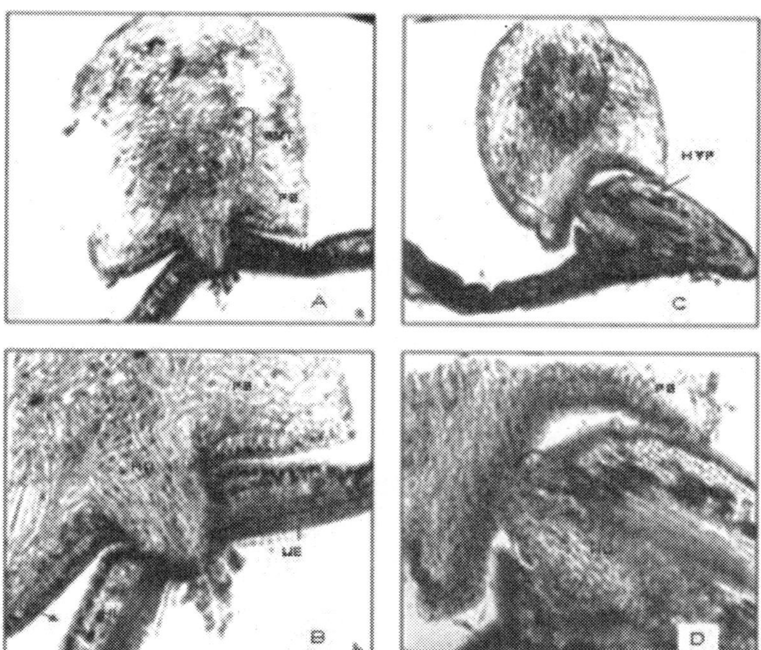


**Fig. 3. A :** Transverse section of petiole of *Pyrus pashia* parasitized by *Cuscuta gigantea* (x 40); **B :** Magnified view of A (x 100); **C :** Transverse section of *Cuscuta* stems depicting haustorial entry into host stem (*Carissa opaca*) showing the ramified haustorial apex and the union of vasculature of haustorium and the host (x 100); **D :** Elongated haustorial cells showing enlarged and hypertrophied nuclei (x 400) (HP – host petiole; HS – host stem; HU – haustorium ; HUC – haustorial cells ; NU – nucleus ; PH – phloem ; PS –parasite stem ; XY – xylem) (Kapoor, 2007)

## HOST-PARASITE RELATIONSHIP OF CUSCUTA

Many species of parasitic plants have the ability to parasitize different host plants species in a community at the same time. This generalist's strategy allows the parasite to invade new areas and become established on host species with which they have not co-evolved. Further, additional preference for certain hosts can lead to parasite-induced changes in plants community structure and diversity (89, 90). The parasitic relationship of *Cuscuta* with its hosts has been the subject of numerous investigations (53, 51, 23, 56, 104, 44). Host morphology and phenology were considered as the two primary factors for which different *Cuscuta* species selected their respective hosts. *Cuscuta* seedling produces stems that grow spirally around susceptible host plants in upward direction and forms haustoria, which ultimately penetrate the host tissue and connect to the phloem and xylem of the host (70, 15). However, the dodders preferably twirl on hosts of high nutritional status and grow away or reject hosts of poor quality. In fact, the behaviour of acceptance and rejection is exhibited before taking up any food from the host (61). In addition, they are more benefited from parasitizing diverse host species (62).

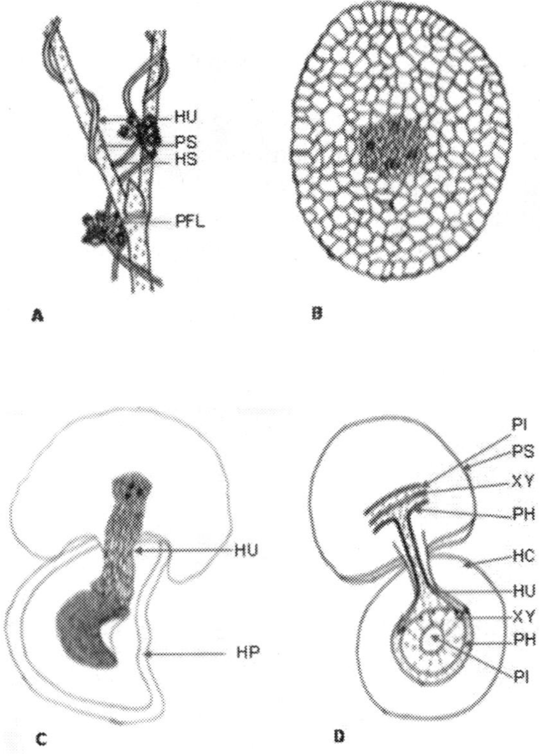


**Fig. 4.** *Cuscuta gigantea*

**A :** Habit; **B :** Transverse section of the stem (x 40); **(C&D) :** Schematic diagrams of the transverse section pf host petiole and stem respectively showing the penetration of haustorium of the parasite (HC- host cortex; HP- host petiole; HS-host stem; HU- haustorium; PFL- parasite flower; PH- phloem; PI- pith; PS- parasite stem; XY- xylem).(Kapoor, 2007)

## NATURE OF HAUSTORIUM IN *CUSCUTA*

The haustorium is the key organ that physically and physiologically bridges the parasite and the host plant. Host selection by parasitic plants is accomplished through the induction of chemical molecular signals, including germination stimulants and haustoria-inducing factors (HIFs). HIFs are generated by HIF-releasing enzymes from the parasite or triggered by redox cycling between electrochemical states of the inducers. Haustorium attachment is non-specific; however, mucilaginous substances produced by haustorial hairs facilitate the attachment to a host. Nutrients and water are transferred uni-directionally through the haustorium from the vascular system of the host into the parasite. Besides, this organ also facilitates hormonal interactions between the two organisms (53, 72, 71, 51, 123, 116). Synergistic effects of far-red light and tactile stimuli could initiate the haustoria formation in *Cuscuta* seedling efficiently in the absence of any host plant (119). Many researchers have studied the haustorial hyphae and their developmental stages in *Cuscuta europaea* and *Cuscuta odorata* (131, 66, 132). The hyphal walls are perforated by numerous plasmodesmata (107, 35). The dodder, like leafhoppers, places its absorptive structures in such a way as to hit the vascular bundle (75). The haustorial coil provides a strong counterforce so that the intrusive organ is well braced (71). Many investigators have worked out the mechanism of dodder adherence to its hosts (27,121). Host recognition in *Cuscuta* is mediated by phototrophic mechanism (43) besides, chemical molecular signals, including germination stimulants and haustoria-inducing factors HIFs) also play important roles in recognition process. Establishment of infection begins when *Cuscuta* seedling encounters a susceptible host. At the contact site of potential host plant, *Cuscuta* develops haustoria and establishes a connection by secreting enzymes and sticky substances consisting mainly of de-esterified pectins from epidermal cells (121,122). Recently, it has been established that *Cuscuta* infection induced a signal in the host leading to expression of plasma membrane bound cell wall-localized protein (attAGP) which helps in adherence (4). From the tip of the haustorium cone, searching hyphae grow extracellularly through host tissues in a manner similar to phloem tubes. On reaching individual sieve elements of phloem, the terminal hyphae cells differentiate into absorbing hyphae (26). The hyphae making contact with the functional phloem undergo a radical change in behaviour. The first indications of this came from the work of Zender (131), who described that the entry of hyphae into a sieve tube was followed by a "rhizopodial" lobing while Dorr (27) demonstrated that the absorbing hyphae produce finger-like protrusions that surround the sieve element. The efflux of substances occur from sieve tubes into the free space of the host, and the parasite is presumed to be superior to the host tissue in its ability to absorb these phloem mobile substances contained in the free space (126,127).

## DISTRIBUTION OF *CUSCUTA*

Holoparasitic species of *Cuscuta* have attracted the attention of many researchers worldwide because of their strange growth habit and extraordinary behaviour. Representatives of the genus *Cuscuta* are widespread and comprise of about 180 species distributed throughout the temperate to the tropical regions of the globe (80) with the greatest species diversity in subtropical and tropical regions. Yuncker (130) reviewed about 158 species of *Cuscuta* distributed in tropical and subtropical regions in five continents profusely growing and parasitizing a wide variety of hosts on large landmasses, but usually rarely and infrequently on islands. These species exist in varied habitats. Graham *et. al.* (36) reported the distribution of dodder in North America and Europe. Crins (18) studied the presence of six species of *Cuscuta* in Ontario. Silberhorn (111) described the invasion of *Cuscuta indecora* in a tidal brackish water marsh in Virginia. Lei-Simon (76) studied the distribution of four species of dodder in the Amargosa valley of Southern Nevada. The largest number of species has been recorded from the western hemisphere where *Cuscuta* is distributed from Southern Canada to Chile. In Argentina, *Cuscuta pauciflorum* has been reported growing as far south as 47° latitude. *Cuscuta* was reported growing luxuriantly in Syria and Jordan (77, 1). Garcia-Miguel (32,33) recorded *Cuscuta rausii* from Greece and western Mediterranean area while Sagalaev and Matveev (106) recorded twelve species of *Cuscuta* from Russia.

*Cuscuta* has also gained importance because of its increasing incidence worldwide, and its ability to cause severe damage to and loss of yield of the host plant (86). Its species have been reported to be a problematic parasite to crop plants in agriculture systems in Africa and Asia (63, 16, 22, 54, 84, 49). In China, *Cuscuta chinensis, C. japonica,* and *C. australis* have a wide host range (128, 29), whereas *Cuscuta gigantea, C. monogyna, C. pulchella, C. chinensis* and *C. reflexa* are commonly found in Afghanistan on many host plant species (96). Rajput and Tahir (97) have given the distributional pattern of fourteen taxa of *Cuscuta* from Pakistan. These are *Cuscuta hyalina, C. reflexa, C. violacea, C. lupuliformis, C. gigantea, C. monogyna, C. lehmaniana, C. epithymum, C. capitata, C. europaea, C. australis, C. pulchella, C. chinensis* and *C. campestris.* Banerji and Das (6) recorded *Cuscuta santapaui* from East Nepal, while Bhattarai *et al.* (8) studied the host range of *Cuscuta reflexa* in the Kathmandu valley, Nepal. In Sri Lanka, Arulandhy and Padmasiri, (5) recorded severe infestations of *Cuscuta* spp. in *Capsicum annum, Allium cepa,* and *Centella asiatica* while Wijesundara *et al.* (125) accounted for more than one hundred wild and cultivated host species, which were vulnerable to infestation from this holoparasite.

In India, Clarke (17) initially mentioned *Cuscuta reflexa* var *anguina* under the subgenus *Monogynella* in Hooker's flora of British India. Species of *Cuscuta* are now known to be widely distributed throughout India ascending the Himalayas upto an altitude of 8000 feet to the plains of West Bengal, Uttar Pradesh, Andhra Pradesh and Punjab extending up to Uttarakhand and Kashmir that widen species distribution to the farthest limit of the Himalayas in the west (Table 1). Few workers have explored the incidence of *Cuscuta* and its hosts from Jammu and Kashmir also. Dhar and Kachroo (25) reported thirteen species of *Cuscuta* from Kashmir out of which eight reach alpine to sub alpine level of Western Himalaya. Three species namely *Cuscuta approximata, C. brevistyla* and *C. europaea* are more pervasive, while three others, *Cuscuta reflexa, C. capitata* and *C. lehmanniana* occur in Eastern Himalaya. The remaining species are distributed with in the areas near to Pakistan and Afghanistan (25). However, from Jammu and Udhampur districts, only *Cuscuta reflexa* (109, 118, 60) and *Cuscuta campestris* have been reported so far on very few host plants (9, 60).

**Table 1.** Host plants parasitized by *Cuscuta* species reported from different parts of India

| *Cuscuta* species | Host species parasitized | Place of occurrence | References |
|---|---|---|---|
| *Cuscuta reflexa* | *Anacardium occidentale* | Uttar Pradesh | |
| | *Bougainvillea spectabilis* | Uttaranchal | 114 |
| | *Brassica campestris* | Andhra Pradesh | 3 |
| | *Camellia sinensis* | Karnataka | 65 |
| | *Casuarina equisetifolia* | Assam | 42 |
| | *Citrus* spp. | Delhi | 85 |
| | *Cleodendrum inerme* | Gujarat | 115 |
| | *Codium variegatum* | Punjab | 99 |
| | *Digitaria ciliaris* Hedges | Maharashtra | 109 |
| | *Jasminium absimile* | Jammu & Kashmir | 25 |
| | *Litchi sinensis* | Rajasthan | 113 |
| | *Medicago sativa* | Jharkhand | 39,40 |
| | Ornamentals | WestBengal | 80 |
| | *Petunia hybrida* | | 118 |
| | *Solanum nigrum* | | 24 |
| | *Spathodea campanulata* | | |
| | *Tabebuia argentea* | | |
| | *Thevetia peruviana* | | |
| | *Ziziphus mauritiana* | | |
| *Cuscuta campestris* | *Cannabis sativa* | Uttar Pradesh | 114 |
| | *Ipomoea palmata* | Jammu and | |
| | *Lantana camara* | Kashmir | 13 |
| | *Medicago sativa* | | 25 |
| | *Parthenium hysterophorus* | | 9 |
| *Cuscuta chinensis* | *Capsicum annum* | Andhra Pradesh | |
| | *Eleusine coracana* | Karnataka | |
| | *Medicago sativa* | Uttar Pradesh | 98 |
| | *Momordica charantia* | | 73 |
| | *Polygonum plebejum* | | |
| | *Vigna sinensis* | | |
| *Cuscuta santapaui* | *Viburnum* sp. | Uttaranchal | |
| | *Indigofera* sp. | | |

| | | | |
|---|---|---|---|
| *Cuscuta indecora* | *Petunia hybrida* | Uttar Pradesh | |
| *Cuscuta appro-ximata* | Many hosts | Kashmir (J&K) | 25 |
| *Cuscuta australis* | Many hosts | Kashmir (J&K) | 25 |
| *Cuscuta brevistyla* | Many hosts | Kashmir (J&K) | 25 |
| *Cuscuta capitata* | Many hosts | Kashmir (J&K) | 25 |
| *Cuscuta europaea* | Many hosts | Kashmir (J&K) | 25 |
| *Cuscuta lehmaniana* | Many hosts | Kashmir (J&K) | 25 |
| *Cuscuta hyalina* | Many hosts | Tamilnadu | |

## HOST RANGE OF *CUSCUTA*

Host range varies widely among species of parasitic plants. Unlike most parasitic weeds that are selective on only several species of plants, dodders can invade host plants ranging from algae (Characeae), pteridophytes (Polypodiaceae), gymnosperms, and a wide variety of angiosperms including wild and cultivated plants (31, 22). Prominent among the angiospermic families are Rosaceae, Solanaceae, Rutaceae, Acanthaceae, Leguminosae, Myrtaceae, Verbenaceae, Asteraceae, Rhamnaceae etc. *Cuscuta* species also demonstrate specialized opportunistic behaviour in their degree of selection and preference for specific individual host species irrespective of host abundance. Host preference refers to the choice of the most desirable host for optimal growth of the parasite. *Cuscuta* optimally forages by a positive relationship between investment and reward. It exhibits a pre-uptake response that allows the parasite to discriminate among hosts and then invest the most on the best host (61). Dicotyledons are more likely to be found its hosts while monocot hosts have been reported as resistant to *Cuscuta* infestation. This may be because dicots provide spatial comforts for the parasite proliferation and have much developed and organized vascular bundles from where the parasites can derive photosynthate, water and minerals. Other advantages to the parasite can be its dispersion over larger distances and surviving for longer period on these hosts. Other workers have also made similar observations (63,88,43,87, 47).

Inability of these parasites to colonize monocot hosts, probably, is due to either pre-existing unfavourable anatomical conditions (e.g. sclerenchymatous tissues and scattered vascular bundles) creating physical barriers, such as a layer of lignified tissues, or a direct defense response applied by the host as denoted by Forstreuter and Weber (30) and Dawson *et al.* (22).These direct defense responses used by the host may include the formation of wall material against the haustorial hyphae, phytoalexin production or production of inhibitors. Hypersensitive defense reactions have also been implicated in imparting resistance in monocots against *Cuscuta* infestation. Many other researchers have also advocated the failure of *Cuscuta* to attack most monocots due to the low efficacy of its enzymes involved in the breakdown of monocotyledonous tissues during penetration by the parasite (50, 22, 43, 112, 11, 60, 103).

## CROP LOSSES DUE TO *CUSCUTA* INFESTATION

*Cuscuta* are known to cause direct and indirect damage to plantations through infestation of the aerial parts of many wild and cultivated plants. These plant parasites often cause dramatic reduction in host growth and reproduction (83, 48, 87). Crop losses ranging from 23 to 100% due to *Cuscuta* spp. have been reported previously (56,105). In semi-humid and semi-arid areas of Africa and Asia, *Cuscuta* species are of immense economic importance since they cause severe yield losses of crops. The United States Department of Agriculture (USDA) has kept dodder in "World's Top Ten Weeds" list. Dodder infested patches in the field continue to enlarge during the growth season and in perennial plants these expand every year. The infected host plants turn weak, their vigour declines and they produce poor yields. Under green house conditions, *Cuscuta* infestation can decrease the biomass of a single host individual by as much as 50% (68,69). The parasitization result in stunted growth of all the host plants and the production of new foliage is greatly suppressed (99). In addition, the reduction in flowering and fruiting in host plants has been reported (83). In severe infections, many hosts are even killed. Thus, due to their ability to infest, impose severe damage and reproduce prolifically on a wide range of angiosperms of great economic importance, *Cuscuta* presents an alarming threat to the diversity and health of our vegetation. Several major crop plants such as *Allium cepa, Anacardium occidentale, Beta vulgaris, Capsicum annum, C. frutescens, Daucus carota, Dolichos lablab, Guizotia abyssinica, Linum usitattisimum, Lycopersicon esculentum, Momordica charantia, Phaseolus aureus, Solanum melongena, S. tuberosum, Trifolium alexandrianum,* and oil seeds crops are known to be some of its hosts in different regions. The fruit plants such as *Ziziphus mauritiana, Prunus persica,* and *Litchi chinensis* also harbour these organisms (80). Besides, *Cuscuta* species are known to exist on woody trees, ornamentals, shrubs and other hosts of economic importance (60).

In India, among the various reported species of this holoparasite, *Cuscuta reflexa* has been recorded as leading destructive parasite on most economically important hosts from different regions of (Table 1). It prefers dicotyledonous hosts, while most of the monocots have been reported as resistant to *Cuscuta* infestation (22). Hosts on which *Cuscuta reflexa* has been described include *Camellia sinensis* from Assam (3), *Bougainvillea spectabilis, Cleodendrum inerme, Casuarina equisetifolia, Thevetia peruviana* and citrus plants in Punjab and Gujarat (42,85). Ramaiah et al. (99) recorded *Cuscuta reflexa* on various hosts such as *Spathodea campanulata*, *Jasminium absimile*, *Codium variegatum Anacardium occidentale* and *Tabebuia argentea* from Karnataka, while in Maharashtra the same species occurred on sixty eight plants (124). Misra and Prasad (80) recorded fifteen plant species belonging to twelve different families including fruit plants such as *Ziziphus mauritiana, Prunus persica, Litchi chinensis*, ornamentals, hedges and others as hosts of *Cuscuta reflexa* from Bihar. Dey and Pati (24) have recorded the parasitism of *Cuscuta reflexa* on *Digitaria ciliaris* in West Bengal. Few workers have investigated the colonization of host plants by *Cuscuta chinensis* from many parts of the country on crops like *Vigna sinensis, Capsicum annum, Momordica charantia, Eleusine coracana* (98,73). Other investigations have revealed that *Brassica campestris, Medicago sativa, Solanum nigrum* also serve as a potential host to various dodder species (115,13,39,40,113).

## Status of *Cuscuta* in Jammu and Kashmir

Three different species of *Cuscuta* viz., *Cuscuta reflexa, C. campestris* and *C. gigantea* are known to exist in Jammu and Kashmir The morphological details studied and taxonomic description of these three species is described herewith:

### *Cuscuta reflexa* Roxb.

*Cuscuta reflexa* Roxb., commonly called as giant dodder, is an annual or facultative perennial shoot holoparasite with branched, light brown to dark brown, thick and succulent stem with inconspicuous scale-like leaves; flowers arranged in paniculate cymes, white to cream, pedicellate, 5.5-7.5 mm long, bracts leaf-like; calyx lobes 5, fleshy, margin scarious, rounded, ovate-obtuse, deeply divided, overlapping at the base, 0.7-2.5 x 1.2-2.2 mm; corolla lobes 5, ivory white, obtuse, upright spreading, 1–1.5 x 0.5–0.7mm, corolla tube 2-3 mm long, scales scarious, reaching the middle of the corolla tube, 1mm long; stamens 5, filaments linear, 2–3 mm long, fused, anthers oblong, basifixed, 1–1.5 mm long; ovary conical, fleshy, 1–2mm long, style 1, thick, inconspicuous with 2 dark brown, unequal and elongated stigmas, 0.7– 1.0 mm long; ovules 4, oblong – elliptical, 0.3 x 0.2 mm, capsule 1-3 x 2 – 3 mm, globose-conical, brown, not circumscissile with a definite line of cleavage; seeds typically 2- 2.5 mm long.

### *Cuscuta campestris* Yuncker

*Cuscuta campestris* Yuncker, referred to as golden dodder, is characterized by thin, slender, green to yellow-brown, branched herbaceous stem with minute scale-like leaves, inflorescence usually of cymose clusters, flowers actinomorphic, bisexual, white ; calyx lobes 5, oval to orbicular, covering more than half of corolla ; corolla lobes 2– 2.5 mm long, oblong- elliptical, with acute to slightly obtuse apex; stamens 5, filaments as long as or slightly longer than anthers; styles 2, linear; ovary globular, fleshy, brown 1-1.2 mm long; stigma capitate or rounded, 0.2mm, brown; ovules 4, triangular or ovoid; capsule globose, membranous, not definitely circumscissile; seeds 4, reddish brown, ovoid or somewhat triangular, usually flattened.

**Cuscuta gigantea Griff.**

*Cuscuta gigantea* Griff., closely related to *C. reflexa,* is an annual or perennial stem holoparasite with rudimentary scales, mature stem brown to dark-brown, thick, twining around the host stem and petioles ; inflorescence racemose; flowers sessile, 6.5– 9 mm long, actinomorphic, bisexual, white; calyx lobes 5, ovate-obtuse, 2–3 x 1–2 mm, overlapping at the middle, thick at the base; corolla lobes 5.4–6.0 x 0.8–1.2mm, oblong to obtuse, inflexed, corolla tube 3.5–5mm long; stamen 5, sessile, free, filaments fused prominently in the petals; anthers oblong, 1.2–1.6mm long ; ovary globose; style 4mm long; stigma 2, oblong with light green tips, 1.0–1.2mm long; ovules 2, orbicular, brown, 0.7–1mm in diameter; capsule 5-6 x 4-5mm, globose-conical; seeds 2, brown-black, smooth, 1.5-1.7mm in diameter.

The presence of these three different species of *Cuscuta* shows their wide adaptation to the varied climatic characteristics of the state.

**Comparative analysis of recorded *Cuscuta* species**

Comparative analysis of the three species of *Cuscuta* viz., *C. reflexa*, *C. campestris* and *C. gigantea* has been presented (Table 3). These stem holoparasites depicts marked differences with regard to their distribution, vegetative morphology, phenology and number of host plants parasitized.

In the field, thin branching stems appears leafless with the leaves reduced to minute scales. Due to the absence of chlorophyll pigments, the stem of *Cuscuta* does not appear fully green and changes its colour from yellow to brown. Tissues of *Cuscuta* plants have low chlorophyll content and it has low accessory photosynthetic pigments such as alpha and beta-carotenes and xanthophylls.

*Cuscuta reflexa* showed profuse growth and fast spreading habit and predominantly attacks woody trees and shrubs. On the other hand, *Cuscuta campestris* stem is more delicate in structure and favours herbaceous hosts, while *C. gigantea* usually parasitizes shrubs and softer parts of trees. With respect to its distribution *C. reflexa is* widely distributed in Jammu and Kashmir (109,25,118,58,59).

*Cuscuta campestris* shows narrow distributional range and is confined to some secluded places in Kashmir and Jammu district of Jammu and Kashmir (25,9,58,59). However, *Cuscuta gigantea* is observed with moderate distribution in different locations of Rajouri district of the Jammu and Kashmir only (60)

The *Cuscuta* allocates its resources much to accomplish sexual reproduction. The reproductive effort occurs at the expense of vegetative growth. The expected average reproductive effort is greater for the annual growing herbs. This reproductive effort increases with decreasing length of growing season and the parasite invests more in reproductive effort as it has very less time for vegetative growth. Vegetative proliferation in these species occurs during the months of February to April.

**Table 2.** Phenology of *Cuscuta* species recorded from Jammu province

| ***Cuscuta* species** | | **Feb** | **Mar** | **Apr** | **May** | **Jun** | **Jul** | **Aug** | **Sept** | **Oct** | **Nov** | **Dec** | **Jan** |
|---|---|---|---|---|---|---|---|---|---|---|---|---|---|
| *Cuscuta reflexa* | Flowering | | | | ■ | ■ | ■ | ■ | ■ | ■ | ■ | ■ | |
| | Fruiting | | | | | | | | ▒ | ▒ | ▒ | ▒ | ▒ |
| *Cuscuta campestris* | Flowering | | | | | ■ | ■ | ■ | ■ | | | | |
| | Fruiting | | | | | | | ▒ | ▒ | ▒ | ▒ | ▒ | |
| *Cuscuta gigantea* | Flowering | | | | | ■ | ■ | ■ | ■ | ■ | ■ | ■ | |
| | Fruiting | | | | | | | | ▒ | ▒ | ▒ | ▒ | ▒ |

Phenological observations reveal that flowering and fruiting periods varied in each of these three dodder species (60) (Table 2 & 3). In early summer, *Cuscuta* produced numerous flowers in loose clusters and their colour varied from white to creamish yellow. Commencement of flowering was noticed in the summer months (May-June) and lasted for 4 to 8 months duration up to December. Similarly, fruit formation began in August-September after 2 to 4 months of initiation of reproductive phase and went upto the month of January. However, an overlapping flowering and fruiting period of four months (June-September) and three months (September-November) was observed respectively in these parasites, while flowering and fruiting was observed from May to January. Because of much longer period of their reproductive growth, abundant seeds are produced. These seeds are much viable and are, therefore, responsible for their propagation in long terms.

**Table 3.** Comparative analysis of three species of *Cuscuta* recorded from the study area

| *Cuscuta* species | Cuscuta reflexa | Cuscuta campestris | Cuscuta gigantea |
|---|---|---|---|
| Habit | Vigorous, typically attacks woody plants. | Species is more delicate in structure and favours herbaceous hosts. | The species usually attacks shrubs and softer parts of trees. |
| Stem | Thread like, light brown to dark brown. | Green when young, intermingled, brown at maturity. | Stem thick, brown to dark brown, not interlaced, making many spirals around the host stem. |
| Distinguishing Taxonomic characters | Scales inside the petals present, capsule not circumscissile with a definite line of cleavage i.e. dehiscing irregularly, style 1,stigma 2,insignificant. | Petals oblong-elliptical; sepals lobes oval to orbicular, not forming carina near the tips, 1.2-1.5mm long, bracts not keeled, filament subulate, style 2, stigma 2. | Stem and flowers without violet spots, corolla tube not forming a sub-globose to urceo-late covering around the capsule, petals 4-6mm long, style1, stigma 2. |
| Flowering | May to December | June to September | June to December |
| Fruiting | September to January | August to November | September to January |
| Number of host plants infested | 34 | 15 | 35 |
| Distribution | Abundant | Narrow | Moderate |

*Cuscuta reflexa* entered flowering in the month of May and extended up to December, and the fruiting started from September until January. On the other hand, *Cuscuta campestris* had comparatively shorter period and flowers started appearing from June to September while fruit formation was observed during August to November. However, in *Cuscuta gigantea*, the flowering (June to December) and fruiting (September-January) almost coincided with *Cuscuta reflexa* (Table 2 & 3). The seeds produced in all the three species were yellow, brown and black, nearly round and had a fine rough surface with a hard seed coat. The variation in the phenology in these species may be due to differential accessibility of the host nutrients.

## ECONOMIC IMPORTANCE OF *CUSCUTA*

The local abundance and availability of some parasitic plants led to their potential use by the humans. Throughout the world, many parasites are utilized in a variety of ways most of which are unrelated to the parasitism of the plants. Ethno-botanical information including traditional applications and modern pharmacognosy of these plants have been given due attention by many workers world over.

Plants of *Cuscuta* have versatility in their use, in both traditional as well as modern medicine. The dry seeds of *Cuscuta chinensis* Lam. constitute an important traditional Chinese herbal medicine widely used for liver and kidney tonic and aphrodisiac to treat impotence and seminal emission (128, 29). In Africa, the dodder is commonly known as widow wort and only widows may collect it for use in tribal medicine due to its abortifacient, contraceptive and oestrogenic properties. Several groups in USA use stem of this parasite as emetic after brewing and then drinking the compound to induce vomiting. Similarly, in United Kingdom and other parts of Europe, the dodder is considered a valuable remedy for liver and gall bladder problems, besides a powerful purgative and diuretic. Lacille et al. (74) have reviewed the use of *Cuscuta* plants as medicine in many different forms in different parts of the globe.

In India, *Cuscuta* species are well known for their medicinal value and used as an herbal remedy for thousand of years. Various parts of parasitic dodder, particularly *Cuscuta reflexa,* were used in tribal medicine for diseases like fits, melancholy, and insanity (2). The stem extract have been found to possess anti-fertility effect (100, 41) and is useful externally against itch and internally in protracted fevers (67,14). A decoction of stem is used in the treatment of respiratory, urinary, spleen, jaundsciatica and scorbutic complaints. The plant is also used in a bath for the treatment of tuberculosis. Yadav et al. (129) have detailed the anti-oxidant activity of *Cuscuta reflexa* while *Cuscuta chinensis* in Indian traditional medicine, is considered to have an anti-tumor effect (81). Das *et. al.* (21) reviewed the ethno-medicinal importance of *Cuscuta reflexa* by tribals of Orissa state. There, the holoparasite extracts have been used against diarrhoea, the most prevalent disease of the state, rheumatism, skin infections, antidote for poisoning, leucorrhoea, eye complaints, fever etc. Tribal of terai region of Uttar Pradesh and West Bengal have been using *Cuscuta reflexa* for curing common ailments like headache, hair fall and rheumatism (64, 117). Local women in several rural areas of Udhampur and Rajouri district of Jammu and Kashmir state use dodder stems as hair tonic to get rid of lice and dandruff (60). Decoction of whole plant of *Cuscuta reflexa* mixed with bark of *Holarrhaena pubescens* and seeds of *Piper nigrum* are used for curing puerperal fever of female cow (7). Moreover, plants are reported to be used against dermatological problems, generalized bodyaches, bone diseases and joint pains (57).

## REFERENCES

1. Abu-Irmaileh B and Thahab S 1998 Infestation by parasitic flowering plants in Jordan *Cuscuta* and *Orobranche, Dir Asat Agri Sciences* **25** 439-455
2. Agarwal RR and Dutt S 1935 Chemical examination of *Cuscuta reflexa* Roxb; The constituents, *Journal of Indian Chemical Society* **1** 384-388
3. Ahlawat YS and Sardar KK 1973 Insect and dodder transmission of tea rose yellow mosaic virus, *Currenmt Science* **42** 181
4. Albert M, Werner M, Proksch P, Fry SC, and Kaldenhoff R 2006 The cell wall-modifying xyloglucan endotransglycosylase/hydrolase LeXTHI is expressed during the defense reaction of tomato against the plant parasite *Cuscuta reflexa, Plant Biology* **6** 402 – 407
5. Arulandhy V and Padmasiri IS 1987 Threat of *Cuscuta* parasite, *Chilli Krishi* **10** 52-54
6. Banerji ML and Das S 1965 A new *Cuscuta* from Nepal, *Journal of the Arnold Arboretum* **46** 86-89
7. Basu R 2002 Ethnobotany Avishkar Publishers, Jaipur, 349-355 pages
8. Bhattarai T, Bhandary H, and Shrestha P 1989 Host range of *Cuscuta reflexa* Roxb

in the Kathmandu valley, Nepal, *Plant ProtQ* **4** 78-80

9. Bhellum BL and Mangotra R 1996 *Cuscuta campestris* Yuncker - A new record for the flora of Jammu and Kashmir state, *Indian Journal of Forestry* **19** 103-104

10. Bouwmeester H, Matusova R, and Beale MH 2003 Secondary metabolite signaling in host-parasite plant interactions, *Current Opinion in Plant Biology* **6** 358-364

11. Bringmann G, Schlauer J, Ruckert M, Wiesen B, Ehrenfeld K, Proksch P, and Czygan FC 1999 Host-derived acetogenins involved in the incompatible parasitic relationship between *Cuscuta reflexa* (Convulvulaceae) and *Ancitrocladus heyneanus* (Ancistrocladaceae), *Plant Biology* **1** 581-584

12. Butchart SHM, Ackakaya HR, Kennedy E, and Hilton-Taylor C 2006 Biodiversity indicators based on trends in conservation status strengths of the IUCN Red list index, *Conservation Biology* **20** 579-581

13. Charles DJ, Singh M, and Sanwal GG 1982 Metabolic activity of the resting seeds of *Cuscuta campestris* and its host alfalfa, *Indian Journal of Biochemistry and Biophysics* **19** 106-110

14. Chopra RN, Nayer SL, and Chopra IC 1992 Glossary of Indian medicinal plants CSIR publication, New Delhi, 85 page

15. Christensen NM, Dorr I, Hansen M, van der Kooij TAW, and Schulz A 2003 Development of *Cuscuta* species on a partially incompatible host Induction of xylem transfer cells, *Protoplasma* **220** 131-142

16. Chrtek J and Osbornova J 1991 Notes on the synanthropic plants of Egypt *Grammica campestris* and other species of family Cuscutaceae, *Folia Geobot Phytotaxon* **26** 287-314

17. Clarke CB 1883 *Cuscuta* In *Fl BritInd* **42** 25-228,

18. Crins WJ and Ford BA 1988 The parasitic dodders (*Cuscuta* cuscutaceae) in Ontario, *The Canadian Field Naturalist* **102** 209-215

19. Culpeper N 1652 @ 2002–2006 herbs 2000 com

20. Darwin E 1825 The Botanic Garden, etc Jones and Co London

21. Das S, Dash SK, and Padhy SN 2003 Ethno-medicinal information from Orissa state, India, A review, *Journal of Human Ecology* **14** 165-227

22. Dawson JH, Musselman LJ, Wolswinkel P, and Dorr I 1994 Biology and control of *Cuscuta*, *Rev Weed Sci* **6** 265-317

23. De Bock F and Fer A 1992 Effects of abscisic acid on the transfer of sucrose from host, *Pelargonium zonale* (L) Aiton, to a phanerogamic parasite, *Cuscuta reflexa* Roxb *Australian, Journal of Plant Physiology* **19** 679-691

24. Dey DK and Pati BR 1998 Records of parasitism of *Cuscuta reflexa* Roxb on *Digitaria ciliaris* Pers, *Journal of Economic and Taxonomic Botany* **22** 235-236

25. Dhar U and Kachroo P 1983 Alpine flora of Kashmir Himalaya Scientific Publishers, Jodhpur, India

26. Dorr I 1969 Fine structure of intracellular growing *Cuscuta* hyphae, *Protoplasma* **67** 123-137

27. Dorr I 1972 The contact of *Cuscuta* hyphae with the sieve tubes of its host plants, *Protoplasma* **75** 167-184

28. Dorr I 1990 Sieve elements in haustoria of parasitic angiosperms, in Sieve elements comparative structure, induction and development (eds Behnke,HDand Sjolund, RD) Springer, New York, 239-253 pages

29. Du XM, Kohinata T, Guo YT, and Kazumoto, M 1998 Components of the ether-insoluble resin glycoside-like fraction from *Cuscuta chinensis, Phytochemistry* **48** 843-850

30. Forstreuter W and Weber HC 1984 *Zum Parasitismus von Cuscuta auf Euphorbia-Wirten Beitr Biol Pflanzen* **59** 31-54

31. Gaertner EE 1950 Studies of seed germination, seed identification and host relationships in dodders, *Cuscuta* sp, *Cornell University Agrcultural Experimental Station Mem* 294

32. Garcia-Miguel A 1998 *Cuscuta raussii* (Convolvulaceae) a new species from Greece, *Annales Botanici Fennici* **20** 171-174

33. Garcia-Miguel A 2001 A new western mediterranean species of *Cuscuta* (convolvulaceae) confirms the presence of holocentric chromosomes in subgenus *Cuscuta, Botanical Journal of the Linnean Society* **135** 169-178

34. Gerarde J 1633 The Herball or Generall Historie of Plantes , London

35. Glockenmann C and Kollmann R 1975 Electron microscopic photos of plasmodesmata @ Peter V Sengbusch-Impressum

36. Graham JH, Forsheiser FT, Stuteville DL, and Erwin DC 1979 A compendium of alfalfa diseases, *The American Phytopathological Society*

37. Graves JD, Press MC, Smith S, and Stewart GR 1992 The carbon canopy economy of the

association between cowpea and the parasitic angiosperm *Striga gesnerioides, Plant Cell Environment* **15** 283– 288

38. Guettard 1744 Memoire sur I'adherence de la Cuscute aux autres plantes *Histde I'AcadRoySci* Paris 170-190 pages
39. Gupta A and Singh M 1983 Mechanism of parasitization by *Cuscuta reflexa* RNase activity in the haustorial region of *Cuscuta* and infected host tissues (*Lantana camara*), *Plant Physiology* **58** 523-526
40. Gupta A and Singh M 1985 Mechanism of parasitism by *Cuscuta reflexa* Distribution of cytokinin in different regions of the parasite vine, *Plant Physiology* **63** 76-78
41. Gupta M, Mazumder UK, Pal DK, and Bhattacharya S 2003 Anti-steroidogenic activity of methanolic extract of *Cuscuta reflexa* Roxb stem and *Corchorus olitorius* L seed in mouse ovary, *Indian Journal of Experimental Biology* **4** 1641-644
42. Gupta SC and Rajeswari VM 1974 Histochemical localiz-ation of DNA and RNA in *Cuscuta reflexa* and its host, *Current Science* **43** 799-800
43. Haidair MA, Orr GL, and Westra P 1997 Effects of light and mechanical stimulation on coiling and prehaustoria formation in *Cuscuta* spp, *Weed Research* **37** 219-228
44. Haupt S and Neumann St 1996 Transfer of assimilates and xenobiotics from host plants to the parasite *Cuscuta reflexa* Roxb In Advances in parasitic plant research (eds Moreno, MT, Cubero, JI, Berner, D, Joel, D, Musse-lman, LJ and Parker, C), 353-364 pages
45. Haupt S, Oparka K J, Sauer N, and Neumann St 2001 Macromolecular trafficking between *Nicotiana tabacum* and the holoparasite *Cuscuta reflexa, Journal of Experi-mental Botany* **52** 173-177
46. Hibberd JM, Bungard RA, Press MC, Jeschke WD, Scholes JD, and Quick WP 1998 Localization of photo-synthetic metabolism in the parasitic angiosperm *Cuscuta reflexa, Planta* **205** 506-513
47. Holm L, Doli J, Panch J, and Harberger J 1997 World weeds natural histories and distribution John Wiley and Sons, New York, USA
48. Holmes B 1993 Evolution's neglected superstars, *New Scientist* **140** 30-33
49. Hyde and Mand Worsten B 2006 http/wwwzimbabwefloracozw/ species data/genusphp
50. Ihl B, Tutakhil N, Hagen A, and Jacob F 1988 Studies on *Cuscuta reflexa* Roxb 7

Defence mechanisms of *Lycopersicon esculentum* Mill, *Flora* **181** 383-393

51. Israel S, Dorr I, and Kollmann R 1980 Das Phloem der Haustorien von *Cuscuta, Protoplasma* **103** 309-321
52. IUCN 2006 Recent list of plants @wwwgooglecom
53. Jacob F and Neumann S 1968 Studies an *Cuscuta reflexa* Roxb I Zur Funktion der Haustorien bei der Aufnahme von Saccharose, *Flora AbtA* **159** 191-203
54. Jayakody DPP 1995 Some biological aspects of *Cuscuta chinensis* Lam In Second weed science congress Weed Science Society of Sri Lanka, Peradeniya
55. Jeschke WD and Hilbert A 1997 Sink- stimulated photosynthesis and sink-dependent increase in nitrate uptake nitrogen and carbon relations of the parasitic association *Cuscuta reflexa-Ricinus communis, Plant Cell Environment* **20** 47-56
56. Jeschke WD, Baumel P, Rath N, Cyzgan FC, and Proksch P 1994 Modelling the flow and partitioning of carbon and nitrogen in the holoparasitic *Cuscuta reflexa* Roxb and its host *Lupinus albus* L 2 Flows between host and parasite and within the parasitized host, *Journal of Experimental Botany* **45** 791-800
57. Kala CP, Farooquee NA, and Dhar U 2004 Prioritization of medicinal plants on the basis of available knowledge, existing practices and use value status in Uttaranchal, India, *Biodiversity and Conservation* **13** 453-469
58. Kapoor V and Sharma YP 2007 Host Range, Severity and Intensity of *Cuscuta campestris* Yuncker Infestations in Jammu Province of Jammu and Kashmir, *Indian Journal of Weed Science* **39** 146-148
59. Kapoor V and Sharma YP 2008 Host Range of *Cuscuta reflexa* Roxb in Jammu Province of Jammu and Kashmir State, India, *Indian Journal of Weed Science* **40** 98-100
60. Kapoor V 2007 Studies on *Cuscuta* infestation in some areas of Jammu MPhil Dissertation University of Jammu, Jammu
61. Kelly CK 1992 Resource choice in *Cuscuta europaea, Proceedings of National Academy of Science* USA **89** 12194-12197
62. Kelly CK and Horning K 1999 Acquisition order and resource value in *Cuscuta attenuate, Proceedings of National Academy of Science* USA **96** 13219-13222
63. Kelly CK, Venable DL, and Zimmerer K 1988 Host specialization in *Cuscuta*

*costaricensis* An assessment of host use relative to host availability, *Oikos* **53** 315-320

64. Khanna KK 2002 Ethnobotany Aavishkar Publishers, Jaipur, 128-145 pages
65. Khanna SK, Khanna R, and Sanwal GG 1975 Studies on alfalfa grown after cutting dodder (*Cuscuta*) infection changes in content and nature of phenolics, *Journal of Experimental Biology* **13** 407-408
66. Kindermann A 1928 Haustorial studien an *Cuscuta* Arten, *Planta* **5** 769-783
67. Kirtikar KR and Basu BD 2001 Indian medicinal plants, 3rd ed Vol 8 Sri Satguru Publication, Delhi
68. Koskela T, Pnustinen S, Salonen V, and Mutikainen P 2002 Resistance and tolerance in a host plant-holoparasitic plant interaction genetic variation and costs, *Evolution* **56** 899-908
69. Koskela T, Salonen V, and Mutikainen P 2000 Local adaptation of a holoparasitic plant, *Cuscuta europaea* variation among populations, *Journal of Evolutionary Biology* **13** 749-755
70. Kuijt J 1969 Biology of parasitic flowering plants University of California Press, USA, 246 page
71. Kuijt J 1977 Haustoria of phanerogamic parasites, *Annual Review of Phytopathology* **17** 91-118
72. Kuijt J and Toth R 1976 Ultrastructure of angiosperm haustoria A review, *Annals of Botany* **401** 121-1130
73. Kumaraswamy AS, Syamasunder J, Lingappa S, and Jayaram G 1978 *Cuscuta chinensis* Lam on *Eleusine coracana* Gaertn (Ragi)- A new record, *Current Science* **47** 469-469
74. Lacaille DMA, Franck U, and Wagner H 2001 Search for potential angiotensin converting enzyme (ACE) - inhibitors from plants, *Phytomedicine* **8** 47-52
75. Lackey CF 1953 Attraction of dodder and beet leafhopper to vascular bundles in the sugarbeet as affected by curlytop, *American Journal of Botany* **40** 221-225
76. Lei-Simon A 2001 Diversity of parasitic *Cuscuta* and their host plant species in a *Larrea-Atriplex* ecotone, *Bulletin of Southern California Academy of Science* **100** 36-43
77. Linke KH 1993 Parasitic plants in Syria, *Candollea* **48** 83-99
78. Machado MA and Zetsche K 1990 A structural, functional and molecular analysis of plastids of the holoparasites *Cuscuta reflexa* and *Cuscuta europaea, Planta* **181** 91-96

79. Mirande M 1900 Recherches physiologiques et anatomiques sur les Cuscutaceae Thesis, *Fac Sci*, Paris

80. Misra B and Prasad SM 1988 Natural hosts of *Cuscuta reflexa, Indian Phytopathology* **41** 154-155

81. Miyahara K, Du XM, Watanabe M, Sugimura C, Yahara S, and Nohara T 1996 Resin glycoside from the seeds of *Cuscuta chinensis, Chemical and Pharmaceutical Bulletin* **44** 481-485

82. Molau U 1995 Reproductive ecology and biology In Parasitic plants (eds) Press, MC and Graves, JD, Chapman and Hall, London, 141-176 pages

83. Musselman LJ 1987 Parasitic weeds in agriculture Vol I *Striga* CRC Press Boca Raton

84. Musselman LJ and Press MC 1995 Introduction to parasitic plants In Parasitic Plants (eds) Press, MC and Graves, JD, Chapman and Hall, London, 1-13 pages

85. Naik BI, Mehta KK, and Srivastava SK 1976 Changes in polyamine levels on infection of plants by *Cuscuta reflexa, Indian Journal of Biochemistry and Biophysics* **13** 306-307

86. Nun NB and Mayer AM 1999 Culture of pectin methylesterase and polyphenol oxidase in *Cuscuta campestris, Phytochemistry* **50** 719-727

87. Parker C and Riches CR 1993 Parasitic weeds of the world Biology and Control Wallingford, UK, CAB International

88. Pate JS, Davidson NJ, Kuo J, and Milburn JA 1990 Water relations of the root hemiparasite *Olax phyllanthi* (Labill) RBr (Olacaceae) and its multiple hosts, *Oecologia* **84** 186-193

89. Pennings SC and Callaway RM 1996 Impact of a parasitic plant on the structure and dynamics of salt marsh vegetation, *Ecology* **77** 1410-1419

90. Pennings SC and Callaway RM 2002 Parasitic plants Parallels and contrast with herbivores, *Oecologia* **131** 479-489

91. Prather LA and Tyrl RJ 1993 The biology of *Cuscuta attenuata* Waterfall (Cuscutaceae), *Proceedings of the Oklahoma Academy of Science* **73** 7-13

92. Press MC 1998 Dracula or Robin Hood? A functional role for root hemiparasites in nutrient poor ecosystems, *Oikos* **82** 609-611

93. Press MC and Graves JD 1995 Parasitic flowering plants Chapman and Hall, London

94. Press MC and Phoenix GK 2005 Impact of parasitic

plants on natural communities, *New Phytologist* **166** 737-751

95. Press MC, Scholes JD and Watling JR 1999 Parasitic plants physiological and ecological interactions with their hosts In Physiological plant ecology (eds) Press, MC, Scholes, JD and Barker, MG Oxford, UK, Blackwell Science, 175-197 pages
96. Rajput MTM and Tahir SS 1988 Flora of West Pakistan cuscutaceae **189** 1-24
97. Rajput MTM and Tahir SS 1990 Taxonomic and biogeographic notes on cuscutaceae from Pakistan and Kashmir, *Pakistan Journal of Botany* **22** 69-74
98. Raju VS and Rao PN 1976 China dodder (*Cuscuta chinensis* Lam) on cowpea- a new record, *Current Science* **45** 839-840
99. Ramaiah SK, Syamasunder J, Chandrashekar M, and Chandrashekaraiah SC 1981 New record of *Cuscuta reflexa* on tulip, jasmine, croton, cashew nut and *Tabebuia, Indian Journal of Mycology and Plant Pathology* **11** 112-113
100. Rao VSN, Dasaradhan P, and Krishnaiah KS 1979 Antifertility effect of some indigenous plants, *Indian Journal of Medical Research* **701** 517-520
101. Raynal-Roques A and Pare J 1998 Biodiversite des phanerogames parasites Lev place dans la classification systematique, *Adansonia* **20** 313-322
102. Riopel JL 1979 Experimental studies in induction of haustoria in *Agalini purpurea* In Proceedings of the second international symposium onparasitic weed North Carolina State University, Raleigh, NC, 173-1656 pages
103. Rispail N, Dita MA, Gonzalez-Verdejo C, Perez-deLuque A, Castillejo MA, Prats E, Roman B, Jorrin J, and Rubiales D 2007 Plant resistance to parasitc plants; molecular approaches to an old foe, *New Phytologist* **173** 703-712
104. Rothe K, Diettrich B, Rahfeld B, and Luckner M 1999 Uptake of phloem specific cardenolides by *Cuscuta* spp growing on *Digitalis lunata* and *Digitalis purpurea, Phytochemistry* **51** 357-361
105. Sadler HA, Else MJ, and Sutherland M 1997 Application of sand for inhibition of swamp dodder (*Cuscuta gronovii*) seedling emergence and survival on cranberry (*Vaccinium macrocarpon* Bogs), *Weed Technology* II 318-323
106. Sagalaev VA and Matveev DE 2000 Floristic findings in the Middle Don *Botanicheskii-*

*Zhurnal-St-Petersburg* **85** 115-119

107. Schumacher W and Halbsguth W 1968 Veber de Anschluss einiger hoheren Parasiten an die Siebrohren der Wiertspflanzen, *Jahrb. Wiss. Bot.* **87** 324-355

108. Seel WE, Parsons AN, and Press MC 1993 Do inorganic solutes limit growth of the facultative hemiparasitic *Rhinanthus minor* L in the absence of a host? *New Phytologist* **124** 283-289

109. Sharma BM and Kachroo P 1981 Flora of Jammu and plants of neighbourhood, Vol 1 and 2 Bishen Singh Mahendra Pal Singh, Dehradun

110. Sherman TD, Pettigrew WT, and Vaughn KC 1999 Structural and immunological characterization of the *Cuscuta pentagona* chloroplast, *Plant Cell Physiology* **40** 592-603

111. Silberhorn GM 1998 Invasion of *Cuscuta indecora* Choisy (Convulvulaceae) in a tidal brackish water marsh in Virginia, *Castanea* **63** 190-191

112. Singh A and Singh M 1997 Incompatibility of *Cuscuta* haustoria with the resistant hosts- *Ipomoea batatas* L and *Lycopersicon esculentum* Mill, *Journal of Plant Physiology* **150** 592-596

113. Singh BP and Brar NS 1985 *Journal of Bombay Natural History Society* **81** 596-599

114. Singh DV 1972 Influence of shading on starch and nucleic acid metabolism of *Cuscuta* on *Medicago sativa, Indian Journal of Experimental Biology* **10** 143-146

115. Singh M and Sanwal GG 1978 A source of error in the paper chromatographic identification of phytate in the presence of inorganic polyphosphate, *Indian Journal of Experimental Biology* **16** 257-258

116. Stewart GR and Press MC 1990 The physiology and biochemistry of parasitic angiosperms, *Annual Review of Plant Physiology and Plant Molecular Biology* **41** 127-151

117. Sur PR and Halder AC 2002 Ethnobotany Avishkar Publisher, Jaipur, 146-168 pages

118. Swami A and Gupta BK 1998 Flora of Udhampur Bishen Singh Mahendra Pal Singh, Dehradun

119. Tada Y, Sugai M, and Furuhashi K 1996 Haustoria of *Cuscuta japonica*, a holoparasitic flowering plant, are induced by the cooperative effects of far-red light and tactile stimuli, *Plant Cell Physiology* **37** 1049-1053

120. van der Kooij TAW, Krause K, Dorr I, and Krupinska K 2000 Molecular, functional and ultrastructural characterization of plastids from six species of the parasitic flowering plant genus, *Cuscuta Planta* **210** 701-707

121. Vaughn KC 2003 Dodder hyphae invade the host a structural and immunocytochemical characterization *Protoplasma* **220** 189-200

122. Vaughn KC 2002 Attachment of the parasitic weed dodder to the host, *Protoplasma* **219** 227-237

123. Visser JH and Dorr I 1986 The haustorium In Parasitic weeds in agriculture1 *Striga* (ed Musselman, L J) CRC Press Inc Boca Raton

124. Vyanjane NT and Mali VR 1981 Occurrence of alfalfa mosaic virus on brinjal in India, *Indian Phytopathology* **34** 164-168

125. Wijesundara DSA, Jayasinghe C, Marambe B, and Tennakoon KU 2001 Host-parasitic associations of *Cuscuta chinensis* Lam in Sri Lanka, *Ann Sri Lanka Depart Agric* **3** 343-351

126. Wolswinkel P 1978 Phloem unloading in stem parts parasitized by *Cuscuta* The release of Ca and K to the free space at 0°C and 25°C, *Plant Physiology* **42** 167-172

127. Wolswinkel P, Ammerlaan A, and Peters HFC 1984 Phloem unloading of amino acids at the site of attachment of *Cuscuta europaea, Plant Physiology* **75** 13-20

128. Xiong YB and Zhou CH 1994 The effect of extracts from herba epimedii and semen *Cuscutae* on the function of male reproduction, *Chinese Pharmaceutical Journal* **29** 89-91

129. Yadav SB, Tripathi V, Singh RK, and Pandey HP 2000 Anti-oxidant activity of *Cuscuta reflexa* stems, *Indian Journal of Pharmaceutical Science* **6** 477-478

130. Yuncker TG 1932 The genus *Cuscuta Mem Torrey Club* **18** 109-331

131. Zender J 1924 Les haustoriums de la *Cuscuta* et les reactions de I' hote, *Bull Soc Bot Geneve Ser* **16** 189-264

132. Zhuk AV 1997 Haustoria morphogenesis and origin in *Cuscuta* species (Cuscutaceae), *Botani Cheskii Zhurnal* **82** 1-15